T0259328

# GAP JUNCTIONS

*Volume 30* • 2000

ADVANCES IN MOLECULAR AND CELL BIOLOGY

# GAP JUNCTIONS

## ADVANCES IN MOLECULAR AND CELL BIOLOGY

*Series Editor:* E. EDWARD BITTAR
*Department of Physiology*
*University of Wisconsin—Madison*
*Madison, Wisconsin*

*Guest Editor:* ELLIOT L. HERTZBERG
*Departments of Neuroscience and of Anatomy*
*and Structural Biology*
*Albert Einstein College of Medicine*
*Bronx, New York*

VOLUME 30 • 2000

JAI PRESS INC.
*Stamford, Connecticut*

*ISBN: 0-7623-0599-1*

*Printed and bound by CPI Antony Rowe, Eastbourne*
*Transferred to digital printing 2006*

# CONTENTS

LIST OF CONTRIBUTORS     vii

PREFACE
*Elliot L. Hertzberg*     xi

GAP JUNCTION GENES AND THEIR REGULATION
*Eric C. Beyer and Klaus Willecke*     1

STRUCTURE AND BIOCHEMISTRY OF GAP
JUNCTIONS
*Mark Yeager and Bruce J. Nicholson*     31

POST-TRANSCRIPTIONAL EVENTS IN THE EXPRESSION
OF GAP JUNCTIONS
*Dale W. Laird and Juan C. Sáez*     99

GAP JUNCTIONS CHANNELS: PERMEABILITY AND
VOLTAGE GATING
*Vytas K. Verselis and Richard Veenstra*     129

GAP JUNCTIONS IN DEVELOPMENT
*Cecilia W. Lo and Norton B. Gilula*     193

GAP JUNCTIONS DURING NEOPLASTIC
TRANSFORMATION
*Mark J. Neveu and John Bertram*     221

GAP JUNCTIONS FUNCTION
*Paolo Meda and David C. Spray*     263

GAP JUNCTIONS AND CONNEXINS IN THE
MAMALIAN CENTRAL NERVOUS SYSTEM
*James I. Nagy and Rolf Dermietzel*     323

INDEX     397

# LIST OF CONTRIBUTORS

*John Bertram*

Cancer Research Center of Hawaii
University of Hawaii at Manoa
Honolulu, Hawaii

*Eric C. Beyer*

Section of Pediatric Hematology/Oncology
Department of Pediatrics
University of Chicago
Chicago, Illinois

*Rolf Dermietzel*

Department of Neuroanatomy and
    Molecular Brain Research
Ruhr University Bochum
Bochum, Germany

*Norton B. Gilula*

Department of Cell Biology
The Scripps Research Institute
La Jolla, California

*Dale W. Laird*

Department of Anatomy and Cell Biology
University of Western Ontario
London, Ontario, Canada

*Cecilia W. Lo*

Biology Department
University of Pennsylvania
Philadelphia, Pennsylvania

*Paolo Meda*

Department of Morphology
University of Geneva Medical School
Geneva, Switzerland

*James I. Nagy*

Department of Physiology
University of Manitoba
Winnipeg, Manitoba, Canada

*Mark J. Neveu*
Pfizer Central Research
Departments of Molecular Genetics and
    Cancer
Groton, Connecticut

*Bruce J. Nicholson*
Department of Biological Sciences
State University of New York—Buffalo
Buffalo, New York

*Juan C. Sáez*
Departmento de Ciencias Fisiológicas
Pontificia Universidad Católica De Chile
Santiago, Chile

*David Spray*
Departments of Neuroscience and Medicine
Albert Einstein College of Medicine
Bronx, New York

*Richard Veenstra*
Department of Pharmacology
State University of New York—Syracuse
Syracuse, New York

*Vytas K. Verselis*
Department of Neuroscience
Albert Einstein College of Medicine
Bronx, New York

*Klaus Willecke*
Institut fur Genetik
Universitat Bonn
Bonn, Germany

*Mark Yeager*
Department of Cell Biology
Research Institute of Scripps Clinic
La Jolla, California

# PREFACE

## FORTY YEARS OF GAP JUNCTIONS

The field of gap junction research arose at the confluence of cell physiology, especially early efforts involving electrophysiology, and the study of cell ultrastructure, which then was starting to utilize electron microscopy for the study of biological samples. Electrophysiological analyses of the nervous system were a key element in the effort to determine the mechanism(s) of synaptic transmission. The finding of electrotonic (electrical) coupling between some neurons was soon complemented by the finding of appropriately specialized regions of cell-cell contact now called gap junctions. While first observed in the nervous system, gap junctions were soon identified in muscle and soon thereafter in many cell types. It was quickly established that gap junctions are present in most if not all contacting cells in metazoans. An especially fine early interpretation of freeze/fracture electron microscopy significantly advanced the concept of cell–cell channels assembled in macular arrays often containing thousands of intercellular channels. By this time, the correlation between the presence of ultrastructurally identifiable gap junctions with the ability of cells to directly transmit not only current, but hydrophilic dyes and low molecular weight metabolites, was well-accepted as the basis of this structure–function correlation of intercellular communication.

From the earliest days, a major thrust of the field was the determination of the role of gap junctional communication in cell and tissue function, with a special emphasis on the nervous system. Soon thereafter, it was appreciated by several groups that the semicrystalline arrays of gap junction particles might serve as a good model system for the study of the structure of integral membrane proteins much as was being developed for analysis of bacteriorhodopsin and the acetylcholine receptor.

The presence of direct intercellular communication soon led to hypotheses for a role in development where it was suggested that gap junctions would be logical sites for transmission of then-elusive morphogens as well as for the regulation of the diffusion of these molecules. Several groups undertook analysis of the role of gap junctions in cell–cell interactions and in the regulation of tissue function, including its extension into the domain of regulation of cell growth the aberration of which was suggested to serve as the molecular basis for at least one mechanism for cell transformation and carcinogenesis.

Hence, by about the early 1970s, both conceptual and experimental framework for future studies of gap junctions was becoming reasonably well developed. By about 1980, most of the early books and conferences centering on gap junctions were divided more or less formally along these lines. Indeed, this current volume is organized on these principles with the addition of a chapter on the molecular biology of the connexin gene family as well as one considering trafficking and assembly of connexins into gap junctions and their covalent modifications. These topics are, in fact, but a logical outgrowth of our effort to understand the regulation of gap junctional at all levels from synthesis to turnover and so fall well within the classic characterization of key categories of gap junction studies.

However, in order to establish molecular mechanisms underlying gap junctional communication and the mechanisms by which it could be regulated, it became essential to identify the proteins constituting gap junctions, an effort which served as my own introduction to the field as a postdoctoral fellow with Dr. Bernie Gilula at the Rockefeller University. A key attraction of entering the gap junction field remains the overwhelming extent to which one must keep up with so many different aspects of biology since it has often turned out that findings in one discipline (e.g., biochemistry) would require modifications of hypotheses and models being developed in the other subfields and vice versa. Out of the success of several groups in isolating unproteolyzed liver and heart gap junctions sprang characterization of their proteins (now called connexins) and the genes encoding them, discovery of a connexin gene family, and access to a plethora of techniques for their study in physiology and pathophysiology.

Perhaps not surprisingly, human genetic diseases based on connexin mutations were found. Quite a surprise, though, was that the first such disease, a peripheral neuropathy termed X-linked Charcot-Marie-Tooth disease that arises from mutations in connexin32, had not been predicted by those of us working with this protein—we had failed to detect its presence in myelin!

Activity in the field, as assessed by literature citations as a function of year of publication (below), has proceeded with noticeable spurts. Prior to about 1975

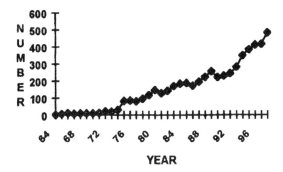

was the era of basic structural identification of gap junctions, the beginnings of an understanding of their functional properties, and the early phases of protein isolation. This was followed by a rather linear rise in publication rate as antibody probes to connexins were developed and most of the connexin genes were identified and as suitable probes were developed and applied to their study. The rapid increase in publication rate after about 1993 reflects a large influx of investigators into the field with the predominant usage of more sophisticated tools and analytical techniques. Whether the blip between 1997 and 1998 reflects a new phase in growth remains to be seen. Clearly, more papers analyzing connexin dysfunction in disease appeared in this year than all of the papers published prior to 1975! We still know little about which specific properties of individual connexins are, presumably, necessary for their tissue-specific patterns of expression. Some recent knock-out and knock-in experiments suggest that at least some overlap of function does exist. Surprises, no doubt, remain.

In considering whether yet another tome on gap junctions was necessary for the field and for those wanting an indepth immersion into our reality, this volume clearly represents an affirmative response. My objective in editing this volume was twofold: to provide a reasoned overview of the field as well as to furnish one that provided this overview within the context of the intellectual boundaries of those who initially attempted to define the purview of gap junction research. The latter objective has been realized by selecting the topics for review in this volume. The former objective was achieved by securing the cooperation of leaders in their fields as chapter co-authors.

In recent years, books focusing on gap junctions have represented two genres: proceedings of conferences and, increasingly, those that focus in greater detail on but one aspect of the field. In light of the burgeoning of citations in the biomedical databases relating to gap junctions (there were perhaps 50-odd papers to assimilate when I began my own efforts in the field—but now there are more than 500 per year), it most certainly will become exceedingly difficult to attempt another such comprehensive overview of this field of research. Should this volume represent the last such endeavor remains to be determined. Should it achieve my objective of providing a thorough overview of our thoughts and endeavors, it will have served its purpose.

As all who have presumed that they could achieve more as the editor of a volume such as this than their colleagues could in putting such a volume together, I must primarily acknowledge the efforts and contributions of those who authored the chapters. My ongoing interactions with the core gap junction group at The Albert Einstein College of Medicine, Drs. Bennett, Spray, Verselis, and Bargiello, has certainly broadened my perspective and aided me in conceiving and executing this project. A special thanks is also due to Dr. Bernie Gilula who introduced me to the field, its players, and its own sets of techniques and considerations. Last, but certainly not least, I thank those efforts of understanding achieved by my family. The longer I remain in academic science, the more I appreciate how heartfelt such acknowledgments truly are.

Elliot L. Hertzberg
*Guest Editor*

# GAP JUNCTION GENES AND THEIR REGULATION

Eric C. Beyer and Klaus Willecke

|     |                                                                                    |     |
| --- | ---------------------------------------------------------------------------------- | --- |
| I.  | Connexin Genes                                                                     | 2   |
|     | A. Introduction                                                                    | 2   |
|     | B. Molecular Cloning of Gap Junction Proteins: The Connexins                       | 2   |
|     | C. Connexin Primary Structure and Topology                                         | 8   |
|     | D. Connexin Gene Structure and Chromosomal Locations                               | 8   |
|     | E. Nonconnexin Components of Gap Junctions                                          | 11  |
| II. | Transcriptional Control and Expression Patterns of Connexin Genes                  | 12  |
|     | A. Analysis of Gene Control Regions                                                 | 12  |
|     | B. Tissue-specific Expression Pattern of Connexin Genes                            | 14  |
|     | C. Cell Type–Specific Expression of Connexin Genes                                  | 16  |
|     | D. Expression of Connexins in Transfected Cells                                     | 18  |
|     | E. Transcriptional and Post-Transcriptional Control of Connexin Gene Expression    | 19  |
| III.| Conclusions and Outlook                                                            | 20  |

Advances in Molecular and Cell Biology, Volume 30, pages 1-30.
ISBN: 0-7623-0599-1

# I.  CONNEXIN GENES

## A.  Introduction

New insights into the regulation of gap junctions and intercellular communication have resulted from the molecular cloning of numerous subunit gap junction proteins and the use of those clones to develop specific probes for gap junction nucleic acid sequences and proteins.

In the 1970s, procedures were developed for the isolation of gap junctions from liver (Evans and Gurd, 1972; Goodenough and Stoekenius, 1972). These isolated gap junctions facilitated study of some of their molecular structure. Makowski and colleagues (Makowski et al., 1976; Caspar et al., 1977) used x-ray diffraction and electron microscopy to develop a low-resolution (25 Å) structural model. The model shows that each hemichannel is composed of a hexameric structure (connexon) containing six apparently identical integral membrane subunits (connexins), which surround a central pore. The connexon joins with another connexon in the plasma membrane of the adjacent cell.

The isolated liver gap junctions are composed primarily of a 27-kD polypeptide and a 21-kD polypeptide (Henderson et al., 1979; Hertzberg and Gilula, 1979). Similarly prepared myocardial gap junctions contain a number of different sized polypeptides, but predominantly one of 43 to 47 kD and its degradation products (Manjunath et al., 1987). Preparations enriched in lens fiber gap junctions contain several prominent proteins, including one of 70 kD, termed MP70 (Kistler et al., 1985). N-terminal sequencing of these proteins by Edman degradation has shown that the liver 27 kD and 21 kD, the heart 43 to 47 kD, and the lens 70 kD (Nicholson et al., 1985; Nicholson et al., 1987; Kistler et al., 1988) are homologous proteins. More details about the isolation of these gap junction proteins and their biochemistry are provided in the next chapter by Yeager and Nicholson.

## B.  Molecular Cloning of Gap Junction Proteins: The Connexins

cDNA and genomic clones encoding these biochemically identified gap junction proteins were isolated by a number of different strategies, including (1) the use of anti-gap junction antibodies to screen bacteriophage expression libraries, (2) hybridization screening of libraries with synthetic oligonucleotide probes based on known amino acid sequences, and (3) low stringency hybridization of phage libraries with previously identified gap junction cDNAs. Paul (1986) and Heynkes and co-workers (1986) cloned a cDNA for the rat liver 27-kD protein; Kumar and Gilula (1986) cloned a cDNA for its human counterpart. These cDNAs encode a polypeptide of 32 kD. Beyer and associates (1987) isolated a homologous rat heart cDNA, which codes for a polypeptide of 43 kD that contains 43% identical amino acids to the protein cloned from rat liver. Zhang and Nicholson (1989) cloned a cDNA encoding a 26-kD polypeptide, which corresponds to

***Table 1.*** Cloned Connexin Genes

| Connexin Nomenclature[a,b] | Greek Letter Nomenclature | Species From Which Cloned | cDNA Source |
|---|---|---|---|
| Cx26 | $\beta_2$ | Rat, mouse, human | Liver, breast |
| Cx30 | $\beta_6$ | Mouse | Brain, skin |
| Cx30.3 | $\beta_5$ | Mouse | Skin |
| Cx31 | $\beta_3$ | Rat, mouse | Placenta, skin, F9 cells |
| Cx31.1 | $\beta_4$ | Rat, mouse | Skin, F9 cells |
| Cx32 | $\beta_1$ | Rat, human, *Xenopus*, mouse | Hepatocytes, F9 cells |
| Cx32.2 | | Atlantic croaker | Ovary |
| Cx32.7 | | Atlantic croaker | Ovary |
| Cx33 | $\alpha_7$ | Rat | Testis |
| Cx34.7 | $\gamma$ or $\delta$ | Perch | Retina |
| Cx35 | $\gamma$ or $\delta$ | Skate, perch | Retina |
| Cx36 | $\gamma$ or $\delta$ | Mouse | Brain |
| Cx37 | $\alpha_4$ | Mouse, rat, human, pig | Lung, endothelium |
| Cx38 | $\alpha_2$ | *Xenopus* | Oocyte, early embryo |
| Cx40/Cx42 | $\alpha_5$ | Rat, dog, mouse, chicken, human | Lung, vascular smooth muscle, myocardium |
| Cx41 | | *Xenopus* | Ovary |
| Cx43 | $\alpha_1$ | Rat, mouse, human, Cow, chicken, *Xenopus* | Myocardium, myometrium, smooth muscle, F9 cells |
| Cx45/Cx43.4 | $\alpha_6$ ($\gamma$) | Chicken, dog, mouse, human, zebrafish | Myocardium, F9 cells, |
| Cx46/Cx56/Cx44 | $\alpha_3$ | Rat, chicken, bovine | Lens fibers |
| Cx46.6 | $\alpha$?, $\gamma$? | Human | Unknown |
| Cx50/Cx45.6 | $\alpha_8$ | Mouse, chick | Lens |
| Cx60 | | Pig | Ovary |

***Notes:*** [a]The connexin nomenclature assigns names for each protein based on its predicted molecular mass in kiloDaltons. The Greek letter system is based on Kumar and Gilula (1992). cDNA source indicates the tissue used to prepare libraries from which connexins were cloned. For connexins cloned only from genomic DNA, the major location of mRNA expression is indicated. The *Xenopus* homolog of rat/human Cx32 is *Xenopus* Cx30. Cx40 in rat, dog, and mouse appear to be homologous to chick Cx42. Therefore, these connexins are considered together. Rat Cx50 is immunologically related to the ovine lens fiber junction protein MP70. The fish Cx34.7 and Cx35 found in retinal neurons are homologs of mouse Cx36, found in brain neurons; they are more closely related to each other than to members of the a and b groups and are therefore included in a new group ($\gamma$ or $\delta$). Cx45 and human Cx46.6 could also constitute a subfamily.

[b]References for cDNA cloning are: Cx26 (Zhang and Nicholson, 1989; Hennemann et al., 1992b), *Xenopus* Cx30 (Gimlich et al., 1988), Cx30.3 (Hennemann et al., 1992a), Cx31 (Hoh et al., 1991; Hennemann et al., 1992c), Cx31.1 (Haefliger et al., 1992; Hennemann et al., 1992a), Cx32 (Heynkes et al., 1986; Kumar and Gilula, 1986; Paul, 1986; Hennemann et al., 1992b), Atlantic Croaker Cx32.2 and Cx32.7 (Yoshizaki et al., 1994; Bruzzone et al., 1995), Cx33 (Haefliger et al., 1992), perch Cx 34.7 and Cx35 (O'Brien et al., 1998), skate Cx35 (O'Brien et al., 1996), Cx36 (Condorelli et al., 1998; Sohl et al., l998), Cx37 (Willecke et al., 1991b; Haefliger et al., 1992; Reed et al., 1993), *Xenopus* Cx38 (Ebihara et al., 1989; Gimlich et al., 1990), Cx40 (Beyer et al., 1992; Haefliger et al., 1992; Hennemann et al., 1992d; Jiang et al., 1994; Kanter et al., 1992), *Xenopus* Cx41 (Yoshizaki and Patino, 1995), chick Cx42 (Beyer, 1990), Cx43 (Beyer et al., 1987; Fishman et al., 1990; Gimlich et al., 1990; Lash et al., 1990; Musil et al., 1990; Beyer and Steinberg, 1991; Lang et al., 1991; Nishi et al., 1991; Hennemann et al., 1992d), Cx44 (Gupta et al., 1994), Cx45 (Beyer, 1990; Hennemann et al., 1992c; Jiang et al., 1994; Kanter et al., 1992) (Essner et al., 1996), Cx46 (Paul et al., 1991), Cx50 (White et al., 1992; Church et al., 1995) Cx45.6 (Jiang et al., 1994), Cx56 (Rup et al., 1993) and pig Cx60 (Itahana et al., 1998).

the liver 21-kD protein. White and colleagues (1992a) have cloned DNA encoding a 50-kD mouse polypeptide that is immunologically related to the sheep lens 70-kD protein.

Cloning of these sequences has confirmed previous suggestions that there is a family of related gap junction proteins. The gap junction protein sequences contain many identical amino acid residues. Many of these proteins are not uniquely expressed in a single tissue or cell type. The mobilities of these proteins on SDS-PAGE may vary with electrophoresis conditions (Green et al., 1988) and their gel mobilities often differ from the sizes predicted from the sequences. Therefore, previous descriptions of gap junction proteins based on tissue of origin or electrophoretic mobility have been abandoned and replaced by an operational nomenclature using the generic term *connexin* (abbreviated Cx) for the protein family, with an indication of species (as necessary) and a numeric suffix designating the predicted molecular mass in kilodaltons (Beyer et al., 1987, 1990). Thus, the 27-kD protein from rat liver is termed *rat connexin32* (Cx32), the 43-kD protein from rat heart is termed *rat connexin43* (Cx43). An alternative nomenclature has been suggested (Gimlich et al., 1990; Kumar and Gilula, 1992), which divides the connexin gene family into α and β classes based on sequence similarities. For better comparison of the original references, both nomenclatures of connexin genes are listed in Table 1.

Further members of the connexin family have been cloned from cDNA libraries and from genomic DNA by a number of strategies. The different cloned connexins are listed in Table 1. Currently, 15 different connexins have been identified in rat or mouse (most have also been demonstrated to be expressed in other mammals including humans). Several additional connexins without identified rodent orthologs have been cloned from other species such as *Xenopus,* pig, and human (Ebihara et al., 1989; Itahana et al., 1998). Different connexins within a single species contain approximately 40% identical amino acids.

Some connexins have been cloned from different species, and are highly conserved. Cx43 or its close homolog has been cloned from *Xenopus* (Gimlich et al., 1990), chick (Musil et al., 1990), human (Fishman et al., 1990), mouse (Beyer and Steinberg, 1991; Nishi et al., 1991; Hennemann et al., 1992d), and cow (Lash et al., 1990) as well as rat (Beyer et al., 1987). The sequences are extremely similar: the mammalian Cx43 proteins show more than 97% amino acid identity; the chick protein has 92% identical amino acids to the rat; and the *Xenopus* protein has 87% identical amino acids to the rat. The neuronal Cx 34.7 and Cx 35 found in perch and skate retina (O'Brian et al., 1996, 1998) are apparent orthologs of neuronal Cx36 found in mouse and rat brain and retina (Condorelli et al., 1998; Sohl et al., 1998). Many of the substitutions are conservative. The amphibian homolog of rat Cx32 (*Xenopus* Cx30) contains 71% identical amino acids to the rat or human proteins (Gimlich et al., 1988). However, other connexins are less well conserved. The Cx40 protein is only about 86% identical between mouse and dog (Hennemann et al., 1992d; Kanter et al., 1992); chick Cx42 (Beyer, 1990) likely is the

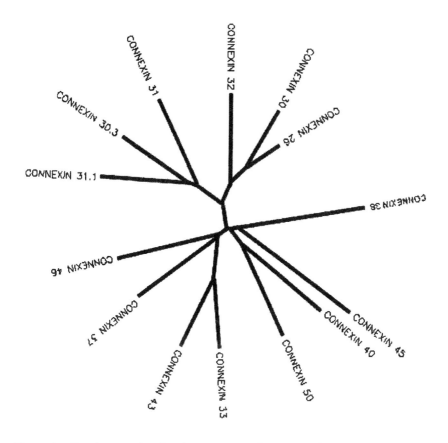

**Figure 1.** Dendrogram showing the sequence relation and phylogeny of the derived connexin polypeptide sequences, based on the cloned connexin cDNAs and genomic DNAs. The connexin proteins listed in Table 1 were compared using the CLUSTAL-W program (Higgins and Sharp, 1988). Where both mouse and rat homologs had been sequenced only the rat connexin is shown, as in all cases they were extremely similar. The relative height of a branch indicates the similarity of adjacent sequences. Similar analysis shows that some connexins (e.g., Cx43) are extremely well conserved between many different species, whereas others (e.g., Cx40/Cx42) are less well conserved. Abbreviations: XEN, Xenopus; HUM, human; CHK, chicken.

ortholog of mammalian Cx40, but it shares only approximately 67% identical amino acids. The closest mammalian counterpart of chicken Cx56 is rat Cx46 (Rup et al., 1993). No mammalian or avian counterpart has been identified for *Xenopus* Cx38, which was cloned from an amphibian embryo library (Ebihara et al., 1989; Gimlich et al., 1990).

## Connexin Primary Structure

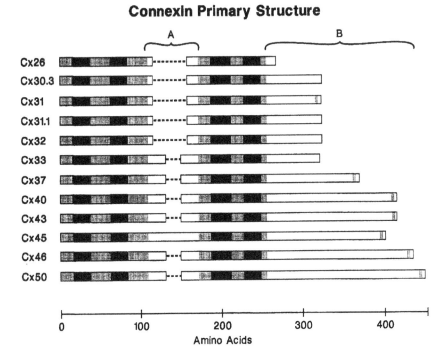

***Figure 2.*** Comparison of the primary amino acid sequences of 12 cloned mammalian connexins. Regions containing many identical amino acids are shaded, those that are unique are not shaded. The black shading represents conserved hydrophobic regions predicted to be transmembrane domains. Dashes represent spaces added to optimize alignment. Each connexin contains two stretches of approximately 100 amino acids that are highly conserved, and two unique regions marked A and B. Several connexins also contain a short serine-rich region near the carboxyl terminus, indicated by the small shaded area.

A dendrogram comparing the degree of relatedness of different connexins is shown in Figure 1. One of the conclusions from this analysis is that the connexins appear to fall into several groups: one (the β group, according to Kumar and Gilula, 1992) includes Cx26, Cx30, Cx30.3, Cx31, Cx31.1, and Cx32; the α group includes Cx33, Cx37, Cx38, Cx40, Cx43, Cx45, Cx46, Cx50, and Cx56. Newly discovered connexins that are expressed in neurons, Cx34.7, Cx35, and Cx36 (O'Brien et al., 1998; Condorelli et al., 1998 and Sohl et al., 1998), appear more closely related to each other than to members of the α or β group; they may be referred to as γ or δ. Cx45 and human Cx46.6 might also constitute a subfamily. Although grouping of connexins by sequence features has previously been noted by several authors (Gimlich et al., 1990; Bennett et al., 1991; Hoh et al., 1991; Haefliger et al., 1992), its biological significance remains unclear.

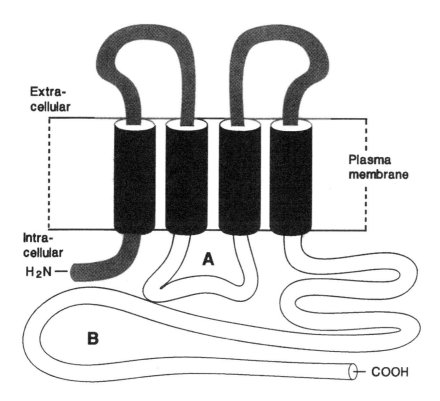

**Figure 3.** Topological model of connexin orientation within the junctional plasma membrane. Shaded regions represent sequences containing many identical amino acids in all connexins. The unshaded, predicted cytoplasmic domains (A and B), correspond to the unique, connexin specific regions shown in Figure 2. The folding shown for cytoplasmic regions is arbitrary, not based on sequence information. The putative third transmembrane region contains charged amino acid residues at conserved positions in all connexin sequences analyzed, followed by hydrophobic amino acid residues (reviewed in Willecke et al., 1991a). Thus, the third transmembrane region in each connexin subunit has been hypothesized to line the aqueous pore of the gap junction channel, which is presumably composed of 12 connexin subunits.

## C.  Connexin Primary Structure and Topology

All of the connexins contain many identical and conserved amino acids. Most of these residues fall in two regions of homology, each containing about 100 amino acids: one at the beginning of the protein and one in the middle. The other portions of the connexins are unique in each connexin, varying both in length and sequence. Figure 2 shows a representation comparing the primary sequences of 13 of the currently identified mammalian connexin polypeptides. A similar comparison of Cx31, Cx31.1, and Cx30.3 amino acid sequences showed that these three genes are most closely related to each other (Hennemann et al., 1992a).

The amino acid sequences derived from the cloned cDNAs have been used to predict the structures of the connexins. Each connexin contains four regions which are predicted to be very hydrophobic; two hydrophobic domains are located within each of the conserved regions. A topology model for the relation of the connexin polypeptides to the junctional plasma membrane has been constructed by assuming that each hydrophobic domain represents a transmembrane segment of the molecule (Beyer et al., 1987; Zimmer et al., 1987). This model has been confirmed for rat Cx32, Cx43, and Cx26 proteins by proteolysis and immunochemistry experiments (reviewed by Yeager and Nicholson and Yeager in this volume). This connexin model is illustrated in Figure 3.

The connexin topology models suggest that the hydrophobic/transmembrane domains and the extracellular regions correspond to the homologous sequences shared between connexins. These are represented as the shaded regions in Figures 2 and 3. Comparison of the primary sequences of all the known connexins reveals that the two predicted extracellular domains are the most conserved regions, each containing three invariant cysteines. The transmembrane domains are somewhat less well conserved. The cytoplasmic domains, with the exception of the short N-terminal region, differ markedly between connexins both in sequence and in length. This is summarized in Figures 2 and 3, in which the two cytoplasmic domains are labeled A and B.

## D.  Connexin Gene Structure and Chromosomal Locations

The organization of connexin genes is becoming clarified by genomic cloning studies and chromosomal mapping. Some characterization and genomic sequence have been published concerning the genes for Cx32, Cx26, Cx43, and other connexins (Miller et al., 1988; Fishman et al., 1991a; Hennemann et al., 1992b; Sullivan et al., 1993. These studies show that these connexin genes all have a rather similar structure, which is illustrated in Figure 4. Each of these connexin genes is composed of two exons: the first is small and contains only 5'-untranslated sequence; the second contains the entire coding sequence and all of the 3'-untranslated regions. Although the genes for other connexins have not been as well characterized, they appear to have similar structures, as Southern blots, PCR

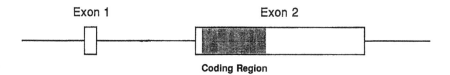

**Figure 4.** Connexin gene structure, based on studies of Cx26, Cx32, and Cx43. Each connexin gene has a short first exon containing only 5'-untranslated sequence. The coding sequence is not interrupted by introns. Where it has been examined, the entire translated sequence and 3'-untranslated sequence are contained within a second exon. Cx36 and its fish counterparts are exceptions to this structure, as the beginning of the Cx36 coding region is interrupted by a 1.1-kb intron.

experiments, and genomic DNA sequencing suggest that their coding sequences, with the exception of Cx 36, are not interrupted by introns (Zhang and Nicholson, 1989; Beyer and Steinberg, 1991; Hoh et al., 1991; Willecke et al., 1991b; Haefliger et al., 1992; Hennemann et al., 1992a, 1992b, 1992d; Kanter et al., 1992; Rup et al., 1993; Church et al., 1995; Dahl et al., 1996). In Cx36, and its fish orthologs, the coding region is interrupted by a 1.14-kilobase (kb) intron (Sohl et al., 1998). Analysis of the human Cx32 gene has demonstrated that it contains two different first exons which are alternatively spliced to generate mRNAs with different 5'-untranslated regions; the entire coding sequence for human Cx32 is in the last exon (Neuhaus et al., 1996). Another interesting result of the genomic studies has been the demonstration of the presence of a processed pseudogene for human connexin43 (Fishman et al., 1990; Willecke et al., 1990; Fishman et al., 1991a). Analysis of the regulatory elements within connexin genes is considered below.

Chromosomal mapping of several connexin genes has been performed (Willecke et al., 1990, Hsieh et al., 1991; Fishman et al., 1991a; Haefliger et al., 1992, Schwarz et al., 1992, 1994; Church et al., 1995; Van Camp et al., 1995; Dahl et al., 1996; Mignon et al., 1996; Geyer et al., 1997; Wenzel et al., 1998). These studies place human Cx26 and Cx46 on chromosome 13; Cx32 on chromosome X; Cx43 on chromosome 6; Cx37, Cx40, and Cx50 on chromosome 1; and the Cx43 pseudogene on chromosome 5. Hsieh and co-workers (1991) report that the homologous mouse loci were assigned to regions of known conserved syntenic groups. Several mouse connexin genes are located on the same chromosome. Cx31.1 and Cx30.3 are adjacent genes on mouse chromosome 4, within 3.4 kb (Hennemann et al., 1992a). Cx37 is linked to Cx31.1 on mouse chromosome 4;

*Table 2.*   Chromosomal localization

| Connexin | Mouse Gene Location[b] | Human Gene Location |
|---|---|---|
| Cx26 | 14[1,2,3] 14D1-E1[4] | 13[1,5];13q11-q12[4] |
| Cx30 | 14[6] | |
| Cx30.3 | 4[7] | |
| Cx31 | 4[3] | 1p34-p36[15] |
| Cx31.1 | 4[2,7] | |
| Cx32 | XD-F4[1], X[2,3] | Xq13.1[9] |
| Cx33 | X[2,3,10] | |
| Cx37 | 4[2,3] | 1p35.1[11] |
| Cx40 | 3[2,3] | 1pter-q12[5] |
| Cx43 | 10[1,2,3] | 6q22.3[12] |
| ΨCx43[a] | | 5[1,5,8] |
| Cx45 | 11[3] | 17[16] |
| Cx46 | 14[1,2] 14D1-E1[4] | 13[1] 13q11-q12[4] |
| Cx50 | 3[10] | 1q21.1[14] |

**Notes:**   [a]ΨCx43 is a pseudogene of human Cx43.

[b]References for chromosomal assignment studies cited: [1]Hsieh et al., 1991; [2]Haefliger et al., 1992 ; [3]Schwarz et al., 1992; [4]Mignon et al., 1996; [5]Willecke et al., 1990; [6]Dahl et al., 1996; [7]Hennemann et al., 1992c; [8]Fishman et al., 1991a; [9]Corcos et al., 1992; [10]Schwarz et al., 1994; [11]Van Camp et al., 1995; [12]Kato et al., 1997; [13]Church et al., 1995; [14]Geyer et al., 1997; [15]Wenzel et al., 1998; [16]Beyer (unpublished).

Cx26 and Cx46 are linked on mouse chromosome 14. Cx40 is not linked to other connexin genes on mouse chromosome 3 (Haefliger et al., 1992; Schwarz et al., 1992) although Cx50 is also present on this chromosome (Schwarz et al., 1994). The mouse Cx32 and Cx33 genes are both located on the X chromosome, but not in close proximity (Haefliger et al., 1992). The chromosomal localization studies are summarized in Table 2.

Several inherited disorders have been associated with defined defects in connexin genes. Most extensively studied is X-linked Charcot-Marie-Tooth disease (CMTX), a peripheral demyelinating neuropathy arising from mutations in the coding or presumptive regulatory sequences of Cx32 (Bergoffen et al., 1993; Fairweather et al., 1994; Palau and Sevilla, 1995; Ionasescu et al., 1996; Ainsworth et al., 1998). Although Cx32 has been immunolocalized to the paranodal myelin loops and Schmidt-Lanterman incisures of myelinating Schwann cells (Bergoffen et al., 1993), the mechanism(s) by which defects of Cx32 lead to the disease is not yet clear. Possibilities include interruption of a signaling pathway necessary for maintenance of myelin structure or disruption of the pathway for efficient transfer of low molecular weight metabolites within the compact myelin structure. A slowing of nerve conduction velocity in brain stem auditory evoked response in a CMTX patient (Nicholson and Corbett, 1996) is consistent with Cx32 serving a function in myelinating oligodendrocytes in the central nervous system similar to that for Schwann cells in the peripheral nervous system. Surprisingly, no other clinical manifestations of CMTX are known, even though Cx32 is expressed at high levels in other tissues, including the liver and endocrine pancreas. Alterations of liver response to sympathetic nerve stimulation in Cx32-knockout mice have been reported (Nelles et al., 1996), although the presence of similar subclinical defects in humans have yet to be demonstrated.

Mutations in Cx43 (especially of serines, which are possible phosphorylation sites) were reported in several patients with heart malformations associated with visceroatrial heterotaxia (Britz-Cunningham et al., 1995). However, the validity of these findings is questionable, as other studies of a large number of such patients were unable to demonstrate any Cx43 mutations (Katsuya et al., 1995; Gebbia et al., 1996; Splitt et al., 1997).

Several mutations of Cx26 have recently been reported in families with nonsyndromic sensorineural deafness (Kelsell et al., 1997). Autosomal dominant inherited cataracts have been mapped to the genes for Cx50 and Cx46 (Mackay et al., 1997; Shiels et al., 1998).

It is quite likely that mutations in other connexins will be demonstrated as the basis for yet other diseases in the coming years.

## E. Nonconnexin Components of Gap Junctions

This chapter has concentrated on the connexin family of proteins, several of which have been definitively demonstrated to form cell-cell channels. However, it is important to consider briefly the possibility that other proteins may also be components of gap junctions or form similarly functioning intercellular channels.

Several studies have suggested that plants might contain connexin-related proteins. Anti-Cx32 and anti-Cx43 antisera have been shown to react with plant polypeptides, and the anti-Cx43 immunoreactivity localizes to plasmodesmata, structures specialized for cellular communication (Meiners and Schindler, 1987; Yahalom et al., 1991). Schulz and colleagues (1992) have localized anti-Cx26 immunoreactivity on the surface of protoplasts prepared from leaves of *Helianthus annuus* and *Vicia faba*. The protein antigenically related to Cx32 has been cloned from *Arabidopsis* (Meiners et al., 1991). Its DNA sequence shows no relation to the connexins; its amino acid sequence shares a number of identical amino acids within the cytoplasmic loop region of Cx32, while there is little similarity to the extracellular or transmembrane domains shared in all connexins. It has been suggested that the plant protein with limited sequence homology to Cx32 may be a protein kinase (Mushegian and Koonin, 1993).

The lens protein MP26, which shows no sequence similarity to the connexins (Gorin et al., 1984), was once widely believed to be a lens gap junction protein, because its abundance in lens membranes parallels the abundance of gap junction structures in that tissue (Bloemendal et al., 1972); but, immunocytochemical studies have not localized it to conventional gap junction structures. However, this protein does form transmembrane channels (Zampighi et al., 1982; Ehring et al., 1990). One suggested role has been that of volume regulation, but it is also possible that by forming transmembrane channels in adjacent cells that are narrowly separated, it could still function to allow cell-cell communication. Another non-connexin lens protein, MP19, which has been isolated and cloned, also localizes to gap junctions (Galvan et al., 1989; Louis et al., 1989; Gutekunst et al., 1990). No functional information about the ability of this protein to form channels has yet been obtained; it could be an additional channel protein, or could be an accessory junctional component. Finally, Finbow and colleagues (1983, 1984) have studied a 16-kD polypeptide, termed ductin, which they find co-enriches with gap junction structures; however, its protein sequence suggests that it is in fact a component of the vacuolar adenosine triphosphatase (ATPase) (Dermietzel et al., 1989). Even though ductin can be resolved from gap junction structures containing connexins (Finbow and Meagher, 1992), Finbow and colleagues maintain that ductin is a constituent of at least some gap junction channels (Finbow and Pitts, 1993; Dunlop et al., 1995).

## II. TRANSCRIPTIONAL CONTROL AND EXPRESSION PATTERNS OF CONNEXIN GENES

### A. Analysis of Gene Control Regions

Analysis of promoter/enhancer regions of connexin genes has, thus far, been most extensive for the earliest isolated and cloned connexins: Cx32, Cx26 and Cx43. Hennemann and co-workers (1992b) have compared about 600-bp sequences of mouse Cx32 and Cx26 genes, upstream of the corresponding start sites of transcription. These sequences did not contain TATA boxes. The promoter of the mouse Cx26 gene is located in a very GC-rich region, which is reminiscent of promoters of housekeeping genes. Putative consensus sequences for a metal response element and the transcription factor NFκB were found. A more recent analysis of 4.8 kb upstream of the human Cx26 coding sequence indicated many similar features but also included elements such as a YY1-like binding site at well as a consensus sequence for a mammary gland factor binding site as is observed in the mouse β-casein gene (Kiang et al., 1997). The promoter region of the mouse Cx32 gene contains two putative binding sites for the transcription factor HNF-1 and consensus motifs for NF-1 as well as NFκB. The transcription factor HNF-1 is expressed in mouse liver, intestine, stomach, kidney and embryonic carcinoma cells, but not in ovary, spleen, heart, brain, and lung (Kuo et al., 1990). This pattern is largely consistent with the reported expression of Cx32 mRNA. Only putative binding sites for the transcription factor NFκB and the fat-specific element were identified in the 5' flanking region of mouse Cx32 as well as Cx26 gene. NFκB has previously been characterized as a pleiotropic mediator of inducible and time-specific gene control, possibly involved in acute phase inducibility of certain genes expressed in liver (Ron et al., 1990). From these studies it becomes apparent that the Cx32 and Cx26 promoter regions are distinct, although there is some overlap in the pattern of transcript expression in different tissues.

Following up on a preliminary study (Miller et al., 1988), Bai and colleagues have characterized several aspects of the rat Cx32 promoter. The basal promoter region was mapped to a region −179 to −34 upstream of the first exon. Using nuclear extracts from rat liver and HuH-7 cells, a human liver-derived cell line, in DNA mobility shift assay they identified three DNA-binding activities in extracts of rat liver nuclei. Two of them are related to the ubiquitous transcription factor SP1. Three DNase I hypersentitive regions were also found, two of which behaved as silencer elements for both native promoter and the heterologous promoter in HuH-7 cells (Bai et al., 1993). In a subsequent study (Bai et al., 1995), a 60-kD protein termed B2 was partially purified from rat liver nuclei based on its activity in the mobility shift assay. B2-binding was initially suggested for the region −152 to −127 based on methylation interference footprinting and was confirmed using a DNA probe. The Cx32-B2 complex is thought to be an essential element in the basal promoter and may be a factor in liver-specific expression of Cx32.

Expression of Cx32 in rodent nerve tissue has been shown to utilize a promoter that differs from the one used in liver. Analysis of the pattern of expression of Cx32 transgene constructs truncated in the coding region and encoding luciferase was undertaken in transgenic mice (Neuhaus et al., 1995). One construct, (including 2.5 kb of sequence upstream from the promoter, exon 1, the intron and modified protein-encoding exon 2) was expressed in liver and pancreas as anticipated. In another construct in which the 5'-sequence was deleted and the intron was truncated to include only 1.8 kb of its 3' end, expression was restricted to the nervous system. Confirmation of expression of the nerve-specific transcript was obtained by RT-PCR using rat nerve RNA. A similar finding was made when Cx32 cDNA was isolated from a rat sciatic nerve library (Sohl et al., 1996). The transcription initiation site for this mRNA was localized within the intron at 444 bp upstream of the rat exon 2. Transcription of this mRNA was stimulated upon nerve crush with kinetics similar to that observed for myelin genes, supporting a role for Cx32 in myelination.

Cx43 appears to have only a single transcriptional start (Sullivan et al., 1993). Several transcriptional regulatory elements in the Cx43 gene have been identified. Cx43 transcription can be augmented by treatment of animals with pharmacological doses of estrogens, and a Cx43 promoter-reporter gene construct was shown to be estrogen responsive when expressed in HeLa cells (Yu et al., 1994). Lye and colleagues (Lefebvre et al., 1995) have suggested that the estrogen effects may be indirect through activation of the immediate early genes Fos or Jun, or both, which may bind to AP-1 sites within the promoter. Geimonen and co-workers (1996) implicated an AP-1 site in transcriptional regulation in response to activation of protein kinase C. Chen and colleagues (1995) have identified regions that contribute to myometrial-specific regulation of Cx43 expression.

Analysis of the genomic sequence of Cx40 has recently demonstrated the presence of a TATA box preceding exon 1, as well as potential binding sites for several transcription factors, including AP-1, AP-2, SP1, TRE, and p53 (Seul et al., 1997; Groenewegen et al., 1998). The basal promoter was localized to a region 300 bp upstream of the transcription start site as well as a strong negative regulatory element between +100 and +297 (Seul et al., 1997). No difference in the site of initiation of transcription in different tissues and cell types is apparent (Groenewegen et al., 1998).

## B.  Tissue-specific Expression Pattern of Connexin Genes

After cloning of the different connexin cDNAs, they were first used for characterization of the corresponding transcripts by Northern blot hybridization or nuclease protection analysis of RNA isolated from whole organs. These studies are now being extended with experiments using in situ hybridization for localization of connexin transcripts within cross-sectioned tissue (Micevych and Abelson, 1991; Kanter et al., 1993). In a few cases, immunolocalization studies employing connexin-specific antisera have demonstrated the connexin-containing cell type within that organ. These studies have demonstrated that some connexins are widely expressed, whereas others are found in much more restricted locations, such as skin or lens.

Cx43 appears to be the most abundant connexin mRNA, both in terms of the number of tissues where it is expressed and in terms of its concentration in many tissues. Cx43 mRNA is particularly abundant in heart (Beyer et al., 1987) but is found in many other tissues, including ovary, lens epithelium, uterus, and kidney.

Cx26 mRNA and Cx32 mRNA are coexpressed in liver, kidney, intestine, lung, spleen, stomach, testes, and brain (Zhang and Nicholson, 1989). However, either gene can also be expressed without the other as shown in leptomeningeal cells (Spray et al., 1991) and pineal cells (expression of Cx26 and Cx43) (Sáez et al., 1991). Naus and colleagues (1990) reported that Cx32 mRNA shows high level expression in hindbrain, whereas Cx43 mRNA is homogeneously distributed throughout different regions of the brain.

The tissue pattern of transcript expression has been investigated for Cx37 and Cx40 in mouse (Willecke et al., 1991b, Hennemann et al., 1992d) as well as in rat (Beyer et al., 1992; Haefliger et al., 1992). Both mRNAs are maximally expressed in lung, but are also found in other adult tissues, for example kidney, heart, and skin. Reed and associates (1993) showed that Cx37 mRNA is also expressed in rat uterus and ovary and human blood vessel endothelium.

Beyer (1990) showed that chick Cx45 mRNA dropped in expression in chick heart by 90% after day 16 of embryonic development and stayed low in the adult animal. Mouse Cx45 mRNA has been found at highest abundance in lung, but has also been detected in heart, embryonic brain, embryonic skin, and embryonic kidney. Furthermore, this transcript decreased in kidney and brain dramatically at day 4 after birth (Hennemann et al., 1992d).

Recently, other connexins have been described that show a rather restricted pattern of tissue expression. Cx31, Cx31.1, and Cx30.3 transcripts were found in embryonic as well as in adult skin (Hoh et al., 1991, Haeflinger et al., 1992, Hennemann et al., 1992a, 1992c). Cx31 mRNA was also detected in placenta and Harderian gland (Hoh et al., 1991). Rat Cx33 mRNA has so far only been found in testes (Haefliger et al., 1992) where a low amount of Cx31.1 mRNA was also detected. Other connexin transcripts, Cx46 (Paul et al., 1991) and Cx50 mRNA (White et al., 1992a), are maximally expressed in lens, but Cx46 is also detected in heart and kidney. Chick Cx56 mRNA (a sequence closely related to Cx46 and Cx50), was only found in lens (Rup et al., 1993). The recent identification of Cx36 as a gap junction protein expressed in mouse neurons (Sohl et al., 1998; Condorelli et al., 1998), and for which homologs (Cx35 and Cx34.7) exist in fish (O'Brien et al., 1998) raises intriguing possibilities for characterization of the roles of electrotonic synapses in neuronal function.

### Connexin Expression in Pregnancy

Transcript expression of Cx43, Cx32, and Cx26 is modulated during pregnancy around in the rat (Risek et al, 1990). A five- to sixfold increase in Cx43 mRNA was observed 1 day before parturition in uterine myometrium and ovary. The transcript elevation may lead to the formation of gap junctions required for synchronizing the contractility of myometrial cells during parturition. In contrast, Cx26 mRNA decreases in endometrium about fourfold, 1 day before parturition. No specific changes of these connexin transcript levels were seen in heart and liver during pregnancy. These findings suggest cell-specific modulation of gap junction expression in two regions of the uterus and in the ovary during pregnancy and differential regulation of Cx43 gene expression in different organs (i.e., heart and uterus), presumably as a result of hormonal control. Winterhager and associates (1991) showed that the Cx26 and Cx43 mRNAs and proteins are expressed in rat uterus during early pregnancy, during decidualization, and at term, whereas Cx32 and Cx37 mRNAs are much less abundant. Cx26 and Cx43 show different spatial and temporal expression in rat endometrium during trophoblast invasion (Winterhager et al., 1993) as well as in late stages of pregnancy (Risek and Gilula, 1991).

### Connexin Expression in Mouse Embryos

Nishi and co-workers (1991) have studied connexin gene expression in pre- and postimplantation mouse embryos as well as during organogenesis. The Cx43 mRNA was the earliest connexin transcript detected in the 4-cell stage embryo; its abundance increased during subsequent development. Valdimarsson and colleagues (1991) also found Cx43 mRNA in 4-cell stage mouse embryos and concluded that its zygotic expression may be critical for preimplantation morphogenesis. Barron and associates (1989) had found Cx32 protein but no Cx32 mRNA during preimplantation development of the zygote to the late morula. Thus, they suggested that the Cx32 protein may be inherited as an oogenetic product through implantation development in the mouse. This hypothesis is disputed by Nishi and co-workers (1991), who could not detect Cx32, Cx26, or Cx43 mRNA in unfertilized oocytes stripped of cumulus cells.

### Connexin Expression in Xenopus Embryos

Gimlich and colleagues (1988, 1990) have studied the differential regulation of Cx43, Cx38, and Cx30 mRNA in *Xenopus* embryos. *Xenopus* Cx38 mRNA is abundant in oocytes, then decreases to levels below detection by 18 hours and is not detected in a number of adult organs, except in the ovary. The *Xenopus* Cx43 mRNA appears during organogenesis, is found in fully grown oocytes, and is rapidly degraded upon oocyte maturation.

In summary, it is evident from the recent literature that connexin gene expression is regulated according to the differentiation program of the different cell types.

## C.   Cell Type–Specific Expression of Connexin Genes

In several cases, expression of connexin transcripts has been analyzed in primary cells or established cell lines. These results further support the conclusion that expression of certain connexins is coordinately regulated by the differentiation program of a cell type. However, these studies also show that the type of connexin expressed in established cell lines is not necessarily representative of the connexin pattern expressed *in vivo* or in parental primary cells.

### Hepatocytes

Previously it had been shown that murine hepatocytes express Cx32 and Cx26 (Nicholson et al., 1987). In the most detailed study to date, Stutenkemper and associates (1992) have compared expression of Cx32, Cx26, and Cx43 mRNA in primary embryonic hepatocytes, immortalized hepatocytes, and liver-derived cell lines. Primary mouse hepatocytes and highly differentiated immortalized derivatives contain relatively large amounts of Cx32 and Cx26 mRNA but only traces of Cx43 transcripts. When these cells dedifferentiate in culture spontaneously or by shift of the culture conditions, they gradually decrease expression of Cx32 and Cx26, in parallel to the decrease in dye coupling. In contrast, the amount of Cx43 mRNA increases in dedifferentiated cells and is predominant in several liver-derived cell lines, established in serum-containing culture media. Dedifferentiation in culture and the expression pattern of connexin transcripts can be reversed.

### Cardiovascular Cells

Cardiac myocytes express Cx40, Cx43, and low amounts of Cx45 mRNA and proteins (Kanter et al., 1992). Analysis by immunocytochemistry using monospecific antisera to different connexins and by in situ hybridization have shown that specialized myocytes from different areas of the heart show different patterns of connexin expression (Bastide et al., 1993; Kanter et al., 1993; Davis et al., 1994; Delorme et al., 1995). Cx43 is a major component of the gap junctions in working myocytes of the atrium and ventricle. Cx40 is either not present or expressed only at low levels by ventricular myocytes, but is a major component of junctions between cells of the conducting system. Cx40 is abundantly expressed by atrial myocytes of at least some species. Cx45 appears to be a minor component of gap junctions in all cardiac regions.

Cx40 is abundantly detected in many types of endothelial cells *in vivo* (Bastide et al., 1993; Bruzzone et al., 1993). Levels of Cx40 appear to fall rapidly on adapting these cells to tissue culture.

Larson and colleagues (1990) showed that cultured cells derived from multiple elements of the blood vessel wall (endothelium, smooth muscle, and pericytes) all expressed Cx43. Beyer and associates (1992) showed that rat Cx40 mRNA is expressed in primary cells of aortic smooth muscle, where it is coexpressed with Cx43 transcripts. Coexpression of Cx40 and Cx43 was also found in the rat smooth muscle cell line A7r5 but not in other cell lines tested. Reed and co-workers (1993) found that primary human or bovine endothelial cells express Cx37 mRNA. The authors did not detect this transcript in primary smooth muscle cells or established A7r5 cells. It appears that endothelial cells from blood vessels in different organs may express different levels of Cx40, Cx37, and Cx43 mRNA. Pepper and associates (1992) found that the extent of gap junctional coupling correlates with the amounts of Cx43 mRNA and protein in bovine microvascular and large vessel endothelial cells.

### Keratinocytes

Hennemann and associates (1992a, 1992c) have shown that Cx31 and Cx31.1 as well as Cx30.3 mRNA are expressed in the highly differentiated Hel-30 cell line derived from mouse keratinocytes. Thus, keratinocytes express at least five different connexins, as Cx43 and Cx26 mRNA have also been found in these cells (Brissette et al., 1991). At present it is not known whether the expression of these connexin transcripts is differently modulated during keratinocyte differentiation.

## F9 Embryonic Carcinoma Cells

Cx31 and Cx45 mRNA were detected in mouse embryonic carcinoma F9 cells, in addition to Cx43 mRNA, and a low level of Cx32 transcripts (Hennemann et al., 1992a). Cx31.1 and Cx30.3 mRNA have also been found in F9 cells. As no quantitative comparison of the different connexin mRNAs and proteins has been reported, we do not know at present to what extent these transcripts are translated into functional protein.

## Other Cell Types

Cx43 mRNA has previously been shown to be expressed in fibroblast cells (Beyer et al., 1987). It is coexpressed with Cx26 mRNA in leptomeningeal cells (Spray et al., 1991). Dermietzel and associates (1991) and Giaume and colleagues (1991) reported that Cx43 mRNA and protein are expressed in rat astrocytes. Beyer and Steinberg (1991) found Cx43 mRNA and protein expressed in mouse macrophages. Berthoud and co-workers (1992) have studied expression of Cx43 protein in MDCK cells, a canine kidney epithelium-derived cell line. Pancreatic islet cells contain Cx43, whereas exocrine pancreatic cells contain Cx32 (Meda et al. 1991). Civitelli and colleagues (1993) have shown that primary human osteoblasts contain Cx43 and Cx45 mRNA.

### D.    Expression of Connexins in Transfected Cells

In order to study functional expression of cloned connexin sequences, they were injected as *in vitro*–transcribed cRNA into *Xenopus* oocytes (Dahl et al., 1987; Swenson et al., 1989; Werner et al., 1989), and measurements of electrical conductance between oocyte pairs were obtained. More recently, these studies have been complemented by transfecting cloned connexin sequences into mammalian cells defective in dye and electrical communication. Cx32 and Cx43 have been functionally expressed in SK-Hep1 cells, a human hepatocarcinoma cell line. (Eghbali et al., 1990; Fishman et al., 1990). C6 rat glioma cells have been transfected with rat Cx43 cDNA (Zhu et al., 1991). The connexin-transfected cells show a reduced frequency of tumorigenesis (Eghbali et al., 1991) or grow more slowly in culture than the nontransfected parental cells (Zhu et al., 1991). This suggests that cell proliferation may be at least partially controlled by intercellular communication through functional gap junctions (see the chapter by Neveu and Bertram in this volume).

Hennemann and associates (1992d) have expressed mouse Cx40 in SK-Hep1 cells and in HeLa cells, a human cervix carcinoma cell line. Beyer and associates (1992), Veenstra and coleagues (1992), Reed and colleagues (1993), and Rup and associates (1993) have transfected rat Cx40; chick Cx42, Cx43, and Cx45; human Cx37; and chick Cx56 DNA into communication- deficient N2A cells, a mouse neuroblastoma cell line. Because electrical coupling is absent or very low in the parental cells, the unitary conductance of single channels measured in transfectants can be assigned to the exogenous connexin. In this way, homotypic gap junction channels composed of homogeneous connexins can be characterized in transfected cells and compared with results from primary cells, which frequently express more than one type of connexin. The formation of heterotypic channels, which have been demonstrated between pairs of *Xenopus* oocytes injected with rat Cx26 and Cx32 cRNA (Barrio et al., 1991) or with mouse Cx37 and Cx40 cRNA (Hennemann et al., 1992d), can also be studied in the mammalian transfection systems. Thus, one can expect that the electrophysiological characterization of homomeric, homotypic, and heterotypic gap junction channels should make rapid progress using cells in which gap junctional communication has been restored by transfection with exogenous connexin DNA.

### E.  Transcriptional and Post-transcriptional Control of Connexin Gene Expression

So far only a few reports have been published in which regulatory effects on the total level of connexin mRNAs have been ascribed to initiation or stabilization of the corresponding mRNAs, or both. Sáez and associates (1990) have shown that 8Br-cAMP increased the stability of Cx32 mRNA at least twofold in primary hepatocytes, but did not lead to an increase of transcription rate. Rosenberg and colleagues (1992) have found that Cx26 mRNA is 2.8 times more abundant in periportal than in pericentral rat hepatocytes, whereas Cx32 mRNA is equally distributed in both zones of the liver. Cx26 mRNA is transcribed at a faster rate than Cx32 mRNA in nuclei isolated from both cellular fractions. Furthermore, the rate of Cx26 mRNA transcription is 3.9-fold faster in nuclei from periportal than from pericentral cells. The authors conclude that Cx26 mRNA synthesis is controlled transcriptionally by "zonation" of hepatocytes in the liver, whereas the relative abundance of connexin mRNAs is determined by post-transcriptional regulatory mechanisms. Stutenkemper and associates (1992) have shown that addition of hydrocortisone to serum free culture medium of immortalized mouse embryonic hepatocytes caused a twofold decrease in Cx26 mRNA, no change in the level of Cx32 mRNA, but a twofold increase of Cx43 mRNA. These experiments do not differentiate between synthesis and stability of the corresponding mRNAs.

Mehta and colleagues (1992) have observed that Cx43 mRNA (but not Cx32 or Cx26 mRNA) in the rat Morris hepatoma cell line is increased by intracellular cyclic adenosine monophosphate (cAMP) 15- to 40-fold, although the rate of Cx43 transcription is only stimulated sixfold. Thus, in addition to its effects on initiation of Cx43 mRNA synthesis, cAMP appears to stabilize Cx43 mRNA. Maximum communication in these cells was reached 12 hours after addition of cAMP although Cx43 mRNA peaked already after 6 hours. The authors attribute this effect to other stages in functional expression of Cx43 mRNA. When the cAMP stimulus was removed, communication in these cells persisted for another 28 hours after the decrease of Cx43 mRNA. As the half-life of Cx43 protein is between 1 and 3 hours (Musil et al., 1990; Laird et al., 1991), the prolonged maintenance of communication could indicate that the half-life of Cx43 protein in functional gap junction channels is considerably longer in the presence of cAMP. The Morris hepatoma cells used in these studies are derived from hepatocytes that express Cx32 and Cx26, but only traces of Cx43; thus, the effects of cAMP on induction of Cx43 should also be analyzed with primary cells expressing Cx43, to verify their physiological relevance.

In addition to the effects of cAMP, Cx43 gene expression can also be regulated by retinoids that exert their effects by binding to nuclear receptor proteins. At very low concentrations ($10^{-10}$ to $10^{-9}$ M) retinoic acid inhibits gap junctional communication, whereas at $10^{-9}$ to $10^{-7}$ M, it enhances intercellular coupling in 10T1/2 or 3T3 mouse fibroblastoid cells (Mehta et al., 1992). Rogers and colleagues (1990) have shown, using 10T1/2 cells, that gap junctional communication and Cx43 mRNA rose after treatment with retinoids. In this cell system, the action of retinoids on cellular growth correlates with their action on gap junctional communication (Mehta et al., 1989, Mehta and Loewenstein, 1990). As mentioned earlier in this chapter, it is not clear that "cAMP response elements" or consensus-binding regions for retinoid receptors exist in the control region of the Cx43 gene.

Lee and associates (1992) discovered that human mammary tumor cell lines lack expression of Cx26 and Cx43 mRNA, in contrast to normal mammary epithelial cells. They suggested that Cx26 may be a class II tumor suppressor gene; that is, downregulated in tumor cells at the transcriptional level and not by loss or mutation of the coding sequence (Lee et al., 1991). Interestingly, human Cx26 is a cell cycle regulated gene, expressed in mammary epithelial cells at a moderate level during G1 and S and strongly upregulated in late S and G2 (Lee et al., 1992). More information about the involvement of gap junctional communication in growth control and cellular transformation can be found in the chapter by Neveu and Bertram later in this volume.

## III.  CONCLUSIONS AND OUTLOOK

The central subject of this chapter is the surprising heterogeneity of connexin genes. For reasons not understood, at least 15 connexin genes have been generated during evolution of mammalian species. Possibly, additional connexin genes are still to be discovered. Several of the known connexin genes show broad patterns of tissue expression; others appear to be restricted in expression to one tissue or even one cell type. Many, if not all, cell types express more than one type of connexin. In view of these differences in transcriptional regulation, it is unlikely that the different connexin genes fulfill exactly the same function in intercellular communication.

The connexin gene products may differ in regulation of cellular coupling, junctional permeability, and channel gating. This may be dependent on the central cytoplasmic loop and the C-terminal region, which are different between all connexins analyzed.

Additional gap junction physiological diversity could be generated by the mixing of different connexins either in two adjacent cells or within a single cell. Heterotypic connexin channels have been demonstrated between pairs of *Xenopus* oocytes injected with rat Cx26 and Cx32 cRNA (Barrio et al., 1991) or mouse Cx37 and Cx40 cRNA (Hennemann et al., 1992d). However, there is also evidence that certain connexins do not form heterotypic channels, as exemplified by Cx43 and Cx40, (Bruzzone et al., 1993; Haubrich et al., 1996) or Cx43 and Cx50 (White et al., 1992b), which do not couple to one another when expressed in *Xenopus* oocytes. These results could explain selective uncoupling among cells expressing different connexins. It has also been shown that two connexins expressed together in the same cell (Cx32/Cx26, Cx46/Cx50, Cx37/Cx43) may form hemichannels comprised of different subunits (Jiang and Goodenough, 1996; Brink et al., 1997; Bevans et al., 1998). Formation of these heteromeric channels may yield additional functional diversity.

Because different connexin channels exhibit unique unitary conductances, it is likely that their velocity of impulse conduction, molecular permeability, or selectivity may also differ. Possibly within an organism, the coordinated programming and maintenance of different pathways of cellular differentiation requires a delicate balance in the concentration of ions and second messenger molecules that is facilitated by the multiplicity of differentially expressed connexins. Future studies of connexin diversity will undoubtedly concentrate on functional differences between connexins and their role in intercellular communication during development as well as their contributions to cell type specific functions. The cloning of connexins has allowed molecular perturbations of their gene expression. As described elsewhere in this volume, several connexin knockout mice have been generated. The consequences of specific knockouts range from but minor abnormalities in mature animals to early embryonic lethality.

## REFERENCES

Ainsworth, P.J., Bolton, C.F., Murphy, B.C., Stuart, J.A., & Hahn, A.F. (1998). Genotype/phenotype correlation in affected individuals of a family with a deletion of the entire coding sequence of the connexin 32 gene. Hum. Genet. 103, 242-424.

Bai, S., Schoenfeld, A., Pietrangelo, A., & Burk, R.D. (1995). Basal promoter of the rat connexin 32 gene: identification and characterization of an essential element and its DNA-binding protein. Mol. Cell Biol. 15, 1439-1445.

Bai, S., Spray, D.C., & Burk, R.D. (1993). Identification of proximal and distal regulatory elements of the rat connexin32 gene. Biochim. Biophys. Acta 1216, 197-204.

Barrio, L.C., Suchyna, T., Bargiello, T., Xu, L.X., Roginsky, R.S., Bennett, M.V.L., & Nicholson, B.J. (1991). Gap juncitons formed by connexins26 and 32 alone and in combination are differently affected by applied voltage. Proc. Natl. Acad. Sci. USA 88, 8410-8414.

Barron, D.J., Valdimarsson, G., Paul, D.L., & Kidder, G.M. (1989). Connexin32, a gap junction protein, is a persistent oogenetic product during preimplantation development in the mouse. Dev. Genet. 10, 318-324.

Bastide, B., Neyses, L., Ganten, D., Paul, M., Willecke, K., & Traub, O. (1993). Gap junction protein connexin40 is preferentially expressed in vascular endothelium and conductive bundles of rat myocardium and is increased under hypertensive conditions. Circ. Res. 73, 1138-1149.

Bennett, M.V., Barrio, L.C., Bargiello, T.A., Spray, D.C., Hertzberg, E., & Sáez, J.C. (1991). Gap junctions: new tools, new answers, new questions. Neuron 6, 305-320.

Bergoffen, J., Scherer, S.S., Wang, S., Scott, M.O., Bone, L.J., Paul, D.L., Chen, K., Lensch, M.W., Chance, P.F., & Fischbeck, K.H. (1993). Connexin mutations in X-linked Charcot-Marie-Tooth disease. Science 262, 2039-2042.

Berthoud, V.M., Ledbetter, M.L.S., Hertzberg, E.L., & Sáez, J.C. (1992). Connexin43 in MDCK cells: regulation by a tumor promoting phorbol ester and $Ca^{2+}$. Eur. J. Cell Biol. 57, 40-50.

Bevans, C.G., Kordel, M., Rhee, S.K., & Harris, A.L. (1998). Isoform composition of connexin channels determines selectivity among second messengers and uncharged molecules. J. Biol. Chem. 273, 2808-2816.

Beyer, E.C. (1990). Molecular cloning and developmental expression of two chick embryo gap junction proteins. J. Biol. Chem. 265, 14439-14443.

Beyer, E.C., Paul, D.L., & Goodenough, D.A. (1987). Connexin43: a protein from rat heart homologous to a gap junction protein from liver. J. Cell Biol. 105, 2621-2629.

Beyer, E.C., Paul, D.L., & Goodenough, D.A. (1990). Connexin family of gap junction proteins. J. Membr. Biol. 116, 187-194.

Beyer, E.C., Reed, K.E., Westphale, E.M., Kanter, H.L., & Larson, D.M. (1992). Molecular cloning and expression of rat connexin40, a gap junction protein expressed in vascular smooth muscle. J. Membr. Biol. 127, 69-76.

Beyer, E.C. & Steinberg, T.H. (1991). Evidence that the gap junction protein connexin-43 is the ATP-induced pore of mouse macrophages. J. Biol. Chem. 266, 7971-7974.

Bloemendal, H.., Zweers, A., Vermorken, F., Dunia, I., & Benedetti, E.L. (1972). The plasma membrane of eye lens fibers. Biochemical and structural characterization. Cell Differ. 1, 91-106.

Brink, P.R., Cronin, K., Banach, K., Peterson, E., Westphale, E.M., Seul, K.H., Ramanan, S.V., & Beyer, E.C. (1997). Evidence for heteromeric gap junction channels formed from rat connexin43 and human connexin37. Am. J. Physiol. (Cell Physiol.) 273, C1386-C1396.

Brissette, J.L., Kumar, N.M., Gilula, N.B., & Dotto, G.P. (1991). The tumor promoter 12-0-tetradecanoylphorbol-13-acetate and the ras oncogene modulate expression and phosphorylation of gap junction proteins. Mol. Cell Biol. 11, 5364-5371.

Britz-Cunningham, S.H., Shah, M.M., Zuppan, C.W., & Fletcher, W.H. (1995). Mutations of the Connexin43 gap-junction gene in patients with heart malformations and defects of laterality. N. Engl. J. Med. 332, 1323-1329.

Bruzzone, R., Haefliger, J.A., Gimlich, R.L., & Paul, D.L. (1993). Connexin40, a component of gap junctions in vascular endothelium, is restricted in its ability to interact with other connexins. Mol. Biol. Cell 4, 7-20.

Bruzzone, R., White, T.W., Yoshizaki. G., Patino, R., & Paul, D.L. (1995). Intercellular channels in teleosts: functional characterization of two connexins from Atlantic croaker. FEBS Lett. 358, 301-304.

Caspar, D.L.D., Goodenough, D.A., Makowski, L., & Phillips, W.P. (1977). Gap junction structures. I. Correlated electron microscopy and X-ray diffraction. J. Cell Biol. 74, 605-628.

Chen, Z.-Q., Lefebvre, D., Bai, X., Reaume, A., Rossant, J., & Lye, S.J. (1995) Identification of two regulatory elements within the promoter region of the mouse connexin43 gene. J.Biol.Chem. 270, 3863-3868.

Church, R.L., Wang, J.H., & Steele, E. (1995). The human lens intrinsic membrane protein MP70 (Cx50) gene: clonal analysis and chromosome mapping. Curr. Eye Res. 14, 215-221.

Civitelli, R., Beyer, E.C., Warlow, P.W., Robertson, A.J., Geist, S.T., & Steinberg, T. (1993). Connexin43 mediates direct intercellular communication in human osteoblast cell networks. J. Clin. Invest. 91, 1888-1896.

Condorelli, D.F., Parenti, R., Spinella, F., Salinaro, A.T., Belluardo, N., Cardile, V., & Cicirata, F. (1998). Cloning of a new gap junction gene (Cx36) highly expressed in mammalian brain neurons. Eur. J. Neurosci. 10, 1202-1208.

Corcos, I.A., Lafreniere, R.G., Begy, C.R., Loch Caruso, R., Willard, H.F., & Glover, T.W. (1992). Refined localization of human connexin32 gene locus, GJB1, to Xq13.1. Genomics 13, 479-480.

Dahl, E., Manthey, D., Chen, Y., Schwarz, H.J., Chang, Y.S., Lalley, P.A., Nicholson, B.J., & Willecke, K. (1996). Molecular cloning and functional expression of mouse connexin-30, a gap junction gene highly expressed in adult brain and skin. J. Biol. Chem. 271, 17903-17910.

Dahl, G., Miller, T., Paul, D.L., Voellmy, D., & Werner, R. (1987). Expression of functional cell-cell channels from cloned rat liver gap junction complementary DNA. Science 236, 1290-1293.

Davis, L.M., Kanter, H.L., Beyer, E.C., & Saffitz, J.E. (1994). Distinct gap junction phenotypes in cardiac tissues with disparate conduction properties. J. Am. Coll. Cardiol. 24, 1124-1132.

Delorme, B., Dahl, E., Jarry-Guichard, T., Marics, I., Briand, J., Willecke, K., Gros, D., & Theveniau-Ruissy. (1995). Developmental regulation of connexin40 gene expression in mouse heart correlates with differentiation of the conducting system. Dev. Dynam. 204, 35-71.

Dermietzel, R., Hertzberg, E.L., Kessler, J.A., & Spray, D.C. (1991). Gap junctions between cultured astrocytes: immunochemical, molecular, and electrophysiological analysis, J. Neurosci. 11, 1421-1432.

Dermietzel, R., Volker, M., Hwang, T.K., Berzborn, R.J., & Meyer, H.E. (1989). A 16 kDa protein co-isolating with gap junctions from brain tissue belonging to the class of proteolipids of the vacuolar HATPases. FEBS Lett. 253, 1-5.

Dunlop, J., Jones, P.C., & Finbow, M.E. (1995). Membrane insertion and assembly of ductin: a polytopic channel with dual orientations. EMBO J. 14, 3609-3616.

Ebihara, L., Beyer, E.C., Swenson, K.I., Paul, D.L., & Goodenough, D.A. (1989). Cloning and expression of a Xenopus embryonic gap junction protein. Science 243, 1194-1195.

Eghbali, B., Kessler, J.A., Reid, L.M., & Spray, D.C. (1991) Involvement of gap junctions in tumorgenesis: transfection of tumor cells with connexin32 cDNA retards growth *in vivo*. Proc. Nat. Acad. Sci. USA 88, 10701-10705.

Eghbali, B., Kessler, J.A., & Spray, D.C. (1990). Expression of gap junctional channels in communication incompetent cells after stable transfection with cDNA encoding connexin32. Proc. Natl Acad. Sci. USA, 87, 1328-1331.

Ehring, G.R., Zampighi, G.A.., Horwitz, J., Bok, D., & Hall, J.E. (1990). Properties of channels reconstituted from the major intrinsic protein of lens fiber membrane. J. Gen. Physiol. 96, 631-664.

Essner, J.J., Laing, J.G., Beyer, E.C., Johnson, R.G., & Hackett, P.B., Jr. (1996). Expression of zebrafish connexin43.4 in the notochord and tail bud of wild-type and mutant no tail embryos. Dev. Biol. 177, 449-462.

Evans, W.H. and Gurd, J.W. (1972). Preparation and properties of nexuses and lipid enriched vesicles from mouse liver plasma membranes. Biochem. J. 128, 691-700.

Fairweather, N., Bell, C., Cochrane, S., Chelly, J., Wang, S., Mostacciuolo, M.L., Monaco, A.P., & Haites, N.E. (1994). Mutations in the connexin 32 gene in X-linked dominant Charcot-Marie-Tooth disease (CMTX1). Hum. Mol. Genet. 3, 29-34.

Finbow, M.E., Buultjens, T.E.J., Lane, N.J., Shuttleworth, J., & Pitts, J.D. (1984). Isolation and char-
    acterization of arthropod gap junctions. EMBO J. 3, 2271-2278.
Finbow, M.E., & Meagher, L. (1992). Connexins and the vacuolar proteolipid-like 16-kDa protein are
    not directly associated with each other but may be components of similar or the same gap junc-
    tional complexes. Exp. Cell Res. 203, 280-284.
Finbow, M.E., & Pitts, J.D. (1993). Is the gap junction channel--the connexon—made of connexin or
    ductin? J. Cell Sci. 106, 463-471.
Finbow, M.E., Shuttleworth, J., Hamilton, A.E., & Pitts, J.D. (1983). Analysis of vertebrate gap junc-
    tion protein. EMBO J. 2, 1479-1486.
Fishman, G.I., Eddy, R.L., Shows, T.B., Rosenthal, L., & Leinwand, L.A. (1991a). The human con-
    nexin gene family of gap junction proteins: distinct chromosomal locations but similar struc-
    tures. Genomics 10, 250-256.
Fishman, G.I., Moreno, A.D., Spray, D.C., & Leinwand, L.A. (1991b). Functional analysis of human
    cardiac gap junction channel mutants. Proc. Nat. Acad. USA 88, 3525-3529.
Fishman, G.I., Spray, D.C., & Leinwand, L.A. (1990). Molecular characterization and functional
    expression of the human cardiac gap junction channel. J. Cell Biol. 111, 589-598.
Galvan, A., Lampe, P.D., Hur, K.C., Howard, J.B., Eccleston, E.D., Arneson, M., & Louis, C.F.
    (1989). Structural organization of the lens fiber plasma membrane protein MP18. J. Biol.
    Chem. 264, 19974-19978.
Gebbia, M., Towbin, J.A., & Casey, B. (1996). Failure to detect connexin43 mutations in 38 cases of
    sporadic and familial heterotaxy. Circulation 94, 1909-1912.
Geimonen, E., Jiang, W., Ali, M., Fishman, G.I., Garfield, R.E., & Andersen, J. (1996) Activation of
    protein kinase C in human uterine smooth muscle induces connexin-43 gene transcription
    through an AP-1 site in the promoter sequence. J. Biol. Chem. 271, 23667-23674.
Geyer, D.D., Church, R.L., Steele, E.C., Jr., Heinzmann, C., Kojis, T.L., Klisak, I., Sparkes, R.S., &
    Bateman, J.B. (1997). Regional mapping of the human MP70 (Cx50; connexin 50) gene by flu-
    orescence in situ hybridization to 1q21.1. Mol. Vis. 3, 13
Giaume, C., Fromaget, C., el Aoumari, A., Cordier, J., Glowinski, J., & Gros, D. (1991). Gap junctions
    in cultured astrocytes: singel channel currents and characterization of the channel-forming pro-
    tein. Neuron 6, 133-143.
Gimlich, R.L., Kumar, N.M., & Gilula, N.B. (1988). Sequence and developmental expression of
    mRNA coding for a gap junction protein in Xenopus. J. Cell Biol. 107, 1065-1073.
Gimlich, R.L., Kumar, N.M., & Gilula, N.B. (1990). Differential regulation of the levels of three gap
    junction mRNAs in Xenopus embryos. J. Cell Biol. 110, 597-605.
Goodenough, D.A., & Stoekenius, W. (1872). The isolation of mouse hepatocyte gap junctions. Pre-
    liminary chemical characterization and X-ray diffraction. J. Cell Biol. 110, 597-605.
Gorin, M.B., Yancey, S.B., Cline, J., Revel, J.P., & Horwitz, J. (1984). The major intrinsic protein
    (MIP) of the bovine lens fiber membrane: characterization and structure based on cDNA clon-
    ing. Cell 39, 49-59.
Green, C.R., Harfst, E., Gourdie, R.G., & Severs, N.J. (1988). Analysis of the rat liver gap junction
    protein: clarification of anomalies in its molecular size. Proc. R. Soc. Lond. [Biol]. 233,
    165-174.
Groenewegen, W.A., van Veen, T.A., van der Velden, H.M., & Jongsma, H.J. (1998). Genomic orga-
    nization of the rat connexin40 gene: identical transcription start sites in heart and lung. Cardio-
    vasc. Res. 38, 463-471.
Gupta, V.K., Berthoud, V.M., Atal, N., Jarillo, J.A., Barrio, L.C., & Beyer, E.C. (1994). Bovine
    connexin44, a lens gap junction protein: molecular cloning, immunologic characterization, and
    functional expression. Invest. Ophthalmol. Vis. Sci. 35, 3747-3758.
Gutekunst, K.A., Rao, G.N., & Church, R.L. (1990). Molecular cloning and complete nucleotide
    sequence of the cDNA encoding a bovine lens intrinsic membrane protein (MP19). Exp. Eye
    Res. 9, 955-961.

Haefliger, J.A., Bruzzone, R., Jenkins, N.A., Gilbert, D.J., Copeland, N.G., & Paul, D.L. (1992). Four novel members of the connexin family of gap junction proteins. Molecular cloning, expression, and chromosome mapping. J. Biol. Chem. 267, 2057-2064.

Haubrich,S., H.J., Schwarz, F., Bukauskas, H., Lichtenberg- Frate, O., Traub,R.Weingart, & Willecke, K. (1996). Incompatibility of connexin40 and -43 hemichannels in gap junctions between mammalian cells is determined by intracellular domains. Mol. Biol. Cell 7, 1995-2006.

Henderson, D., Eibl, H., & Weber, K. (1979). Structure and biochemistry of mouse hepatic gap junctions. J. Mol. Biol. 132, 193-218.

Hennemann, H., Dahl, E., White, J.B., Schwarz, H.J., Lalley, P.A., Chang, S., Nicholson, B.J., & Willecke, K. (1992a). Two gap junction genes, connexin 31.1 and 30.3, are closely linked on mouse chromosome 4 and preferentially expressed in skin. J. Biol. Chem. 267, 17225-17233.

Hennemann, H., Kozjek, G., Dahl, E., Nicholson, B., & Willecke, K. (1992b). Molecular cloning of mouse connexins26 and -32: similar genomic organization but distinct promoter sequences of two gap junction genes. Eur. J. Cell Biol. 58, 81-89.

Hennemann, H., Schwarz, H.J., & Willecke, K. (1992c). Characterization of gap junction genes expressed in F9 embryonic carcinoma cells: molecular cloning of mouse connexin31 and -45 cDNAs. Eur. J. Cell Biol. 57, 51-58.

Hennemann, H., Suchyna, T., Lichtenberg Frate, H., Jungbluth, S., Dahl, E., Schwarz, J., Nicholson, B.J., & Willecke, K. (1992d). Molecular cloning and functional expression of mouse connexin40, a second gap junction gene preferentially expressed in lung. J. Cell Biol. 117, 1299-1310.

Hertzberg, E.L., & Gilula, N.B. (1979). Isolation and characterization of gap junctions from rat liver. J. Biol. Chem. 254, 2138-2147.

Heynkes, R., Kozjek, G., Traub, O., & Willecke, K. (1986). Identification of a rat liver cDNA and mRNA coding for the 28 kDa gap junction protein. FEBS Lett. 205, 56-60.

Higgins, D.G., & Sharp, P.M. (1988). CLUSTAL: a package for performing multiple sequence alignment on a microcomputer. Gene 73, 237-244.

Hoh, J.H., John, S.A., & Revel, J.P. (1991). Molecular cloning and characterization of a new member of the gap junction gene family, connexin-31. J. Biol. Chem. 266, 6524-6531.

Hsieh, C.L., Kumar, N.M., Gilula, N.B., & Francke, U. (1991). Distribution of genes for gap junction membrane channel proteins on human and mouse chromosomes. Somat. Cell Mol. Genet. 17, 191-200.

Ionasescu, V., Ionasescu, R., & Searby, C. (1996). Correlation between connexin 32 gene mutations and clinical phenotype in X-linked dominant Charcot-Marie-Tooth neuropathy. Am. J. Med. Genet. 63, 486-491.

Itahana, K., Tanaka, T., Morikazu, Y., Komatu, S., Ishida, N., & Takeya, T. (1998). Isolation and characterization of a novel connexin gene, Cx-60, in porcine ovarian follicles. Endocrinology 139, 320-329.

Jiang, J.X., & Goodenoughm D.A. (1996). Heteromeric connexons in lens gap junction channels. Proc. Natl. Acad. Sci. USA 93, 1287-1291.

Jiang, J.X., White, T.W., Goodenough, D.A., & Paul, D.L. (1994). Molecular cloning and functional characterization of chick lens fiber connexin 45.6. Mol. Biol. Cell 5, 363-373.

Kanter, H.L., Saffitz, J.E., & Beyer, E.C. (1992). Cardiac myocytes express multiple gap junction proteins. Circ. Res. 70, 438-444.

Kanter, H.L., Laing, J.G., Beau, S.L., Beyer E.C., & Saffitz, J.E. (1993). Distinct patterns of connexin expression in canine Purkinje fibers and ventricular muscle. Circ. Res. 72, 1124-1131.

Kiang, D.T., Jin, N., Tu, Z.J., & Lin, H.H. (1997). Upstream genomic sequence of the human connexin26 gene. Gene 199, 165-171.

Kistler, J., Christie, D., & Bullivant, S. (1988). Homologies between gap junction proteins in lens, heart and liver. Nature 331, 721-723.

Kistler, J., Kirkland, B., & Bullivant, S. (1985). Identification of a 70,000-D protein in lens membrane junctional domains. J. Cell Biol. 101, 28-35.

Kumar, N.M., & Gilula, N.B. (1986). Cloning and characterization of human and rat liver cDNAs coding for a gap junction protein. J. Cell Biol. 103, 767-776.

Kumar, N.M., & Gilula, N.B. (1992). Molecular biology and genetics of gap junction channels. Semin. Cell Biol. 3, 3-16.

Kuo, C.J., Conley, B.P., Hsieh, C.L., Francke, U., & Crabtree, B.R. (1990). Molecular cloning, functional expression, and chromosomal localization of mouse hepatocyte nuclear factor 1. Proc. Natl. Acad. Sci. USA 87, 9838-9842.

Laird, D.W., Puranam, K.L., & Revel, J.-P. (1991). Turnover and phosphorylation dynamics of connexin43 gap junction protein in cultured cardiac myocytes, Biochem. J. 273, 67-72.

Lang, L.M., Beyer, E.C., Schwartz, A.L., & Gitlin, J.D. (1991). Molecular cloning of a rat uterine gap junction protein and analysis of gene expression during gestation. Am. J. Physiol. 260, E787-E793.

Larson, D.M., Haudenschild, C.C., & Beyer, E.C. (1990). Gap junction messenger RNA expression by vascular wall cells. Circ. Res. 66, 1074-1080.

Lash, J.A., Critser, E.S., & Pressler, M.L. (1990). Cloning of a gap junctional protein from vascular smooth muscle and expression in two-cell mouse embryos. J. Biol. Chem. 265, 13113-13117.

Lee, S., Tomasetto, W.C., Paul, D., Keymarsi, K., & Sager, R. (1992) Transcriptional downregulation of gap junction proteins blocks junctional communication in human mammary tumor cell lines. J. Cell Biol. 118, 1213-1221.

Lee, S., Tomasetto, W.C., & Sager, R. (1991) Positive selection of candidate tumor suppressor genes by subtractive hybridization. Proc. Natl. Acad. Sci. USA 88, 2825-2829.

Lefebvre, D.L., Piersanti, M., Bai, X.-H., Chen, Z.-Q., & Lye, S.J. (1995). Myometrial transcriptional regulation of gap junction gene, connexin-43. Reprod.Fertil.Dev. 7, 603-611.

Louis, C.F., Hur, K.C., Galvan, A.C., TenBroek, E.M., Jarvis, L.J., Eccleston, E.D., & Howard, J.B. (1989). Identification of an 18,000-Dalton protein in mammalian lens fiber cell membranes. J. Biol. Chem. 264, 19967-19973.

Mackay, D., Ionides, A., Berry, V., Moore, A., Bhattacharya, S., & Shiels, A. (1997). A new locus for dominant "zonular pulverulent" cataract, on chromosome 13. Am. J. Hum. Genet. 60, 1474-1478.

Makowski, L., Caspar, D.L.D., Phillips, W.C., & Goodenough, D.A. (1976). Gap junction structures. II. Analysis of the X-ray diffraction data. J. Cell Biol. 74, 629-645.

Manjunath, C.K., Nicholson, B.J., Teplow, D., Hood, L., Page, E., & Revel, J.P. (1987). The cardiac gap junction protein (Mr 47,000) has a tissue- specific cytoplasmic domain of Mr 17,000 at its carboxy-terminus. Biochem. Biophys. Res. Commun. 142, 228-234.

Meda, P., Chanson, M., Pepper, M., Giordano, E., Bosco, D., Traub, O., Willecke, K., El Aoumari, A., Gros, D., Beyer, E.C., Orci, L., Spray, D.C. (1991). *In vivo* modulation of connexin43 gene expression and junctional coupling of pancreatic B-cells. Exp. Cell Res. 192, 469-480.

Mehta, P.P., & Loewenstein, W.R. (1990). Differential regulation of communication by retinoid acid in homologous and heterologous junctions between normal and transformed cells. J. Cell Biol. 113, 371-379.

Mehta, P.P., Yamamoto, M., & Rose, B. (1992). Transcription of the gene for the gap junctional protein connexin43 and expression of functional cell-to-cell channels are regulated by cAMP. Mol. Biol. Cell 3, 839-850.

Meiners, S., & Schindler, M. (1987). Immunological evidence for gap junction polypeptide in plant cells. J. Biol. Chem. 262, 951-953.

Meiners, S., Xu, A., & Schindler, M. (1991). Gap junction protein homologue from *Arabidopsis thaliana:* evidence for connexins in plants. Proc. Natl. Acad. Sci. USA 88, 4119-4122.

Micevych, P.E., & Abelson, L. (1991). Distribution of mRNAs coding for liver and heart gap junction proteins in the rat central nervous system. J. Comp. Neurol. 305, 96-118.

Mignon, C., Fromaget, C., Mattei, M.G., Gros, D., Yamasaki, H., & Mesnil, M. (1996). Assignment of connexin 26 (GJB2) and 46 (GJA3) genes to human chromosome 13q11-->q12 and mouse chromosome 14D1-E1 by in situ hybridization. Cytogenet. Cell Genet. 72, 185-186.

Miller, T., Dahl, G., & Werner, R. (1988). Structure of a gap junction gene: rat connexin-32. Biosci. Rep. 8, 455-464.

Mushegian, A. R., & Koonin, E. V. (1993). The proposed plant connexin is a protein kinase-like protein. Plant Cell 5, 998-999.

Musil, L.S., Beyer, E.C., & Goodenough, D.A. (1990). Expression of the gap junction protein connexin43 in embryonic chick lens: molecular cloning, ultrastructural localization, and post-translational phosphorylation. J. Membr. Biol. 116, 163-175.

Naus, C.C.G., Belliveau, D.J., & Bechberger, J.F. (1990) Regional differences in connexin32 and connexin43 messenger RNAs in rat brain. Neurosci. Lett. 111, 297-302.

Nelles, E., Butzler, C., Jung, D., Temme, A., Gabriel, H.D., Dahl, U., Traub, O., Stumpel, F., Jungermann, K., Zielasek, J., Toyka, K.V., Dermietzel, R., & Willecke, K. (1996). Defective propagation of signals generated by sympathetic nerve stimulation in the liver of connexin32-deficient mice. Proc. Natl. Acad. Sci. USA 93, 9565-9570.

Neuhaus, I.M., Bone, L., Wang, S., Ionasescu, V., & Werner, R. (1996). The human connexin32 gene is transcribed from two tissue-specific promoters. Biosci. Rep. 16, 239-248.

Neuhaus, I.M., Dahl, G., & Werner, R. (1995). Use of alternate promoters for tissue-specific expression of the gene coding for connexin32. Gene 158, 257-262.

Nicholson, B., Dermietzel, R., Teplow, D., Traub, O., Willecke, K., & Revel, J.P. (1987). Two homologous protein components of hepatic gap junctions. Nature 329, 732-734.

Nicholson, B.J., Gros, D.B., Kent, S.B.H., Hood, L.E., & Revel, J.P. (1985). The Mr 28,000 gap junction proteins from rat heart and liver are different but related. J. Biol. Chem. 260, 6514-6517.

Nicholson, G., & Corbett, A. (1996). Slowing of central conduction in X-linked Charcot-Marie-Tooth neuropathy shown by brain stem auditory evoked responses. J. Neurol. Neurosurg. Psychiatry 61, 43-46.

Nishi, M., Kumar, N.M., & Gilula, N.B. (1991). Developmental regulation of gap junction gene expression during mouse embryonic development. Dev. Biol. 146, 117-130.

O'Brien, J., al-Ubaidi, M.R., & Ripps, H. (1996). Connexin 35: a gap-junctional protein expressed preferentially in the skate retina. Mol. Biol. Cell 7, 233-243.

O'Brien, J., Bruzzone, R., White, T.W., al-Ubaidi, M.R., & Ripps, H. (1998). Cloning and expression of two related connexins from the perch retina define a distinct subgroup of the connexin family. J. Neurosci. 18, 7625-7637.

Palau, F., & Sevilla, T. (1995). [Genetics of peripheral neuropathies and hereditary ataxias]. Neurologia 10, 32-43.

Paul, D.L. (1986). Molecular cloning of cDNA for rat liver gap junction protein. J. Cell Biol. 103, 123-134.

Paul, D.L., Ebihara, L., Takemoto, L.J., Swenson, K.I., & Goodenough, D.A. (1991). Connexin46, a novel lens gap junction protein, induces voltage-gated currents in nonjunctional plasma membrane of Xenopus oocytes. J. Cell Biol. 115, 1077-1089.

Pepper, M.S., Montesano, R., El Aoumari, A., Gros, D., Orci, L., & Meda, P. (1992). Coupling and connexin43 expression in microvascular and large vessel endothelial cells. Am. J. Physiol. 262, C1246-C1257.

Reed, K.E., Westphale, E.M., Larson, D.M., Wang, H.Z., Veenstra, R.D., & Beyer, E.C. (1993). Molecular cloning and functional expression of human connexin37,an endothelial cell gap junction protein. J. Clin. Invest. 91, 997-1004.

Risek, B., & Gilula, N.B. (1991) Spatiotemporal expression of three gap junction gene products involved in fetomaternal communication during rat pregnancy. Development 113, 165-181.

Risek, B., Guthrie, S., Kumar, N., & Gilula, N.B. (1990). Modulation of gap junction transcript and protein expression during pregnancy in rat. J. Cell. Biol. 110, 269-282.

Rogers, M., Berestecky, J.M., Hossain, M.Z., Guo, H., Kadle, R., Nicholson, B.J., & Bertram, J. (1990). Retinoid-enhanced gap junctional communication is achieved by increased levels of connexin mRNA and protein, Mol. Carcinog. 3, 335-343.

Ron, D., Brasia, A.R., Wright, K.A., Tate, J.E., & Habener, J.F. (1990). An inducible 50-kilodalton NFkappaB-like protein and a constitutive protein both bind the acute phase response element of the angiotensin gene. Mol. Cell. Biol., 10, 1023-1032.

Rosenberg, E., Spray, D.C., & Reid, L.M. (1992). Transcriptional and posttranscriptional control of connexin mRNAs in periportal and pericentral hepatocytes, Eur. J. Cell Biol. 59, 21-26.

Rup, D.M., Veenstra, R.D., Wang, H.Z., Brink, P.R., & Beyer, E.C. (1993). Chick connexin56, a novel lens gap junction protein. J. Biol. Chem. 268, 706-712.

Sáez, J.C., Bennett, M.V.L., & Spray, C. (1990). Hepatocyte gap junctions: metabolic regulation and possible role in liver metabolism. In: *Transduction in Biological Systems* (Hidalgo, C., Bacigalupo, J., Jaimovich, E., & Vergara, J., Eds.), pp. 231-243. Plenum, New York.

Sáez, J.C., Berthoud, V.M., Kadle, R., Traub, O., Nicholson, B.J., Bennett, M.V.L., & Dermietzel, R. (1991). Pinealocytes in rats: connexin identification and increase in coupling caused by norepinephrine. Brain Res. 568, 265-275.

Schulz, M., Traub, O., Knop, M., Willecke, K., & Schnabl, H. (1992). Immunofluorescent localization of connexin26-like protein at the surface of mesophyll protoplasts from *Vicia faba L.* and *Helianthus annuus L.* Bot. Acta 105, 111-115.

Schwarz, H. J., Chang, Y. S., Lalley, P. A., & Willecke, K. (1994). Chromosomal assignments of mouse genes for connexin 50 and connexin 33 by somatic cell hybridization. Somat. Cell Mol. Genet. 20, 243-247.

Schwarz, H.J., SookChang, Y., Hennemann, H., Dahl, E., Lalley, P.A., & Willecke, K. (1992). Chromosomal assignments of mouse connexin genes, coding for gap junctional proteins, by somatic cell hybridization. Somat. Cell Mol. Genet. 18, 351-359.

Seul, K. H., Tadros, P. N. , & Beyer, E. C. (1997). Mouse connexin40: gene structure and promoter analysis. Genomics 46, 120-126.

Shiels, A., Mackay, D., Ionides, A., Berry, V., Moore, A., & Bhattacharya, S. (1998). A missense mutation in the human connexin50 gene (GJA8) underlies autosomal dominant "zonular pulverulent" cataract, on chromosome 1q. Am. J. Hum. Genet. 62, 526-532.

Sohl, G., Degen, J., Teubner, B., & Willecke, K. (1998). The murine gap junction gene connexin36 is highly expressed in mouse retina and regulated during brain development. FEBS Lett. 428, 27-31.

Sohl, G., Gillen, C., Bosse, F., Gleichmann, M., Muller, H.W., & Willecke, K. (1996). A second alternative transcript of the gap junction gene connexin32 is expressed in murine Schwann cells and modulated in injured sciatic nerve. Eur. J. Cell Biol. 69, 267-275.

Splitt, M.P., Tsai, M.Y., Burn, J., & Goodship, J.A. (1997). Absence of mutations in the regulatory domain of the gap junction protein connexin 43 in patients with visceroatrial heterotaxy. Heart 77, 369-370.

Spray, D.C., Moreno, A.P., Kessler, J.A., & Dermietzel, R. (1991). Characterization of gap junctions between cultured leptomeningeal cells. Brain Res. 568, 1-4.

Stutenkemper, R., Geisse, S., Schwarz, H.J., Look, J., Traub, O., Nicholson, B.J., & Willecke, K. (1992). The hepatocyte specific phenotype of murine liver cells correlates with high expression of connexin32 and connexin26 but very low expression of connexine43. Exp. Cell Res., 102, 43-54.

Sullivan, R., Ruangvoravat, C., Joo, D., Morgan, J., Wang, B.L., Wang, X.K., & Lo, C.W. (1993). Structure, sequence and expression of the mouse Cx43 gene encoding connexin 43. Gene 130, 191-199.

Swenson, K.I., Jordan, I.R., Beyer, E.C., & Paul, D.L. (1989). Formation of gap junctions by expression of connexins in Xenopus oocyte pairs, Cell 57, 145-155.

Valdimarsson, G., De Sousa, P.A., Beyer, E.C., Paul, D.L., & Kidder, G.M. (1991). Zygotic expression of the connexin43 gene supplies subunits of gap junction assembly during mouse preimplantation development. Mol. Reprod. Devel. 30, 18-26.

Van Camp, G., Coucke, P., Speleman, F., Van Roy, N., Beyer, E.C., Oostra, B.A., & Willems, P.J. (1995). The gene for human gap junction protein connexin37 (GJA4) maps to chromosome 1p35.1, in the vicinity of D1S195. Genomics 30, 402-403.

Veenstra, R.D., Wang, H.Z., Westphale, E.M., & Beyer, E.C. (1992). Multiple connexins confer distinct regulatory and conductance properties of gap junctions in developing heart. Circ. Res. 71, 1277-1283.

Wenzel, K, Manthey, D., Willecke, K., Grzeschik, K.H., & Traub, O. (1998). Human gap junction protein connexin31: molecular cloning and expression analysis. Biochem. Biophys. Res. Commun. 248, 910-915.

Werner, R., Levine, E., Rabbadam-Diehl, C., & Dahl, G. (1989). Formation of hybrid cell-cell channels. Proc. Natl. Acad. Sci. USA 80, 5380-5384.

White, T.W., Bruzzone, R., Goodenough, D.A., & Paul, D.L. (1992a). Mouse Cx50, a functional member of the connexin family of gap junction proteins, is the lens fiber protein MP70. Mol. Biol. Cell 3, 711-720.

White, T.W., Gimlich, R.L., Paul, D.L., & Goodenough, D.A. (1992b). Reconstitution of lenticular communication pathways by the expression of connexins43, -46, and -50 in paired Xenopus oocytes, Mol. Biol. Cell 3, 293a.

Willecke, K., Hennemann, H., Dahl, E., Jungbluth, S., & Heynkes, R. (1991a). The diversity of connexin genes encoding gap junction proteins. Eur. J. Cell Biol. 56, 1-7.

Willecke, K., Heynkes, R., Dahl, E., Stutenkemper, R., Hennemann, H., Jungbluth, S., Suchyna, T., & Nicholson, B.J. (1991b). Mouse connexin37: cloning and functional expression of a gap junction gene highly expressed in lung. J. Cell Biol. 114, 1049-1057.

Willecke, K., Jungbluth, S., Dahl, E., Hennemann, H., Heynkes, R., & Grzeschik, K.H. (1990). Six genes of the human connexin gene family coding for gap junctional proteins are assigned to four different human chromosomes. Eur. J. Cell Biol. 53, 275-280.

Winterhager, E., Grümmer, R., Jahn, E., Willecke, K., & Traub O. (1993). Spatial and temporal expression of connexin26 and -43 in rat endometrium during trophoblast invasion. Develop. Biol. 157, 399-409.

Winterhager, E., Stutenkemper, R., Traub, O., Beyer, E., & Willecke, K. (1991). Expression of different connexin genes in rat uterus during decidualization and at term. Eur. J. Cell Biol., 55, 133-142.

Yahalom, A., Warmbrodt, R.D., Laird, D.W., Traub, O., Revel, J.P., Willecke, K., & Epel, B.L. (1991). Maize mesocotyl plasmodesmata proteins cross-react with connexin gap junction protein antibodies. Plant Cell 3, 407-417.

Yoshizaki, G., & Patino, R. (1995). Molecular cloning, tissue distribution, and hormonal control in the ovary of Cx41 mRNA, a novel Xenopus connexin gene transcript. Mol. Reprod. Dev. 42, 7-18.

Yoshizaki, G., Patino, R., & Thomas, P. (1994). Connexin messenger ribonucleic acids in the ovary of Atlantic croaker: molecular cloning and characterization, hormonal control, and correlation with appearance of oocyte maturational competence. Biol. Reprod. 51, 493-503.

Yu, W., Dahl, G., & Werner, R. (1994) The connexin43 gene is responsive to estrogen. Proc. R. Soc. Lond. 255, 125-132.

Zampighi, G.A., Simon, S.A., Robertson, J.D., McIntosh, T.J., & Costello, M.J. (1982). On the structural organization of isolated bovine lens fiber junctions. J. Cell Biol. 933, 175-189.

Zhang, J.T., & Nicholson, B.J. (1989). Sequence and tissue distribution of a second protein of hepatic gap junctions, Cx26, as deduced from its cDNA. J. Cell Biol. 109, 3391-3401.

Zhu, D., Caveney, S., Kidder, G.M., & Naus, C.G. (1991). Transfection of C6 glioma cells with connexin43 cDNA: Analysis of expression, intercellular coupling, and cell proliferation. Proc. Natl. Acad. Sci. USA 88, 1883-1887.

Zimmer, D.B., Green, C.R., Evans, W.H., & Gilula, N.B. (1987). Topological analysis of the major
    protein in isolated intact rat liver gap junctions and gap junction-derived single membrane
    structures. J. Biol. Chem. 262, 7751-7763.

# STRUCTURE AND BIOCHEMISTRY OF GAP JUNCTIONS*

Mark Yeager and Bruce J. Nicholson

|       |                                                                      |      |
| ----- | -------------------------------------------------------------------- | ---- |
| I.    | Historical Perspective                                               | 32   |
| II.   | Biochemical Characterization and Composition                         | 35   |
|       | A.   Origins                                                         | 35   |
|       | B.   Current Protocols                                              | 39   |
|       | C.   Protein Components                                             | 40   |
|       | D.   Criteria for Identifying Gap Junction Proteins—Functional Assays | 43   |
|       | E.   The Era of Cloning and the Connexin Family                    | 45   |
|       | F.   Heterotypic and Heteromeric Interactions between Connexins      | 47   |
|       | G.   Post-Translational Modification                               | 54   |
|       | H.   Other Components                                              | 55   |
| III.  | Structure Analysis                                                  | 57   |
|       | A.   Background                                                    | 57   |
|       | B.   One-Dimensional Electron Density Profile by X-ray Diffraction   | 58   |
|       | C.   Low Resolution Two-Dimensional Projection Density Maps          | 59   |
|       | D.   Low Resolution Three-Dimensional Density Maps                   | 61   |
|       | E.   Membrane Topology                                             | 63   |

*Sections of this review have been summarized in Yeager and Nicholson (1996) and Yeager (1997; 1998) and are reproduced here by permission.

**Advances in Molecular and Cell Biology, Volume 30, pages 31-98.**
**Copyright © 2000 by JAI Press Inc.**
**All rights of reproduction in any form reserved.**
**ISBN: 0-7623-0599-1**

F.   Secondary Structure Models.....................................64
G.   Higher Resolution Structure Analysis by Electron Cryo-Crystallography...66
H.   Structure of the Pore.........................................72
I.   Models for Connexon Docking...................................73
J.   Gating Models.................................................75
IV.  Identification of Functional Sites That Regulate Gating ......76
A.   Phosphorylation ..............................................76
B.   Gating by pH..................................................77
C.   Voltage Gating ...............................................78
V.   Conclusion ....................................................79

# I.  HISTORICAL PERSPECTIVE

The field of gap junction research was born from the marriage of two quite disparate areas over 30 years ago. It had long been recognized that cells of multicellular organisms interact with both their immediate neighbors and cells at a distance through a variety of modes of transmembrane signaling. In the selected cases of cardiac muscle (Weidmann, 1952; Barr et al., 1965; Weidmann, 1966) and some neurons (Furshpan and Potter, 1959; Robertson, 1963), there was also evidence for electrical coupling, indicating a direct flow of ions between cells (Potter et al., 1966; Furshpan and Potter, 1968). Subsequently, these observations were broadened to nonexcitable cells by observing the analogous passage of larger metabolites (Subak-Sharpe et al., 1966, 1969) and dyes (Loewenstein and Kanno, 1964). Clearly this process required the kinds of intimate contacts between cells that had been described earlier in other contexts such as tight junctions and desmosomes. The challenge, however, was to identify a specific structure responsible for the phenomenon of electrical coupling. This was complicated by the inadequate distinction at that time between different intercellular contacts such as tight junctions and gap junctions (Karrer, 1960; Dewey and Barr, 1962; Trelstad et al., 1966).

A final resolution was achieved using a membrane impermeable, colloidal form of lanthanum that remained in the extracellular spaces of en bloc fixed tissue. This electron dense stain provided definition of a uniform, approximately 20-Å gap between membranes of apposed cells at focal plaquelike regions (Revel and Karnovsky, 1967). This narrow separation of apposed membranes remains the most consistent and recognizable feature of these "gap" junctional structures, whether *in situ* or after isolation (Figure 1). En face views of these domains revealed a close-packed array of stain-excluding particles, often showing a stain-filled center (Revel and Karnovsky, 1967). This provided a plausible structural correlate for the coupling that had been reported. In fact, this same array had been found using analogous, but less controlled conditions of permanganate precipitation (Robertson, 1963). The nature of these particles was clarified with the advent of freeze-fracture techniques that provided internal views of membrane structure (Kreutziger, 1968; Chalcroft and Bullivant, 1970; Goodenough and Revel, 1970).

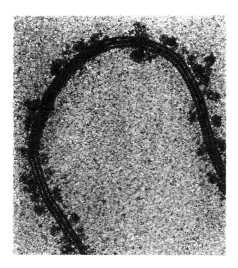

**Figure 1.** Conventional thin-section electron microscopy has served as a keystone for defining the "morphological signature" of gap junctions. The characteristic appearance is septalaminar, with stain exclusion in the hydrophobic domains of the lipid bilayer and stain accumulation in the extracellular gap and on the cytoplasmic faces (Brightman and Reese, 1969). The cardiac gap junction has been labeled by site-specific, peptide antibodies against a synthetic peptide spanning residues 131 to 143 in Cx43. The gold particles are attached to immunoglobulin antibodies and serve as a marker to verify that residues 131 to 143 are localized on the cytoplasmic membrane face. Such experiments, using immunogold electron microscopy, have been used to confirm the folding models shown in Figure 5. Reproduced from Yeager (1998) with permission from Academic Press.

Closely packed arrays of intramembranous particles in the cytoplasmic half of the membrane were aligned with corresponding pits in the extracellular half of the bilayer of the apposing cell in specialized domains of the membrane where the extracellular space narrowed (Figure 2). By utilizing rotary rather than unidirectional shadowing of the fracture replicas, central depressions in the particles could be detected (Hanna et al., 1985). This led to the model of the gap junction, which persists to date, as an array of aqueous channels connecting the cytoplasms of adjacent cells. The oligomeric channels are formed by the end-to-end docking of two hemichannels, termed connexons, each of which is formed by a hexameric cluster of protein subunits, termed connexins.

With the morphological signature of the gap junction thus defined, similar structures were identified in many tissues (Friend and Gilula, 1972; Gilula et al., 1972; Hertzberg and Skibbens, 1984) of virtually every metazoan phylum studied

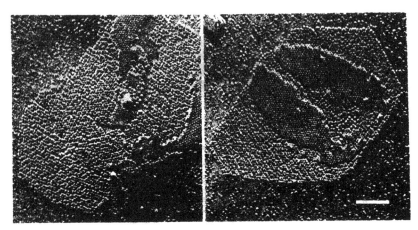

*Figure 2.*   Electron micrographs of freeze-fractured BHK cells expressing either full-length (**A**) or a mutant in which the majority of the C-terminal domain has been deleted from Cx43 (truncation at Lys263; see Figure 5a) (**B**). Both replicas display plaques that exhibit particles on the P fracture face and corresponding depressions on the E face that are typical of mammalian gap junctions. Bar represents 200 nm. Reproduced from Unger et al. (1999a) with permision from Academic Press.

(Loewenstein, 1967; Friend and Gilula, 1972; Flower, 1977; Peracchia, 1980). Indeed, the ubiquitous presence of these structures throughout the multicellular animal kingdom, from the primitive ciliated mesozoan (Revel, 1987) to humans, led to the notion that gap junctions are a prerequisite for multicellularity. Despite their widespread occurrence, gap junctions retained certain characteristic features, with the variations in their structures arising as much from the preparation of the tissue as its source (Gilula and Satir, 1971; Hudspeth and Revel, 1971; Friend and Gilula, 1972). Arthropods proved a notable exception as the gap junctions showed a distinctive freeze-fracture morphology in which the intramembranous particles segregated to the extracellular half of the bilayer, and formed dispersed, randomly arranged aggregates (Flower, 1977).

As previously noted, functional demonstrations of cell-to-cell coupling had also been established in a wide range of systems, using both electrical coupling coefficients and the intercellular transfer of low molecular weight, membrane impermeable, fluorescent dyes or cellular metabolites. Several studies, in particular by Loewenstein and colleagues, utilized an extensive array of fluorescent probes to establish an upper permeability limit for these channels of approximately 1000 D (Flagg-Newton et al., 1979), or approximately 2000 D in the case of insect gap junctions (Schwarzmann et al., 1981). However, even in these early studies there were indications that features other than just size, such as the net charge on the probe, could contribute to the permeability of solutes through gap junctions (Flagg-Newton et al., 1979; Brink and Dewey, 1980). Communication between cells was also monitored by metabolic coupling where the intercellular exchange of cellular metabolites, typically nucleotides, was followed via rescue of a mutant phenotype by co-culture with a complementing cell (reviewed in Hooper and Subak-Sharpe, 1981). The first direct demonstration of the passage of physiologically relevant metabolites and signaling molecules such as cyclic adenosine monophosphate (cAMP) through junctional channels was established in this way (Lawrence et al., 1978). Fluorescence imaging (Dunlap et al., 1987; Sáez et al., 1989) also demonstrated that other secondary messengers, such as IP3 and $Ca^{2+}$, could pass between cells through gap junctions. However, identification of the specific transjunctional permeants that might modify cell behavior under any given physiological condition has proven elusive. One promising approach has been recently described by Goldberg and associates (1998), in which labeled metabolites from a "donor" cell population were transferred to fluorescently tagged "recipient" cells. Fluorescence-activated cell sorting was used to retrieve the "recipient" cells. The initial results demonstrate that adenosine triphosphate (ATP) and adenosine diphosphate (ADP) molecules equilibrate within a cell population in less than 2 hours of cell contact.

In most cases the presence of morphologically identified gap junctions could account for the coupling of cells, and in some cases represented the only identifiable intimate contact between coupled cells (Revel et al., 1971). However, a few exceptions were noted among invertebrates. Based on the restricted extracellular space and their large surface area, septate junctions in coelenterates and insects had been proposed as an alternative means of cell coupling (Gilula et al., 1970). Closer analysis, and a consideration of the different morphologies of the gap junctions in some of these systems, eventually led to the identification of gap junction structures on the basal side of the septate junctions (Hudspeth and Revel, 1971). Septate junctions are now generally considered to be the functional equivalents of tight junctions in these systems. Thus, the working assumption of the field is that structures defined morphologically as gap junctions are the only mediators of the direct exchange of ions and small metabolites between cells.

## II. BIOCHEMICAL CHARACTERIZATION AND COMPOSITION

### A. Origins

After the initial characterization of the morphological features of the gap junction, the identification of biochemical components was difficult, as morphological criteria were the only assay for purity. In fact, no enzymatic or specific binding activity has as yet been associated with gap junctions. Isolation procedures have

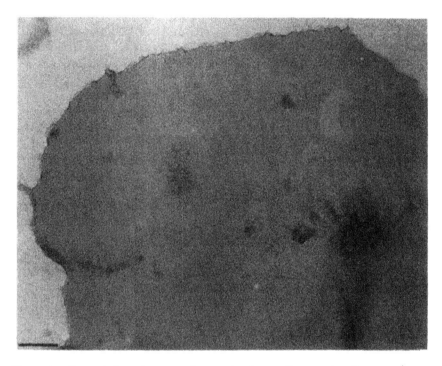

***Figure 3.*** Transmission electron micrograph of a cardiac gap junction membrane plaque negatively stained with uranyl acetate. The membrane sheet is formed by a mosaic of crystalline domains, the largest of which is approximately 0.5 $\mu m^2$. This gap junction preparation has undergone protease cleavage in the region of residues 252 to 271 (Yeager and Gilula, 1992) to release the cytoplasmic C-tail domain. This presumably accounts for the clarity of the connexons compared with images of gap junctions containing uncleaved $\alpha_1$ connexin. Such two-dimensional crystals contain structural information to about 15 Å resolution, which is sufficient to define the molecular boundary and quaternary structure of the connexons. Scale bar = 1000 Å. Reproduced from Yeager (1994) with permission of *Acta Crystallographica*.

relied on the resistance of these closely packed arrays of proteins to treatments with various nonionic detergents that solubilize most cell membranes (Evans and Gurd, 1972; Goodenough and Stoeckenius, 1972). Not only do these detergent treatments leave the protein arrays intact (Figure 3), but they also do not disrupt the pairing of the membranes (see Figure 1). Although this structure seemed quite unique and characteristic, there were potential pitfalls to applying purely morphological criteria to the assessment of purity (see Hertzberg, 1983; Nicholson and

Revel, 1983, for reviews), and this has produced several controversies (see the discussion of "Nonconnexin Proteins Associated with 'Junctional' Structures" in section II.C). Although a variety of detergents have been employed in the isolation of gap junctions, the greatest success was achieved with N-lauryl sarcosine (0.3% to 0.5%, depending on the tissue of origin) (Hertzberg and Gilula, 1979; Gros et al., 1983), and sodium deoxycholate (0.5% to 1.0%) (Henderson et al., 1979). An important modification to this general approach was the substitution of alkali treatment (sonication in ice cold 20 mM NaOH) for the usual detergent extraction of crude membrane fractions (Hertzberg, 1984). Although one might expect some lipid saponification with this process, few free fatty acids were detected in the supernatant of this treatment (Hertzberg and Henderson, personal communication). Furthermore, exogenous hydrophobic agents were not added, and the yield of gap junctional protein was greatly increased.

Aside from the critical importance of the solubility of gap junctions, the second most important step in all isolation procedures has been separation on the basis of density. Most procedures have utilized discontinuous sucrose gradients to isolate plasma membranes prior to detergent or alkali extraction (Culvenor and Evans, 1977; Hertzberg, 1983; Nicholson and Revel, 1983), but the two-phase system has also been effective (Finbow et al., 1980; Nicholson et al., 1981). The conditions used vary with the tissue, but plasma membranes typically sediment with a density between 1.15 and 1.20 g/cm$^3$. Of more specific relevance to gap junctions, density gradients have also been employed following the solubilization and washing of the membranes to separate the gap junctions from the less dense lipid residue and the more dense matrix and cytoskeletal material. These gradients have often included low concentrations of detergents (0.1% N-lauryl sarcosine or 0.3% deoxycholate) or denaturants such as urea (1 to 6 M), or both, to reduce aggregation (reviewed in Hertzberg, 1983, and Nicholson and Revel, 1983). The junctions usually float at an interface spanning the densities of 1.127 (30% [w/v] sucrose) to 1.20 (54% [w/v] sucrose), consistent with an estimated density of 1.165 g/cm$^3$ for gap junction plaques based on equilibrium centrifugation on a continuous gradient (Hertzberg and Gilula, 1979).

Other steps in the various purification schemes described by different laboratories and for different tissues have removed residual or adhering insoluble contaminants such as elements of the extracellular matrix, cytoskeleton, and crystalline deposits of proteins inside cells. Initially, proteases were used as they had no obvious effect on junctional morphology (Goodenough and Stoeckenius, 1972). In fact, the first application of the term *connexin* referred to proteolytic fragments of what we now recognize as the principle protein component of liver gap junctions (Goodenough, 1974), emphasizing the precarious nature of a purely morphological assay. Indeed, this problem resurfaced in the early protocols to isolate cardiac gap junctions in which the principle protein component was thought to have a comparable molecular weight to the liver gap junction protein (i.e., Mr 28,000 [Gros et al., 1983; Nicholson et al., 1985]). Only when stringent efforts were applied to prevent adventitious proteolysis during isolation was the full length 43-kD gap junction protein (or connexin) identified (Manjunath et al., 1985). More recent strategies to displace nonspecifically interacting components have included high salt (Hertzberg and Gilula, 1979; Nicholson et al., 1981) and alkaline pH stripping of membrane fractions (Henderson et al., 1979; Hertzberg and Gilula, 1979; Nicholson et al., 1981), and the inclusion in density gradients of chaotropic agents such as urea (Henderson et al., 1979; Gros et al., 1983) and low levels of detergents such as deoxycholate (Manjunath et al., 1984a) or N-lauryl sarcosine (Gros et al., 1983). Indeed, since the first well-characterized protocols were published, many variations have been applied with little change in the final product other than increased yield and simplification of the isolation procedure.

The efficiency of the various protocols can be assessed by comparing their estimated yields with the techniques that are currently used. Reasonable quantitative interpretation of hepatic morphology suggests that approximately 4% of the lateral surface or approximately 1% of the total plasma membrane is occupied by gap junctions (Yee and Revel, 1978). Based on this interpretation, as well as estimates of the fraction of tissue protein associated with plasma membranes and protein assays at all stages of the isolation (Hertzberg and Gilula, 1979), yields can be deduced for various isolation procedures. The values range from 0.2% in the earlier preparations (Hertzberg and Gilula, 1979) to 1% (Nicholson et al., 1981) or even 10% in later refinements (Hertzberg, 1984). By adding back trace amounts of labeled, purified protein to the crude fractions, the largest losses probably occur during detergent extraction and final density gradient separation. These estimates for the yield refer to assembled gap junction plaques. However, immunoprecipitation experiments have suggested that the pool of intracellular protein (distributed between the Golgi, endoplasmic reticulum, and lysosomes) is equal to or greater than in the plasma membrane (Musil et al., 1990b). Even the surface membrane population may not all be assembled into junctional structures that are morphologically identifiable and resistant to the usual detergent extractions (Musil and Goodenough, 1993). Thus, it is important to realize that all methods for isolating gap junctions select for only a fraction of the total connexin expressed in a cell.

The exact fraction of surface connexins in the form of hemichannels has been a matter of conjecture and is likely to depend on cell and connexin type. This was exemplified in the characterization of connexin46 (Cx46) in the paired *Xenopus* oocyte system. In addition to forming intercellular channels typical of other connexins, Cx46 also formed open hemichannels in the cell membrane, which caused lysis of the oocytes unless elevated $Ca^{2+}$ was used in the medium to induce hemichannel closure (Paul et al., 1991). These hemichannels had very similar electrical properties to intact gap junctions, although the polarity of their response to voltage appears reversed (Ebihara et al., 1995). Interestingly, lens fiber cells, which express Cx46 in abundance, have electrically tight nonjunctional membranes (Mathias and Rae, 1985; Cooper et al., 1989), indicating that hemichannel formation of Cx46 is an aberration of the oocyte system. This may be a consequence of a lack of critical posttranslational modifications reflected in a mobility shift of the protein by SDS–polyacrylamide gel electrophoresis (SDS-PAGE) (Paul et al., 1991). This same rationale could explain the apparent formation of patent hemichannels following reconstitution of purified connexins into lipid bilayers (Harris et al., 1992).

The alternative view, that potentially functional hemichannels may be a widespread feature of cells, is supported by recent observations of Li and colleagues (1996). In the presence of low $Ca^{2+}$, a variety of cultured cells take up extracellular Lucifer yellow in a pattern that was closely correlated with the expression of Cx43. This observation has since been extended to transfected HeLa cells expressing a variety of connexins (Li and Johnson, personal communication). These experiments are consistent with earlier claims that large open channels in the membranes of retinal horizontal cells represent connexon hemichannels (deVries and Schwartz, 1992; Malchow et al., 1993).

## B. Current Protocols

Procedures for the isolation of gap junctions have been established for only a few vertebrate tissues, principally liver (Henderson et al., 1979; Hertzberg and Gilula, 1979; Finbow et al., 1980; Nicholson et al., 1981), heart (Kensler and Goodenough, 1980; Manjunath et al., 1982; Gros et al., 1983; Manjunath et al., 1985), uterus (Zervos et al., 1985) and eye lens (Kistler et al., 1985). Procedures for some invertebrates, mainly arthropods (Buultjens et al., 1988; Ryerse, 1989), and various cell lines have also been reported, but the nature of the principle components is still controversial (see later discussion). Current protocols involve the isolation of a plasma membrane fraction on a discontinuous sucrose gradient (typically at the 40/54 or 35/49% [w/v] interface), followed by extraction with either high pH (0.1 M NaOH, pH 12.5 for 30 minutes on ice) or a nonionic detergent (0.3% to 0.5% N-lauryl sarcosine or 0.5% to 1.0% deoxycholate). The final fraction is then separated from other insoluble components with an additional discontinuous sucrose gradient (20/40/54% [w/v] sucrose) incorporating detergents (0.1% N-lauryl sarcosine or 0.3% deoxycholate) or chaotropic agents (1 M urea or alkaline conditions such as pH 10). Gap junctions are typically collected between the 35% to 40% and 49% to 54% (w/v) sucrose layers. For some samples, the original isolation of membranes can be omitted without compromising the final purity of the fractions (Zhang and Nicholson, 1994). For other tissues, specific steps must be incorporated to deal with particularly abundant insoluble impurities. For example, in cardiac tissue, pretreatment with 0.6 M KI depolymerizes the majority of the contractile apparatus (Kensler and Goodenough, 1980), and in rat liver tissue, high pH treatments are required to dissolve uricase crystals (Hertzberg and Gilula, 1979).

## C.  Protein Components

### Identifying the First Connexins

At first view, the composition of the most enriched gap junction fractions appeared rather heterogeneous. However, it was soon demonstrated that many of the bands seen by SDS-PAGE were a result of degradation by endogenous proteases (Hertzberg and Gilula, 1979; Nicholson et al., 1981; Manjunath et al., 1985) and aggregation into multimers. For example, liver gap junction proteins aggregate in SDS, particularly when heated (Henderson et al., 1979; Nicholson et al., 1981). In contrast, the cardiac gap junction protein does not tend to aggregate when heated (Manjunath et al., 1985). Biochemical characterization of each of the bands (Henderson et al., 1979; Hertzberg and Gilula, 1979; Gros et al., 1983) and microsequencing (Nicholson et al., 1981, 1985; Manjunath et al., 1987) revealed that the preparations from the most extensively studied tissues (i.e., liver and heart) comprise a single major protein, termed *connexin*. Despite original observations that these proteins were of similar size (Gros et al., 1983) and immunologically related (Dermietzel et al., 1984; Hertzberg and Skibbens, 1984), it was soon evident that they differed substantially in both molecular weight (Manjunath et al., 1984a, 1985, 1987) and sequence (Gros et al. 1983; Nicholson et al., 1985). The predominant liver connexin has a relative mobility on SDS-PAGE of 26 to 28 kD (now known as Cx32), whereas the heart protein migrates as a broad band at 43 kD (now known as Cx43), which is the result of different degrees of phosphorylation, both *in vitro* (Crow et al., 1990; Musil et al., 1990a) and in a tissue specific manner *in vivo* (Kadle et al., 1991).

This rather reductionistic biochemical analysis of gap junctions became more complex with the recognition that gap junctions from a single tissue source could contain multiple connexin types (Nicholson et al., 1987). The well-characterized hepatic gap junctions were shown to contain a second component of Mr 21,000 (now known as Cx26) that was related to, yet distinct from, the major Mr 28,000 (i.e., Cx32) form and colocalized to the same gap junctional plaques, visualized by both immunofluorescence (Nicholson et al., 1987) and immunoelectron microscopy (Traub et al., 1989; Kuraoka et al., 1993). Although the relative levels of Cx32 and Cx26 varies greatly between species (Nicholson et al., 1987; Zhang and Nicholson, 1989), in cell location within the hepatic acinus (Traub et al., 1989), and during the early stages of hepatic tumorigenesis (Sakamoto et al., 1992; Neveu et al., 1994), we still do not understand their respective physiological significance. With the advent of cloning (see the next section and the chapter by Laird and Sáez in this volume), it has become evident that multiple connexins are typically coexpressed in the same tissue (Risek and Gilula, 1991; Hennemann et al., 1992a; Kanter et al., 1992; Risek et al., 1992) and are frequently associated in the same junctional structures (Kanter et al., 1993b). In hindsight, the lack of evidence for multiple connexins in the original biochemical analyses is notable. In some cases this may have been the result of the relatively low levels of expression of the additional connexin(s). However, this seems less likely in the heart, where Cx40 is now known to be relatively abundant, although mostly with a different cellular distribution (Bastide et al., 1993; Kanter et al., 1993a; Gros et al., 1994). Similar mobilities of Cx43 and Cx40 on SDS gels, and the heterogeneity introduced by the presence of multiple phosphorylated forms of both proteins would have complicated the issue, but it is nonetheless surprising that sequencing studies did not reveal the heterogeneity in some bands. Aside from this oversight, possibly based on selective blockage of the Cx40 N-terminus, the failure to detect additional connexins in the original biochemical analyses may have resulted, in part, from differences in the detergent solubility of junctions composed of different connexins, as documented in comparisons of hepatic and cardiac gap junctions (Manjunath and Page, 1986).

## Nonconnexin Proteins Associated with "Junctional" Structures

Two other products of the biochemical analyses of gap junctions, MIP26 and ductin, are worthy of discussion. The eye lens is a particularly abundant source of gap junctions (Kuszak et al., 1978), although their detailed structure differs from junctions of most vertebrate tissues in terms of particle packing in freeze-fracture replicas (Zampighi et al., 1982; Lo and Harding, 1986) and membrane spacing in thin sections (Goodenough, 1979; Zampighi et al., 1982). Isolations from this tissue were very straightforward and yielded milligram quantities of the main intrinsic protein (MIP) of 26 kD (Zampighi et al., 1982). However, the primary sequence of MIP26 was shown to be unrelated to any of the other connexins that have since been characterized or cloned (Nicholson et al., 1983). Indeed, MIP26 became the prototype (Gorin et al., 1984) of a novel and extensive gene family, including such proteins as plant TIP (Johnson et al., 1990), nodulin (Jacobs et al., 1987), *Drosophila* big brain (Rao et al., 1990), mammalian CHIP28 (Agre et al, 1993; Verkman, 1993), yeast FPSI (Van Aelst et al., 1991) and *Escherichia coli* GlpF (Muramatsu and Mizuno, 1989). Although specific functions have not been associated with all members of the superfamily, a unifying theme seems to be passive transmembrane transport ranging from water (CHIP28) to large metabolites (e.g., GlpF, a glycerol facilitator) (see Ehring et al., 1993, for review).

Supportive evidence for the role of MIP26 in gap junction function includes immunolabeling of junctional structures (Sas et al., 1985; Fitzgerald et al., 1985) and use of MIP26 antibodies to block intercellular dye spread (Johnson et al., 1988). However, there is not universal agreement on the results (Paul and Goodenough, 1983; Kistler et al., 1985). Also, unlike connexins, expression of MIP26 does not appear to induce intercellular coupling in oocytes (Swenson et al., 1989). Isolation protocols similar to those applied to other tissues have since identified a member of the connexin family, MP70 (now identified as Cx50 [White et al., 1992]), in gap junctional structures of the lens (Gruijters et al., 1987; Kistler et al.,

1985, 1988). Although MIP26 appears to form channels on reconstitution (Girsch and Peracchia, 1985; Zampighi et al., 1985; Chandy et al., 1997) and forms closely packed arrays of proteins on the surface of lens fiber cells, it now seems likely that the structure and functional role of MIP26 is distinct from gap junctions (Zampighi et al., 1989; Konig et al., 1997). Indeed, the MIP family member CHIP28 is a water-selective channel that can be reconstituted into two-dimensional (2D) crystals in membrane bilayers. Electron crystallography reveals that CHIP28 assembles as a tetramer (Jap and Li, 1995; Mitra et al., 1995; Walz et al., 1995), in contrast to connexons, which are hexameric. Three-dimensional (3D) maps have revealed that the CHIP28 monomer is formed by a barrel of six $\alpha$-helices (Cheng et al., 1997; Li et al., 1997; Walz et al., 1997), in contrast to gap junction channels where each connexin subunit is folded as four transmembrane $\alpha$-helices (Unger et al., 1999b) (see section III.G).

Ductin is a second case of a nonconnexin protein that forms structures that morphologically resemble gap junctions. A streamlined preparation in which the initial plasma membrane isolation was omitted (Finbow et al., 1983; Dermietzel et al., 1989b) yielded paired membrane structures that displayed a closely packed array of particles by negative stain electron microscopy. However, they were found to mainly contain a 16-kD protein that has now been identified as a member of the $H^+$/ATPase family (Mandel et al., 1988; Finbow et al., 1992). Similar preparations from a variety of tissues and cell lines all contained this protein (Finbow et al., 1983; Buultjens et al., 1988), including a homologous 18-kD protein in a membrane fraction from arthropods (Finbow et al., 1984; Berdan and Gilula, 1988; Ryerse, 1989). Extensive arguments, including the blocking of cell coupling with anti–16-kD antisera and immunohistochemical localization, have been presented to support a role for ductin in gap junctional structure (see Finbow and Pitts, 1993, for review). Nevertheless, a convincing *in situ* localization of ductin to gap junctional structures at the electron microscope level and a demonstration that its expression in any system promotes intercellular coupling are lacking. Thus, it seems most likely that this represents another example of the limitations imposed by a purely morphological assay in any biochemical isolation. However specialized a structure may appear, analogous structures can be found, or induced to form, under appropriate conditions. In the case of ductin, this may have resulted from the failure to exclude lysosomal or other non–plasma membranes from the original detergent treatment (Finbow and Meagher, 1992), which may induce the formation of these junctionlike structures.

The strongest evidence linking ductin to gap junction structures has been obtained from invertebrate tissues such as the arthropod hepatopancreas. Extensive efforts failed to identify polypeptides homologous with connexins. However, recent studies have indicated that invertebrate gap junctions may be composed of homologous proteins unrelated to vertebrate connexins. Several mutants in *Caenorhabditis elegans* (unc and eat) and *Drosophila melanogaster* (ochre and pas) that are characterized by asynchronous behavior of cells that are normally coupled, have all been linked to defects in a protein family termed *innexins*. This circumstantial evidence was recently made more compelling by the induction of intercellular coupling in oocyte pairs injected with some innexin RNAs (Phelan et al., 1998). Notably, several other members of the innexin family did not induce coupling between oocytes. Nonetheless, these proteins currently present the best candidates for the major structural components of gap junctions in invertebrates. The innexins share some similarities with connexins and members of the invertebrate family of gap junction proteins, such as four transmembrane segments, of which one can be modeled as an amphipathic α-helix, and two loops that contain a conserved pattern of cysteines. However, sequence alignments indicate that these protein families appear to have arisen independently (Krishnan et al., 1993, 1995; Phelan et al., 1998; Shimohigashi and Meinertzhagen, 1998; Starich et al., 1996; Sun and Wyman, 1996).

## D. Criteria for Identifying Gap Junction Proteins—Functional Assays

These observations emphasize the need for more reliable criteria to establish the identity of a protein as a gap junctional component. Clearly, morphological criteria alone no longer suffice, although association with junctional structures *in situ* is a useful screening feature. It would also be unnecessarily exclusionistic to insist that all gap junction components must be related to the known family of connexins. In fact, it is possible some connexins may not form gap junctions (e.g., Cx31.1 [Henneman et al., 1992a] and Cx33 [Bruzzone et al., 1992, Chang et al., 1996]). Ultimately, the best criterion is the assay of function. A "negative" approach is to block intercellular coupling with antibodies demonstrated to be specific to the protein of interest. In the case of gap junctions, this necessitates the introduction of the antibodies into the cell, usually by microinjection. However, it is likely that not all antibodies will bind to regions that result in closure of the channels (Lal et al., 1993). Furthermore, extraneous antibodies could also mediate uncoupling of cells indirectly by interfering with cell adhesion, or disrupting metabolism, leading to an acidification of the cell—a known cause of uncoupling (possibly explaining the results obtained with anti–16-kD antibodies [Finbow and Pitts, 1993]).

A sounder course is to actually demonstrate the formation of intercellular channels capable of passing ions and some larger molecules, although the properties of these channels may vary between different connexins. Effective systems have included the expression of connexins in paired *Xenopus* oocytes (Dahl et al., 1987; Swenson et al., 1989; Nicholson et al., 1993) or in mammalian cells that have low levels of endogenous coupling (Eghbali et al., 1990; Fishman et al., 1990; Reed et al., 1993). In the former, the low levels of coupling found in virtually any system studied to date can be eliminated by injecting the oocytes with antisense oligonucleotides (Barrio et al., 1991; Willecke et al., 1991b). The advantage of using transfected mammalian cells is that single channel properties can be examined. Of course, even a functional assay represents a necessary, but not sufficient requirement, as some systems fail to couple owing to the lack of elements other than components of the gap junction channel itself. For example, the induction of coupling between L cells is dependent on the introduction of cell adhesion molecules that provide a necessary step for the assembly of the gap junctions (Musil et al., 1990b), and antibodies to N-cadherin block formation of gap junctions between cultured lens cells (Frenzel and Johnson, 1996).

These complications can be circumvented in a cell-free system using purified components, and two general strategies for reconstitution have been adopted. Firstly, isolated rodent liver gap junction plaques have been applied to lipid bilayers (Spray et al., 1986; Young et al., 1987). This approach does not represent true reconstitution since the channels are not removed from their native environment, and the starting material is not homogeneous with respect to either protein or lipid. Channel properties that were recorded from such a "reconstitution" of hepatic gap junctions correlated in some respects to those *in situ* as they exhibited conductances of 130 to 160 pS and could be blocked by connexin antibodies (Spray et al., 1986; Young et al., 1987). However, gating characteristics in response to voltage, pH, and $Ca^{2+}$ varied markedly between these preparations and those seen *in situ*.

The second, more rigorous approach is the incorporation into liposomes or lipid bilayers of connexins purified by immunoaffinity techniques in n-octylglucoside (Claassen and Spooner, 1988; Harris et al., 1992; Rhee et al., 1996) or other non-ionic detergent extracts (Stauffer et al., 1991), which appear to leave the hemichannels intact (Stauffer et al., 1991; Harris et al., 1992). In one case, channel activity could be reconstituted from connexin subunits purified by SDS-PAGE (Young et al., 1987). Using both heart (Claassen and Spooner, 1988) and liver (Spray et al., 1986; Young et al., 1987; Harris et al., 1992; Rhee et al., 1996) gap junctions as starting material, liposomes can be produced that are permeable to smaller molecules such as Azure A (Claassen and Spooner, 1988), Lucifer yellow, and sucrose (Harris et al., 1992; Bevans et al., 1998; Rhee et al., 1996) and impermeable to larger molecules such as dextran (Claassen and Spooner, 1988). Most recently, the use of a graded series of probes suggested that hemichannels containing Cx26 have a smaller exclusion limit than those containing predominantly Cx32 (Bevans et al., 1998), a property that also seems consistent with the behavior of intact homomeric channels expressed in *Xenopus* oocytes (Weber and Nicholson, unpublished). Depending on their composition, the reconstituted channels also showed differential permeability for natural metabolites, notably cyclic nucleotides (Bevans et al., 1998).

The single channel properties of reconstituted liver gap junction channels have been variable. Conductances of 50 pS (Harris et al., 1992), 150 pS (Young et al., 1987; Harris et al., 1992; Spray et al., 1986), and 180 to 200 pS (Rhee et al., 1996) have been reported from reconstitutions of hepatic gap junctions containing predominantly Cx32 and a variable contribution from Cx26, depending on the species used. Buehler and co-workers (1995) reconstituted purified Cx26 channels derived from baculovirus infected cells and recorded unitary conductances of approximately 55 pS. In an even more defined system, Falk and associates (1997) recorded channel activity from fused microsomes with bilayers that contain connexins (Cx26, Cx32, or Cx43) produced by cotranslational insertion directly from the respective cRNAs. This strategy offers the potential to compare homogeneous connexin populations in a similar lipid background, although it also suffers from the significant complication of resolving the connexin-induced channels from the significant background of channels seen in microsomal membranes. Nonetheless, initial studies have associated hemichannel conductances of 52 pS, 170 pS, and 35 pS with Cx26, Cx32, and Cx43, respectively.

Varying sensitivities to pH and voltage have been reported in these reconstituted systems, and in one case, the voltage sensitivity was highly asymmetrical (Harris et al., 1992) as would be expected since the incorporation of connexins into single bilayers should yield hemichannels. Unfortunately, the properties and physiological significance of hemichannels *in vivo* remains to be established. Thus, the degree to which the reconstitution of connexins serves to reproduce gap junction channels comparable to those *in situ* has been difficult to assess. Perhaps most perplexing is the apparent lack of correlation between the single channel conductances that have been reported from hemichannels, particularly in reconstituted systems, and the conductances associated with the corresponding connexins in intact gap junctions. However, given the rudimentary state of our current understanding of the pore structure of gap junctions (see section III.H) and how this would influence ionic flux through the channel, there could be many explanations for this apparent discrepancy. Nevertheless, when fully characterized, these experimental approaches should provide unique insights into many currently intractable questions of channel structure and function. Not only will they enhance the electrophysiological analysis of channel permeability now limited by perfusion of whole cell systems, but they will also open new approaches to defined analyses of protein-lipid interactions (see the discussion of "Lipids" in section II.H).

## E. The Era of Cloning and the Connexin Family

The characterization and sequencing of the first connexins heralded the use of antibodies and degenerate oligonucleotides to isolate cDNA clones for other members of the connexin family. Both approaches proved successful (Kumar and Gilula, 1986; Paul, 1986; Zhang and Nicholson, 1989), which then led to low stringency hybridization and polymerase chain reaction (PCR) screens of genomic and cDNA libraries that yielded a plethora of family members from mammals (15 at the latest count), as well as avian (Beyer, 1990) and amphibian (Gimlich et al., 1990) variants. These clones and their relationships have been reviewed by Beyer and associates (1990), Willecke and co-workers (1991a),

Kumar and Gilula (1992, 1996), Goodenough and associates (1996), and Nicholson and Bruzzone (1997), and are discussed in depth in the preceding chapter by Beyer and Willecke in this volume.

In the absence of a clear understanding of the functional differences between connexins, it has been difficult to establish a logical nomenclature for distinguishing the various members. For example, the acetylcholine receptor displayed a clear distinction between subunits required for functional assembly; consequently an $\alpha$, $\beta$, $\gamma$, $\delta$ terminology was utilized. Because gap junctions can clearly assemble from a single type of subunit, such a functional distinction is not applicable. Therefore, one practical approach was to add the molecular weight of each variant as a suffix to the descriptor "connexin" (Beyer et al., 1987) or "Cx." Hence, the 28-kD and 21-kD proteins of liver are now known as Cx32 and Cx26 respectively, based on their predicted formula weights established from the deduced protein sequence. The major connexin in cardiac gap junctions is Cx43 and that in lens preparations (MP70) appears to be Cx50. Although this terminology carries the advantage of making no functional inferences, it ultimately becomes clumsy as the family grows (hence Cx30, 30.3, 31, 31.1, and 32—see the previous chapter by Beyer and Willecke) and can also lead to confusion in comparisons between species. For example, chick Cx46 = rat Cx50 and chick Cx56 = rat Cx46 based on physiological properties and sequence (White et al., 1992). As our functional understanding of these proteins increases, refinements in the nomenclature are likely to become apparent.

Based on the broad division of the currently known connexin gene family into two more closely related groups, an $\alpha$ and $\beta$ terminology (Gimlich et al., 1990; Kumar and Gilula, 1996) (and subsequently, an analogous group II and I classification by Bennett et al., 1991) has also been proposed for the connexins, in which subscripts are arbitrarily used to distinguish group members. The electrophysiological analysis of paired oocytes suggests that, in general, connexins within each class tend to pair with each other (see section II.F). At present, there is a limited range of species available to establish evolutionary ties. With this caveat in mind, the extrapolation of these comparisons of connexin sequences leads to the surprising conclusion that the two major classes, $\alpha$ and $\beta$, diverged sometime near the prokaryotic/eukaryotic split, well before multicellularity arose (Hoh et al., 1991). Tentative support for this conclusion was provided by the identification of a distantly related gene in plants (Meiners et al., 1991). However, the homology of this plant protein with the connexins in terms of both sequence and predicted membrane topology is of marginal significance, and in fact it shares more homology with members of the protein kinase family (Mushegian and Koonin, 1993). As previously discussed, insect gap junctions comprise a family of proteins unrelated to connexins (innexins), further complicating any assignment of plant proteins as connexin homologs. The isolation of a connexin member in an invertebrate would certainly clarify these issues, but this goal remains elusive.

It is intriguing to speculate on the possible origins of connexins. The function of the putative plant connexin is unknown. However, several antibodies to mammalian connexins have identified two proteins localized in different regions of the plasmodesmata of plants (Meiners and Schindler, 1987; Yahalom et al., 1991). Although structurally different from junctions, plasmodesmata nonetheless have a similar function: the flow of metabolites less than approximately 1000 Da between adjacent cells. One possibility is that the plant proteins serve a more adhesive role, which is now a secondary function in mammals. A prediction of these ancient branch points in the evolutionary tree is that single-celled organisms should also have a relative of this family, and if so, the issue of function again arises. It is possible that adhesion may have represented an important early property of this family. Alternatively, regulated hemichannels may have a unique role in single-celled organisms. It is also possible that transient intercellular channels may be required at selected points in the unicellular life cycle.

A notable puzzle is the failure to detect the innexin family in vertebrates or the connexin protein in invertebrates. Did they both coexist at some point, and if so, what was their function and why was one family lost in each of the major evolutionary branches of the metazoa? However, the failure to identify nonvertebrate connexins may indicate wide divergence in the family. This may result from connexins performing different roles in vertebrates and invertebrates.

## F. Heterotypic and Heteromeric Interactions between Connexins

If multiple connexins can be coexpressed in a single cell type, to what extent can different connexins interact within a gap junction plaque? Interactions between connexins can be divided into three general types: homotypic channels formed by the same connexin; heterotypic channels formed by homomeric hemichannels in which the two hemichannels contain different connexins (as encountered when cells with different connexin phenotypes interact); and heteromeric hemichannels in which there are different connexins within the hemichannel (Figure 4; Kumar and Gilula, 1996). Evidence for heterotypic channels has been largely based on the electrophysiological analysis of paired *Xenopus* oocytes (Werner et al., 1989; Barrio et al., 1991; Hennemann et al., 1992c; White et al., 1994, 1995; Dahl et al., 1996; Cao et al., 1998) or transfected cell lines (Suchyna et al. 1993; Elfgang et al., 1995; Cao et al. 1998) expressing different connexins. Through the efforts of several laboratories, a mapping of allowed and disallowed heterotypic interactions within the connexin family has been established (Table 1). Overall, these results reveal that some connexins are highly selective in their communication patterns (e.g., Cx31 and Cx40), whereas others are rather promiscuous in heterotypic coupling (e.g., Cx46). Most connexins, however, show an intermediate behavior, typically interacting with more closely related connexins (e.g., Cx32, Cx26, Cx37, and Cx43). Thus, it appears that connexins within a homology class (i.e., $\alpha$ and $\beta$) couple more frequently than those between classes.

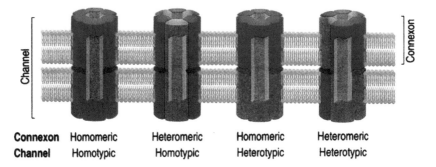

**Figure 4.** Schematic models for the possible arrangements of connexons to form gap junction channels. Connexons, which consist of six connexin subunits (red and blue), may be homomeric (composed of six identical connexin subunits) or heteromeric (composed of more than one connexin isotype). Connexons associate end-to-end to form a double membrane gap junction intercellular channel. The channel may be homotypic (if connexons are identical) or heterotypic (if the two connexons are different). Reproduced from Kumar and Gilula (1996) by permission of *Cell*.

The two notable exceptions to this rule appear to be Cx30.3 (a β connexin that preferentially pairs with alphas) and Cx50 (an α connexin that frequently pairs with betas). Consistent with the earlier implication of the second extracellular loop (E2) as a major determinant of specificity in heterotypic interactions between connexins (White et al., 1994), two residues within E2 were recently found to be responsible for this "switched" pairing preference (Zhu and Nicholson, unpublished results). At least two connexins have also been identified that fail to form

***Table 1.*** Heterotypic Coupling of Connexins

| Cx | 26 $\beta_2$ | 30 $\beta_6$ | 30.3 $\beta_5$ | 31 $\beta_3$ | 31.1 $\beta_4$ | 32 $\beta_1$ | 33 $\alpha_7$ | 37 $\alpha_4$ | 40 $\alpha_5$ | 43 $\alpha_1$ | 45 $\alpha_6$ | 46 $\alpha_3$ | 50 $\alpha_8$ |
|---|---|---|---|---|---|---|---|---|---|---|---|---|---|
| Xe 38 | − | − | + | | − | − | | + | − | + | − | − | − |
| 50 | + | | | | | + | | | − | − | | + | + |
| 46 | + | | | | + | | | | − | + | + | | |
| 45 | − | | − | − | − | − | | + | + | + | + | | |
| 43 | − | | + | − | − | − | − | + | − | + | | | |
| 40 | − | | + | − | − | − | | + | + | | | | |
| 37 | − | | + | − | − | − | − | + | | | | | |
| 33 | | | | | | − | − | | | | | | |
| 32 | + | + | − | − | − | + | | | | | | | |
| 31.1 | − | | − | | − | | | | | | | | |
| 31 | − | | | + | | | | | | | | | |
| 30.3 | − | − | + | | | | | | | | | | |
| 30 | + | + | | | | | | | | | | | |
| 26 | + | | | | | | | | | | | | |

even homotypic interactions (Cx33 [Bruzzone, et al., 1994a] and Cx31.1 [Hennemann et al., 1992a]). Evidence from Chang and associates (1996) suggests that Cx33 may play a unique role as a selective transdominant, negative regulator of Cx37 expression in the testis, without affecting the functional expression of other connexins such as Cx32.

An interesting aspect of these interactions is the degree to which the properties (e.g., voltage gating) of the resulting heterotypic channels are often substantially modified from the parental homotypic form (Barrio et al., 1991; Hennemann et al., 1992b). These changes may arise from allosteric interactions between the hemichannels (Hennemann et al., 1992c; Zhang and Nicholson, 1994) or may reflect the unmasking of hemichannel properties not evident in the symmetrical homotypic case (Suchyna et al., 1994; Verselis et al., 1994; Bukauskas et al., 1995). Although these studies have all been performed in artificially contrived systems, it is notable that competition studies between the homo- and heterotypic interactions have revealed either no significant difference (Barrio et al., 1991; Nicholson et al., 1993) or a preference for the heterotypic interaction (Werner et al., 1989, 1993).

Two separate lines of evidence support the presence of heterotypic junctions *in vivo*. Analysis of intact and split liver gap junctions by scanning transmission electron microscopy (STEM) revealed mass distributions consistent with heterotypic and homotypic, but not heteromeric junctional channels (Sosinsky, 1995) (see the discussion of this topic in section III.C). Measurements of electrical coupling between pairs of hepatocytes have also, in rare cases, revealed rectifying behavior (Verselis et al., 1994) that is characteristic of heterotypic channels formed by Cx32 and Cx26 in paired oocytes (Barrio et al., 1991). Immunoelectron microscopy has demonstrated the existence of asymmetrically labeled gap junctions, for example, between astrocytes, which express Cx43 (Dermietzel et al., 1989a; Yamamoto et al., 1990; Nagy et al., 1992) and Cx30 (Nagy et al., 1997, 1999) and oligodendrocytes, which express Cx32 (Dermietzel et al., 1989a) and Cx45 (Dermietzel et al., 1997; Kunzelmann et al., 1997). It is likely that this heterotypic arrangement of connexons will manifest itself in the properties of cell-cell communication between astrocytes and oligodendrocytes.

The issue of heteromeric interactions (i.e., between different connexin subunits within a hemichannel) has been more difficult to assess as the properties of hemichannels have not been extensively examined and any studies on intact channels or junctions are complicated by potential heterotypic interactions (i.e., between hemichannels). Hepatic gap junctions are the best characterized case of connexins that are colocalized *in situ*. Immunoelectron microscopy demonstrated that Cx32 and Cx26 are intermingled, with no evidence of segregated domains (Kuraoka et al., 1993; Zhang and Nicholson, 1994). However, limits on labeling densities using relatively large immunoglobulin G (IgG) probes did not allow resolution of the oligomeric structure of individual channels.

Biochemical strategies to address this issue have recently become feasible with the ability to examine the composition of detergent-solubilized hemichannels. Some protocols (e.g., 1% Triton X-100 at 4°C.) isolate hemichannels by selecting for nonassembled hemichannels, rather than solubilizing connexons from assembled plaques (Musil and Goodenough, 1993). Other detergent treatments are thought to disrupt the interactions between the two hemichannels within the gap junction, but not those between the subunits of the hemichannels (Stauffer, 1995). Both approaches have been used to isolate connexons from animal tissues (Ghosh et al., 1995; Jiang and Goodenough, 1996), exogenous expression systems (Ghosh et al., 1995; Stauffer, 1995), and even cell-free systems (Falk et al., 1997). Immunoprecipitation and/or Western immunoblots of a 9S species (compatible with connexin hexamers) suggested that heteromeric hemichannels can assemble between the two β connexins, Cx32 and Cx26, or the two α connexins, Cx46 and Cx50. Based on a limited comparison, heteromeric interactions between the α and β connexins (i.e., Cx43 and Cx32) appear to be disfavored (Falk et al., 1997). Without *in situ* cross-linking experiments, it has been difficult to rule out the possibility of subunit exchange between oligomers. For example, such exchange has been shown to rapidly occur in the case of enzymes where the kinetics can be monitored (Milligan and Koshland, 1988). However, in the case of connexins expressed singly or together in an exogenous system, exchange between oligomeric forms did not appear to occur significantly (Ghosh et al., 1995; Falk et al., 1997). Contrary to these biochemical results, STEM analysis of isolated junctions from rat liver (Sosinsky, 1995; described earlier) were interpreted as consistent with heterotypic, but not heteromeric gap junction channels.

A more indirect means of inferring the co-oligomerization of connexins has been provided by the identification of transdominant negative connexins that can presumably be effective only if they interact with wild-type connexins through oligomerization. Several of these constructs were described as mutants associated with Charcot-Marie-Tooth disease (Bruzzone et al., 1994b), whereas wild-type Cx33 also seems to exert selective transdominant negative effects (Chang et al., 1996). Expression of various wild-type connexins with these constructs in oocytes indicates that Cx32 interacts with Cx26 (presumably in heteromeric, although possibly also in heterotypic associations), but not Cx40, whereas Cx33 interacts strongly with Cx37, less well with Cx43, and not all with Cx32. Hence, it would appear that allowed heteromeric interactions between connexins are specifically governed by complex rules that we have only just begun to explore.

Interesting, and perhaps provocative, *in vitro* model systems have been utilized to assess the possibility of heteromeric interactions. In one study, isolated connexins were reconstituted into liposomal membranes. The results suggested that heteromeric Cx32/Cx26 (hemi)channels might have different permeability properties from the homotypic forms (Bevans et al., 1998). In another approach, several connexins were coexpressed in transfected cell lines prior to pairing and measurements of coupling. When cells that expressed rat Cx43 and human Cx37

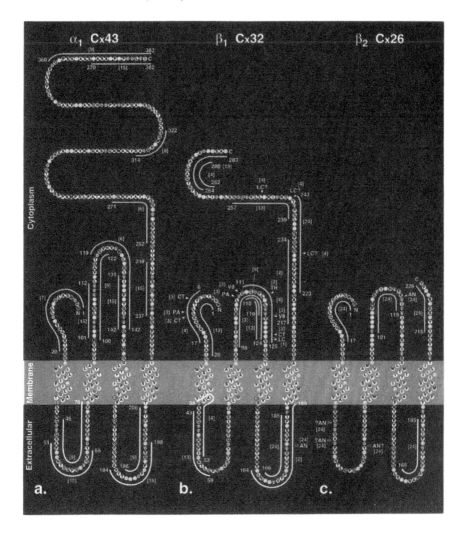

**Figure 5.** Folding models for gap junction proteins: **a**, $\alpha_1$ (Cx43); **b**, $\beta_1$(Cx32); and **c**, $\beta_2$ (Cx26). The amino-acid sequences of $\alpha_1$ (Cx43) (Beyer et al., 1987), $\beta_1$ (Cx32) (Kumar and Gilula, 1986; Paul et al., 1986), and $\beta_2$ (Cx26) (Zhang and Nicholson, 1989) were deduced from cDNA analysis, and the residues are coded as follows: hydrophobic in yellow, acidic in red, basic in blue, and cysteine in green. Hydropathy analysis predicts four membrane-spanning domains, referred to as M1, M2, M3, and M4, proceeding from the N- to the C-terminus. The predicted locations of the extracellular and cytoplasmic regions were confirmed with site-directed antibodies (blue and yellow bars indicate cytoplasmic and extracellular epitopes, respectively) and proteases (solid arrowheads indicate sites accessible from the cytoplasmic face; open arrowheads indicate sites only accessible after separation of the membranes). CT

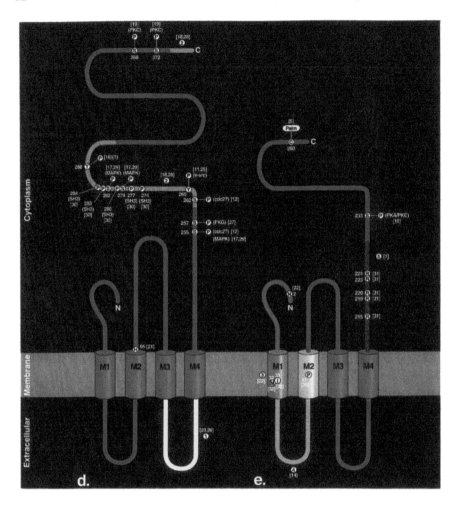

indicates chymotrypsin; PA, proteinase A; V8, staphylococcal V8 protease; TH, thrombin; LC, Lys-C protease; and AN, Asp-N protease. Note the three conserved cysteine residues (shown in green) located in each of the extracellular loops (designated E1 and E2) of the three connexins. **d**, $\alpha_1$ (Cx43) and **e**, $\beta_1$ (Cx32), indicating the locations of various functionally important residues (His95 in $\alpha_1$ [Cx43] and Asn2 and Pro87 in $\beta_1$ [Cx32]) and domains 1 (yellow) and 2 (blue) in $\alpha_1$ (Cx43) and 3 (orange), 4 (green), and 5 (violet) in $\beta_1$ (Cx32) as determined by mutagenesis and chimera studies. Indicated sites of covalent modification are based on consensus sequences, modification of synthetic peptides *in vitro*, and mutagenesis studies. The significance of the functional domains (circled numbers) and specifically mutated residues are as follows: Domain 1 (yellow) is the predominant determinant for specificity of heterotypic interactions between $\alpha_1$ (Cx43), $\alpha_8$ (Cx50), and $\alpha_3$ (Cx46). (This result was based on chimeras of $\alpha_8$ [Cx50] and $\alpha_3$ [Cx46].) Domains 2 (blue) impart high sensitivity to pH gating of $\alpha_1$ (Cx43). Gating by pH is also influenced by the charge on His95. The SH3 domains of v-Src bind to proline-rich motifs and a phosphorylated tyrosine at position 265 of $\alpha_1$ (Cx43). Domain 3 (orange) determines the polarity of the voltage sensor where an Asn2 to Asp2 change from $\beta_1$ (Cx32) to $\beta_2$ (Cx26) changes the gating polarity. Domains 3 (orange) and 4 (green) determine the voltage gating parameters; the first extracellular loop, E1, is important in determining the transjunctional voltage required for closure; Pro87 serves as a critical feature of M2 involved in the transduction of the voltage response. (This observation was based on studies in $\beta_2$ (Cx26) but has not been confirmed for other connexins.) Domain 5 (violet) constitutes potential calmodulin binding sites. Arginine residues have an inhibitory effect on the gating sensitivity of $\beta_1$ (Cx32) to cytosolic acidification by $CO_2$ exposure. The numbered references are as follows: [1]Zimmer et al. (1987); [2]Goodenough et al. (1988); [3]Hertzberg et al. (1988); [4]Milks et al. (1988) [5]Willecke et al. (1988); [6]Beyer et al. (1989); [7]Yancey et al. (1989); [8]El Aoumari et al. (1990); [9]Laird and Revel (1990); [10]Sáez et al. (1990); [11]Swenson et al. (1990); [12]Kennelly and Krebs (1991); [13]Rahman and Evans (1991); [14]Rubin et al. (1992); [15]Yeager and Gilula (1992); [16]Goldberg and Lau (1993); [17]Kanamitsu and Lau (1993); [18]Liu et al. (1993); [19]Sáez et al. (1993); [20]Suchyna et al. (1993); [21]Ek et al., (1994); [22]Verselis et al. (1994); [23]White et al. (1994); [24]Zhang and Nicholson (1994); [25]Loo et al., (1995); [26]White et al. (1995); [27]Kwak et al. (1996); [28]Morley et al., (1996); [29]Warn-Cramer et al. (1996); [30]Kanemitsu et al., (1997); [31]Wang and Peracchia (1997). Modified from Yeager and Nicholson (1996) by permission of Current Biology Ltd. and reproduced from Yeager (1998) with permission of Academic Press.

---

were paired with each other, the single intercellular channel properties were more complex than could be accounted for by homomeric and heterotypic combinations, alone, and were most readily interpretable as indicating the presence of heteromeric channels (Brink et al., 1997). That such channels can occur in model systems raises interesting possibilities, and complications, for *in vivo* studies. The lack of similar observations for cells that endogenously express more than one connexin *in vivo* raises the possibility that assembly pathways might favor homomeric connexons. This idea is exemplified by the novel observation that Cx26 can insert into membranes post-translationally and, unlike Cx32, may not have to traverse the normal endoplasmic reticulum to Golgi route when oligomerization occurs (Zhang et al., 1996). In addition, there is evidence for some separation of

Cx26 and Cx32 pools in hepatocytes *in situ* (Diez et al., 1999; George et al., 1999).

## G.  Post-Translational Modification

Post-translational modification is an important issue when considering the topology of any membrane protein. The most intensively studied post-translational modification of connexins has been phosphorylation, particularly of Cx43 and Cx32. For Cx43, phosphorylation causes upward shifts in mobility on an SDS gel (Crow et al., 1990, Musil et al., 1990b) and appears to occur ubiquitously, albeit to varying degrees in different tissues (Kadle et al., 1991; but see Hossain et al., 1994). The phosphorylation of Cx43 appears to occur relatively late in biosynthesis and has been correlated with assembly of detergent insoluble gap junction plaques (Musil and Goodenough, 1993). However, no cause-and-effect relationship has yet been established. Targets for a variety of kinases (protein kinase C [PKC], cAMP-dependent kinase [PKA], cdc2 kinase, cyclic guanosine monophosphate [cGMP]-dependent kinase (PKG), mitogen-activated protein [MAP] kinase, v-src, etc.) have been identified in Cx43 (Figure 5d) and Cx32 (Figure 5e). Other studies have recently demonstrated a novel mitosis-associated phosphorylation of Cx43 (Xie et al., 1997; Lampe et al., 1998). The potential functional significance of phosphorylation is discussed in section IV.A. and, from a somewhat different perspective, in the chapter by Laird and Sáez in this volume.

Palmitic or myristic acid moieties, or both, can be metabolically incorporated into Cx32 (Willecke et al., 1988). Although incorporation of radioactivity from either acyl chain was observed, it is unclear which form(s) is added since controls for the interconversion of the two forms within the cell were not performed. However, chemical cleavage of the acyl chain was consistent with a thioester-linked palmitic acid residue. Furthermore, the observation that the amino-termini of Cx32, Cx26, and Cx43 are not blocked during sequencing (Nicholson et al., 1981) suggests that quantitative myristilation of the protein in plaques does not occur. A consensus $[C(X)_3C]$ palmitylation site is found at the carboxy-terminus of Cx32. However, its absence from other connexins indicates that such a modification, should it occur, is not a general feature of connexin structure. No studies have as yet examined the functional significance of tethering either the amino-terminus (via myristilation) or carboxy-terminus (via palmitylation) to the membrane. However, it is interesting to speculate that the former could contribute to the relative inaccessibility of the N-terminus to proteases and antibodies. But this could also be attributed to the conserved hydrophobic residues in the N-terminus of connexins, which exhibit a sequence that may fold as an amphipathic $\alpha$-helix and closely associate with either the membrane or other protein domains. Notably, several studies have failed to demonstrate any glycosylation of connexins, consistent with the presence of only one consensus glycosylation site in the sequence that is located on the cytoplasmically disposed N-terminus.

Hertzberg and co-workers reported an interesting phenomenon associated with several connexins (Cx43, Cx32, and Cx26) and even the unrelated MIP26 of lens (Hertzberg, 1995). Isoelectric focusing of these proteins revealed a pI of 6 to 7, in marked contrast to the pI of approximately 9 predicted from the amino acid sequences. The pI shift appeared to be correlated with the assembly of connexins into detergent-insoluble gap junction structures, but could be accounted for by phosphorylation. At present, the specific nature of the modification underlying this significant acidification of the protein remains unknown, but the linkage appears to be alkali labile. This observation suggests that we do not have a full understanding of the range of potentially regulatory modifications of connexins.

## H.  Other Components

### Lipids

In contrast to the attention lavished on the protein components of gap junctions, there is a paucity of information about the lipid milieu. An obstacle to the interpretation of such studies is that the isolation of gap junctions involves membrane extractions with reagents that are likely to remove lipids. Nonetheless, efforts to identify the composition of the remaining lipids have been reviewed by Malewicz and colleagues (1990). It is noteworthy that lipid analyses on lens junctional fractions are likely to also include MIP26-containing membranes. The few analyses of liver gap junctions have yielded somewhat variable results based on the preparation technique (Evans and Gurd, 1972; Goodenough and Stoeckenius, 1972; Hertzberg and Gilula, 1979). However, common features are a relative depletion of sphingomyelin, high phosphatidylcholine levels (Hertzberg and Gilula, 1979) and a cholesterol to phospholipid ratio that is two- to sixfold higher than found in liver plasma membrane (Henderson et al., 1979). In contrast, the ratio of cholesterol to protein is relatively unaffected by the isolation procedure. Attempts to visualize the cholesterol *in situ* with filipin (Severs, 1981; Robenek et al., 1982) have not succeeded. This may not be surprising, given the close packing of connexons in gap junction membranes. In addition, the original studies implied that cholesterol may be tightly bound to the connexins, which may interfere with the formation of complexes with filipin (Feltkamp and Van der Waerden, 1983; Berdan and Shivers, 1985). Cholesterol supplementation in cell culture medium, either directly or via low density lipoprotein, appears to significantly affect the rate of junction assembly, but not the synthesis of connexins directly (Meyer et al., 1990). The mechanism for this effect, other than that it can be blocked with cycloheximide, remains obscure.

A variety of lipid reagents have been reported to influence gap junction function, but effects may not be due to incorporation into junctional membranes (e.g., diacylglycerol [Yada et al., 1985; Aylsworth et al., 1986], prostaglandins [Radu et al., 1982], arachidonic acid, and other lipoxygenases [Massey et al., 1992]). Free unsaturated fatty acids in the range of C14 to C18 have been reported to inhibit cell coupling (Burt et al., 1991). It has been speculated that these fatty acids may incorporate between phospholipids and modulate the fluidity of gap junction membranes. In one case, the blockage of coupling by the lipophilic anesthetic halothane was directly correlated with effects on the fluidity of cholesterol-rich membrane domains and not bulk lipid, consistent with a specific influence on gap junctions (Bastiaanse et al., 1993). Extensive correlation of the structural features of lipophilic agents and their potency as uncouplers has led Burt and colleagues to conclude that disruption toward the center of the lipid bilayer is critical. However, the high degree of specificity and potency of these agents also suggest specific disruption of the lipid/protein interface of connexins rather than effects on the bulk lipid (Burt et al., 1993). More recently, the sleep-inducing lipid oleamide (Cravatt et al. , 1995) was shown to specifically block dye transfer between cultured cells (Guan et al., 1997; Boger et al., 1998). Such lipophilic agents may be useful for examining the role of lipids in gap junction function. New approaches beyond those of traditional biochemical analysis will be necessary. This certainly represents one of the possible applications of reconstituted systems once they are perfected.

### Other Proteins

Are there other proteins that may interact with gap junctions? Little is known, principally because junctions are isolated by a combination of detergents and chaotropic reagents, such as high salt, alkaline pH, and urea, that disrupt noncovalent interactions. Furthermore, connexins and hemichannels are only solubilized by rather stringent conditions. Thus, without employing crosslinking, direct biochemical approaches to identifying accessory elements are not feasible. Nevertheless, two alternative strategies have yielded interesting results.

To investigate the mechanism of $Ca^{2+}$-induced gating of gap junctions, the interaction of calmodulin with Cx32 was studied by the gel overlay technique (Van Eldik et al., 1985). Binding sites were identified in both the C- and N- termini of Cx32 (Zimmer et al., 1987). The relevance of these observations to channel gating remains unclear as the binding appeared to be constitutive and was not modulated by $Ca^{2+}$. In fact, calmodulin "binding" might simply reflect ion-exchange between acidic calmodulin and basic targets on connexins, which have predicted pIs typically greater than 10 (but see section III.G). The gel overlay technique has also been employed to demonstrate an association of guanosine triphosphate (GTP)-binding proteins with a presumed Cx32-like protein in heart membranes (Doucet et al., 1992), but the presence of significant amounts of Cx32, a predominantly liver-associated connexin, in heart gap junctions is puzzling.

Another approach avoids the artifacts of the isolated system and relies on the *in situ* localization of proteins to gap junction structures. In a very early study, pp60v-src was localized at the plasma membrane, in the vicinity of gap junctions, following its activation in NRK cells (Willingham et al., 1980). This association has been directly confirmed by co-precipitation of v-src and Cx43 (Kanemitsu et al., 1997; Zhon et al., 1999) and is consistent with the demonstration of rapid tyrosine phosphorylation of Cx43 in response to v-src expression (Swenson et al., 1990). Both morphological and immunological criteria have also localized basic fibroblast growth factor with cardiac (Kardami et al., 1991) and astrocyte gap junctions that contain Cx43 (Yamamoto et al., 1991). The possibility of connexin interaction with other elements of junctional complexes, such as ZO-1, a peripheral membrane protein associated with tight junctions in epithelial cells and adeherens junctions in cardiac myocytes, may underlie aspects of the organization of junctional complexes *in vivo* (Toyofuku et al., 1998). Steps in the synthesis, assembly and turnover of gap junction channels appear to follow the general secretory pathway for membrane proteins (reviewed in Yeager et al., 1998). The turnover of gap junctions is thought to occur via internalization of complete dodecameric channels in a complex process that involves lysosomal as well as proteasomal pathways (Ginzberg and Gilula, 1979; Risinger and Larsen, 1983; Naus et al., 1983; Rahman et al., 1993; Laing and Beyer, 1995; Laird, 1996; Laing et al., 1997).

Nonetheless, the most striking result of the attempts to identify proteins associated with connexins may be a negative one. Gap junctions are unique among plasma membrane structures involved in interactions between cells in that they appear to lack a connection with the cytoskeleton, as most notably demonstrated in deep-etched, freeze-fracture images of the cytoplasmic surface (Hirokawa and Heuser, 1982). Earlier studies suggested an association of actin with gap junctions (Larsen et al., 1979), and pharmacological disruption of the cytoskeleton with cytochalasin B was found to enhance gap junction formation (Tadvalkar and Pinto Da Silva, 1983). However, neither study demonstrated a compelling attachment between gap junctions and the cytoskeleton. However, they do raise the possibility that hemichannels at the cell surface could be "tethered" to the cytoskeleton, which could regulate gap junction formation or turnover. Studies on gap junction biosynthesis have thus far not revealed such interactions (Musil and Goodenough, 1993; Rahman et al., 1993).

Given our limited knowledge about accessory elements, new strategies will need to be developed to address this important issue, perhaps utilizing both crosslinking and genetic approaches that have been successful in other fields. For example, experiments using the yeast two hybrid system suggest an association between ZO1 and Cx43 (Giepmans and Moolenaar, 1998; Toyofuku et al., 1998). Independent lines of evidence have also recently indicated that pH gating of Cx43 may require as yet ill-defined cytoplasmic factors, although this does not appear to be the case with Cx46 (Cao and Nicholson, unpublished results).

## III.  STRUCTURE ANALYSIS

### A.  Background

Conventional thin-section electron microscopy has served as a keystone for defining the "morphological signature" of gap junctions. Although the nuances of stain distribution may vary with the preparative technique (Brightman and Reese, 1969), the characteristic appearance is septalaminar (see Figure 1) with stain exclusion in the hydrophobic domains of the lipid bilayer and stain accumulation in the extracellular gap and on the cytoplasmic faces. (This can be distinguished from tight junctions, which appear pentalaminar by thin-section electron microscopy [Revel and Karnovsky, 1967; Brightman and Reese, 1969; Friend and Gilula, 1972].) The narrow extracellular gap is a feature that has been particularly useful for delineating the polypeptide topology because macromolecular probes such as proteases and antibodies are excluded. In contrast, the cytoplasmic surfaces are readily accessible as exemplified by the decoration of gold labeled secondary antibodies attached to site-specific peptide antibodies directed against a sequence predicted to reside on the cytoplasmic surface (see Figure 1).

## B.   One-Dimensional Electron Density Profile by X-ray Diffraction

Low angle x-ray scattering of gap junctions in aqueous buffers (Makowski et al., 1977) allowed more precise examination of the distribution of lipid and protein than traditional staining patterns of thin sections (see Figure 1). Specimens were prepared by pelleting gap junctions into stacks to separate the equatorial diffraction arising from the hexagonal packing of the connexons from the meridional diffraction arising from the electron density fluctuations perpendicular to the plane of the gap junction membranes. In order to compute a one-dimensional map, the phases for the meridional diffraction amplitudes had to be determined. The problem was simplified because the centrosymmetry of the specimens restricted the phases to 0 and 180 degrees. Phase assignments were based on the minimum wavelength principle (Sayre, 1952): because the thickness of liver gap junctions is approximately 150 Å, the phases for adjacent peaks in the diffraction pattern must be of opposite sign if the peak separation in the diffraction pattern is less than 2/150 Å. In addition, density maps not consistent with the low resolution images provided by thin-section electron microscopy could be excluded. The computed one-dimensional electron density maps were dominated by four peaks that corresponded to the location of the electron-dense phospholipid headgroups in the two lipid bilayers comprising the gap junction. This map provided precise boundaries for the lipid bilayer domains of liver gap junctions: the headgroups within each bilayer were separated by approximately 42 Å, the hydrophobic portion of the bilayers was approximately 32 Å thick, and the extracellular gap had a thickness of approximately 35 Å. The electron density in the aqueous extracellular gap and cytoplasmic regions could be adjusted by varying the amount of sucrose in the buffer (Makowski et al., 1984). In this way the distribution of protein in gap junctions could be examined. The analysis compared the predicted electron density for different models with the experimental density profiles derived for gap junctions suspended in aqueous buffers with varying electron densities. Although the transmembrane distribution of connexins is now an accepted fact, the x-ray scattering analysis provided the first compelling evidence for such models. The electron density within the center of the hydrophobic region was much higher than would be expected for pure lipid, indicating the transmembrane disposition of the protein. The higher electron density in the extracellular gap compared with the electron density of the buffer confirmed that the connexons span the gap, as expected from thin-section electron micrographs.

## C.  Low Resolution Two-dimensional Projection Density Maps

The paracrystalline packing of the annular connexons in liver gap junctions is a critical feature that has allowed the application of electron crystallography. Electron microscopy and digital image processing remains to date as the approach that has enriched our knowledge of gap junction structure more than any other method. The first studies focused on the analysis of mouse (Caspar et al., 1977) and rat liver (Zampighi and Unwin, 1979) gap junctions. More recently, negatively stained heart (Yeager and Gilula, 1992; Yeager, 1994, 1998), lens (Lampe et al., 1991), liver (Perkins et al., 1997, 1998), and arthropod (Sikerwar et al., 1991) gap junctions have been manipulated in order to grow 2D crystals amenable to electron microscopy and image analysis.

There are several advantages of electron crystallography:

1.  The diffraction pattern is generated by all molecules in the crystal, thereby improving the signal-to-noise ratio.
2.  Noise in the images is evident between the diffraction spots and can be removed by filtering.
3.  The phases for computing the density map by Fourier transformation are directly calculated from the images, unlike x-ray diffraction in which the phase information is lost.
4.  Further averaging is provided by enforcing the inherent symmetry in the packing of the molecules in the crystal lattice.

The 2D projection density map is computed by back Fourier transformation and provides an *in focu* view of the distribution of mass. Conserved features of all of the different gap junctions structures are the annular appearance of the connexon surrounding a central region of negative stain (Yeager, 1995). Based on connexin molecular weight and protein density, the annulus supports a model in which the channel is formed by an oligomeric cluster of connexin subunits. The diffraction patterns of negatively stained specimens extend to a nominal resolution of 15 to 20 Å, and the computed phases to this level of resolution are consistent with plane group symmetry p6. Therefore, in the published density maps, sixfold symmetry has been enforced, and the six domains comprising the annular connexons correspond to a hexamer of subunits.

## Molecular Heterogeneity

Several caveats regarding the method of electron microscopy and image analysis should be noted. The method is an averaging technique, so that any molecular heterogeneity will not be revealed in an averaged map. For instance, liver gap junctions not only contain Cx32 but also Cx26. At 15- to 20-Å resolution, subtle differences in the 2D projection density maps of channels formed by Cx32 and Cx26 connexons would be obscured. In addition, an averaged map at 15-Å resolution would also probably obscure heteromeric connexons in which oligomers are formed by both Cx32 and Cx26 connexins (see section II.F). An example is the acetylcholine receptor that is formed by a heteropentamer of 2α, β, γ, and δ subunits. Projection density maps at 17-Å resolution did not reveal major differences between the homologous but different subunits (Toyoshima and Unwin, 1988). Another type of heterogeneity that would probably be obscured at approximately 15-Å resolution is variability in the stoichiometry of the oligomer. For instance, a small fraction of pentameric or heptameric connexons would probably be obscured in an averaged map at low resolution.

## Atomic Force Microscopy

Atomic force microscopy (AFM) is a technique that may allow examination of molecular heterogeneity in gap junctions. In this method, an extremely fine needle prepared by electron beam deposition is dragged along the surface of the specimen. Deflections of the needle on the order of Ångstroms can be detected by observing the reflection of laser light from the needle. An important advantage of AFM is that fully hydrated specimens can be examined in physiological buffers. Furthermore, the images obtained are point to point and not averaged, so that heterogeneity within the sample is resolved. Although the AFM images are prone to tip-induced artifacts, continued research and improvements in tip design will likely allow recording of the surface topography of membranes with greater fidelity. However, another limitation that limits resolution is the inherent deformability of biological specimens compared with inorganic materials. The best images thus far obtained from liver gap junctions are approaching the quality obtained by conventional transmission electron microscopy of negatively stained specimens (Hoh et al., 1993).

## Scanning Transmission Electron Microscopy

Scanning transmission electron microscopy (STEM) is a second structural approach that has allowed investigation of molecular heterogeneity in gap junction specimens (Sosinsky, 1995). Liver gap junction samples were mixed with a standard such as tobacco mosaic virus (TMV), and the specimen was freeze-dried. The TMV serves as an internal mass standard to calibrate the mass measurements of connexons within a gap junction. Since the lipid in the membranes will affect the measurements, it is best to interpret the mass measurements qualitatively. Histograms for the mass measurements of liver gap junctions yielded three types of distributions: narrow and unimodal, broad and unimodal, or bimodal. The narrow, unimodal distributions and the bimodal distributions were consistent with homotypic channels formed by Cx32/Cx32 or Cx26/Cx26 pairings. Broad unimodal distributions could arise from heterotypic homomeric channels or pairing of heteromeric connexons. This ambiguity was resolved by examining the mass distributions for urea-split liver gap junction plaques. Because the average masses for the split junctions was always half of the mass for the intact junction, the connexons were homomeric. Hence, the broad unimodal distributions were consistent with heterotypic channels formed by the pairing of homomeric Cx26 connexons and homomeric Cx32 connexons.

## Variation in Projection Density Maps

There are differences between the density maps derived in different laboratories that may relate to bona fide chemical differences in structure, differing effects on the structure during isolation, or variability in negative staining. For instance, the maps of mouse liver gap junctions (Baker et al., 1985) displayed density on the threefold symmetry axis that has been interpreted as an arm of the connexin polypeptide (Makowski et al., 1984). Alternatively, this density could represent nonconnexin protein, but electrophoresis of the purest gap junction preparations have not suggested that there are other proteins associated with gap junction plaques. Maps of rat liver (Gogol and Unwin, 1988) and cardiac (Yeager and Gilula, 1992) gap junctions have not shown significant density on the threefold axis. In addition, a map has been derived for liver gap junctions from crude preparations of plasma membranes, thereby minimizing the effects of sarkosyl or alkali treatment and multiple centrifugation steps (Sikerwar and Unwin, 1988). This projection density map displayed the characteristic hexameric connexon as well as an absence of density on the threefold axis. Interestingly, the density map of cardiac gap junctions (Yeager and Gilula, (1992) reveals a larger central channel. This difference may be related to differences in the isolation method or variability in staining. Electron cryo-microscopy of heart and liver gap junctions isolated by the same protocol should allow careful scrutiny of these differences.

## D.  Low Resolution Three-Dimensional Structure

In spite of the limiting resolution, electron microscopy and image analysis have been powerful methods for defining the 3D molecular envelope of the connexons in liver gap junctions. The approach relies on the ability to tilt the membrane crystals via the goniometer stage of the electron microscope (Henderson and Unwin, 1975; Amos et al., 1982). The projected views represent planar sections through the 3D density map. The 2D sections are combined in Fourier space, and the 3D Fourier transform is interpolated in the regions for which sections were not obtained. The 3D density map is then obtained by back Fourier transformation.

Low resolution 3D density maps have been derived for rat liver gap junctions either stained with uranyl acetate (Unwin and Zampighi, 1980; Perkins et al., 1997) or in the frozen hydrated state (Unwin and Ennis, 1984). Specimens imaged by negative stain microscopy are sensitive to beam-induced shrinkage, and the thickness of the liver gap junction was adjusted to match the membrane thickness measured from x-ray scattering of hydrated specimens. In addition, the stain puddles around the perimeter of the specimen so that the images reveal the molecular envelope of the specimen. In the case of electron cryo-microscopy of unstained specimens, the derived maps not only reveal the surface contours of a structure but also the internal organization. Electron cryo-microscopy of unstained samples also minimizes the potential specimen shrinkage and distortion encountered with negative stain microscopy. Nonetheless, the overall maps obtained by negative stain and cryo-electron microscopy are quite similar.

The basic design of the liver gap junction connexon is a cluster of six rodlike subunits oriented roughly perpendicular to the membrane plane. The subunits have a diameter of approximately 25 Å and surround a central channel that has a diameter of approximately 20 Å. At a resolution of 18 Å, the contours of the channel within the hydrophobic interior of the bilayer were not revealed. The connexon extends approximately 20 Å into the extracellular space so that the gap is approximately 40-Å thick. The thickness of the extracellular gap is larger than observed by conventional thin-section electron microscopy, possibly owing to shrinkage during dehydration and embedding. Examinations of the contrast perpendicular to the membranes plane reveals that the negative stain in the center of the connexon in *en face* views primarily resides in an extracellular vestibule with less stain accumulation on the cytoplasmic face of the channel.

Interestingly, the 3D maps only resolve density extending approximately 10 Å into the cytoplasmic space. This is substantially less than would be expected, knowing that the amino terminal loop, the substantially longer carboxy tail, as well as the loop between the second (M2) and third (M3) membrane-spanning domains should all reside on the cytoplasmic face and account for more than one-third of the mass of Cx32, the predominant component in rat liver gap junctions. One possible explanation is that the cytoplasmic domains exhibit conformational flexibility so that their density would be smeared out during crystallographic averaging. Another explanation is that the map is less well resolved in the z direction perpendicular to the membrane because the specimens can only be tilted up to approximately 60 degrees. The resulting "missing cone" of data tends to smear the map in the z direction. A comparison of the x-ray scattering density profiles of native and trypsin-treated mouse liver gap junctions showed that protein extends 85 to 90 Å from the center of the gap (Makowski et al., 1984). The lipid headgroup region on the cytoplasmic leaflet of the bilayer is centered approximately

64 Å from the center of the gap. Assuming a headgroup region approximately 10-Å thick, the cytoplasmic protein extends 15 to 20 Å beyond the membrane surface, which should certainly be sufficient to accommodate the connexin domains predicted to reside on the cytoplasmic face, particularly given the higher complement of Cx26 in mouse liver gap junctions. Image analysis of edge views of cardiac gap junctions (Yeager, 1998) suggests a gap junction thickness of approximately 250 Å so that the protein on the cytoplasmic face may extend 40 to 50 Å from the membrane surface, consistent with the larger size of the Cx43 compound with the Cx32 carboxy tail.

## E. Membrane Topology

The 3D density map determined by electron microscopy and image analysis at 15- to 20-Å resolution defines the molecular boundary of the cylindrical connexin subunits and the hexameric quaternary arrangement in the connexon (Unwin and Ennis, 1984; Perkins et al., 1997). However, the maps are not of sufficient resolution to define the folding of the polypeptide chain. Hydropathy analyses of the complete sequences of the various connexin family members reveal clear patterns of conserved and variable domains and suggest a putative folding pattern within the membrane. Four hydrophobic domains of between 20 and 28 residues suggest that the connexins contain four transmembrane regions (see Figure 5). These domains, and particularly the cysteine-rich hydrophilic loops connecting the first and second spans, and the third and fourth spans, represent the most conserved regions of the protein family (see the preceding chapter by Beyer and Willecke in this volume). The most variable domains, both in length and sequence, are the carboxy-terminal domain and the hydrophilic loop connecting the second and third spans, which have a cytoplasmic location based on early protease protection assays of isolated gap junctions (Nicholson et al., 1981; Gros et al., 1983; Manjunath et al., 1987).

Direct biochemical confirmation of the connexin membrane topology suggested by hydropathy analysis has been greatly aided by the close association of the paired membranes in isolated gap junctions. The narrow extracellular gap of approximately 20 Å does not allow access of proteases or antibodies unless specific treatments are employed to separate the membranes (Manjunath et al., 1984b; Zimmer et al., 1987). In isolated junctions, these treatments utilize strongly denaturing conditions (6 M urea at pH 12) that may alter the folding of the polypeptide chain on the cytoplasmic face, thus complicating interpretation of the results. Detailed topological studies of three members of the connexin family (Cx32, Cx43, and Cx26; references given in Figure 5) have utilized a combination of site-specific, peptide antibodies and treatments with highly specific proteases. Indeed, evidence defining the topology of connexins is perhaps more compelling than for any other polytopic membrane protein other than those for which atomic structures have been determined. In addition to the broad view of folding provided by these studies, a number of more specific conclusions about interactions between subdomains can be gleaned.

Limited accessibility of the 20-residue amino-terminus to both proteases (Hertzberg et al., 1988) and many antibodies suggests that residues 11 to 22 are protected by either folding of the polypeptide or close association with the membrane. Protection of the carboxy-terminal domain of Cx26 was also demonstrated by a proteolytic time course, which showed that the binding of a C-terminal antibody required the cleavage of the cytoplasmic loop. After exposure, this domain could also be cleaved by proteases (Zhang and Nicholson, 1994). This is consistent with a recent model proposed by Wang and Peracchia (1997) in which interactions between the membrane proximal part of the C-terminus and the cytoplasmic loop of Cx32 are thought to affect the gating of the channel in response to reduced pH (see section V).

## F.   Secondary Structure Models

Before 1997, the available density maps provided by x-ray diffraction and electron microscopy of gap junctions were not of sufficient resolution to resolve elements of secondary structure. Therefore, hypothetical working models were based on inferences from (1) the topological mapping of the primary sequence (see Figure 5) combined with the 3D density maps provided by electron image analysis (Milks et al., 1988); (2) x-ray diffraction patterns (Makowski et al., 1984; Tibbits et al., 1990); and (3) circular dichroism spectroscopy (Cascio et al., 1990; Yeager, 1993).

(1) The topological studies described earlier support a folding model with four transmembrane domains of sufficient length so that they could be folded as $\alpha$-helices. By analogy with soluble proteins having four antiparallel $\alpha$-helices, a model was proposed by Milks and colleagues (1988) in which the four transmembrane domains of Cx32 connexin were folded as a four-helix bundle that could be accommodated within the cross-sectional area of each connexin, estimated as 500 $\text{Å}^2$ from 3D density maps of liver gap junctions. Six M3 transmembrane domains from each of the six subunits were proposed to form the boundary of the aqueous channel, as M3 contains a series of polar amino acids spaced so that they could form an amphipathic $\alpha$-helix.

(2) X-ray diffraction patterns recorded from oriented gap junction membranes displayed sharp fringes centered at approximately 4.7 Å on the meridian and diffraction centered at approximately 11 Å on the equator. These patterns were initially interpreted as $\beta$-sheets with the strands running more parallel than perpendicular to the surface. To test this hypothesis, connexon models were built that contained a transmembrane core based on known soluble proteins structures that exhibited $\alpha$-helical bundle and $\beta$-sheet conformations. In fact, the predicted diffraction patterns for purely $\alpha$-helical conformations were in closer agreement with the x-ray diffraction data. For $\alpha$-helices packed perpendicular to the mem-

brane plane, a diffraction fringe at approximately 5 Å would have been predicted. However, in four helix bundle proteins, the α-helices are tilted with respect to each other. This tilting causes a shift in the fringe to approximately 4.7 Å.

(3) From 190 to 240 nm, circular dichroism (CD) spectra of proteins are sensitive to molecular geometry, and are therefore quite useful for determining protein secondary structure and monitoring conformational changes (Johnson, 1988; Fasman, 1993; Bloemendal and Johnson, 1995; Woody, 1995). CD spectroscopy measures the difference in absorption of left and right circularly polarized light. The magnitude of these difference measurements are quite small so that the signal-to-noise ratio is typically <<1. Therefore, a protocol was typically used in which spectra are recorded over several hours by accumulating repetitive scans. In this way, each scan could be statistically compared with all others in order to detect instrumental variations during recording. A second relevant technical aspect was that the signal is dramatically affected by light scattering and results in absorption flattening in the far ultraviolet region that will confound the prediction of secondary structure (Glaeser and Jap, 1985). Light scattering can certainly occur with membrane specimens, and samples were sonicated and centrifuged so that only the smallest vesicles are used for CD measurements. CD spectra recorded from suspensions of rat liver (Cascio et al., 1990) and heart gap junction vesicles (Yeager, 1993) displayed considerable similarity to the spectra for a polypeptide with an α-helical conformation, and the estimated content of α-helix (40% to 50%) is sufficient so that the four transmembrane domains may be folded as α-helices.

Apart from CD spectroscopy, several other useful biophysical techniques include Fourier transform infrared (FTIR), time-resolved fluorescence, electron paramagnetic resonance (EPR) and nuclear magnetic resonance (NMR) spectroscopy. The advantage of FTIR spectroscopy is that spectra can be recorded from stacked membrane pellets that are fully hydrated so that information can be obtained about the angular orientation of α-helices with respect to the membrane surface (Braiman and Rothschild, 1988; Cooper and Knutson, 1995; Haris and Chapman, 1995; Jackson and Mantsch, 1995). Time-resolved fluorescence spectroscopy is a powerful method for examining conformational dynamics (Stryer, 1968; Somogyi and Lakos, 1993; Jiskoot et al., 1995). This approach may be particularly useful to detect conformational changes that occur during gating of membrane channels. Given the advancing expertise in mutagenesis of connexins, it may be possible to replace all but one of the tryptophan residues with phenylalanine (Menezes et al., 1990). The remaining tryptophan residue could then serve as a reporter of local environment and dynamics during channel gating. In addition, energy transfer between donor and acceptor chromophores attached to macromolecules can be used to determine proximity relationships (Wu and Stryer, 1971; Clegg, 1995; Jiskoot et al., 1995). EPR spectroscopy is another powerful method to study protein structure and dynamics (Todd et al., 1989; Hubbell and Altenbach, 1994; Voss et al., 1995; Hubbell et al., 1996, 1998). For instance, cysteine substitution mutagenesis can be used to introduce spin labels at particular sites in membrane proteins, and the EPR spectrum reveals amino acid side chain dynamics, solvent accessibility, polarity, and electrostatic potential. Although proteins of fairly sizable molecular weight can now be examined by NMR spectroscopy, it is still not feasible to attempt to carry out NMR studies of membrane proteins in detergent micelles. The tumbling rate of the protein-detergent complex would be so slow that the spectral lines would exhibit substantial line broadening, precluding interpretation. Nevertheless, small soluble regions of membrane proteins may be amenable to NMR spectroscopy. For instance, NMR spectroscopy has been used to examine the structure of the third cytoplasmic loop and the carboxy-terminal domain of bovine rhodopsin (Yeagle et al., 1995a, 1995b). The synthesized peptides were biologically active, suggesting that the conformation of the peptides is relevant to the native protein structure. For connexins, candidate domains are the extracellular loops, the cytoplasmic loop between M2 and M3, as well as the carboxy-tail.

Needless to say, such soluble domains which could be readily expressed in bacterial or insect systems may also be more amenable to conventional 3D crystallization and examination by x-ray crystallography than the full-length protein. This has certainly been a fruitful strategy in several other systems in which the hydrophilic, water-soluble domains of bitopic membrane proteins were expressed, purified, and examined at atomic resolution (Wang et al., 1990; De Vos, et al., 1992; Shapiro et al., 1995). However, this approach does not allow examination of transmembrane domains. An integrated strategy that combines NMR or crystallographic analyses of expressed water-soluble domains with electron cryo-crystallography and spectroscopic analysis of membrane domains should be of general utility in the higher resolution structure analysis of polytopic membrane proteins.

## G.  Higher Resolution Structure Analysis by Electron Cryo-Crystallography

In order to resolve elements of secondary structure such as $\alpha$-helices and $\beta$-sheet, the diffraction patterns from 2D crystals would have to extend to better than 10-Å resolution. Based on the experience with bacteriorhodopsin (Henderson et al., 1990), porin (Jap et al., 1991), LHCII (Kühlbrandt et al., 1994), and aquaporin I (Cheng et al., 1997; Li et al.; 1997; Walz et al.; 1997), a chemically pure protein is required to grow high resolution 2D crystals. However, the source to date for 2D crystallization studies of gap junctions has been preparations isolated from heart and liver tissue. For example, isolated cardiac gap junctions possess very little inherent crystallinity compared with liver gap junctions. A strategy for *in situ* crystallization is to expose the isolated junctions to low concentrations of nonionic detergents in order to extract lipid and concentrate the protein in the membrane plane.

Examination of biological specimens in the frozen-hydrated state offers the possibility for higher resolution structure analysis. Nevertheless, electron cryomicroscopy of native gap junctions isolated from tissues such as the liver have been limited to 15- to 20-Å resolution (Unwin and Ennis, 1984; Gogol and Unwin, 1988). This limiting resolution is presumably based on molecular heterogeneity that could arise from (1) expression of multiple connexins within cells of the same tissue that could lead to formation of gap junctions that are assembled from different homomeric connexons, formation of heterotypic channels in which homomeric connexons are composed of different connexins, and formation of heteromeric connexons in which the oligomers contain multiple connexin types (Barrio et al., 1991; Elfgang et al., 1995; Sosinsky, 1995; Stauffer, 1995; White et al., 1995; Jiang and Goodenough, 1996); (2) multiple oligomeric states within a connexon population (such as a small fraction of pentamers among the hexamers); (3) inherent flexibility in the native protein; (4) partial denaturation during specimen preparation; (5) the presence of nonconnexin proteins that may disorder the lattice; (6) partial proteolysis; (7) different degrees of post-translational modification; (8) rotational flexibility of the connexons in the lattice; and (9) lipid heterogeneity that may prevent precise chemical interactions at the protein-lipid interface that may be necessary for 2D crystallization of the connexons.

Several of these problems can be overcome by expression of a single recombinant connexin. Progress has certainly been made in the over-expression of Cx32 in insect cells using a baculovirus vector (Stauffer et al., 1991). The ability of over-expressed connexins to self-assemble into recognizable gap junction structures in infected cells is convincing evidence that the recombinant protein exhibits the critical structural elements of the native protein. Although initial 3D crystallization trials were encouraging (Stauffer et al., 1991), crystals suitable for x-ray crystallography have not been forthcoming. Alternatively, detergent-solubilized and purified connexons may be amenable to *in vitro* reconstitution with lipids in order to grow 2D crystals (Jap et al., 1992; Kühlbrandt, 1992; Yeager et al., 1999). However, this approach has not been successful in yielding high resolution 2D crystals thus far.

Recombinant connexins have now been expressed in a stably transfected baby hamster kidney cell line under control of the inducible metallothionin promotor (Kumar et al., 1995). Ultrastructural studies demonstrated that a truncated form of rat Cx43 that lacks most of the large C-terminal domain (truncation at Lys263) assembles into gap junctions having the characteristic septalaminar morphology. Freeze-fracture images revealed that the gap junctions form small 2D crystals (Figure 2), and their crystallinity and purity could be improved by extraction with nondenaturing detergents such as Tween20 and DHPC (1,2-diheptanoyl-sn-phosphocholine) (Unger et al., 1997; Yeager et al., 1999; Unger et al. 1999a), an approach similar to that taken for crystallization of native rat heart gap junctions (Yeager, 1994).

## Projection Density Map

A projection density map derived from negatively stained 2D crystals of a recombinant Cx43 (Figure 6) closely resembled the maps for native cardiac gap

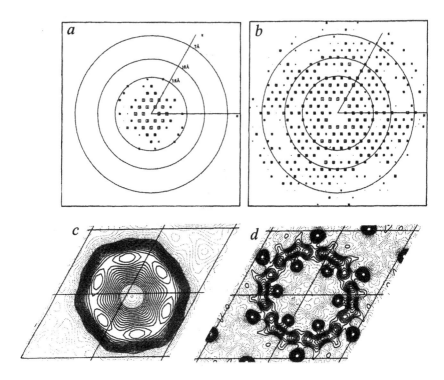

**Figure 6.** Computed diffraction patterns (**a** and **b**) and projection density maps (**c** and **d**) for gap junctions containing a recombinant form of the rat heart Cx43 in which the majority of the C-terminal domain has been deleted (truncation at Lys 263) (Unger et al., 1997). (**a**) Negatively stained crystals display diffraction to approximately 15-Å resolution. (**c**) The projection density map shows the hexameric connexons, similar to maps at comparable resolution from rat heart gap junctions (Yeager and Gilula, 1992; Yeager, 1994). (**b**) When unstained crystals are examined using electron cryo-microscopy, the diffraction patterns extend to approximately 7-Å resolution. (**d**) The puliminary projection density map at 7-Å resolution (Unger et al., 1997) displays three major features: a ring of circular densities centered at a radius of 17-Å interpreted as α-helices that line the channel, a ring of densities centered at a radius of 33-Å interpreted as α-helices that are most exposed to the lipid, and a continuous band of density at 25-Å radius separating the two groups of helices. The hexagonal lattice had parameters $a = b = 79$-Å and $\gamma = 120$-Å. The spacing between grid bars is 40Å. Modified from Yeager and Nicholson (1996) and reproduced by permission of *Current Biology Ltd.*

junction membranes (Yeager and Gilula, 1992; Yeager and Nicholson, 1996). The recombinant connexon had a diameter of approximately 65 Å and was formed by a hexameric cluster of connexin subunits. A projection density map based on the analysis of unstained, frozen-hydrated 2D crystals (Unger et al., 1997) showed that the recombinant connexon has a diameter of approximately 65 Å and is formed by a hexameric cluster of connexin subunits (Figure 6d). At 7-Å resolution, the map of the recombinant channel shows substantially more detail than the map of rat heart gap junctions at 16-Å resolution (Yeager and Gilula, 1992). In particular, at 17-Å radius, the channel is lined by circular densities with the characteristic appearance of transmembrane α-helices that are oriented roughly perpendicular to the membrane plane (Unwin and Henderson, 1975; Kühlbrandt and Downing, 1989; Havelka et al., 1993; Schertler et al., 1993; Jap and Li, 1995; Karrasch et al., 1995; Mitra et al., 1995; Walz et al., 1995; Hebert et al., 1997). A similar appearance for densities at 33-Å radius suggests the presence of α-helices at the interface with the membrane lipids. The two rings of α-helices are separated by a continuous band of density at a radius of 25 Å, which may arise from the superposition of projections of additional transmembrane α-helices and polypeptide density arising from the extracellular and intracellular loops within each connexin subunit.

A notable feature of the projection density map is the 30-degree displacement between the rings of α-helices at 17- and 33-Å radius, which places constraints on possible structural models for the intercellular channel (Figure 7; Unger et al., 1997). As a consequence of the 30-degree displacement between the α-helices that line the channel (at 17-Å radius) and the α-helices that are most exposed to the lipid (at 33-α radius), the two connexons forming the channel are rotationally staggered with respect to each other (as shown by the arcs in Figure 7a and b). The amount of rotational stagger will dictate whether 6 or 12 peaks are resolved in the outer ring of α-helices at 33-α radius. That is, models with less than 30 degrees of rotational stagger between the connexons are not consistent with the map in Figure 6d, as there would be 12 rather than 6 peaks in the outer ring of α-helices (Figure 8b). In addition, the 30-degree displacement between the rings of helices is not consistent with models in which the α-helices are colinear through the center of the channel (see Figure 7c and d). The model shown in Figure 7a is in best agreement with the projection density map (see Figure 6d) that shows superposition of the resolved α helices between the apposed connexins. Note that this model predicts that each subunit in one connexon will interact with two connexin subunits in the apposing connexon. Such an arrangement may confer stability in the docking of the connexons.

Electron microscopy and image analysis of negatively stained gap junction plaques that had been split in the extracellular gap by urea treatment showed that the extracellular surface of each connexon contains six protrusions (Perkins et al., 1997). In order to form a tight seal with the apposed connexon, computer modeling of the map at approximately 18 Å resolution predicted that these protrusions interdigitate in such a way that requires a 30-degree stagger (Perkins et al., 1998),

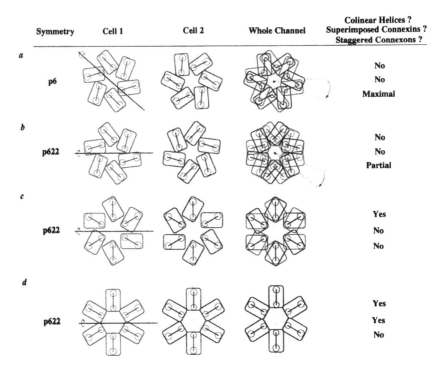

| Symmetry | Cell 1 | Cell 2 | Whole Channel | Colinear Helices ?<br>Superimposed Connexins ?<br>Staggered Connexons ? |
|---|---|---|---|---|
| *a* p6 | | | | No<br>No<br>Maximal |
| *b* p622 | | | | No<br>No<br>Partial |
| *c* p622 | | | | Yes<br>No<br>No |
| *d* p622 | | | | Yes<br>Yes<br>No |

***Figure 7.*** Schematic models for the packing of helices, connexins, and connexons in the gap junction intercellular channel. Each connexin subunit is represented by a rectangle, and the transmembrane α-helices are depicted as circles. The twofold symmetry axes located in the extracellular gap generate the views of the apposed connexons in cell 1 (red) and cell 2 (blue) that form the intercellular channel. The model in (**a**) is in accordance with the observed projection density map (Figure 6d) and predicts that the connexons within the channel will be rotationally staggered, as shown by the dashed lines and the arc. The superposition of the helices and this rotation dictate that the α-helices within one connexin will be superimposed with helices within two connexin subunits in the apposed connexon. The models shown in (**b**), (**c**), and (**d**) are inconsistent with the projection map shown in Figure 6d. The model in (**b**) is representative of a class of structures that involve variable degrees of rotational stagger between the connexons and was generated by a rotation around a twofold symmetry axis that would give rise to strict *p*622 symmetry. A characteristic feature of such models is 12 density peaks at high radius. Depending on the rotational stagger (indicated by the dashed lines and the arc), the 6 peaks at low radius may be broadened or resolved into 12 separate peaks. The models in (**c**) and (**d**) also display *p*622 symmetry, but the connexons are not rotationally staggered. Note that the α-helices are colinear through the center of the channel, as shown by the dotted lines. In model (**d**), there is superposition of the connexin subunits of apposed connexons. Reproduced from Yeager (1998) by permission of Academic Press.

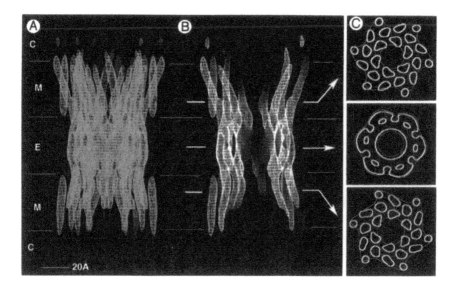

***Figure 8.*** Three-dimensional structure of a recombinant cardiac gap junction channel in which the majority of the C-terminal domain has been deleted from Cx43 (truncation at Lys 263). **A**. A full side view is shown. **B**. The density has been cropped to show the channel interior. The approximate boundaries for the membrane bilayers (M), extracellular gap (E) and the cytoplasmic space (C) are indicated. The white arrows identify the locations of the cross-sections that are parallel to the membrane bilayers (**C**). The red contours in (**C**) are at 1 σ above the mean density and include data to a resolution of 15-Å resolution. These contours define the boundary of the connexon and can be compared to previous low resolution structural studies of liver (Makowski et al., 1977; Unwin and Ennis, 1984) and heart (Yeager and Gilula, 1992) gap junctions. The yellow contours at 1.5 σ above the mean density include data to a resolution of 7.5 Å. The roughly circular shape of these contours within the hydrophobic region of the bilayers is consistent with 24 transmembrane α-helices per connexon. The red asterisk in (B) marks the narrowest part of the channel where the aqueous pore is approximately 15 Å in diameter, not accounting for the contribution of amino acid side chains that are not resolved at the current limit of resolution. The noncrystallographic twofold symmetry that relates the two connexons of a gap junction channel has not been applied to the map. Hence, the similarity of the two connexons provides an independent measure on the quality of the reconstruction. Reproduced from Unger et al. (1999b) with permission of *Science*.

as was predicted from the projection map at 7-Å resolution (Unger et al., 1997). However, it should be noted that some interpretations of the higher resolution 3D map (Figure 4c in Unger et al., 1999b) do not require rotational stagger between the connexons. Ultimately, a 3D map that resolves the connecting loops between the transmembrane $\alpha$-helices will be required to delineate the molecular boundary of each connexin subunit in order to confirm the 30-degree rotational stagger between connexons.

### Three-Dimensional Density Map

To further explore the $\alpha$-helical folding of the connexin polypeptide, a 3D density map was determined at resolutions of approximately 7.5 Å in the membrane plane and 21 Å in the vertical direction by recording and analyzing images of frozen-hydrated, tilted 2D crystals (Unger et al., 1999a,b). A side view of the 3D map (see Figure 8) showed that the recombinant gap junction channel had a thickness of approximately 150 Å. This reduced thickness, in comparison with approximately 250 Å for native cardiac gap junction channels (Yeager, 1998), was consistent with the lack of the 13-kD cytoplasmic carboxy-tail in $\alpha_1$-Cx263T. Consistent with the projection map (Unger et al., 1997), the outer diameter within the membrane region was approximately 70 Å, but the 3D map also revealed that the diameter decreased to approximately 50 Å in the extracellular portion of the channel. A vertically sectioned view of the 3D map (see Figure 8b) revealed that the channel narrowed from approximately 40 to 15 Å (neglecting the contributions of amino acid side chains) in proceeding from the cytoplasmic to the extracellular side of the bilayer. The aqueous pathway then widened again to a diameter of approximately 25 Å within the extracellular vestibule.

Cross-sections of the map within the hydrophobic region of the bilayers (see Figure 8c, top and bottom) displayed roughly circular contours of density that were typical, at the current limit of resolution, for cross-sections of $\alpha$-helices. Each connexon contained 24 circular densities, consistent with the topological model in which each connexin subunit has four transmembrane $\alpha$-helices.

### H.  Structure of the Pore

Identification of the region of any channel protein that contributes directly to formation of the pore is clearly essential to understanding the structural basis for its properties. In the case of gap junctions, M3 has been the de facto choice based on its amphipathic character when modeled as an $\alpha$-helix (see Figure 5a, b, and c). Some patients with Charcot-Marie-Tooth disease have a point mutation in Cx32 at residue 26, from serine to the larger, hydrophobic residue leucine. This mutation within M1 (see Figure 5B) was associated with a decrease in the permeability of the channel to large, neutral polyethylene glycol compounds without any change in single channel conductance (Oh et al., 1996). These results are certainly consistent with a role for M1 in lining the channel, but as with many mutations in the absence of structure, it remains possible that effects could be exerted through conformational change from a distance. In a more direct approach, Zhou and colleagues (1997) used the substituted cysteine accessibility method (SCAM) that has been employed to probe the pores of several channels (Xu and Akabas, 1993; Akabas et al., 1994a, 1994b; Kurz et al., 1995). Channels formed from Cx46 connexons and Cx43 chimeric connexons, containing the Cx32 El Loop, were examined in paired oocytes. The thiol reagent maleimido-butyryl-biocytin was used to probe cysteine replacement mutants in the N-terminal segment M1 and in M3. Two residues in M1, and three in M3, were accessible to aqueous sulfhydryl reagents, leading to a partial block of the channel. By far the greatest effect was seen with two adjacent residues centrally located in M1. It is possible that the reduced labeling of M3 was because the mutated residues reside in a narrow portion of the pore, which may sterically prevent labeling by a cysteine reagent with a molecular weight of 537 Da. An important caveat in the latter analysis is that, in order to have access of the channel by the sulfhydryl reagents, experiments were done exclusively on hemichannels. Thus, caution should be used in extrapolating these results to the whole channel. The development of an intracellular perfusion system for oocytes has recently allowed extension of the SCAM strategy to intact gap junction channels composed of Cx32 (Skerrett et al, 1998). These studies confirm, and extend, the mapping of aqueous accessible sites on one side of the M3 helix along its entire length, strongly implicating this domain as the predominant contributor to the lining of the channel. In general, the SCAM results lead to a model in which multiple α-helices contribute to the wall of the pore. Notably, the 3D map at 7.5-Å resolution (Unger et al., 1999b), showed that two α-helices contribute to lining the pore of the channel (see Figure 8).

## I.  Models for Connexon Docking

The highly conserved extracellular loops (designated E1 and E2) presumably mediate connexon docking and thereby dictate the selectivity of heterotypic connexin interactions. The high degree of structural conservation is reflected by the invariant distribution of three cysteines in each loop (see Figure 5a, b, and c). Because mutation of any cysteine abrogates functional expression in the oocyte system, they are all essential for normal folding or channel function, or for both (Dahl et al., 1991, 1992). The oocyte system has also been used to show that the assembly of functional gap junctions is enhanced under conditions that promote an exchange of disulfides (Dahl et al., 1991), raising the intriguing possibility that different covalent linkage patterns may distinguish the hemichannel from the mature channel of the plaque. Of great value in building models for the interactions between these extracellular loops has been the elegant demonstration, made independently by two groups (John and Revel, 1991; Rahman and Evans, 1991), that the disulfide bridges are exclusively intramolecular, with at least one linking the two loops of a single connexin. Nevertheless, there are clear sequence differences between connexins that are critical. However, in the absence of an atomic resolution model for the conformation of and interactions between these loops, it is very difficult to discern which specific residues these may be.

In the oxidizing extracellular environment, it is likely that all three possible disulfides will form. A mutagenesis strategy has provided an initial mapping of the disulfides (Foote et al., 1998). Rather than deleting the cysteines, the first and third cysteine of each loop were moved within the sequence, both individually and in pairwise combinations, in order to identify the likely partners involved in disulfide bonds. Only certain paired movements of the cysteine residues were compatible with the assembly of functional channels. The pattern of pairings indicated that the first cysteine of each loop paired with the third cysteine of the opposite loop, suggesting that all three disulfides form between the loops, and that the loops are arranged in an antiparallel configuration. Movements two or four residues away from the wild-type position were more effective than one or three in the restoration of functional channels. This periodicity suggested a β-sheet conformation for E1 and E2 that may be stabilized in an antiparallel configuration by three disulfides bridges. Testing a complete set of combinations of the six extracellular cysteine residues should lead to a complete mapping of the disulfide connections within the "docking" loops of connexins. Nevertheless, Foote and associates (1998) have proposed that the interdigitation of protruding extracellular loops could form 24 stranded, "inner" and "outer" antiparallel β barrels. Such a model is similar to bacterial porins (Jap et al., 1991; Weiss et al., 1991), but the barrels would be formed from β strands contributed by different subunits. In contrast, each monomer within a porin trimer folds as a β barrel. Direct structural information is clearly needed. Although the precise secondary structure within the gap was not revealed in the higher resolution 3D map (Unger et al., 1999b), it was nevertheless clear that the interior band of density provided a continuous wall of protein (see Figure 7b), which functions as a tight electrical and chemical seal to exclude the exchange of substances with the extracellular milieu.

Several studies have attempted to determine which domains confer the specificity for heterotypic gap junction formation through the exchange of extracellular loops between connexins. In one case (White et al., 1994), the second extracellular loop (E2) of Cx46, a connexin capable of heterotypic coupling with both Cx43 and 50, was sufficient to confer on Cx50 the ability to couple with Cx43 (a normally forbidden combination). Although this points to a dominant role for E2 in defining heterotypic specificity, White emphasized that in this particular combination, the first extracellular loops (E1) of Cx46 and Cx50 share 92% sequence identity. Nonetheless, further support for the dominant role of E2 was provided by a chimera of Cx32 (N-terminal half, including E1) and Cx43 (C-terminal half, including E2), which showed Cx43-like interactions in pairing with *Xenopus* Cx38 (Bruzzone et al., 1994) and Cx46, but not Cx50 (White et al., 1995). Most recently, preferential pairings within the α and β families were found to be dictated by two specific residues within E2 (Zhu and Nicholson, unpublished results).

Notwithstanding the consistent implication of E2 in defining the specificity of heterotypic interactions between connexins, recent evidence indicates that not only primary sequence, but also tertiary structure of these domains are critical determinants of docking specificity. Strikingly, an unpaired cysteine movement in E2 that disrupted the normal homotypic pairing of Cx32, induced a novel ability of Cx32 to pair with *Xenopus* Cx38 (Foote et al., 1998). Because no significant change in primary sequence was involved, this change in specificity of heterotypic interactions apparently resulted from a distortion in the folding of E2. The same general conclusion can be deduced from the work of Haubrick and colleagues (1996) employing a chimera of Cx43 and Cx40. In this case, the replacement of the extracellular domains of Cx40 with those of Cx43 only partially imparted Cx43 pairing characteristics. In some respects this chimera retained the properties of Cx40, presumably due to the constraints imposed by the transmembrane domains. These results underscore the limitations of the molecular biological approaches for defining functional domains in the absence of an atomic resolution structure.

## J. Gating Models

By comparing changes in the 3D structure in the presence and absence of $Ca^{2+}$, a model was proposed for the gating of liver gap junctions (Unwin and Zampighi, 1980; Unwin and Ennis, 1984). In the transition from the "open" to the "closed" configuration, the protein subunits decrease their angle of tilt by about 5 degrees tangential to the channel axis and slide with respect to one another along their long axis. Such a mechanism is appealing because it would involve sliding of adjacent protein surfaces to accomplish channel closure that would be more energetically favorable than dramatic conformational changes involving refolding of polypeptide chains. Although x-ray diffraction studies by Unwin and Ennis (1983) provided evidence favoring the $Ca^{2+}$-mediated structural transition, Caspar and associates (1988) maintained that possible $Ca^{2+}$-mediated changes in the x-ray diffraction patterns were masked by changes that resulted from variations in lattice constants and connexon packing related to different preparations.

A vertical section through the electron density map derived from x-ray patterns showed density on the cytoplasmic face not seen in the map obtained by electron microscopy (Makowski, 1988). This density in the x-ray map was termed a *gating structure* and was located outside the bilayer in the cytoplasmic domain of the connexon (Caspar et al., 1988; Makowski, 1988). Gating was postulated to involve some kind of displacement of the gating structure. However, this change has not been delineated in separate specimens containing open and closed channels. A higher resolution structure analysis would presumably resolve the discrepancies between the maps provided by electron crystallography and x-ray diffraction. In addition, spectroscopic techniques coupled with site-specific mutagenesis, as described in the next section, may provide an alternative strategy to the investigation of the molecular events associated with channel gating.

## IV. IDENTIFICATION OF FUNCTIONAL SITES THAT REGULATE GATING

As has taken place in other areas as they have matured, the gap junction field has become more introspective by seeking the molecular basis underlying the physiological behavior of these channels. Several approaches have been used, each with strengths and inherent limitations, particularly in the absence of a high resolution structural map of the channel.

### A.  Phosphorylation

The hypervariable C-terminus has been extensively examined by molecular manipulations and has been shown to be a site for differential regulation of connexins. Phosphorylation sites for PKC and PKA have been mapped to the C-terminal region of Cx43 (Sáez et al., 1993) and Cx32 (Sáez et al., 1990), respectively, using comparisons of synthetic phosphopeptides and those derived from phosphorylation of the protein *in situ* (see Figure 5d and e). At least one site for tyrosine phosphorylation in response to pp60v-src has been mapped by mutagenesis to residue 265 of Cx43 (Swenson et al., 1990). Direct, *in vitro* phosphorylation of Cx43 by v-src was demonstrated (Loo et al., 1995), as was a requirement for v-src SH2 and SH3 domains for binding to a proline-rich region of Cx43 and for interaction with phosphorylated tyr265 and a proline-rich region between residues 260 and 280, respectively (Kanemitsu et al., 1997). However, both the binding and direct phosphorylation of Cx43 by v-src were recently shown to not be required for acute closure of the gap junction channels, although longer-term effects on biosynthesis were likely (Zhou et al., 1999). Instead, gating in response to v-src was mediated by the action of MAP kinase on a cluster of sites on the C-terminus (see Figure 5d) that had previously been shown to be a substrate for MAP kinase (Warn-Cramer et al., 1996). Furthermore, the mechanism of channel closure was shown, like pH gating (Morley et al., 1996), to utilize a "ball and chain" mechanism where the C-terminus can act as an independent domain (Zhou et al., 1999), thereby linking these results to the cascade of growth regulatory signals initiated by PDGF and, EGF, as well as the pathogenic effects of oncogenes such as v-src and other factors (Kanemitsu and Lau, 1993).

This MAP kinase phosphorylation site has also been mapped within the C-terminal domain (see Figure 5d) using synthetic peptides and comparisons with peptide maps from *in situ* phosphorylated Cx43 (Warn-Cramer et al., 1996). A similar approach has identified sites on Cx43 for phosphorylation by p34cdc2/

cyclin B, a process that may result in diminished cell-cell coupling during mitosis (Kanemitsu et al., 1998). To date, only the action of MAPK tyrosine kinases have been directly associated with the closure of gap junction channels. However, phosphorylation by other kinases has been correlated with modulation of channel conductance (Moreno et al., 1992, 1994), assembly into detergent-resistant structures at the cell surface (Musil and Goodenough, 1993), and susceptibility of the protein to proteases, indicating a possible role in degradation. In addition, truncation of the C-terminal domain of Cx43 modified both the conductance of the channel (Fishman et al., 1991) and the sensitivity of gating to pH (Liu et al., 1993), thus directly implicating this domain in the regulation of channel function.

## B.  Gating by pH

In an elegant series of studies, Delmar, Taffett, and colleagues showed that the C-terminal domain could still effect pH-induced gating of a truncated form of Cx43 when expressed as a separate polypeptide in *Xenopus* oocytes (Ek-Vitorin et al., 1996; Morley et al., 1996, 1997; Calero et al., 1998). This result is analogous to that demonstrating the "ball and chain" inactivation mechanism for voltage-gated $K^+$ channels. Critical sequences contributing to the "ball" were further identified by deletion mutagenesis (Morley et al., 1996; see Figure 5d). A similar mechanism was also implicated for the insulin-induced uncoupling of cells expressing Cx43 (Homma et al., 1998) and in v-src induced closure of Cx43 channels (Zhou et al., 1999). Both chimeric and mutagenic analyses have also suggested that the cytoplasmic M2-M3 loop plays a role in pH-mediated gating. Specifically, the higher pH sensitivity of *Xenopus* Cx38 was found to be associated with the C-terminal half of the cytoplasmic loop in chimeric constructs with Cx32 (Wang et al., 1996; Wang and Peracchia, 1996). However, one histidine at the N-terminus of this loop in Cx43 appeared to represent a titratable component of the gating machinery (Ek et al., 1994; see Figure 5d). Furthermore, several basic residues in the membrane proximal part of the C-terminal domain seemed to greatly reduce the sensitivity of Cx32 to gating by reduced pH (Wang and Peracchia, 1997). These studies suggested that (1) gating involves complex interactions between the C-terminus and the cytoplasmic loop domains; and, (2) pH can directly mediate gating of gap junctional channels. However, recent studies by Cao and Nicholson (unpublished) have implicated an as yet unidentified cytoplasmic component(s) as being required for the pH gating of both Cx43 and Cx32. A wealth of possibilities exist, including kinases and synergistic effectors such as $Ca^{2+}$. With respect to the latter, it is important to note that calmodulin binding sites have been mapped to the C-terminal domain of Cx32 (Zimmer et al., 1987; see Figure 5e) and Cx43 (He and Shalloway, unpublished results). These results serve to reinforce the importance of identifying other factors that may associate with connexins (see the earlier discussion of "Other Proteins" in section II.H).

## C.  Voltage Gating

The mechanism behind gating of gap junctional channels, specifically in response to voltage, has received more attention, principally due to the very successful application of chimeric approaches between two "compatible" connexins—Cx26 and Cx32 (Rubin et al., 1992; Verselis et al., 1994). By swapping the extracellular loops of Cx32 and Cx26, Verselis and colleagues determined that E1 plays a significant role in establishing the level of transjunctional voltage required for half-maximal closure of the channel (i.e., V0). However, the overall voltage response appears to be a complex function of interactions between several parts of the protein sequence located between the N-terminus and the carboxyl end of M2 (Rubin et al., 1992). In addition, studies on chimeras between Cx32 and *Xenopus* Cx38 suggest that the cytoplasmic loop between M2 and M3 may also define these properties (Wang et al., 1996). The polarity of the voltage sensor is reversed between Cx26 and Cx32, (with Cx26 hemichannels closing in response to transjunctional voltages which are positive at their cytoplasmic face, while Cx32 hemichannels respond to negative voltages [Verselis et al., 1994]). Additional chimeric combinations indicated that polarity was determined by the M1 and N-terminal domains (Verselis et al., 1994), and was particularly sensitive to charges at position 2 in the sequence (Asn in Cx32; Asp in Cx26). The polarity is also influenced by the relative distribution of charges within the membrane, as reversing their polarity can also reverse the polarity of the voltage response of Cx26 (Yox et al., 1993).

A quandary facing all channels studied to date, is how channel closure caused by movements of the pore-lining domain (putatively M3 in gap junctions) or adjacent cytoplasmic domains is mediated by movements of a distant voltage sensor (apparently associated with M1 [Verselis et al., 1994]). The 3D map at 7.5-Å resolution (see Figure 7) shows that each subunit of the channel is comprised of four transmembrane α-helices. A movement of one α-helix (e.g., in response to voltage) could most readily be transmitted to its neighbors if one of the α-helices in the bundle were distorted in some way. Consistent with such a model is the observation that the substitution of a highly conserved proline in M2 of Cx26 leads to a reversal of the gating response, in that the channels now open rather than close in response to voltage (Suchyna et al., 1993). The importance of this proline in the directional distortion of M2 is further supported by recent findings (Smith et al., 1998) indicating that the original closing behavior of the channel can be retrieved in this mutant by reintroducing the proline into M2, seven residues away from its original position (corresponding to exactly two turns of an α-helix). These results also illustrate a useful tactic in point mutagenesis studies where one frequently learns more from a rescue of the original phenotype than one does from the initial disruption. Although these results are self-consistent for Cx26, the failure to see the same results for analogous mutations in Cx32 (Versailis et al., 1994) raises questions as to the generality of this mechanism (Suchyna et al., 1993).

# V. CONCLUSION

In spite of detailed differences in the projection density maps of gap junctions isolated from different tissues, there are important conserved features. For instance, the hexameric connexon with a central channel is clearly a conserved motif in the design of gap junction membrane channels. The conduit for intercellular communication is formed by connexons in register between closely apposed cells. Gap junction channels can be viewed as having a modular design. Homology in the transmembrane sequences (see Figure 5) suggests that the bundle of 24 α-helices (see Figure 8) that form each connexon will likely be a common architecture for the transmembrane channel. Similarly, sequence homology in the extracellular domains suggests similar mechanisms for connexon-connexon interactions. However, it is clear that differences in the extracellular loops appear to limit the promiscuity allowed in the pairing of heterotypic connexons (see section II.F). The cytoplasmic loop between M2 and M3 and especially the carboxy tail domain are the most divergent regions between different connexins (see Figure 5) and presumably confer unique functional properties for different connexins. However, these regions may be recalcitrant to structural determination, as a consequence of polypeptide flexibility that would lead to loss of resolution in any method utilizing averaging.

The identification of specific functional sites or domains is critical for understanding the regulation of gap junction channels. Instead of blindly accumulating site mutations, a better strategy may be the generation of chimeric connexins where a loop domain of one connexin can be substituted for another. Unfortunately, with some notable exceptions of a series of swaps between Cx32 and Cx26 (see section IV.C), such chimeric connexins have proven distressingly uninstructive as they fail to form functional channels (Bruzzone et al., 1994a; Nicholson, unpublished observations). This may reflect a requirement for strict interactions of different connexins that are only satisfied when "compatible" swaps are made between connexins that interact *in vivo*.

A cautionary note in all functional studies on gap junctions is that by definition they are performed in exogenous systems, and at times the results may vary from that which might be seen in the native environment of the channel. Interpretation of the results will also remain limited as long as we lack an experimentally determined atomic resolution structure of the channel. The isolation of apparently intact hemichannels using appropriate detergents (Stauffer et al., 1991; Harris et al., 1992), and their expression in large quantities in the baculovirus system (Stauffer et al., 1991), raises hopes for the eventual generation of highly ordered 2D or 3D connexon crystals. To date, 2D crystals that are ordered to 7-Å resolution have been generated from recombinant, C-terminal-truncated Cx43 (Unger et al., 1997, 1999a,b). Electron cryo-crystallography has for the first time confirmed the widely accepted model that the transmembrane domains of each connexin are folded as four α-helices. However, crystals that are ordered to better than 4-Å resolution would be required to visualize amino acid side chains.

Currently, there are numerous areas for continued exploration of the structural and functional properties of gap junction channels: delineation of the precise interactions between different regions in the channel (e.g., the carboxy-tail, the M2-M3 loop and the amino-tail), the precise folding of the extracellular loops that confers stability but also specificity in connexon-connexon pairing, the identification and orientation of the transmembrane α-helices in the 3D map, and specific mechanisms for gating. The regulation of gap junction channels is particularly fascinating as they manifest properties of both ligand-gated channels (see sections IVA and B) and voltage-gated channels (see section IV.C.). In addition, the lipid environment can regulate the gating state (see the discussion of "Lipids" in section II.H). One can imagine that gating involves multiple conformational switches such as movements of the C-tail during pH gating, movements of the N-tail during voltage gating or changes in α-helical packing resulting from lipophilic molecules that affect the α-helices on the perimeter of the channel. Such multiple triggers may converge on a final common conformational change that presumably involves a site in the bilayer near the extracellular gap, as this is the narrowest portion of the channel (see Figure 8). The exploration of these mechanisms will require a strategy that integrates molecular biology, crystallography, and spectroscopic methods with functional studies using electrophysiology and transport assays.

Several human diseases have now been related to connexin mutations, such as the X-linked form of Charcot-Marie-Tooth disease (Bergoffen et al., 1993; Bone et al., 1997); the most common form of nonsyndromic neurosensory autosomal recessive deafness (Zelante et al., 1997; White et al., 1998); and possibly, some patients with developmental anomalies of the cardiovascular system (Britz-Cunningham et al., 1995). In addition, transgenic mice lacking $\alpha_1$ Cx43, the principal gap junction protein in the heart, die soon after birth and exhibit developmental malformations (Reaume et al., 1995; Ewart et al., 1997). Heterozygotes are viable but display slowed ventricular conduction (Guerrero et al., 1997). The potential to correlate the structure and biochemistry of gap junction channels with human disease makes this a particularly exciting area of research.

# REFERENCES

Agre, P., Preston, G.M., Smith, B.L., Jung, J.S., Raina, S., Moon. C., Guggino, W.B., & Nielsen, S. (1993). Aquaporin CHIP: the archetypal molecular water channel. Am. J. Physiol. 265, F463-F476.

Akabas, M.H., Kaufmann, C., Archdeacon, P., & Karlin, A. (1994a). Identification of acetylcholine receptor channel-lining residues in the entire M2 segment of the α subunit. Neuron 13, 919-927.

Akabas, M.H., Kaufmann, C., Cook, T.A., & Archdeacon, P. (1994b). Amino acid residues lining the chloride channel of the cystic fibrosis transmembrane conductance regulator. J. Biol. Chem. 269, 14865-14868.

Amos, L. A., Henderson, R., & Unwin, P.N.T. (1982). Three-dimensional structure determination by electron microscopy of two-dimensional crystals. Prog. Biophys. Mol. Biol. 39, 183-231.

Aylsworth, C.F., Trosko, J.E., & Welsch, C.W. (1986). Influence of lipids on gap-junction-mediated intercellular communication between Chinese hamster cells *in vitro*. Cancer Res. 46, 4527-4533.

Baker, T.S., Sosinsky, G.E., Casper, D.L.D., Gall, C., & Goodenough, D.A. (1985). Gap junction structures. VII. Analysis of connexon images obtained with cationic and anionic negative stains. J. Mol. Biol. 184, 81-98.

Barr, L., Dewey, M.M., & Berger W. (1965). Propagation of action potentials and the structure of the nexus in cardiac muscle. J. Gen. Physiol. 48, 797-823.

Barrio, L.C., Suchyna, T., Bargiello, T., Xu, L.X., Roginski, R.S., Bennett, M.V.L., & Nicholson, B.J. (1991). Gap junctions formed by connexins 26 and 32 alone and in combination are differently affected by applied voltage [published erratum appears in Proc Natl. Acad. Sci. USA (1992) 89, 4220]. Proc. Natl. Acad. of Sci. USA 88, 8410-8414.

Bastiaanse, E.M.L., Jongsma, H.J., van der Laarse, A., & Takens-Kwak, B.R. (1993). Heptanol-induced decrease in cardiac gap junctional conductance is mediated by a decrease in the fluidity of membranous cholesterol-rich domains. J.Membr. Biol. 136, 135-145.

Bastide, B., Neyses, L., Ganten, D., Paul, M., Willecke, K., & Traub, O. (1993). Gap junction protein connexin40 is preferentially expressed in vascular endothelium and conductive bundles of rat myocardium and is increased under hypertensive conditions. Circ. Res. 73, 1138-1149.

Bennett, M.V.L., Barrio, L.C., Bargiello, T.A., Spray, D.C., Hertzberg, E., & Sáez, J.C. (1991). Gap junctions: new tools, new answers, new questions. Neuron 6, 305-320.

Berdan, R.C. and Gilula, N.B. (1988). The arthropod gap junction and pseudo-gap junction: isolation and preliminary biochemical analysis. Cell Tissue Res. 251, 257-274.

Berdan, R.C., & Shivers, R.R. (1985). Filipin-cholesterol complexes in plasma membranes and cell junctions of *Tenebrio molitor* epidermis. Tissue Cell 17, 177-187.

Bergoffen, J, Scherer, S.S., Wang, S., Scott, M.O., Bone, L.J., Paul, D.L., Chen, K., Lensch, M.W., Chance, P.F., & Fischbeck, K.H. (1993). Connexin mutations in X-linked Charcot-Marie-Tooth disease. Science 262, 2039-2042.

Bevans, C. G., Kordel, M., Rhee, S. K., & Harris, A. L. (1998). Isoform composition of connexin channels determines selectivity among second messengers and uncharged molecules. J. Biol. Chem. 273, 2808-2816.

Beyer, E.C. (1990). Molecular cloning and developmental expression of two chick embryo gap junction proteins. J. Biol. Chem. 265, 14439-14443.

Beyer, E.C., Kistler, J., Paul, D.L., & Goodenough, D.A. (1989). Antisera directed against connexin43 peptides react with a 43-kD protein localized to gap junctions in myocardium and other tissues. J. Cell Biol. 108, 595-605.

Beyer, E.C., Paul, D.L., & Goodenough, D.A. (1987). Connexin 43: a protein from rat heart homologous to a gap junction protein from liver. J. Cell Biol. 105, 2621-2629.

Beyer, E.C., Paul, D.L., & Goodenough, D.A. (1990). Connexin family of gap junction proteins. J Membr. Biol. 116, 187-194.

Bloemendal, M., & Johnson, W.C., Jr. (1995). Structural information on proteins from circular dichroism spectroscopy—possibilities and limitations. Pharm. Biotechnol. 7, 65-100.

Boger, D.L., Patterson, J.E., Guan, X., Cravatt, B.F., Lerner, R.A., & Gilula, N.B. (1998). Chemical requirements for inhibition of gap junction communication by the biologically active lipid oleamide. Proc. Natl. Acad. Sci. USA 95, 4810-4815.

Bone, L.J., Deschenes, S.M., Balice-Gordon, R.J., Fischbeck, K.H., & Scherer, S.S. (1997). Connexin32 and X-linked Charcot-Marie-Tooth disease. Neurobiol. Dis. 4, 221-230.

Braiman, M.S., & Rothschild, K.J. (1988). Fourier transform infrared techniques for probing membrane protein structure. Ann. Rev. Biophys. Biophys. Chem. 17, 541-570.

Brightman, M.W., & Reese, T.S. (1969). Junctions between intimately apposed cell membranes in the vertebrate brain. J. Cell Biol. 40, 648-677.

Brink, P.R., Cronin, K., Banach, K., Peterson, E., Westphale, E.M., Seul, K.H., Ramanan, S.V., & Beyer, E.C. (1997). Evidence for heteromeric gap junction channels formed from rat connexin43 and human connexin37. Am. J. Physiol. 273, C1386-C1396.

Brink, P.R., & Dewey, M.M. (1980). Evidence for fixed charge in the nexus. Nature 285, 101-102.

Britz-Cunningham, S.H., Shah, M.M., Zuppan, C.W., & Fletcher, W.H. (1995). Mutations of the *connexin43* gap-junction gene in patients with heart malformations and defects of laterality. N Engl. J. Med. 332, 1323-1329.

Bruzzone, R., White, T.W., & Paul, D.L. (1994a). Expression of chimeric connexins reveals new properties of the formation and gating behavior of gap junction channels. J. Cell Sci. 107, 955-967.

Bruzzone, R., White, T.W., Scherer, S.S., Fischbeck, K.H., & Paul, D.L. (1994b). Null mutations of connexin32 in patients with X-linked Charcot-Marie-Tooth disease. Neuron 13, 1253-1260.

Buehler, L.K., Stauffer, K.A., Gilula, N.B. and Kumar, N.M. (1995). Single channel behavior of recombinant $\beta_2$ gap junction connexons reconstituted into planar lipid bilayers. Biophys. J. 68, 1767-1775.

Bukauskas, F.F., Elfgang, C., Willecke, K., & Weingart, R. (1995). Heterotypic gap junction channels (connexin26–connexin32) violate the paradigm of unitary conductance. Pflügers Arch. Eur. J. Physiol. 429, 870-872.

Burt, J.M., Massey, K.D., & Minnich, B.N. (1991). Uncoupling of cardiac cells by fatty acids: structure-activity relationships. Am. J. Physiol. 260, C439-C448.

Burt, J.M., Minnich, B.N., Massey, K.D., Ovadia, M., Moore, L.K., & Hirschi, K.K. (1993). Influence of lipophilic compounds on gap-junction channels. In: *Progress in Cell Research,* vol. 3: *Gap Junctions* (Hall, J.E., Zampighi, G.A., & Davis, R.M., Eds.), pp. 113-120. Elsevier, Amsterdam.

Buultjens, T.E.J., Finbow, M.E., Lane, N.J., & Pitts, J.D. (1988). Tissue and species conservation of the vertebrate and arthropod forms of the low molecular weight (16-18000) proteins of gap junctions. Cell Tissue Res. 251, 571-580.

Calero, G., Kanemitsu, M., Taffet, S.M., Lau, A.F., & Delmar, M. (1998). A 17mer peptide interferes with acidification-induced uncoupling of connexin43. Circ. Res. 82, 929-935.

Cao, F., Eckert, R., Elfgang, C., Nitsche, J.M., Snyder, S.A., Hülser, D.F., Willecke, K., & Nicholson, B.J. (1998). A quantitative analysis of connexin-specific permeability differences of gap junctions expressed in HeLa transfectants and *Xenopus* oocytes. J. Cell Sci. 111, 31-43.

Cascio, M., Gogol, E., & Wallace, B.A. (1990). The secondary structure of gap junctions. Influence of isolation methods and proteolysis. J. Biol. Chem. 265, 2358-2364.

Caspar, D.L.D., Goodenough, D.A., Makowski, L., & Phillips, W.C. (1977). Gap junction structures I. Correlated electron microscopy and x-ray diffraction. J. Cell Biol. 74, 605-628.

Caspar, D.L.D., Sosinsky, G.E., Tibbitts, T.T., Phillips, W.C., & Goodenough, D.A. (1988). Gap junction structure. In: *Modern Cell Biology,* vol. 7: *Gap Junctions* (Hertzberg, E.L., & Johnson, R.G., Eds.), pp. 117-133. Alan R. Liss, New York.

Chalcroft, J.P., & Bullivant, S. (1970). An interpretation of liver cell membrane and junction structure based on observation of freeze-fracture replicas of both sides of the fracture. J. Cell Biol. 47, 49-60.

Chandy, G., Zampighi, G.A., Kreman, M., & Hall, J.E. (1997). Comparison of the water transporting properties of MIP and AQP1. J. Membr. Biol. 159, 29-39.

Chang, M., Werner, R., & Dahl, G. (1996). A role for an inhibitory connexin in testis? Develop. Biol. 175, 50-56.

Cheng, A., van Hoek, A.N., Yeager, M., Verkman, A.S., & Mitra, A.K. (1997). Three-dimensional organization of a human water channel. Nature 387, 627-630.

Claassen, D.E., & Spooner, B.S. (1988). Reconstitution of cardiac gap junction channeling activity into liposomes: a functional assay for gap junctions. Biochem. Biophys. Res. Comm. 154, 194-198.

Clegg, R.M. (1995). Fluorescence resonance energy transfer. Curr. Opin. Biotechnol. 6, 103-110.

Cooper, E.A., & Knutson, K. (1995). Fourier transform infrared spectroscopy investigations of protein structure. Pharm. Biotechnol. 7, 101-143.

Cooper, K., Rae, J.L., & Gates, P. (1989). Membrane and junctional properties of dissociated frog lens epithelial cells. J. Membr. Biol. 111, 215-227.

Cravatt, B.F., Prospero-Garcia, O., Siuzdak, G., Gilula, N.B., Henriksen, S.J., Boger, D.L., & Lerner, R.A. (1995). Chemical characterization of a family of brain lipids that induce sleep. Science 268, 1506-1509.

Crow, D.S., Beyer, E.C., Paul, D.L., Kobe, S.S., & Lau, A.F. (1990). Phosphorylation of connexin43 gap junction protein in uninfected and Rous sarcoma virus-transformed mammalian fibroblasts. Mol. Cell. Biol. 10, 1754-1763.

Culvenor, J.G., & Evans, W.H. (1977). Preparation of hepatic gap (communicating) junctions. Identification of the constituent polypeptide subunits. Biochem. J. 168, 475-481.

Dahl, E., Manthey, D., Chen, Y., Schwarz, H.-J., Chang, Y.S., Lalley, P.A., Nicholson, B.J., & Willecke, K. (1996). Molecular cloning and functional expression of a mouse connexin-30 gap junction gene highly expressed in adult brain and skin. J. Biol. Chem. 271, 17903-17910.

Dahl, G., Levine, E., Rabadan-Diehl, C., & Werner, R. (1991). Cell/cell channel formation involves disulfide exchange. Eur. J. Biochem. 197, 141-144.

Dahl, G., Miller, T., Paul, D., Voellmy, R., & Werner, R. (1987). Expression of functional cell-cell channels from cloned rat liver gap junction complementary DNA. Science 236, 1290-1293.

Dahl, G., Werner, R., Levine, E., & Rabadan-Diehl, C. (1992). Mutational analysis of gap junction formation. Biophys. J. 62, 172-180.

Dermietzel, R., Farooq, M., Kessler, J.A., Althaus, H., Hertzberg, E.L., & Spray, D.C. (1997). Oligodendrocytes express gap junction proteins connexin32 and connexin45. Glia 20, 101-114.

Dermietzel, R., Leibstein, A., Frixen, U., Janssen-Timmen, U., Traub, O., & Willecke, K. (1984). Gap junctions in several tissues share antigenic determinants with liver gap junctions. EMBO J. 3, 2261-2270.

Dermietzel, R., Traub, O., Hwang, T. K., Beyer, E., Bennett, M. V. L., Spray, D. C., & Willecke, K. (1989a). Differential expression of three gap junction proteins in developing and mature brain tissues. Proc. Natl. Acad. Sci. USA 86, 10148-10152.

Dermietzel, R., Völker, M., Hwang, T.-K., Berzborn, R.J., & Meyer, H.E. (1989b). A 16 kDa protein co-isolating with gap junctions from brain tissue belonging to the class of proteolipids of the vacuolar $H^+$-ATPases. FEBS Lett. 253, 1-5.

De Vos, A.M., Ultsch, M., & Kossiakoff, A.A. (1992). Human growth hormone and extracellular domain of its receptor: crystal structure of the complex. Science 255, 306-312.

DeVries, S.H., & Schwartz, E.A. (1992). Hemi gap junction channels in solitary horizontal cells of the catfish retina. J. Physiol. 445, 201-230.

Dewey, M.M., & Barr, L. (1962). Intercellular connection between smooth muscle cells: the nexus. Science 137, 670-672.

Diez, J.A., Ahmad, S., & Evans (1999). Assembly of heteromeric connexons in guinea-pig liver enroute to the Golgi apparatus, plasma membrane and gap junctions. Eur. J. Biochem. 262, 142-148.

Doucet, J.-P., Pierce, G.N., Hertzberg, E.L., & Tuana, B.S. (1992). Low molecular weight GTP-binding proteins in cardiac muscle. Association with a 32-kDa component related to connexins. J. Biol. Chem. 267, 16503-16508.

Dunlap, K., Takeda, K., & Brehm, P. (1987). Activation of a calcium-dependent photoprotein by chemical signalling through gap junctions. Nature 325, 60-62.

Dupont, E., El Aoumari, A., Roustiau-Sévère, S., Briand, J.P., & Gros, D. (1988). Immunological characterization of rat cardiac gap junctions: Presence of common antigenic determinants in heart of other vertebrate species and in various organs. J. Membr. Biol. 104, 119-128.

Ebihara, L., Berthoud, V.M., & Beyer, E.C. (1995). Distinct behavior of connexin56 and connexin46 gap junctional channels can be predicted from the behavior of their hemi-gap-junctional channels. Biophys. J. 68, 1796-1803.

Eghbali, B., Kessler, J.A., & Spray, D.C. (1990). Expression of gap junction channels in communication-incompetent cells after stable transfection with cDNA encoding connexin 32. Proc. Natl. Acad. Sci. USA 87, 1328-1331.

Ehring, G.R., Zampighi, G.A., & Hall, J.E. (1993). Does MIP play a role in cell-cell communication? In: Progress in Cell Research, vol.3: Gap Junctions (Hall, J.E., Zampighi, G.A., & Davis, R.M., Eds.), pp. 153-162. Elsevier, Amsterdam.

Ek, J.F., Delmar, M., Perzova, R., & Taffet, S.M. (1994). Role of Histidine 95 in pH gating of the cardiac gap junction protein connexin43. Circ. Res. 74, 1058-1064.

Ek-Vitorín, J.F., Calero, G., Morley, G.E., Coombs, W., Taffet, S.M., & Delmar, M. (1996). pH regulation of connexin43: molecular analysis of the gating particle. Biophys. J. 71, 1273-1284.

El Aoumari, A. Fromaget, C., Dupont, E., Reggio, H., Durbec, P., Briand, J.-P., Böller, K., Kreitman, B., & Gros, D. (1990). Conservation of a cytoplasmic carboxy-terminal domain of connexin 43, a gap junctional protein, in mammal heart and brain. J. Membr. Biol. 115, 229-240.

Elfgang, C., Eckert, R., Lichtenberg-Fraté, H., Butterweck, A., Traub, O., Klein, R.A., Hülser, D.F., & Willecke, K. (1995). Specific permeability and selective formation of gap junction channels in connexin-transfected HeLa cells. J. Cell Biol. 129, 805-817.

Evans, W.H., & Gurd, J.W. (1972). Preparation and properties of nexuses and lipid-enriched vesicles from mouse liver plasma membranes. Biochem. J. 128, 691-700.

Ewart, J.L., Cohen, M.F., Meyer, R.A., Huang, G.Y., Wessels, A., Gourdie, R.G., Chin, A.J., Park, S.M.J., Lazatin, B.O., Villabon, S., & Lo, C.W. (1997). Heart and neural tube defects in transgenic mice overexpressing the Cx43 gap junction gene. Development 124, 1281-1292.

Falk, M.M., Buehler, L.K., Kumar, N.M., & Gilula, N.B. (1997). Cell-free synthesis and assembly of connexins into functional gap junction membrane channels. EMBO J. 16, 2703-2716.

Fasman, G.D. (1993). Distinguishing transmembrane helices from peripheral helices by circular dichroism. Biotechnol. Appl. Biochem. 18, 111-138.

Feltkamp, C.A., & Van der Waerden, A.W.M. (1983). Junction formation between cultured normal rat hepatocytes. An ultrastructural study on the presence of cholesterol and the structure of developing tight-junction strands. J. Cell Sci. 63, 271-286.

Finbow, M.E., Buultjens, T.E.J., Lane, N.J., Shuttleworth, J., & Pitts, J.D. (1984). Isolation and characterization of arthropod gap junctions. EMBO J. 3, 2271-2278.

Finbow, M.E., Eliopoulos, E.E., Jackson, P.J., Keen, J.N, Meagher, L., Thompson, P., Jones, P., & Findlay, J.B.C. (1992). Structure of a 16 kDa integral membrane protein that has identity to the putative proton channel of the vacuolar $H^+$-ATPase. Protein Eng. 5, 7-15.

Finbow, M.E., & Meagher, L. (1992). Connexins and the vacuolar proteolipid-like 16-kDa protein are not directly associated with each other but may be components of similar or the same gap junctional complexes. Exp. Cell Res. 203, 280-284.

Finbow, M.E., & Pitts, J.D. (1993). Is the gap junction channel—the connexon—made of connexin or ductin? J.Cell Sci. 106, 463-472.

Finbow, M.E., Shuttleworth, J., Hamilton, A.E., & Pitts, J.D. (1983). Analysis of vertebrate gap junction protein. EMBO J. 2, 1479-1486.

Finbow, M., Yancey, S.B., Johnson, R., & Revel, J.-P. (1980). Independent lines of evidence suggesting a major gap junctional protein with a molecular weight of 26,000. Proc. Natl. Acad. Sci. USA 77, 970-974.

Fishman, G.I., Moreno, A.P., Spray, D.C., & Leinwand, L.A. (1991). Functional analysis of human cardiac gap junction channel mutants. Proc. Natl. Acad. USA 88, 3525-3529.

Fishman, G.I., Spray, D.C., & Leinwand, L.A. (1990). Molecular characterization and functional expression of the human cardiac gap junction channel. J. Cell Biol. 111, 589-598.

FitzGerald, P.G., Bok, D., & Horwitz, J. (1985). The distribution of the main intrinsic membrane polypeptide in ocular lens. Current Eye Res. 4, 1203-1218.

Flagg-Newton, J., Simpson, I., & Loewenstein, W.R. (1979). Permeability of the cell-to-cell membrane channels in mammalian cell junctions. Science 205, 404-407.

Flower, N.E. (1977). Invertebrate gap junctions. J. Cell Sci. 25, 163-171.

Foote, C.I., Zhou, L., Zhu, X., & Nicholson, B.J. (1998). The pattern of disulfide linkages in the extracellular loop regions of connexin 32 suggests a model for the docking interface of gap junctions. J. Cell Biol. 140, 1187-1197.

Friend, D.S., & Gilula, N.B. (1972). Variations in tight and gap junctions in mammalian tissues. J. Cell Biol. 53, 758-776.

Furshpan, E.J., & Potter, D.D. (1959). Transmission at the giant motor synapses of the crayfish. J. Physiol. 145, 289-325.

Furshpan, E.J., & Potter, D.D. (1968). Low-resistance junctions between cells in embryos and tissue culture. Curr. Top. Dev. Biol. 3, 95-127.

George, C.H., Kendall, J.M., & Evans, W.H. (1999). Intracellular trafficking pathways in the assembly of connexins into gap junctions. J. Biol. Chem. 274, 8678-8685.

Ghosh, S., Safarik, R., Klier, G., Monosov, E., Gilula, N.B., & Kumar, N.M. (1995). Evidence for gap junction hetero-oligomers. Mol. Biol. Cell Suppl. 6, 189a.

Ghoshroy, S., Goodenough, D.A., & Sosinsky, G.E. (1995). Preparation, characterization and structure of half gap junctional layers split with urea and EGTA. J. Membr. Biol. 146, 15-28.

Giepmans, B.N. & Moolenaar, W.H. (1998). The gap junction protein connexin43 interacts with the second PDZ domain of the zonula occludens-1 protein. Current Biol. 8, 931-934.

Gilula, N.B., Branton, D., & Satir, P. (1970). The septate junction: a structural basis for intercellular coupling. Proc. Natl. Acad. Sci.USA 67, 213-220.

Gilula, N.B., Reeves, O.R., & Steinbach, A. (1972). Metabolic coupling, ionic coupling and cell contacts. Nature 235, 262-265.

Gilula, N.B., & Satir, P. (1971). Septate and gap junctions in molluscan gill epithelium. J. Cell Biol. 51, 869-872.

Gimlich, R.L., Kumar, N.M., & Gilula, N.B. (1990). Differential regulation of the levels of three gap junction mRNAs in *Xenopus* embryos. J. Cell Biol. 110, 597-605.

Ginzberg, R.D., & Gilula, N.B. (1979). Modulation of cell junctions during differentiation of the chicken otocyst sensory ephithelium. Dev. Biol. 68, 110-129.

Girsch, S.J., & Peracchia, C. (1985). Lens cell-to-cell channel protein: I. Self-assembly into liposomes and permeability regulation by calmodulin. J. Membr. Biol. 83, 217-225.

Glaeser, R.M., & Jap, B.K. (1985). Absorption flattening in the circular dichroism spectra of small membrane fragments. Biochemistry 24, 6398-6401.

Gogol, E., & Unwin, N. (1988). Organization of connexons in isolated rat liver gap junctions. Biophys J. 54, 105-112.

Goldberg, G.S., Lampe, P.D., Gibson, S.P., Harris, W.P., Nicholson, B.J., & Naus, C.C.G. (1995). Capturing and identification of transjunctional molecules from cells in culture. Mol. Biol. Cell Suppl. 6, 188a.

Goldberg, G.S., Lampe, P.D., Sheedy, D., Stewart, C.C., Nicholson, B.J., & Naus, C.C.G. (1998). Direct isolation and analysis of endogenous transjunctional ADP from Cx43 transfected C6 glioma cells. Exp. Cell Res. 239, 82-92.

Goldberg, G.S., & Lau, A.F. (1993). Dynamics of connexin43 phosphorylation in pp60$^{v-src}$-transformed cells. Biochem. J. 295, 735-742.

Goodenough, D.A. (1974). Bulk isolation of mouse hepatocyte gap junctions. Characterization of the principal protein, connexin. J. Cell Biol. 61, 557-563.

Goodenough, D.A. (1979). Lens gap junctions: a structural hypothesis for nonregulated low-resistance intercellular pathways. Invest. Ophthalmol. Vis. Sci. 18, 1104-1122.

Goodenough, D.A., Goliger, J.A., & Paul, D.L. (1996). Connexins, connexons, and intercellular communication. Ann. Rev. Biochem. 65, 475-502.

Goodenough, D.A., Paul, D.L., & Jesaitis, L. (1988). Topological distribution of two connexin32 antigenic sites in intact and Split rodent hepatocyte gap junctions. J. Cell Biol. 107, 1817-1824.

Goodenough, D.A., & Revel, J.-P. (1970). A fine structural analysis of intercellular junctions in the mouse liver. J. Cell Biol. 45, 272-290.

Goodenough, D.A., & Stoeckenius, W. (1972). The isolation of mouse hepatocyte gap junctions. Preliminary chemical characterization and x-ray diffraction. J. Cell Biol. 54, 646-656.

Gorin, M.B., Yancey, S.B., Cline, J., Revel, J.-P., & Horwitz, J. (1984). The major intrinsic protein (MIP) of the bovine lens fiber membrane: characterization and structure based on cDNA cloning. Cell 39, 49-59.

Gros, D., Jarry-Guichard, T., Ten Velde, I., de Maziere, A., van Kempen, M.J.A., Davoust, J., Briand, J.P., Moorman, A.F.M., & Jongsma, H.J. (1994). Restricted distribution of connexin40, a gap junctional protein, in mammalian heart. Circ. Res. 74, 839-851.

Gros, D.B., Nicholson, B.J., & Revel, J.-P. (1983). Comparative analysis of the gap junction protein from rat heart and liver: is there a tissue specificity of gap junctions? Cell 35, 539-549.

Gruijters, W.T.M., Kistler, J., Bullivant, S., & Goodenough, D.A. (1987). Immunolocalization of MP70 in lens fiber 16-17-nm intercellular junctions. J. Cell Biol. 104, 565-572.

Guan, X., Cravatt, B.F., Ehring, G.R., Hall, J.E., Boger, D.L., Lerner, R.A., & Gilula, N.B. (1997). The sleep-inducing lipid oleamide deconvolutes gap junction communication and calcium wave transmission in glial cells. J. Cell Biol. 139, 1785-1792.

Guerrero, P.A., Schuessler, R.B., Davis, L.M., Beyer, E.C., Johnson, C.M., Yamada, K.A., & Saffitz, J.E. (1997). Slow ventricular conduction in mice heterozygous for a connexin43 null mutation. J.Clin.Invest. 99, 1991-1998.

Hanna, R.B., Ornberg, R.L., & Reese, T.S. (1985). Structural details of rapidly frozen gap junctions. In: Gap Junctions. (Bennett, M.V.L., & Spray, D.C., Eds.), pp. 23-32. Cold Spring Harbor Laboratory, Cold Spring Harbor, NY.

Harfst, E., Severs, N.J., & Green, C.R. (1990). Cardiac myocyte gap junctions: evidence for a major connexon protein with an apparent relative molecular mass of 70 000. J. Cell Sci. 96, 591-604.

Haris, P.I., & Chapman, D. (1995). The conformational analysis of peptides using Fourier transform IR spectroscopy. Biopolymers 37, 251-263.

Harris, A.L., Walter, A., Paul, D., Goodenough, D.A., & Zimmerberg, J. (1992). Ion channels in single bilayers induced by rat connexin32. Mol. Brain Res. 15, 269-280.

Haubrich, S., Schwarz, H.-J., Bukauskas, F., Lichtenberg-Fraté, H., Traub, O., Weingart, R., & Willecke, K. (1996). Incompatibility of connexin 40 and 43 hemichannels in gap junctions between mammalian cells is determined by intracellular domains. Mol. Biol. Cell 7, 1995-2006.

Havelka, W.A., Henderson, R., Heymann, J.A.W., & Oesterhelt, D. (1993). Projection structure of halorhodopsin from Halobacterium halobium at 6Å resolution obtained by electron cryomicroscopy. J. Mol. Biol. 234, 837-846.

Hebert, H., Schmidt-Krey, I., Morgenstern, R., Murata, K., Hirai, T., Mitsuoka, K., & Fujiyoshi, Y. (1997). The 3.0Å projection structure of microsomal glutathione transferase as determined by electron crystallography of p $2_12_12$ two-dimensional crystals. J. Mol. Biol. 271, 751-758.

Henderson, D., Eibl, H., & Weber, K. (1979). Structure and biochemistry of mouse hepatic gap junctions. J. Mol. Biol. 132, 193-218.

Henderson, R., Baldwin, J.M., Ceska, T.A., Zemlin, F., Beckmann, E., & Downing, K.H. (1990). Model for the structure of bacteriorhodopsin based on high-resolution electron cryo-microscopy. J. Mol. Biol. 213, 899-929.

Henderson, R., & Unwin, P.N.T. (1975). Three-dimensional model of purple membrane obtained by electron microscopy. Nature 257, 28-32.

Hennemann, H., Dahl, E., White, J.B., Schwarz, H.-J., Lalley, P.A., Chang, S., Nicholson, B.J., & Willecke, K. (1992a). Two gap junction genes, connexin 31.1 and 30.3, are closely linked on mouse chromosome 4 and preferentially expressed in skin. J. Biol. Chem. 267, 17225-17233.

Hennemann, H., Schwarz, H.-J., & Willecke, K. (1992b). Characterization of gap junction genes expressed in F9 embryonic carcinoma cells: molecular cloning of mouse connexin31 and -45 cDNAs. Eur. J. Cell Biol. 57, 51-58.

Hennemann, H., Suchyna, T., Lichtenberg-Fraté, H., Jungbluth, S., Dahl, E., Schwarz, J., Nicholson, B.J., & Willecke, K. (1992c). Molecular cloning and functional expression of mouse connexin40, a second gap junction gene preferentially expressed in lung. J. Cell Biol. 117, 1299-1310.

Hertzberg, E.L. (1983). Isolation and characterization of liver gap junctions. Methods Enzymol. 98, 501-510.

Hertzberg, E.L. (1984). A detergent-independent procedure for the isolation of gap junctions from rat liver. J. Biol. Chem. 259, 9936-9943.

Hertzberg, E.L. (1995). Isoelectric focusing of gap junction proteins indicates that covalent modification(s) other than phosphorylation alter pI and may be involved in gap junction assembly. Mol. Biol. Cell. 6, 189a.

Hertzberg, E.L., & Chan, T.C. (1996). Connexins, the unglycosylated integral membrane proteins comprising gap junctions, exhibit multiple, unanticipated pI forms. (Submitted for publication.)

Hertzberg, E.L., Disher, R.M., Tiller, A.A., Zhou, Y., & Cook, R.G. (1988). Topology of the $M_r$ 27,000 liver gap junction protein. Cytoplasmic localization of amino- and carboxyl termini and a hydrophilic domain which is protease-hypersensitive. J. Biol. Chem. 263, 19105-19111.

Hertzberg, E.L., & Gilula, N.B. (1979). Isolation and characterization of gap junctions from rat liver. J. Biol. Chem. 254, 2138-2147.

Hertzberg, E.L., & Skibbens, R.V. (1984). A protein homologous to the 27,000 dalton liver gap junction protein is present in a wide variety of species and tissues. Cell 39, 61-69.

Hirokawa, N., & Heuser, J. (1982). The inside and outside of gap-junction membranes visualized by deep etching. Cell 30, 395-406.

Hoh, J.H., John, S.A., & Revel, J.-P. (1991). Molecular cloning and characterization of a new member of the gap junction gene family, connexin-31. J. Biol. Chem. 266, 6524-6531.

Hoh, J.H., Sosinsky, G.E., Revel, J.-P., & Hansma, P.K. (1993). Structure of the extracellular surface of the gap junction by atomic force microscopy. Biophys. J. 65, 149-163.

Homma, N., Alvarado, J.L., Coombs, W., Stergiopoulos, K., Taffet, S.M., Lau, A.F., & Delmar, M. (1998). A particle-receptor model for the insulin-induced closure of connexin43 channels. Circ. Res. 83, 27-32.

Hooper, M.L., & Subak-Sharpe, J.H. (1981). Metabolic cooperation between cells. Int. Rev. Cytol. 69, 45-104.

Hossain, M.Z., Murphy, L.J., Hertzberg, E.L., & Nagy, J.I. (1994). Phosphorylated forms of connexin43 predominate in rat brain: demonstration by rapid inactivation of brain metabolism. J. Neurochem. 62, 2394-2403.

Hubbell, W.L., & Altenbach, C. (1994). Investigation of structure and dynamics in membrane proteins using site-directed spin labeling. Curr. Opin. Struct. Biol. 4, 566-573.

Hubbell, W.L., Gross, A., Langen, R., & Lietzow, M.A. (1998). Recent advances in site-directed spin labeling of proteins. Curr. Opin. Struct. Biol. 8, 649-656.

Hubbell, W.L., Mchaourab, H.S., Altenbach, C., & Lietzow, M.A. (1996). Watching proteins move using site-directed spin labeling. Structure 4, 779-783.

Hudspeth, A.J., & Revel, J.-P. (1971). Coexistence of gap and septate junctions in an invertebrate epithelium. J. Cell Biol. 50, 92-101.

Jackson, M., & Mantsch, H.H. (1995). The use and misuse of FTIR spectroscopy in the determination of protein structure. Crit. Rev. Biochem. Mol. Biol. 30, 95-120.

Jacobs, F.A., Zhang, M., Fortin, M.G., & Verma, D.P.S. (1987). Several nodulins of soybean share structural domains but differ in their subcellular locations. Nucleic Acids Res. 15, 1271-1280.

Jap, B.K., & Li, H. (1995). Structure of the osmo-regulated H₂O-channel, AQP-CHIP, in projection at 3.5Å resolution. J. Mol. Biol. 251, 413-420.

Jap, B.K., Walian, P.J., & Gehring, K. (1991). Structural architecture of an outer membrane channel as determined by electron crystallography. Nature 350 , 167-170.

Jiang, J.X., & Goodenough, D.A. (1996). Heteromeric connexons in lens gap junction channels. Proc. Natl. Acad. Sci. USA 93, 1287-1291.

Jiskoot, W., Hlady, V., Naleway, J.J., & Herron, J.N. (1995). Application of fluorescence spectroscopy for determining the structure and function of proteins. Pharm. Biotechnol. 7, 1-63.

John, S.A., & Revel, J.-P. (1991). Connexon integrity is maintained by non-covalent bonds: intramolecular disulfide bonds link the extracellular domains in rat connexin-43. Biochem. Biophys. Res. Comm. 178, 1312-1318.

Johnson, K.D., Höfte, H., & Chrispeels, M.J. (1990). An intrinsic tonoplast protein of protein storage vacuoles in seeds is structurally related to a bacterial solute transporter (GlpF). Plant Cell 2, 525-532.

Johnson, R.G., Klukas, K.A., Tze-Hong, L., & Spray, D.C. (1988). Antibodies to MP28 are localized to lens junctions, alter intercellular permeability, and demonstrate increased expression during development. In: Modern Cell Biology, vol.7: Gap Junctions (Hertzberg, E.L., & Johnson, R.G., Eds.), pp. 81-98. Alan R. Liss, New York.

Johnson, W.C., Jr. (1988). Secondary structure of proteins through circular dichroism spectroscopy. Ann. Rev. Biophys. Biophys. Chem. 17, 145-166.

Kadle, R., Zhang, J.T., & Nicholson, B.J. (1991). Tissue-specific distribution of differentially phosphorylated forms of Cx43. Mol. Cell. Biol. 11, 363-369.

Kanemitsu, M.Y., Jiang, W., & Eckhart, W. (1998). Cdc2-mediated phosphorylation of the gap junction protein, connexin43, during mitosis. Cell Growth Differ. 9, 13-21.

Kanemitsu, M.Y., & Lau, A.F. (1993). Epidermal growth factor stimulates the disruption of gap junctional communication and connexin43 phosphorylation independent of 12-0-tetradecanoylphorbol 13-acetate-sensitive protein kinase C: the possible involvement of mitogen-activated protein kinase. Mol. Biol. Cell 4, 837-848.

Kanemitsu, M.Y., Loo, L.W.M., Simon, S., Lau, A.F., & Eckhart, W. (1997). Tyrosine phosphorylation of connexin 43 by v-Src is mediated by SH2 and SH3 domain interactions. J. Biol. Chem. 272, 22824-22831.

Kanter, H.L., Laing, J.G., Beau, S.L., Beyer, E.C., & Saffitz, J.E. (1993a). Distinct patterns of connexin expression in canine Purkinje fibers and ventricular muscle. Circ. Res. 72, 1124-1131.

Kanter, H.L., Laing, J.G., Beyer, E.C., Green, K.G., & Saffitz, J.E. (1993b). Multiple connexins colocalize in canine ventricular myocyte gap junctions. Circ. Res. 73, 344-350.

Kanter, H.L., Saffitz, J.E., & Beyer, E.C. (1992). Cardiac myocytes express multiple gap junction proteins. Circ. Res. 70, 438-444.

Kardami, E., Stoski, R.M., Doble, B.W., Yamamoto, T., Hertzberg, E.L., & Nagy, J.I. (1991). Biochemical and ultrastructural evidence for the association of basic fibroblast growth factor with cardiac gap junctions. J. Biol. Chem. 266, 19551-19557.

Karrasch, S., Bullough, P.A., & Ghosh, R. (1995). The 8.5Å projection map of the light-harvesting complex I from Rhodospirillum rubrum reveals a ring composed of 16 subunits. EMBO J. 14, 631-638.

Karrer, H.E. (1960). The striated musculature of blood vessels. II. Cell interconnections and cell surface. J. Biophys. Biochem. Cytol. 8, 135-150.

Kennelly, P.J., & Krebs, E.G. (1991). Consensus sequences as substrate specificity determinants for protein kinases and protein phosphatases. J. Biol. Chem. 266, 15555-15558.

Kensler, R.W., & Goodenough, D.A. (1980). Isolation of mouse myocardial gap junctions. J. Cell Biol. 86, 755-764.

Kistler, J., Christie, D., & Bullivant, S. (1988). Homologies between gap junction proteins in lens, heart and liver. Nature 331, 721-723.

Kistler, J., Kirkland, B., & Bullivant, S. (1985). Identification of a 70,000-D protein in lens membrane junctional domains. J. Cell Biol. 101, 28-35.

König, N., Zampighi, G.A., & Butler, P.J.G. (1997). Characterisation of the major intrinsic protein (MIP) from bovine lens fibre membranes by electron microscopy and hydrodynamics. J. Mol. Biol. 265, 590-602.

Kreutziger, G.O. (1968). Freeze etching of intercellular junctions of mouse liver. In: *26th Proceedings of the Electron Microscopy Society of America,* p. 138. Claitor's, Baton Rouge, LA.

Krishnan, S.N., Frei, E., Schalet, A.P., & Wyman, R.J. (1995). Molecular basis of intracistronic complementation in the Passover locus of *Drosophila.* Proc. Natl. Acad. Sci. USA 92, 2021-2025.

Krishnan, S.N., Frei, E., Swain, G.P., & Wyman, R.J. (1993). *Passover:* a gene required for synaptic connectivity in the giant fiber system of *Drosophila.* Cell 73, 967-977.

Kühlbrandt, W. (1992). Two-dimensional crystallization of membrane proteins. Quart. Rev. Biophysics 25, 1-49.

Kühlbrandt, W., & Downing, K.H. (1989). Two-dimensional structure of plant light-harvesting complex at 3.7Å resolution by electron crystallography. J. Mol. Biol., 207, 823-828.

Kühlbrandt, W., Wang, D.N., & Fujiyoshi, Y. (1994). Atomic model of plant light-harvesting complex by electron crystallography. Nature 367, 614-621.

Kumar, N.M., Friend, D.S., & Gilula, N.B. (1995). Synthesis and assembly of human $\beta_1$ gap junctions in BHK cells by DNA transfection with the human $\beta_1$ cDNA. J. Cell Sci. 108, 3725-3734.

Kumar, N.M., & Gilula, N.B. (1986). Cloning and characterization of human and rat liver cDNAs coding for a gap junction protein. J. Cell Biol. 103, 767-776.

Kumar, N.M., & Gilula, N.B. (1992). Molecular biology and genetics of gap junction channels. Sem. Cell Biol. 3, 3-16.

Kumar, N.M., & Gilula, N.B. (1996). The gap junction communication channel. Cell 84, 381-388.

Kunzelmann, P., Blümcke, I., Traub O., Dermietzel, R., & Willecke, K. (1997). Coexpression of connexin45 and -32 in oligodendrocytes of rat brain. J. Neurocytol. 26, 17-22.

Kuraoka, A., Iida, H., Hatae, T., Shibata, Y., Itoh, M., & Kurita, T. (1993). Localization of gap junction proteins, connexins 32 and 26, in rat and guinea pig liver as revealed by quick-freeze, deep-etch immunoelectron microscopy. J. Histochem. Cytochem. 41, 971-980.

Kürz, L.L., Zühlke, R.D., Zhang, H.-J., & Joho, R.H. (1995). Side-chain accessabilities in the pore of a $K^+$ channel probed by sulfhydryl-specific reagents after cysteine-scanning mutagenesis. Biophys. J. 68, 900-905.

Kuszak, J., Maisel, H., & Harding, C.V. (1978). Gap junctions of chick lens fiber cells. Exp. Eye Res. 27, 495-498.

Kwak, B.R., Sáez, J.C., Wilders, R., Chanson, M., Fishman, G.I., Hertzberg, E.L., Spray, D.C., & Jongsma, H.J. (1995). Effects of cGMP-dependent phosphorylation on rat and human connexin43 gap junction channels. Pflügers Arch. 430, 770-778.

Laing, J.G., & Beyer, E.C. (1995). The gap junction protein connexin43 is degraded via the ubiquitin proteasome pathway. J. Biol. Chem. 270, 26399-26403.

Laing, J.G., Tadros, P.N., Westphale, E.M., & Beyer, E.C. (1997). Degradation of connexin43 gap junctions involves both the proteasome and the lysosome. Exp. Cell Res. 236, 482-492.

Laird, D.W. (1996). The life cycle of a connexin: gap junction formation, removal, and degradation. J. Bioenerg. Biomembr. 28, 311-318.

Laird, D.W., & Revel, J.-P. (1990). Biochemical and immunochemical analysis of the arrangement of connexin43 in rat heart gap junction membranes. J. Cell Sci. 97, 109-117.

Lal, R., Laird, D.W., & Revel, J.-P. (1993). Antibody perturbation analysis of gap-junction permeability in rat cardiac myocytes. Pflügers Arch. Eur. J. Physiol. 422, 449-457.

Lampe, P.D., Kistler, J., Hefti, A., Bond, J., Müller, S., Johnson, R.G., & Engel, A. (1991). *In vitro* assembly of gap junctions. J. Struct. Biol., 107, 281-290.

Lampe, P.D., Kurata, W.E., Warn-Cramer, B.J., & Lau, A.F. (1998). Formation of a distinct connexin43 phosphoisoform in mitotic cells is dependent upon p34$^{cdc2}$ kinase. J. Cell Sci. 111, 833-841.

Larsen, W.J., Tung, H.-N., Murray, S.A., & Swenson, C.A. (1979). Evidence for the participation of actin microfilaments and bristle coats in the internalization of gap junction membrane. J. Cell Biol. 83, 576-587.

Lawrence, T.S., Beers, W.H., & Gilula, N.B. (1978). Transmission of hormonal stimulation by cell-to-cell communication. Nature 272, 501-506.

Li, H., Lee, S., & Jap, B.K. (1997). Molecular design of aquaporin-1 water channel as revealed by electron crystallography. Nat. Struct. Biol. 4, 263-265.

Li, H., Liu, T.-F., Lazrak, A., Peracchia, C., Goldberg, G.S., Lampe, P.D., & Johnson, R.G. (1996). Properties and regulation of gap junctional hemichannels in the plasma membranes of cultured cells. J. Cell Biol. 134, 1019-1030.

Liu, S., Taffet, S., Stoner, L., Delmar, M., Vallano, M.L., & Jalife, J. (1993). A structural basis for the unequal sensitivity of the major cardiac and liver gap junctions to intracellular acidification: The carboxyl tail length. Biophys. J. 64, 1422-1433.

Lo, W.-K., & Harding, C.V. (1986). Structure and distribution of gap junctions in lens epithelium and fiber cells. Cell Tissue Res. 244, 253-263.

Loewenstein, W.R. (1981). Junctional intercellular communication: The cell-to-cell membrane channel. Physiol. Rev. 61, 829-913.

Loewenstein, W.R. (1967). On the genesis of cellular communication. Devel. Biol. 15, 503-520.

Loewenstein, W.R., & Kanno, Y. (1964). Studies on an epithelial (gland) cell junction. I. Modifications of surface membrane permeability. J. Cell Biol. 22, 565-586.

Loo, L.W.M., Berestecky, J.M., Kanemitsu, M.Y., Lau, A.F. (1995). pp60$^{src}$-mediated phosphorylation of connexin 43, a gap junction protein. J. Biol. Chem. 270, 12751-12761.

Makowski, L. (1988). X-ray diffraction studies of gap junction structure. Adv. Cell Biol. 2, 119-158.

Makowski, L., Caspar, D.L.D., Phillips, W.C., & Goodenough, D.A. (1977). Gap junction structures II. Analysis of the x-ray diffraction data. J. Cell Biol. 74, 629-645.

Makowski, L., Caspar, D.L.D., Phillips, W.C., & Goodenough, D.A. (1984). Gap junction structures V. Structural chemistry inferred from x-ray diffraction measurements on sucrose accessibility and trypsin susceptibility. J. Mol. Biol. 174, 449-481.

Malchow, R.P., Qian, H., & Ripps, H. (1993). Evidence for hemi-gap junctional channels in isolated horizontal cells of the skate retina. J. Neurosci. Res. 35, 237-245.

Malewicz, B., Kumar, V.V., Johnson, R.G., & Baumann, W.J. (1990). Lipids in gap junction assembly and function. Lipids 25, 419-427.

Mandel, M., Moriyama, Y., Hulmes, J.D., Pan, Y.-C.E., Nelson, H., & Nelson, N. (1988). cDNA sequence encoding the 16-kDa proteolipid of chromaffin granules implies gene duplication in the evolution of H$^+$-ATPases. Proc. Natl. Acad. Sci. USA 85, 5521-5524.

Manjunath, C.K., Goings, G.E., & Page, E. (1982). Isolation and protein composition of gap junctions from rabbit hearts. Biochem. J. 205, 189-194.

Manjunath, C.K., Goings, G.E., & Page, E. (1984a). Cytoplasmic surface and intramembrane components of rat heart gap junctional proteins. Am. J. Physiol. 246, H865-H875.

Manjunath, C.K., Goings, G.E., & Page, E. (1984b). Detergent sensitivity and splitting of isolated liver gap junctions. J. Membr. Biol. 78, 147-155.

Manjunath, C.K., Goings, G.E., & Page, E. (1985). Proteolysis of cardiac gap junctions during their isolation from rat hearts. J. Membr. Biol. 85, 159-168.

Manjunath, C.K., Nicholson, B.J., Teplow, D., Hood, L., Page, E., & Revel, J.-P. (1987). The cardiac gap junction protein ($M_r$ 47,000) has a tissue-specific cytoplasmic domain of $M_r$ 17,000 at its carboxy-terminus. Biochem. Biophys. Res. Comm. 142, 228-234.

Manjunath, C.K., & Page, E. (1986). Rat heart gap junctions as disulfide-bonded connexon multimers: Their depolymerization and solubilization in deoxycholate. J. Membr. Biol. 90, 43-57.

Massey, K.D., Minnich, B.N., & Burt, J.M. (1992). Arachidonic acid and lipoxygenase metabolites uncouple neonatal rat cardiac myocyte pairs. Am. J. Physiol. 263, C494-C501.

Mathias, R.T., & Rae, J.L. (1985). Transport properties of the lens. Am. J. Physiol. 249, C181-C190.

McNutt, N.S., & Weinstein, R.S. (1970). The ultrastructure of the nexus. A correlated thin-section and freeze-cleave study. J. Cell Biol. 47, 666-688.

Meiners, S., & Schindler, M. (1987). Immunological evidence for gap junction polypeptide in plant cells. J. Biol. Chem. 262, 951-953.

Meiners, S., Xu, A., & Schindler, M. (1991). Gap junction protein homologue from *Arabidopsis thaliana:* Evidence for connexins in plants. Proc. Natl. Acad. Sci. U.S.A. 88, 4119-4122.

Menezes, M.E., Roepe, P.D., & Kaback, H.R. (1990). Design of a membrane transport protein for fluorescence spectroscopy. Proc. Natl. Acad. Sci. USA 87, 1638-1642.

Meyer, R., Malewicz, B., Baumann, W.J., & Johnson, R.G. (1990). Increased gap junction assembly between cultured cells upon cholesterol supplementation. J. Cell Sci. 96, 231-238.

Milks, L.C., Kumar, N.M., Houghten, R., Unwin, N., & Gilula, N.B. (1988). Topology of the 32-kD liver gap junction protein determined by site-directed antibody localizations. EMBO J. 7, 2967-2975.

Milligan, D.L., & Koshland, D.E., Jr. (1988). Site-directed cross-linking. Establishing the dimeric structure of the aspartate receptor of bacterial chemotaxis. J. Biol. Chem. 263, 6268-6275.

Mitra, A.K., van Hoek, A.N., Wiener, M.C., Verkman, A.S., & Yeager, M. (1995). The CHIP28 water channel visualized in ice by electron crystallography. Nat. Struct. Biol. 2, 726-729.

Moreno, A.P., Fishman, G.I., & Spray, D.C. (1992). Phosphorylation shifts unitary conductance and modifies voltage dependent kinetics of human connexin43 gap junction channels. Biophys. J. 62, 51-53.

Moreno, A.P., Sáez, J.C., Fishman, G.I., & Spray, D.C. (1994). Human connexin43 gap junction channels. Regulation of unitary conductances by phosphorylation. Circ. Res. 74, 1050-1057.

Morley, G.E., Ek-Vitorín, J.F., Taffet, S.M., & Delmar, M. (1997). Structure of connexin43 and its regulation by pHi. J. Cardiovasc. Electrophysiol. 8, 939-951.

Morley, G.E., Taffet, S.M., & Delmar, M. (1996). Intramolecular interactions mediate pH regulation of connexin43 channels. Biophys. J. 70, 1294-1302.

Muramatsu, S., & Mizuno, T. (1989). Nucleotide sequence of the region encompassing the *glpKF* operon and its upstream region containing a bent DNA sequence of *Escherichia coli.* Nucleic Acids Res. 17, 4378.

Mushegian, A.R., & Koonin, E.V. (1993). The proposed plant connexin is a protein kinase-like protein. Plant Cell 5, 998-999.

Musil, L.S., Beyer, E.C., & Goodenough, D.A. (1990a). Expression of the gap junction protein connexin43 in embryonic chick lens: molecular cloning, ultrastructural localization, and post-translational phosphorylation. J. Membr. Biol. 116, 163-175.

Musil, L.S., Cunningham, B.A., Edelman, G.M., & Goodenough, D.A. (1990b). Differential phosphorylation of the gap junction protein connexin43 in junctional communication-competent and -deficient cell lines. J. Cell Biol. 111, 2077-2088.

Musil, L.S., & Goodenough, D.A. (1993). Multisubunit assembly of an integral plasma membrane channel protein, gap junction connexin43, occurs after exit from the ER. Cell 74, 1065-1077.

Nagy, J.I., Ochalski, P.A.Y., Li, J., & Hertzberg, E.L. (1997). Evidence for the co-localization of another connexin with connexin 43 at astrocytic gap junctions in rat brain. Neuroscience 78, 533-548.

Nagy, J.I., Patel, D., Ochalski, P.A.Y., & Stelmack, G.L. (1999). Connexin30 in rodent, cat and human brain: selective expression in gray matter astrocytes, co-localization with connexin43 at gap junctions and late developmental appearance. Neuroscience 88, 447-468.

Nagy, J.I., Yamamoto, T., Sawchuk, M.A., Nance, D.M., & Hertzberg, E.L. (1992). Quantitative immunohistochemical and biochemical correlates of connexin43 localization in rat brain. Glia 5, 1-9.

Naus, C.C., Hearn, S., Zhu, D., Nicholson, B.J., & Shivers, R.R. (1993). Ultrastructural analysis of gap junctions in C6 glioma cells transfected with connexin43 cDNA. Exp. Cell Res. 206, 72-84.

Neveu, M.J., Hully, J.R., Babcock, K.L., Hertzberg, E.L., Nicholson, B.J., Paul, D.L., & Pitot, H.C. (1994). Multiple mechanisms are responsible for altered expression of gap junction genes during oncogenesis in rat liver. J. Cell Sci. 107, 83-95.

Nicholson, B., Dermietzel, R., Teplow, D., Traub, O., Willecke, K., & Revel, J.-P. (1987). Two homologous protein components of hepatic gap junctions. Nature 329, 732-734.

Nicholson, B.J., Gros, D.B., Kent, S.B.H., Hood, L.E., & Revel, J.-P. (1985). The $M_r$ 28,000 gap junction proteins from rat heart and liver are different but related. J. Biol. Chem. 260, 6514-6517.

Nicholson, B.J., Hunkapiller, M.W., Grim, L.B., Hood, L.E., & Revel, J.-P. (1981). Rat liver gap junction protein: properties and partial sequence. Proc. Natl. Acad. Sci. USA 78, 7594-7598.

Nicholson, B.J., & Revel, J.-P. (1983). Gap junctions in liver: isolation, morphological analysis, and quantitation. Methods Enzymol. 98, 519-537.

Nicholson, B.J., Suchyna, T., Xu, L.X., Hammernick, P., Cao, F.L., Fourtner, C., Barrio, L., & Bennett, M.V.L. (1993). Divergent properties of connexins expressed in Xenopus oocytes. In: Progress in Cell Research, vol.3: Gap Junctions (Hall, J.E., Zampighi, G.A., & Davis, R.M., Eds.), pp. 3-13. Elsevier, Amsterdam.

Nicholson, B.J., Takemoto, L.J., Hunkapiller, M.W., Hood, L.E., & Revel, J.-P. (1983). Differences between liver gap junction protein and lens MIP 26 from rat: implications for tissue specificity of gap junctions. Cell 32, 967-978.

Nicholson, S.M., & Bruzzone, R. (1997). Gap junctions: getting the message through. Curr. Biol. 7, R340-R344.

Oh, S., Ri, Y., Bennett, M.V.L., Trexler, E.B., Verselis, V.K. and Bargiello, T.A. (1997). Changes in permeability caused by connexin 32 mutations underlie X-linked Charcot-Marie-Tooth Disease. Neuron 19, 927-938.

Paul, D.L. (1986). Molecular cloning of cDNA for rat liver gap junction protein. J. Cell Biol. 103, 123-134.

Paul, D.L., Ebihara, L., Takemoto, L.J., Swenson, K.I., & Goodenough, D.A. (1991). Connexin46, a novel lens gap junction protein, induces voltage-gated currents in nonjunctional plasma membrane of Xenopus oocytes. J. Cell Biol. 115, 1077-1089.

Paul, D.L., & Goodenough, D.A. (1983). Preparation, characterization, and localization of antisera against bovine MP26, an integral protein from lens fiber plasma membrane. J. Cell Biol. 96, 625-632.

Peracchia, C. (1980). Structural correlates of gap junction permeation. Int. Rev. Cytol. 66, 81-146.

Perkins, G.A., Goodenough, D.A., & Sosinsky, G.E. (1998). Formation of the gap junction intercellular channel requires a 30° rotation for interdigitating two apposing connexons. J. Mol. Biol. 277, 171-177.

Perkins, G., Goodenough, D., & Sosinsky, G. (1997). Three-dimensional structure of the gap junction connexon. Biophys. J. 72, 533-544.

Pfahnl, A., Zhou, X.-W., Tian, J., Werner, R., & Dahl, G. (1996). Mapping of the pore of gap junction channels by cysteine scanning mutagenesis. Biophys. J. 70, A31.

Phelan, P., Stebbings, L.A., Baines, R.A., Bacon, J.P., Davies, J.A., & Ford, C. (1998). Drosophila Shaking-B protein forms gap junctions in paired Xenopus oocytes. Nature 391, 181-184.

Potter, D.D., Furshpan, E.J., & Lennox, E.S. (1966). Connections between cells of the developing squid as revealed by electrophysiological methods. Proc. Natl. Acad. Sci. USA 55, 328-336.

Preston, G.M., & Agre, P. (1991). Isolation of the cDNA for erythrocyte integral membrane protein of 28 kilodaltons: Member of an ancient channel family. Proc. Natl. Acad. Sci. USA 88: 11110-11114.

Radu, A., Dahl, G., & Loewenstein, W.R. (1982). Hormonal regulation of cell junction permeability: upregulation by catecholamine and prostaglandin $E_1$. J. Membr. Biol. 70, 239-251.

Rahman, S., Carlile, G., & Evans, W.H. (1993). Assembly of hepatic gap junctions. Topography and distribution of connexin 32 in intracellular and plasma membranes determined using sequence-specific antibodies. J. Biol. Chem. 268, 1260-1265.

Rahman, S., & Evans, W.H. (1991). Topography of connexin32 in rat liver gap junctions. Evidence for an intramolecular disulphide linkage connecting the two extracellular peptide loops. J. Cell Sci. 100, 567-578.

Rao, Y., Jan, L.Y., & Jan, Y.N. (1990). Similarity of the product of the *Drosophila* neurogenic gene *big brain* to transmembrane channel proteins. Nature 345, 163-167.

Reaume, A.G., de Sousa, P.A., Kulkarni, S., Langille, B.L., Zhu, D., Davies, T.C., Juneja, S.C., Kidder, G.M., & Rossant, J. (1995). Cardiac malformation in neonatal mice lacking connexin43. Science 267, 1831-1834.

Reed, K.E., Westphale, E.M., Larson, D.M., Wang, H.-Z., Veenstra, R.D., & Beyer, E.C. (1993). Molecular cloning and functional expression of human connexin37, an endothelial cell gap junction protein. J. Clin. Invest. 91, 997-1004.

Revel, J.-P. (1988). The oldest multicellular animal and its junctions. In: *Modern Cell Biology,* vol 7: *Gap Junctions* (Hertzberg, E.L., & Johnson, R.G., Eds.), pp. 135-149. Alan R. Liss, New York.

Revel, J.-P., & Karnovsky, M.J. (1967). Hexagonal array of subunits in intercelluluar junctions of the mouse heart and liver. J. Cell Biol. 33, C7-C12.

Revel, J.-P., Yee, A.G., & Hudspeth, A.J. (1971). Gap junctions between electrotonically coupled cells in tissue culture and in brown fat. Proc. Natl. Acad. Sci. USA 68, 2924-2927.

Rhee, S.K., Bevans, C.G., & Harris, A.L. (1996). Channel-forming activity of immunoaffinity-purified connexin32 in single phospholipid membranes. Biochemistry 35, 9212-9223.

Risek, B., & Gilula, N.B. (1991). Spatiotemporal expression of three gap junction gene products involved in fetomaternal communication during rat pregnancy. Development 113, 165-181.

Risek, B., Guthrie, S., Kumar, N., & Gilula, N. B. (1990). Modulation of gap junction transcript and protein expression during pregnancy in the rat. J. Cell Biol. 110, 269-282.

Risek, B., Klier, F.G., & Gilula, N.B. (1992). Multiple gap junction genes are utilized during rat skin and hair development. Development 116, 639-651.

Risinger, M.A., & Larsen, W.J. (1983). Interaction of filipin with junctional membrane at different stages of the junction's life history. Tissue Cell 15, 1-15.

Robenek, H., Jung, W., & Gebhardt, R. (1982). The topography of filipin-cholesterol complexes in the plasma membrane of cultured hepatocytes and their relation to cell junction formation. J. Ultrastructure Res. 78, 95-106.

Robertson, J.D. (1963). The occurrence of a subunit pattern in the unit membranes of club endings in Mauthner cell synapses in goldfish brains. J. Cell Biol. 19, 201-221.

Rohlmann, A., Laskawi, R., Hofer, A., Dobo, E., Dermietzel, R., & Wolff, J.R. (1993). Facial nerve lesions lead to increased immunostaining of the astrocytic gap junction protein (connexin 43) in the corresponding facial nucleus of rats. Neurosci. Lett. 154, 206-208.

Rubin, J.B., Verselis, V.K., Bennett, M.V.L., & Bargiello, T.A. (1992). A domain substitution procedure and its use to analyze voltage dependence of homotypic gap junctions formed by connexins 26 and 32. Proc. Natl. Acad. Sci. USA 89, 3820-3824.

Ryerse, J.S. (1989). Isolation and characterization of gap junctions from *Drosophila melanogaster.* Cell Tissue Res. 256, 7-16.

Sáez, J.C., Connor, J.A., Spray, D.C., & Bennett, M.V.L. (1989). Hepatocyte gap junctions are permeable to the second messenger, inositol 1,4,5-trisphosphate, and to calcium ions. Proc. Natl. Acad. Sci. USA 86, 2708-2712.

Sáez, J.C., Nairn, A.C., Czernik, A.J., Spray, D.C., & Hertzberg, E.L. (1993). Rat connexin43: regulation by phosphorylation in heart. In: *Progress in Cell Research,* vol.3: *Gap Junctions* (Hall, J.E., Zampighi, G.A., & Davis, R.M., Eds.), pp. 275-281. Elsevier, Amsterdam.

Sáez, J.C., Nairn, A.C., Czernik, A.J., Spray, D.C., Hertzberg, E.L., Greengard, P., & Bennett, M.V.L. (1990). Phosphorylation of connexin 32, a hepatocyte gap-junction protein, by cAMP-dependent protein kinase, protein kinase C and $Ca^{2+}$/calmodulin-dependent protein kinase II. Eur. J. Biochem. 192, 263-273.

Sakamoto, H., Oyamada, M., Enomoto, K., & Mori, M. (1992). Differential changes in expression of gap junction proteins connexin 26 and 32 during hepatocarcinogenesis in rats. Jpn. J. Cancer Res. 83, 1210-1215.

Sas, D.F., Sas, M.J., Johnson, K.R., Menko, A.S., & Johnson, R.G. (1985). Junctions between lens fiber cells are labeled with a monoclonal antibody shown to be specific for MP26. J. Cell Biol. 100, 216-225.

Sayre, D. (1952). Some implications of a theorem due to Shannon. Acta Cryst. 5: 843.

Schertler, G.F.X., Villa, C., & Henderson, R. (1993). Projection structure of rhodopsin. Nature 362,770-772.

Schwarzmann, G., Wiegandt, H., Rose, B., Zimmerman, A., Ben-Haim, D., & Loewenstein, W.R. (1981). Diameter of the cell-to-cell junctional membrane channels as probed with neutral molecules. Science 213, 551-553.

Severs, N.J. (1981). Plasma membrane cholesterol in myocardial muscle and capillary endothelial cells. Distribution of filipin-induced deformations in freeze-fracture. Eur. J. Cell Biol. 25, 289-299.

Shapiro, L., Fannon, A.M., Kwong, P.D., Thompson, A., Lehmann, M.S., Grübel, G., Legrand, J.-F., Als-Nielsen, J., Colman, D.R., & Hendrickson, W.A. (1995). Structural basis of cell-cell adhesion by cadherins. Nature 374, 327-337.

Shibata, Y., Manjunath, C.K., & Page, E. (1985). Differences between cytoplasmic surfaces of deep-etched heart and liver gap junctions. Am. J. Physiol. 249, H690-H693.

Shimohigashi, M., & Meinertzhagen, I.A. (1998). The shaking B gene in Drosophila regulates the number of gap junctions between photoreceptor terminals in the lamina. J. Neurobiol. 35, 105-117.

Sikerwar, S.S., Downing, K.H., & Glaeser, R.M. (1991). Three-dimensional structure of an invertebrate intercellular communicating junction. J. Struct. Biol. 106, 255-63.

Sikerwar, S.S., & Unwin, N. (1988). Three-dimensional structure of gap junctions in fragmented plasma membranes from rat liver. Biophys. J. 54, 113-119.

Skerrett, M., Aronowitz, J., Cymes, G., Kasperek, E., & Nicholson, B.J. (1998). Identification of amino acids lining the gap junction pore. Mol. Biol. Cell 9, 94a.

Smith, J.F., Xu, L.X., & Nicholson, B.J. (1998). Significance of the location of proline in the M2 domain of Cx 26 for gap junction function and gating. Mol. Biol. Cell 9, 93a.

Somogyi, B., & Lakos, Z. (1993). Protein dynamics and fluorescence quenching. J. Photochem. Photobiol.B. 18, 3-16.

Sosinsky, G. (1995). Mixing of connexins in gap junction membrane channels. Proc. Natl. Acad. Sci., USA 92, 9210-9214.

Spray, D.C., Sáez, J.C., Brosius, D., Bennett, M.V.L., & Hertzberg, E.L. (1986). Isolated liver gap junctions: gating of transjunctional currents is similar to that in intact pairs of rat hepatocytes. Proc. Natl. Acad. Sci. U.S.A. 83, 5494-5497.

Starich, T.A., Lee, R.Y.N., Panzarella, C., Avery, L., & Shaw, J.E. (1996). eat-5 and unc-7 represent a multigene family in Caenorhabditis elegans involved in cell-cell coupling. J. Cell Biol. 134, 537-548.

Stauffer, K.A. (1995). The gap junction proteins $\beta_1$-connexin (connexin-32) and $\beta_2$-connexin (connexin-26) can form heteromeric hemichannels. J. Biol. Chem. 270, 6768-6772.

Stauffer, K.A., Kumar, N.M., Gilula, N.B., & Unwin, N. (1991). Isolation and purification of gap junction channels. J. Cell Biol. 115, 141-150.

Stryer, L. (1968). Fluorescence spectroscopy of proteins. Science 162, 526-533.

Subak-Sharpe, H., Bürk, R.R., & Pitts, J.D. (1966). Metabolic co-operation by cell to cell transfer between genetically different mammalian cells in tissue culture. Heredity 21, 342-343.

Subak-Sharpe, H., Bürk, R.R., & Pitts, J.D. (1969). Metabolic co-operation between biochemically marked mammalian cells in tissue culture. J. Cell Sci. 4, 353-367.

Suchyna, T.M., Veenstra, R., Chilton, M., & Nicholson, B.J. (1994). Different ionic permeabilities of connexins 26 and 32 produce rectifying gap junction channels. Mol. Biol. Cell Suppl. 5, 199a.

Suchyna, T.M., Xu, L.X., Gao, F., Fourtner, C.R., & Nicholson, B.J. (1993). Identification of a proline residue as a transduction element involved in voltage gating of gap junctions. Nature 365, 847-849.

Sun, Y.-A., & Wyman, R.J. (1996). Passover eliminates gap junctional communication between neurons of the giant fiber system *in Drosophila*. J. Neurobiol. 30, 340-348.

Swenson, K.I., Jordan, J.R., Beyer, E.C., & Paul, D.L. (1989). Formation of gap junctions by expression of connexins in Xenopus oocyte pairs. Cell 57, 145-155.

Swenson, K.I., Piwnica-Worms, H., McNamee, H., & Paul, D.L. (1990). Tyrosine phosphorylation of the gap junction protein connexin43 is required for the $pp60^{v-src}$-induced inhibition of communication. Cell Regulation 1, 989-1002.

Tadvalkar, G., & Pinto Da Silva, P. (1983). *In vitro*, rapid assembly of gap junctions is induced by cytoskeleton disruptors. J. Cell Biol. 96, 1279-1287.

Tibbits, T.T., Caspar, D.L.D., Philips, W.C., & Goodenough, D.A. (1990). Diffraction diagnosis of protein folding in gap junction connexons. Biophys. J. 57, 1025-1036.

Todd, A.P., Cong, J., Levinthal, F., Levinthal, C., & Hubbell, W.L. (1989). Site-directed mutagenesis of colicin E1 provides specific attachment sites for spin labels whose spectra are sensitive to local conformation. Proteins 6, 294-305.

Toyofuku, T., Yabuki, M., Otsu, K., Kuzuya, T., Hori, M., & Tada, M. (1998). Direct association of the gap junction protein connexin-43 with ZO-1 in cardiac myocytes. J. Biol. Chem. 273, 12725-12731.

Toyoshima, C., & Unwin, N. (1988). Ion channel of acetylcholine receptor reconstructed from images of postsynaptic membranes. Nature 336, 247-250.

Traub, O., Look, J., Dermietzel, R., Brümmer, F., Hülser, D., & Willecke, K. (1989). Comparative characterization of the 21-kD and 26-kD gap junction proteins in murine liver and cultured hepatocytes. J. Cell Biol. 108, 1039-1051.

Trelstad, R., Revel, J.-P., & Hay, E.D. (1966). Tight junctions between cells in the early chick embryo as visualized by electron microscopy. J. Cell Biol. 31, C6-C10.

Unger, V.M., Kumar, N.M., Gilula, N.B., & Yeager, M. (1997). Projection structure of a gap junction membrane channel at 7Å resolution. Natl. Struct. Biol. 4, 39-43.

Unger, V.M., Kumar, N.M., Gilula, N.B., & Yeager, M. (1999a). Expression, two-dimensional crystallization and electron cryo-crystallography of recombinant gap junction membrane channels. J. Struct. Biol. 128, 98-105.

Unger, V.M., Kumar, N.M., Gilula, N.B., & Yeager, M. (1999b). Three-dimensional structure of a recombinant gap junction membrane channel. Science 283, 1176-1180.

Unwin, P.N.T., & Ennis, P.D. (1983). Calcium-mediated changes in gap junction structure: evidence from the low angle X-ray pattern. J. Cell Biol. 97, 1459-1466.

Unwin, P.N.T., & Ennis, P.D. (1984). Two configurations of a channel-forming membrane protein. Nature 307, 609-613.

Unwin, P.N., & Henderson, R. (1975). Molecular structure determination by electron microscopy of unstained crystalline specimens. J. Mol. Biol. 94, 425-440.

Unwin, P.N.T., & Zampighi, G. (1980). Structure of the junction between communicating cells. Nature 283, 545-549.

Van Aelst, L., Hohmann, S., Zimmermann, F.K., Jans, A.W.H., & Thevelein, J.M. (1991). A yeast homologue of the bovine lens fibre MIP gene family complements the growth defect of a *Sac-*

*charomyces cerevisiae* mutant on fermentable sugars but not its defect in glucose-induced RAS-mediated cAMP signalling. EMBO J. 10, 2095-2104.

Van Eldik, L.J., Hertzberg, E.L., Berdan, R.C., & Gilula, N.B. (1985). Interaction of calmodulin and other calcium-modulated proteins with mammalian and arthropod junctional membrane proteins. Biochem. Biophys. Res. Comm. 126, 825-832.

Verkman, A.S. (1993). *Water Channels: A Volume in Molecular Biology Intelligence Series.* R.G. Landes, Austin, TX.

Verselis, V.K., Ginter, C.S., & Bargiello, T.A. (1994). Opposite voltage gating polarities of two closely related connexins. Nature 368, 348-354.

Voss, J., Hubbell, W.L., & Kaback, H.R. (1995). Distance determination in proteins using designed metal ion binding sites and site-directed spin labeling: Application to the lactose permease of *Escherichia coli.* Proc. Natl. Acad. Sci. USA 92, 12300-12303.

Walz, T., Hirai, T., Murata, K., Heymann, J.B., Mitsuoka, K., Fujiyoshi, Y., Smith, B.L., Agre, P., & Engel, A. (1997). The three-dimensional structure of aquaporin-1. Nature 387, 624-627.

Walz, T., Typke, D., Smith, B.L., Agre, P., & Engel, A. (1995). Projection map of aquaporin-1 determined by electron crystallography. Nature Struct. Biol. 2, 730-732.

Wang, J., Yan, Y., Garrett, T.P.J., Liu, J., Rodgers, D.W., Garlick, R.L., Tarr, G.E., Husain, Y., Reinherz, E.L., & Harrison, S.C. (1990). Atomic structure of a fragment of human CD4 containing two immunoglobulin-like domains. Nature 348, 411-418.

Wang, X.G., & Peracchia, C. (1996). Connexin 32/38 chimeras suggest a role for the second half of inner loop in gap junction gating by low pH. Am. J. Physiol. 271, C1743-1749.

Wang, X.G., & Peracchia, C. (1997). Positive charges of the initial C-terminus domain of Cx32 inhibit gap junction gating sensitivity to $CO_2$. Biophysical J. 73, 798-806.

Wang, X., Li, L., Peracchia, L.L., & Peracchia, C. (1996). Chimeric evidence for a role of the connexin cytoplasmic loop in gap junction channel gating. Pflügers Arch. 431, 844-852.

Warn-Cramer, B.J., Lampe, P.D., Kurata, W.E., Kanemitsu, M.Y., Loo, L.W.M., Eckhart, W., Lau, A.F. (1996). Characterization of the mitogen activated protein kinase phosphorylation sites on the connexin 43 gap junction protein. J. Biol. Chem. 271, 3779-3786.

Weidmann, S. (1952). The electrical constants of purkinje fibers. J. Physiol. 118, 348-360.

Weidmann, S. (1966). The diffusion of radiopotassium across intercalated disks of mammalian cardiac muscle. J. Physiol. 187, 323-342.

Weiss, M.S., Abele, U., Weckesser, J., Welte, W., Schiltz, E., & Schulz, G.E. (1991). Molecular architecture and electrostatic properties of a bacterial porin. Science 254, 1627-1630.

Werner, R., Levine, E., Rabadan-Diehl, C., & Dahl, G. (1989). Formation of hybrid cell-cell channels. Proc. Natl. Acad. Sci. USA 86, 5380-5384.

Werner, R., Rabadan-Diehl, C., Levine, E., & Dahl, G. (1993). Affinities between connexins. In: *Progress in Cell Research,* vol.3: *Gap Junctions* (Hall, J.E., Zampighi, G.A., & Davis, R.M., Eds.), pp. 21-24. Elsevier, Amsterdam.

White, T.W., Bruzzone, R., Goodenough, D.A., & Paul, D.L. (1992). Mouse Cx50, a functional member of the connexin family of gap junction proteins, is the lens fiber protein MP70. Mol. Biol. Cell 3, 711-720.

White, T.W., Bruzzone, R., Wolfram, S., Paul, D.L., & Goodenough, D.A. (1994). Selective interactions among the multiple connexin proteins expressed in the vertebrate lens: the second extracellular domain is a determinant of compatibility between connexins. J. Cell Biol. 125, 879-892

White, T.W., Deans, M.R., Kelsell, D.P., & Paul, D.L. (1998). Connexin mutations in deafness. Nature 394, 630.

White, T.W., Paul, D.L., Goodenough, D.A., & Bruzzone, R. (1995). Functional analysis of selective interactions among rodent connexins. Mol. Biol. Cell 6, 459-470.

Willecke, K., Hennemann, H., Dahl, E., Jungbluth, S., & Heynkes, R. (1991a). The diversity of connexin genes encoding gap junctional proteins. European J. Cell Biol. 56, 1-7.

Willecke, K., Heynkes, R., Dahl, E., Stutenkemper, R., Hennemann, H., Jungbluth, S., Suchyna, T., & Nicholson, B.J. (1991b). Mouse connexin37: cloning and functional expression of a gap junction gene highly expressed in lung. J. Cell Biol. 114, 1049-1057.

Willecke, K., Traub, O., Look, J., Stutenkemper, R., & Dermietzel, R. (1988). Different protein components contribute to the structure and function of hepatic gap junctions. In: *Modern Cell Biology*, vol. 7: *Gap Junctions* (Hertzberg, E.L., & Johnson, R.G., Eds.), pp. 41-52. Alan R. Liss, New York.

Willingham, M.C., Pastan, I., Shih, T.Y., & Scolnick, E.M. (1980). Localization of the *src* gene product of the Harvey strain of MSV to plasma membrane of transformed cells by electron microscopic immunocytochemistry. Cell 19, 1005-1014.

Woody, R.W. (1995). Circular dichroism. Methods Enzymol. 246, 34-71.

Wu, C.-W. and Stryer, L. (1972). Proximity relationships in rhodopsin. Proc. Natl. Acad. Sci. USA 69, 1104-1108.

Xie, H.-Q., Laird, D.W., Chang, T.-H., & Hu, V.W. (1997). A mitosis-specific phosphorylation of the gap junction protein connexin43 in human vascular cells: biochemical characterization and localization. J. Cell Biol. 137, 203-210.

Xu, M., & Akabas, M.H. (1993). Amino acids lining the channel of the $\gamma$-aminobutyric acid type A receptor identified by cysteine substitution. J. Biol. Chem. 268, 21505-21508.

Yada, T., Rose, B., & Loewenstein, W.R. (1985). Diacylglycerol downregulates junctional membrane permeability. TMB-8 blocks this effect. J. Membr. Biol. 88, 217-232.

Yahalom, A., Warmbrodt, R.D., Laird, D.W., Traub, O., Revel, J.-P., Willecke, K., & Epel, B.L. (1991). Maize mesocotyl plasmodesmata proteins cross-react with connexin gap junction protein antibodies. Plant Cell 3, 407-417.

Yamamoto, T., Kardami, E., & Nagy, J.I. (1991). Basic fibroblast growth factor in rat brain: localization to glial gap junctions correlates with connexin43 distribution. Brain Res. 554, 336-343.

Yamamoto, T., Ochalski, A., Hertzberg, E.L., & Nagy, J.I. (1990). LM and EM immunolocalization of the gap junctional protein connexin 43 in rat brain. Brain Res. 508, 313-319.

Yancey, S.B. (I), John, S.A. (II), Lal, R. (III), Austin, B.J., & Revel, J.-P. (1989). The 43-kD polypeptide of heart gap junctions: Immunolocalization (I), topology (II), and functional domains (III). J. Cell Biol. 108, 2241-2254.

Yeager, M. (1993). Structure and design of cardiac gap-junction membrane channels. In: *Progress in Cell Research,* vol.3: *Gap Junctions* (Hall, J.E., Zampighi, G.A., & Davis, R.M., Eds.), pp. 47-55. Elsevier, Amsterdam.

Yeager, M. (1994). *In situ* two-dimensional crystallization of a polytopic membrane protein: the cardiac gap junction channel. Acta Cryst. D50, 632-638.

Yeager, M. (1995). Electron microscopic image analysis of cardiac gap junction membrane crystals. Microsc. Res. Tech. 31, 452-466.

Yeager, M. (1997). Structure of cardiac gap junction membrane channels: Progress toward a higher resolution model. In: *Discontinuous Conduction in the Heart.* (Spooner, P.M., Joyner, R.W., & Jalife, J, Eds.), pp. 161-184. Futura, Armonk, NY.

Yeager, M. (1998). Structure of cardiac gap junction intercellular channels. J. Struct. Biol. 121, 231-245.

Yeager, M., & Gilula, N.B. (1992). Membrane topology and quaternary structure of cardiac gap junction ion channels. J. Mol. Biol. 223, 929-948.

Yeager, M., & Nicholson, B.J. (1996). Structure of gap junction intercellular channels. Curr. Opin. Struct. Biol. 6, 183-192.

Yeager, M., Unger, V.M., & Falk, M.M. (1998). Synthesis, assembly and structure of gap junction intercellular channels. Curr. Opin. Struct. Biol. 8, 517-524.

Yeager, M., Unger, V.M., & Mitra, A.K. (1999). Three-dimensional structure of membrane proteins determined by two-dimensional crystallization, electron cryomicroscopy, and image analysis. Methods Enzymol. 294, 135-180.

Yeagle, P.L., Alderfer, J.L., & Albert, A.D. (1995a). Structure of the carboxy-terminal domain of bovine rhodopsin. Nat. Struct. Biol. 2, 832-834.

Yeagle, P.L., Alderfer, J.L., & Albert, A.D. (1995b). Structure of the third cytoplasmic loop of bovine rhodopsin. Biochemistry 34, 14621-14625.

Yee, A.G., & Revel, J.-P. (1978). Loss and reappearance of gap junctions in regenerating liver. J. Cell Biol. 78, 554-564.

Young, J.D., Cohn, Z.A., & Gilula, N.B. (1987). Functional assembly of gap junction conductance in lipid bilayers: demonstration that the major 27 kd protein forms the junctional channel. Cell 48, 733-743.

Yox, D., Rosinski, C., Xu, L.X., & Nicholson, B.J. (1993). Switching of charged amino acids in membrane spanning segments of connexin 26 reverses the polarity of voltage gating of their gap junctional channels. Mol. Biol. Cell. Suppl. 4, 328a.

Zampighi, G.A., Hall, J.E., Ehring, G.R., & Simon, S.A. (1989). The structural organization and protein composition of lens fiber junctions. J. Cell Biol. 108, 2255-2275.

Zampighi, G.A., Hall, J.E., & Kreman, M. (1985). Purified lens junctional protein forms channels in planar lipid films. Proc. Natl. Acad. Sci. USA 82, 8468-8472.

Zampighi, G., Simon, S.A., Robertson, J.D., McIntosh, T.J., & Costello, M.J. (1982). On the structural organization of isolated bovine lens fiber junctions. J. Cell Biol. 93, 175-189.

Zampighi, G., & Unwin, P.N.T. (1979). Two forms of isolated gap junctions. J. Mol. Biol. 135, 451-464.

Zelante, L., Gasparini, P., Estivill, X., Melchionda, S., D'Agruma, L., Govea, N., Milá, M., Monica, M.D., Lufti, J., Shohat, M., Mansfield, E., Delgrosso, K., Rappaport, E., Surrey, S., & Fortina, P. (1997). Connexin26 mutations associated with the most common form of non-syndromic neurosensory autosomal recessive deafness (DFNB1) in Mediterraneans. Hum. Mol. Genet. 6, 1605-1609.

Zervos, A.S., Hope, J., & Evans, W.H. (1985). Preparation of a gap junction fraction from uteri of pregnant rats: the 28-kD polypeptides of uterus, liver, and heart gap junctions are homologous. J. Cell Biol. 101, 1363-1370.

Zhang, J.-T., & Nicholson, B.J. (1989). Sequence and tissue distribution of a second protein of hepatic gap junctions, Cx26, as deduced from its cDNA. J. Cell Biol. 109, 3391-3401.

Zhang, J.-T., & Nicholson, B.J. (1994). The topological structure of connexin 26 and its distribution compared to connexin 32 in hepatic gap junctions. J. Membr. Biol. 139, 15-29.

Zhang, J.-T., Chen, M., Foote, C.I., & Nicholson, B.J. (1996). Membrane integration of in vitro-translated gap junctional proteins: co- and post-translational mechanisms. Mol. Biol. Cell 7, 471-482.

Zhou, L., Kasperek, E.M., & Nicholson, B.J. (1999). Dissection of the molecular basis of pp60[v-src] induced gating of connexin 43 gap junction channels. J. Cell Biol. 144, 1033-1045.

Zhou, X.-W., Pfahnl, A., Werner, R., Hudder, A., Llanes, A., Luebke, A., & Dahl, G. (1997). Identification of a pore lining segment in gap junction hemichannels. Biophys. J. 72, 1946-1953.

Zhu, H., Ciubotaru, M., & Nicholson, B.J. (1998). Two residues determine docking specificity of heterotypic interactions between connexins of different subfamilies. Mol. Biol. Cell 9, 94a.

Zimmer, D.B., Green, C.R., Evans, W.H., & Gilula, N.B. (1987). Topological analysis of the major protein in isolated intact rat liver gap junctions and gap junction-derived single membrane structures. J. Biol. Chem. 262, 7751-7763.

# POST-TRANSCRIPTIONAL EVENTS IN THE EXPRESSION OF GAP JUNCTIONS

Dale W. Laird and Juan C. Sáez

| | | |
|---|---|---|
| I. | Introduction . . . . . . . . . . . . . . . . . . . . . . . . . . . . . . . . . . . . . . . . . . . . . . . . . . | 100 |
| II. | Regulation of the Steady State Levels of Connexin Transcripts . . . . . . . . . . . . | 100 |
| III. | Connexin Trafficking and Gap Junction Assembly. . . . . . . . . . . . . . . . . . . . . . | 102 |
| | A. *In Vitro* Translation of Connexins . . . . . . . . . . . . . . . . . . . . . . . . . . . . . | 102 |
| | B. Connexins in the Golgi Apparatus. . . . . . . . . . . . . . . . . . . . . . . . . . . . . . | 102 |
| | C. Oligomerization of Connexins. . . . . . . . . . . . . . . . . . . . . . . . . . . . . . . . | 103 |
| | D. Intercellular Connexon Pairing and Plaque Formation . . . . . . . . . . . . . . . | 105 |
| | E. Role of Cell Adhesion Molecules . . . . . . . . . . . . . . . . . . . . . . . . . . . . . | 106 |
| IV. | Internalization and Degradation of Gap Junctions . . . . . . . . . . . . . . . . . . . . . | 107 |
| V. | Post-Translational Regulation . . . . . . . . . . . . . . . . . . . . . . . . . . . . . . . . . . . | 110 |
| | A. Phosphorylation of Connexins and Apparent Molecular Weight Shifts . . . | 110 |
| | B. Kinases Known to Phosphorylate Connexins, Consensus Sequences, and Phosphorylation Sites . . . . . . . . . . . . . . . . . . . . . . . . . | 111 |
| | C. Role of Phosphatases. . . . . . . . . . . . . . . . . . . . . . . . . . . . . . . . . . . . . . | 115 |
| | D. Functional Role of Phosphorylation . . . . . . . . . . . . . . . . . . . . . . . . . . . | 115 |
| | E. Other Modifications to Connexins . . . . . . . . . . . . . . . . . . . . . . . . . . . . . | 117 |
| VI. | Summary and Prospective for the Future . . . . . . . . . . . . . . . . . . . . . . . . . . . . | 118 |

**Advances in Molecular and Cell Biology, Volume 30, pages 99-128.**
**Copyright © 2000 by JAI Press Inc.**
**All rights of reproduction in any form reserved.**
**ISBN: 0-7623-0599-1**

# INTRODUCTION

Fundamental to understanding the biological role(s) that gap junctions play is a detailed molecular analysis of post-transcriptional events that regulate the constituent protein of gap junctions, the connexins. Before probes, such as antibodies and cDNAs, to specific connexins were developed it was clear that the abundance of gap junctions observed morphologically and the degree of gap junctional intercellular communication were correlated and under strict cell regulation. During the past decade partial molecular explanation for gap junction diversity has been obtained for only a few of the 15 different connexins that have thus far been cloned from rodents. This chapter presents the current understanding of different post-transcriptional events, including the stability of connexin transcripts, intracellular connexin trafficking, post-translational modifications, and gap junction formation and removal. From the compiled information it is tempting to propose that homologous connexin domains give rise to similar features in different gap junctions. However, this generalization does not explain that the combination of some connexins fails to lead to the formation of heterotypic gap junctions. On the other hand, there is evidence that the more diverse protein domains among connexins confer the needed structure to allow for differential regulation. Future details of the post-transcriptional events that occur to connexins are expected to enhance our understanding of the diverse regulatory mechanism that govern each gap junction type and its involvement in cell or tissue functions.

# II. REGULATION OF THE STEADY STATE LEVELS OF CONNEXIN TRANSCRIPTS

Although the abundance of mRNA is determined by its rate of transcription and its rate of degradation, in this chapter we comment only on post-transcriptional events. The regulation of transcription is discussed elsewhere.

Changes in the abundance of connexin mRNA have been observed in different cell types and tissues under the influence of different components of the extracellular matrix, hormones, or second messengers. In some of these studies, changes in the levels of connexin transcript are associated with changes in the rate of transcription of its mRNA, but in most of them it remains unknown at which level (i.e., changes in transcription rate, half-life of the transcripts, or half-life of the protein) the abundance of the junctional protein is controlled.

The disappearance of cell-cell communication and connexin32 between cultured adult rat hepatocytes is delayed by increased intracellular levels of cyclic adenosine monophosphate (cAMP) without changes in the rate of transcription of connexin32 mRNA (Sáez et al., 1989). This effect is associated with a delayed loss of connexin32 transcripts, suggesting that the half-life of the transcripts is increased. In the same system the levels of connexin32 transcripts are increased by either actinomycin D or cycloheximide, blockers of mRNA and protein synthesis, respectively. Therefore, the involvement of a factor, presumably a protein, that participates in the degradation of connexin32 was suggested (Sáez et al., 1989). This factor might be related to the labile *trans*-acting protein factors that are proposed to regulate the steady state levels of certain mRNAs (Morello et al., 1990; Klausner et al., 1993). Similarly, it has been reported that cycloheximide inhibited the loss of connexin26 transcripts while connexin32 mRNA was unaffected (Kren et al., 1993). Although the differences of the cycloheximide effects on the levels of connexin32 mRNA described earlier remain unknown, it might be related to changes in post-transcriptional regulation between *in vivo* and *in vitro* conditions.

During liver inflammation induced by endotoxin, degradation of connexin32 mRNA is prevented by actinomycin D but not by cycloheximide, suggesting that a factor responsible for connexin32 degradation is already present in an inactive form within hepatocytes and is activated upon stimulation by extracellular agents or the transcription of other genes, leading to the degradation of connexin32 (Gingalewski et al., 1996). Nonetheless, the sequence of connexin32 does not present conventional destabilizing elements. Therefore, further studies will be required to understand the molecular mechanism of connexin32 mRNA degradation under those conditions.

Studies of transcriptional rate and steady state levels of connexin26 and connexin32 mRNAs in periportal and pericentral hepatocytes also indicate that the abundance of hepatic connexins are controlled post-transcriptionally (Rosemberg et al., 1992). Differential regulation of gap junctions transcripts has also been detected during early stages of development of *Xenopus* (Gimlich et al., 1990) and mouse embryos (Nishi et al., 1991). Similarly, changes in levels of connexins mRNAs has been detected during organogenesis of mouse kidney (Nishi et al., 1991) and chick heart (Beyer, 1990). In all these examples, post-transcriptional regulation of connexin mRNAs may be relevant, although transcriptional changes are also possible. In other systems, such as a communication-deficient rat Morris hepatoma cell line, the expression of functional gap junctions induced by cAMP is the result of an increase in the rate of connexin43 transcription (Mehta et al., 1992). However, in a mouse tumor cell line (Atkinson et al., 1995) and in human malignant prostate cell lines (Mehta et al., 1996), cAMP increased gap junctional communication but did not change connexin43 mRNA levels. Nevertheless, it appears that the expression of connexins can be regulated either by changes in the transcription rate or the stability of the transcripts. Whether both mechanisms could simultaneously contribute to a change in the expression of connexins remains to be determined.

# III. CONNEXIN TRAFFICKING AND GAP JUNCTION ASSEMBLY

It is likely that the only functional site for connexins is at the plasma membrane. In the process of reaching the plasma membrane, connexins must pass through different cellular compartments and assemble in a highly ordered fashion. During connexin trafficking, connexins form intramolecular disulfide linkages and oligomerize into connexons or "hemichannels." Once connexons reach the cell surface they dock precisely with connexons from a neighboring cell to form gap junction channels. Gap junction channels, in turn, cluster to form well-defined plaques that may exceed 1 μm in diameter. Although details of these events are not well resolved, a basic understanding is beginning to unfold.

## A. *In Vitro* Translation of Connexins

Until recently, it was thought that all connexins insert cotranslationally into the endoplasmic reticulum. Consistent with this hypothesis, Chen and associates (1990) demonstrated that connexin26 and connexin32 are cotranslationally inserted into reticulocyte cell-free microsomes. In 1996, however, Zhang and colleagues published a report that showed, unlike most other class III membrane proteins, connexin26 also has the novel ability of being able to posttranslationally insert into microsomes with native topology (Zhang et al., 1996). In other *in vitro* studies of several connexins, Falk and co-workers (1994) found that connexins inserted into pancreatic endoplasmic reticulum microsomes had their cytoplasmically exposed amino terminals cleaved by signal peptidase. Similar cleavage sites were also identified in cells that overexpressed connexins, suggesting that an unidentified factor must be present under normal *in vivo* conditions to prevent this unusual processing (Falk et al., 1994). In later studies, Falk and colleagues used a cell free translation system to conclude that the N-terminal of connexins plays an important role in regulating heteromeric connexin assembly (Falk et al., 1997). Although several questions remain unanswered, the mechanistic details of connexin insertion into the endoplasmic reticulum will be elucidated more fully as the insertion of other connexin mutants are evaluated. To this end, Zhang and co-workers used this approach to show that the adenosine triphosphate (ATP)-dependent mechanism of connexin26 post-translational insertion into membranes was not related to the length of the C-terminal domain (Zhang et al., 1996). Whether connexin26 can post-translationally insert into other membranes in live cells remains to be tested.

## B. Connexins in the Golgi Apparatus

After de novo synthesis, connexins are believed to traffic from the endoplasmic reticulum to the Golgi apparatus. Distribution of connexins in several intracellular compartments has been observed by light microscopy in a variety of cell types (Musil et al., 1990a, 1990b; Zhu et al., 1991; Berthoud et al., 1992; De Sousa et al., 1993; Naus et al., 1993; Laird et al., 1993). Double-labeling for connexin43 and constituent proteins of the Golgi apparatus has been used to conclusively localize connexin43 to the Golgi apparatus of cardiac myocytes (Puranam et al., 1993), a mammary tumor cell line (Laird et al., 1995), and cells of the myometrium (Hendrix et al., 1992). Subcellular fractionation studies of the rat liver revealed that the largest amount of connexin32 is found in the lateral plasma membranes followed by the Golgi apparatus, sinusoidal plasma membranes, and lysosomes (Rahman et al., 1993). All of these studies represent compelling evidence that at least two connexins enter the Golgi apparatus. However, no electron microscopic localization of connexins to the Golgi apparatus exists to confirm these light microscopic and biochemical results.

It is unclear what, if any, post-transcriptional modifications occur to all connexins in the Golgi apparatus. There is no evidence that the glycosylating enzymes that are associated with Golgi membranes act on connexins (Rahman et al., 1993; Wang et al., 1995). In fact, connexins have no appropriate extracellular consensus sites for N-linked glycosylation. Another possibility is that one or more connexins are phosphorylated early in the secretory pathway. Crow and colleagues (1990) found that phosphorylation of connexin43 occurred within 15 minutes of de novo synthesis in vole fibroblasts, suggesting that the first stages of phosphorylation may occur within the endoplasmic reticulum, intermediate compartment or the Golgi apparatus. In support of this hypothesis, an alkaline phosphatase–sensitive isoform of connexin43 was identified in monensin-treated neonatal cardiac myocytes (Laird et al., 1993; Puranam et al., 1993) and in brefeldin A–treated rat mammary tumor cells (Laird et al., 1995), further suggesting that connexin43 is phosphorylated in early compartments of the secretory pathway at least in some cell types. Although the nonphosphorylated form of connexin43 has been detected at the plasma membrane (Musil and Goodenough, 1991), it is unclear whether it is the dephosphorylation product of a previously phosphorylated form of connexin43 or a nonphosphorylated form newly inserted into the plasma membrane. Protein phosphorylation is not an obligatory feature of all connexins since at least connexin26 appears not to be phosphorylated (Traub et al., 1989; Sáez et al., 1990). Thus, proposed phosphorylation of connexins early in the secretory pathway is likely not an absolute requirement for connexin trafficking to the plasma membrane. Whether phosphorylation of connexins affects the intracellular kinetic transport of connexins to the plasma membrane has not been reported.

## C. Oligomerization of Connexins

The oligomerization of connexins into gap junction hemichannels or to what has commonly been referred to as connexons may in theory occur in the (1) endoplasmic reticulum, (2) intermediate compartment, (3) Golgi apparatus, or (4) plasma membrane. As a first approximation, the fact that the assembly of T-cell receptor (Lippincott-Schwartz et al., 1989), mannose 6-phosphate receptor (Hille et al., 1990), and influenza hemagglutinin (Broakman et al., 1991) occur in the endoplasmic reticulum suggests that this may also be the location for connexin oligomerization. Musil and Goodenough (1993) addressed this question *in vitro* by taking a biochemical approach. Using inhibitors of protein transport (i.e., Brefeldin A, carbonyl cyanide m-chlorophenylhydrazone, low temperatures), chemical crosslinking, and sedimentation velocity centrifugation of detergent-solubilized material, they established that the multisubunit assembly of connexin43 occurs after leaving the endoplasmic reticulum, most likely in the trans-Golgi network. The oligomerization of connexins into connexons late in the secretory pathway may prevent unintentional pairing of connexons and gap junction formation within the endoplasmic reticulum. Moreover, these intracellular hemichannels must be "closed" in order to prevent leakage of cytosolic factors into the lumen of the trans-Golgi network. What prevents connexin oligomerization in the endoplasmic reticulum is not well understood. It is possible that connexin interaction with chaperone(s) obstruct connexon assembly in the endoplasm reticulum; however, no connexin chaperone has been identified to date. It also appears likely that, under defined conditions, connexins can oligomerize in the endoplasmic reticulum or endoplasmic reticulum–derived membranes. Kumar and associates (1995) observed connexin32 connexons within the endoplasmic reticulum of baby hamster kidney cells that express high levels of connexin32. More recently, Falk and co-workers (1997) established that connexin43, connexin32 and connexin26 could selectively oligomerize into apparent hexamers in endoplasmic reticulum–derived membrane vesicles. These authors went on to show that when connexin32 or connexin26 were reconstituted into lipid bilayers single channel conductances were obtained similar to those found in *in vivo* systems.

Consistent with molecular models of gap junctions proposed by Makowski and colleagues (1977), further detailed structural analysis of rat liver gap junction channels revealed that most liver connexins are arranged in hexamers (Cascio et al., 1995). Furthermore, it became apparent that connexins can oligomerize into both homomeric and heteromeric connexons (Sosinsky 1995; Stauffer 1995). The existence of heteromeric connexons was later supported by Jiang and Goodenough (1996) when they observed that connexin46 and connexin50 formed heteromeric hemichannels in the lens. The mere ability of cells to make heteromeric connexons vastly increases the combination of connexons that may form and further complicates the task of identifying the role of each gap junction channel.

Another means for tissues to establish selective communication with neighboring cells is to sort connexins of one type to one intercellular surface and a second connexin to a distinct intercellular domain. At present this concept has not been well established. However, intracellular or plasma membrane sorting of at least two different connexins has been documented. Guerrier and associates (1995) examined polarized thyroid epithelial cells that express both connexin43 and connexin32 and found that these connexins were localized in gap junction plaques at distinct membrane domains. Furthermore, the observation by Simon and co-workers (1997) that connexin43 was found between adjacent granulosa cells whereas connexin37 was the predominant connexin at the oocyte-granulosa junction would support the possibility that connexin sorting may occur within granulosa cells. It will be interesting to determine if connexin sorting mechanisms are utilized to establish selective gap junction communication in other heterotypic cell populations.

### D. Intercellular Connexon Pairing and Plaque Formation

Once intracellular oligomerization of connexins has occurred, they are thought to undergo vesicular transport to the cell surface. Ultrastructural evidence shows that small particle aggregates are found within particle-poor, flattened regions of the membrane as junctions are beginning to form between reaggregating cells. These aggregates have been referred to as formation plaques (Johnson et al., 1974), formation zones (Porvaznik et al., 1979), or formation bands (Montesano, 1980). However, formation plaques were not found in reaggregated blastomeres of Fundulus (Ne'eman et al., 1980) or in hepatocytes during rat liver regeneration (Yee and Revel, 1978; Yancey et al., 1979), both known to be active in gap junction formation.

It is still unresolved whether gap junction plaques increase in size by using a common mechanism. Gap junction growth might occur by coalescence of two or more small particle aggregates, recruitment of additional intramembrane particles to small particle aggregates, or fusion of the small particle aggregates to larger junctional plaques (Johnson et al., 1974; Larsen and Risinger, 1985). Alternatively, they could all be steps of a single sequence of events. Using an *in vitro* system, Kistler and colleagues (1993) showed that the lens fiber gap junction assembly can occur as a two-step process in which molecular interactions between adjacent pore structures are particularly important for gap junctions to grow beyond 10 pore complexes. The significance of these results *in vivo* remains to be demonstrated.

In several cultured cell systems, the existence of regulatable hemichannels in nonjunctional membranes has been reported (Beyer and Steinberg, 1991; Paul et al., 1991; DeVries and Schwartz, 1992; Ebihara et al., 1995; Li et al., 1996). Beyer and Steinberg (1991) demonstrated that J774 macrophages can, in response to ATP, open pores that allow fura-2 to enter the cell, suggesting that these cells have hemichannels. A thorough demonstration that hemichannels can exist on the surface of several mammalian cell types was provided by Li and associates (1996). In these studies, the authors were able to show that connexin hemichannels could be opened by depleting the extracellular environment of calcium, allowing for the uptake of fluorescent dyes. However, it is important to note that cell surface hemichannels would be expected to remain "closed" until appropriate pairing with a hemichannel from a neighboring cell, otherwise they would be incompatible with the maintenance of a membrane potential and cell viability.

The noncovalent forces that occur when one connexon (hemichannel) aligns precisely with a connexon from a neighboring cell (Ghoshroy et al., 1995) results in a sealed gap junction channel. The amino acid residues involved in generating this intercellular connexon interaction is beginning to be understood. Dahl and co-workers observed that synthetic peptides that represent either the first or second extracellular loop region of connexin32 were capable of reducing intercellular coupling in paired oocytes. Site-directed mutagenesis of any of the six cysteines contained in the extracellular loops to serines prevented gap junction channel formation (Dahl et al., 1991, 1992) whereas mutation of amino acid residues of the extracellular loops that differ amongst the connexins, altered, but did not prevent the formation of functional channels (Werner et al., 1993). An extensive mutagenesis approach demonstrated positions of intramolecular disulfides and is consistent with β-sheet structure in these extracellular domains (Foote et al., 1998). These studies suggest that intramolecular disulfide bonds are likely involved in common and basic features of recognition, whereas extracellular loop amino acid residues that are not conserved amongst the connexins may play a part in determining the affinity of hemichannel interaction. Haubrich and colleagues (1996) demonstrate that murine connexin43 hemichannels expressed in HeLa cells were unable to dock and form functional heterotypic channels with murine connexin40, and this selectivity was conveyed not only by the extracellular loops of the connexins but also by the cytoplasmic loop. A conceivable molecular mechanism that may also define the specificity of interaction between connexins is that the less homologous sequence domains between connexins affect the spatial conformation of the highly homologous extracellular loops.

Assembly of gap junctions can be inhibited or enhanced by number of cell treatments. Lampe (1994) showed in Novikoff hepatoma cells that the phorbol ester TPA, which activates protein kinase C, dramatically inhibited gap junction assembly. Conversely, the second messenger cyclic AMP (cAMP) increased the assembly of connexin43 into gap junctions and increased intercellular communication in a mouse mammary tumor cell line without increasing connexin43 mRNA or protein levels (Atkinson et al., 1995). Wang and Rose (1995) determined that cAMP-induced clustering of gap junction channels in rat Morris hepatoma H5123 cells transfected with connexin43 was enhanced by inhibition of glycosylation and dependent on intact microfilaments. Earlier Meyer and associates (1991) showed that Low Density Lipoprotein (LDL) treatment of Novikoff cells resulted in increased gap junction assembly and dye transfer in the absence of an increase in total connexin43 protein. In all of these studies, intercellular communication is regulated by the amount of gap junction assembly that is taking place.

## E.   Role of Cell Adhesion Molecules

The involvement of cell adhesion molecules in gap junction formation has received support from different perspectives. Several years ago, antibodies against calcium-dependent or calcium-independent cell adhesion molecules were shown to be capable of disrupting intercellular coupling between epithelial cells (Imhof et al., 1983) and neuroblasts (Keane et al., 1988), respectively. In later studies, Musil and colleagues (1990b) showed that when communication-deficient S180 mouse sarcoma cells were transfected with the cDNA for L-CAM (E-cadherin), gap junction plaque formation was greatly enhanced, suggesting that calcium-dependent cell adhesion is a necessary factor in plaque formation. This was supported by Jongen and co-workers. (1991), who showed that intercellular communication in mouse epidermal cells was regulated by E-cadherin. The mechanism of E-cadherin–mediated gap junction formation in early mouse embryos does not require phosphorylation of E-cadherin itself (Aghion et al., 1994). Meyer and associates (1992) demonstrated that the formation of connexin43-containing gap junction channels in Novikoff cells can be inhibited by Fab' antibodies to A-CAM (N-cadherin). Consistent with these results N-cadherin/catenin complexes were required in advance of connexin43 gap junction formation in rat cardiomyocytes (Hertig et al., 1996) and for gap junctions to form in regenerating hepatocytes (Fujimoto et al., 1997). Thus, it appears that calcium-dependent cell-cell adhesion molecules, in particular, play a direct role in gap junction plaque formation in most cell types studied although expression of cadherins in mouse L cells has been reported to decrease gap junction intercellular communication (Wang and Rose, 1997). A more thorough examination of the role calcium-independent cell adhesion molecules play in gap junction formation or retention has yet to be completed.

One explanation for the role of cell adhesion molecules is that their cell-cell interaction may act to position apposing membranes for gap junction channel formation. The large extracellular domains of cadherins would likely prevent their actual inclusion in the gap junction plaque; however, cadherins that are positioned near the site of a forming gap junction may be critical. Whether gap junction channels or plaques, or both, can form in the absence of cell adhesion molecules remain unclear, and whether other yet undefined molecules play a part in docking hemichannels is also unresolved. Biochemical and morphological analysis of purified preparations of both liver (Hertzberg, 1984) and heart (Manjunath et al., 1984) gap junctions, suggest that no other major tightly associated proteins are isolated with gap junction plaques. However, if such proteins do exist, they may have been removed by the harsh treatments used during the isolation of the gap junction plaques.

## IV.  INTERNALIZATION AND DEGRADATION OF GAP JUNCTIONS

Gap junction proteins undergo a continual renewal process. Connexin43 turnover in cardiac myocytes, normal rat kidney (NRK) cells, primary cultures of chick lens epithelial cells, and communication-incompetent cells is similar with half-lives of 1.0 to 2.5 hours (Musil et al.,1990a, 1990b; Laird et al., 1991). Likewise, connexin32 and connexin26 have half-lives of 1.3 to 3.0 hours in cultured embryonic mouse hepatocytes (Traub et al., 1987, 1989). Moreover, the pulse chase data obtained to date generally fits a monophasic turnover for connexins. *In vivo* turnover studies were done in the early 1980s, wherein detergent-resistant liver membranes (plaques) were found to have a half-life of 5 hours in mouse (Fallon and Goodenough, 1981) and 19 hours in rat (Yancey et al., 1981). The latter value of gap junction turnover is likely to be an overestimate owing to radioisotope reutilization. Also, it is now clear that both of these groups were monitoring the turnover of connexin32 and connexin26. The surprising conclusion from these studies is that connexins appear to turn over faster than the average membrane protein (Hare, 1990) and connexin43, for instance, does not obey the N-end rule (Bachmair et al., 1986) which predicts that it will have a half-life of more than 20 hours. The dynamic nature of connexins suggests that gap junction formation and removal may play a part in regulating cell-cell communication in response to physiological stimuli.

The mechanism of gap junction internalization and degradation remains poorly understood. It has been frequently suggested that gap junctions are internalized by endocytosis of either entire or fractions of a gap junctional plaque, generating a cytoplasmic vesiclelike structures termed "annular" gap junctions. These structures, first described by Bjorkman (1962), are double-membrane structures that contain the septilaminar appearance of intact gap junctions. Annular profiles of gap junctions have been identified by many investigators within the cytoplasm of several cell types (Bjorkman, 1962; Ginzberg and Gilula, 1979; Yancey et al., 1979; Larsen 1983; Mazet et al., 1985; Pelletier, 1988; Risley et al., 1992; Naus et al., 1993). Immunogold labeling of these annular profiles with anticonnexin antibodies confirmed that they contain the structural subunit of gap junctions (Dermietzel et al., 1991; Risley et al., 1992; Naus et al., 1993).

In order to determine if annular gap junctions were formed from the internalization of cell surface gap junctions, we followed the fate of immunolabeled gap junctions in live NRK cells by fluorescent confocal microscopy (Laird and Chodock, unpublished data). Gap junctions that were tagged with microinjected anticonnexin43 antibodies were observed to be removed into the cytoplasm of a contacting cell. These results provide compelling evidence that connexin43 from one cell was internalized into the cytoplasm of an apposing cell. Although microinjected cells were not examined at the electron microscope level, it is feasible that the antibody-tagged gap junctions are being internalized as classical annular junctions.

An early report described small patches of clathrinlike bristles associated with invaginations of the junctional membrane in granulosa cells (Larsen et al., 1979), suggesting that gap junctions could be internalized in endosomelike vesicles. In some cases these regions appear connected to the surface membrane by uncoated extensions. Once the gap junctional plaque vesicle is in the cytoplasm it has been reported to be surrounded by a halo of 4-7 nm filaments believed to be actin (Larsen et al., 1979).

The degradation of gap junctions has recently become a subject of discussion. Several early reports have identified gap junctions, fragments of gap junctions, or connexins in electron dense compartments, lysosomes, or phagolysosomes in both normal and tumor cells (Larsen and Tung, 1978; Ginzberg and Gilula, 1979; Murray et al., 1981; Larsen and Risinger, 1985; Naus et al., 1993). Consistent with these findings, gap junction fragments were identified by anticonnexin43 immunogold labeling techniques in electron dense compartments (i.e., lysosomes) of rat mammary tumor and iris epithelial cells (Chodock, Igdoura, Hand, Morales, and Laird, unpublished data). Moreover, mammary tumor cells accumulated large numbers of connexin43-positive structures when lysosomal enzymes were inhibited with leupeptin or ammonium chloride (Chodock and Laird, unpublished data). Nevertheless, it is difficult to conclude from these qualitative studies that the majority of gap junctions are destined for lysosomal degradation. In the teleost fish retina few annular gap junctions were found to contain acid phosphatase, making it unclear whether fusion of annular gap junctions with lysosomes is a subsequent step or whether gap junctions are subjected to a different degradative pathway (Vaughan and Lasater, 1992).

An alternate pathway for gap junction degradation is through the action of proteasomes. Laing and Beyer (1995) showed that connexin43 is accumulated and its half-life is prolonged in BWEM cells treated with the proteasomal inhibitor N-acetyl-L-leucyl-L-leucinyl-norleucinal. These authors further demonstrate that the degradation of connexin43 is impaired in E36 Chinese hamster ovary cells that have a thermolabile ubiquitin-activating enzyme, E1, suggesting that connexin43 is subject to ubiquitin-mediated degradation. These results represent one of the first examples of proteasomal degradation of a class III integral membrane protein. Nonlysosomal pathways for connexin degradation had been suggested in earlier *in vitro* work by Elvira and co-workers (1993), in which the calcium-dependent neutral proteases, calpains, were found to be capable of cleaving connexin32 but not connexin26. Calpain proteolysis of connexin32 was prevented when this connexin was phosphorylated by protein kinase C (Elvira et al., 1993). The possibility that both proteosomal and lysosomal pathways act in concert, perhaps in a sequential manner, is suggested by a study using inhibitors specific to each pathway, coupled with localization of connexin43 by immunohistochemistry (Laing et al. 1997).

Together these results suggest that there may be multiple pathways for connexin degradation. It is difficult to deduce how connexins fit within the catalytic pocket of a proteasome for degradation and how the proteasome degrades the transmembrane and extracellular domains of a connexin. However, it is clear that we do not fully understand how proteasomes function. It is possible that proteasomes act in tandem with lysosomes by partially degrading connexins. Finally, it will be interesting to determine if different pathways for connexin degradation exist in various cell types, perhaps to accommodate a broader range of gap junction turnover than originally expected.

In addition to the above mechanisms of gap junction degradation, studies in metamorphosing insects suggest that gap junctions may uncouple and disperse through the membrane as opposed to being internalized and degraded (Lane and Swales, 1980). Moreover, Zampighi and co-workers (1989) found that junctional particles in plaques disperse upon acidification of crayfish lateral axons. Immunofluorescent and immunogold-electron microscopy studies on regenerating hepatocytes were used to show that gap junctions may disperse into smaller aggregates prior to removal from the cell surface (Fujimoto et al., 1997). These authors also propose that gap junction precursors may remain on the cell surface. Thus, the rapid establishment of electrical coupling and appearance of gap junctions after dispersion and reaggregation of some cells might also be explained through reutilization of gap junction channels at the level of the plasma membrane (Loewenstein, 1967; Ito and Loewenstein, 1969; Johnson et al., 1974; Zampighi et al., 1989). Alternatively, a pool of gap junctions channels might be available in the cytoplasm for rapid insertion and clustering, perhaps associated with microfilaments as observed in Morris hepatoma H5123 cells (Wang and Rose, 1995). Conversely, drugs that disrupted the cytoskeletal system have been reported to cause rapid gap junction formation (Tadvalkar and Pinto da Silva, 1983). The existence of a pool of gap junction channels either in the plasma membrane or in association with other cell compartments may also explain the formation of gap junctions in the absence of protein synthesis (Katcher and Reese, 1985; Epstein et al., 1977; Ne'eman et al., 1980; Laird et al., 1995).

## V.  POST-TRANSLATIONAL REGULATION

### A.  Phosphorylation of Connexins and Apparent Molecular Weight Shifts

Shortly after the techniques for isolating connexin32 had been established, Sáez and associates (1986, 1990), Traub and colleagues (1987, 1989), and Takeda and co-workers (1987, 1989) reported that this connexin was phosphorylated in primary cultures of rodent hepatocytes. As the list of cloned and sequenced connexins grew over the subsequent years, it was speculated that all connexins would be phosphoproteins. However, this hypothesis was rebutted when attempts to phosphorylate the second connexin identified from hepatocytes, connexin26, were unsuccessful (Traub et al., 1989; Sáez et al., 1990). In the early 1990s, connexin43 was shown to be a phosphoprotein in vole fibroblasts (Crow et al., 1990), embryonic chick lens (Musil et al., 1990a), adult (Laird and Revel, 1990) and neonatal (Laird et al., 1991) cardiac myocytes, and several cell lines (Hertzberg et al., 1989; Musil et al., 1990a, 1990b; Brissette et al., 1991; Musil and Goodenough, 1991; Oh et al., 1991; Berthoud et al., 1992; Reynhout et al.,

1992; Berthoud et al., 1993). In all of these studies, the phosphorylation of connexin43 was accompanied by apparent molecular weight shifts on SDS-poly-acrylamide gels, a phenomena not apparent when connexin32 (Sáez et al., 1986; Traub et al., 1987; Takeda et al., 1989), connexin46 (Jiang et al., 1993), or connexin45 (Laing et al., 1994) are phosphorylated.

Alkaline phosphatase digestion of isolated cardiac gap junctions increased the mobility of the broad spectrum of bands representing connexin43, implying that the heterogeneity in cardiac connexin43 is indeed a result of phosphorylation (Laird and Revel, 1990). More recently, alkaline phosphatase treatment of connexin44 and connexin56 reduced their apparent molecular weights strongly suggesting that they also exist as phosphorylated proteins (Gupta et al., 1994; Berthoud et al., 1997). In both adult rat heart (Laird and Revel, 1990) and brain tissue (Hossain et al., 1994), the phosphorylated species of connexin43 are more predominant than the nonphosphorylated species of the protein. Furthermore, no change in the mobility of the connexin43 fragments that lack the carboxy terminal is seen, suggesting that the carboxy terminal of connexin43 contains the majority of the phosphorylated residues (Laird and Revel, 1990) but the possibility that other sites exist in the N-terminal or intracellular loop of the protein has not been completely ruled out. Connexin56, on the other hand, has been shown to be phosphorylated both in the intracellular loop connecting the second and third transmembrane segments and the carboxy terminal (Berthoud et al., 1997).

In cultured cells, immunoprecipitation analysis shows that connexin43 is detected predominantly as two phosphorylated species (connexin43-$P_1$ and connexin43-$P_2$) of slower electrophoretic mobility than the unphosphorylated species (connexin43-NP) (Crow et al., 1990; Musil et al., 1990a, 1990b; Laird et al., 1991; Lau et al., 1991). Alkaline phosphatase digestion of $^{35}$S-Met–labeled connexin43 removes the phosphate residues from the slow migrating species of connexin43 with a quantitative recovery of connexin43-NP (Musil et al., 1990a; Laird et al., 1991). Phosphorylated species of connexin43 that appear to migrate slower than the $P_2$ form have been reported (Crow et al., 1992; Lau et al., 1992; Moreno et al., 1994), as well as species that migrate below the $P_1$ form (Crow et al., 1990; Budunova et al., 1993; Puranam et al., 1993; Laird et al., 1995). A species of connexin43 that migrates slower than connexin43-$P_2$ was identified in mitotic human vascular cells (Xie et al., 1997; Lampe et al., 1998). The molecular basis of the phosphorylation-induced shifts in apparent molecular weights of connexin43 remains unknown. It is conceivable that conformational changes induced by phosphorylation of different connexin43 amino acid residues alters the binding of SDS.

## B. Kinases Known to Phosphorylate Connexins, Consensus Sequences, and Phosphorylation Sites

Connexin32 has been demonstrated to be phosphorylated by cAMP-dependent protein kinase at the consensus sequence $^{230}$Lys-Arg-Gly-$^{233}$Ser (Sáez et al.,

1990) located in the C-terminal of connexin32. Protein kinase C has also been demonstrated to phosphorylate connexin32 in intact cells and isolated gap junctions (Takeda et al., 1989) at serine 233 (Sáez et al., 1990). Two-dimensional tryptic maps of protein kinase C phosphorylated connexin32 in isolated junctions or *in vivo* after stimulating protein kinase C show several phosphopeptides besides that containing the serine 233 (Sáez et al., 1990). Consistently, connexin32 phosphorylated by cAMP-dependent protein kinase can be further phosphorylated by protein kinase C, but this does not occur if during phosphorylation the order of the protein kinases is reversed (Takeda et al., 1989). Thus, other sites of phosphorylation of connexin32 by protein kinase C remain to be identified. In rat hepatocytes, norepinephrine, an indirect activator of protein kinase C, stimulated phosphorylation of connexin32 and staurosporine, a potent inhibitor of protein kinase C, inhibited this stimulatory effect (Takeda et al., 1989). Analysis of protease-digested and cyanogen bromide cleaved fragments of phosphorylated connexin32 showed that the amino acid residues phosphorylated by protein kinase C are located in the C-terminal region of the protein (Takeda et al., 1989). Although not a good substrate, connexin32 in isolated gap junctions from liver can also be phosphorylated by $Ca^{2+}$/calmodulin-dependent protein kinase II (Sáez et al., 1990) or by epidermal growth factor receptor tyrosine kinase (Diez et al., 1995) but is not phosphorylated by tyrosine kinase $pp60^{v-src}$ (Nairn, A.C. and Sáez, J.C., unpublished observation).

The polypeptide product of mouse connexin50 (White et al., 1992), MP70, was the second gap junctional protein shown to be phosphorylated *in vitro* by cAMP-dependent protein kinase (Voorter and Kistler, 1989). This protein is abundant in gap junctions isolated from young lens fiber cells and presents a consensus sequence for phosphorylation in the C-terminal as the proteolytic product of MP70, a 38-kD protein, found in older fibers (lens nucleus) lacks the C-terminal region and is not phosphorylated by cAMP-dependent protein kinase (Voorter and Kistler, 1989). Connexin50 is also phosphorylated by a cAMP-independent, $Ca^{2+}$-independent but $Mg^{2+}$-dependent endogenous lens fiber protein kinase (Arneson et al., 1995).

The kinase(s) or sites involved in the phosphorylation of the chick heart connexin45 have not been determined. However, it is interesting that 12-O-tetradecanoylphorbol-13-acetate (TPA) activation of protein kinase C was found to eliminate the phosphorylation of connexin45 in marked contrast to its effect on connexin43 (Laing et al., 1994). The differential effect of TPA on these two connexins is not understood.

The most well-studied phosphorylated connexin is connexin43. Preliminary studies of cyanogen bromide fragments of connexin43 suggest that there are multiple phosphorylated serines and possibly a phosphorylated threonine (Lau et al., 1991; Goldberg and Lau, 1993). However, unlike connexin32 and connexin50, a consensus sequence does not exist on connexin43 for cAMP-dependent protein kinase phosphorylation. In Madin-Darby canine kidney (MDCK) and clone 9 cells, the electrophoretic mobility of connexin43 is not affected by membrane permeant forms of cAMP or cyclic guanosine monophosphate (cGMP), suggesting that protein kinases activated by these second messengers may not be phosphorylating connexin43 (Berthoud et al., 1992, 1993). In addition, in a gap junction–deficient cell line transfected with human connexin43, cGMP does not affect the incorporation of $^{32}P_i$ or its immunoblot pattern (Takens-Kwak et al., 1995). Nevertheless, in the same cell line transfected with rat connexin43, an increase in cGMP leads to changes in the ratio of $^{32}P_i$ incorporation into connexin43 although it does not affect the amount of each isoform detected by immunoblotting (Takens-Kwak et al., 1995). This differential effect of phosphorylation on connexin43 from different species may be due to their difference in the amino acid residue 257, which in the rat form is a serine (Beyer et al., 1987) but in human form is an alanine (Fishman et al., 1990).

Increased connexin43 phosphorylation as a result of treatment with TPA has been observed in WB cells, keratinocytes, MDCK cells, lens epithelial cells, neonatal cardiac myocytes, clone 9 cells, BWEM cells, and Novikoff cells (Brissette et al., 1991; Oh et al., 1991; Berthoud et al., 1992; Reynhout et al., 1992; Sáez et al., 1993; Berthoud et al., 1993; Budunova et al., 1993; Laing et al., 1994; Lampe, 1994; Matesic et al., 1994). In most of these cell types, the unphosphorylated isoform of connexin43 is predominant and activation of protein kinase C induces phosphorylation of connexin43. Similar to several other connexins (Hoh et al., 1991), consensus sites for protein kinase C exist on the carboxy-tail of connexin43. Two dimensional tryptic phosphopeptides maps of different phosphorylated isoforms revealed that TPA induced phosphorylation of the unphosphorylated and phosphorylated connexin43 forms on serine residues while connexin43 lacked several other phosphopeptides seen in control cells (Berthoud et al., 1993). Currently, the exact sites of protein kinase C phosphorylation of connexin43 are being mapped. As part of this process, synthetic peptides representing primary sequences of connexin43 have been used as tools to determine potential protein kinase C phosphorylation sites. A synthetic peptide with sequence corresponding to residues 360 to 375, located in the carboxy terminal of connexin43 (Beyer et a., 1987; Yancey et al., 1989), is phosphorylated by protein kinase C on serine residues 368 and 372 (Sáez et al., 1993). Whether these sites are phosphorylated by protein kinase C *in vivo* remains to be resolved.

Early work by Maldonado and associates (1988) showed that growth factors were capable of modulating junctional cell-cell communication. This observation was extended by Lau and co-workers who found that epidermal growth factor (EGF) not only disrupted gap junctional communication, but that this correlated with enhanced serine phosphorylation and the appearance of a novel phosphorylated form of connexin43 (Lau et al., 1992). Furthermore, the EGF stimulation of connexin43 phosphorylation was found to be independent of TPA-sensitive protein kinase C (Kanemitsu and Lau, 1993). Phosphorylation of serine residues different from those phosphorylated under resting conditions might explain the

EGF-induced effect on cell-cell communication. Lau and colleagues used site-directed mutagenesis and phosphotryptic peptide analysis to identify the mitogen-activated protein (MAP) kinase phosphorylation sites on connexin43 to be the consensus sites at $Ser^{255}$, $Ser^{279}$, and $Ser^{282}$ (Warn-Cramer et al., 1996). Subsequent studies extended this observation to demonstrate that EGF is acting to stimulate MAP kinase, with the phosphorylation at $Ser^{279}$ and $Ser^{282}$ sufficient to inhibit gap junctional communication (Warn-Cramer et al., 1998). Since virtually all mitogens lead to activation of serine/threonine protein kinases or extracellular signal-regulated kinases, it will be of interest to study the possible participation of mitogen-activated protein (MAP) kinases as mediators of growth factor effects on connexin43 containing gap junctions.

In addition to containing consensus sites for protein kinase C and MAP kinase phosphorylation, the carboxy terminal of rat connexin43 emerging from the lipid bilayer into the cytoplasm has a consensus sequence for $p34^{cdc2}$ kinase (S/T-P-X-R/K) (Kennelly and Krebs, 1991), which is involved in cell cycle control. As predicted, a synthetic peptide containing this sequence (amino acids 254 to 266 of rat connexin43) is phosphorylated by purified $p34^{cdc2}$ kinase (Sáez et al., 1997). A similar approach demonstrated $p34^{cdc}$/cyclin B phosphorylation of another synthetic peptide (amino acids 241 to 264) and demonstrated that the sites phosphorylated *in vitro* were also phosphorylated *in vivo* (Kanemitsu et al., 1998). Consistent with this finding, the GST-connexin43 fusion protein was found to be a substrate for $p34^{cdc2}$ protein kinase (Lampe et al., 1996). Lampe and associates (1998) subsequently demonstrated that, while phosphorylation at $Ser^{255}$ occurs *in vitro*, the major phosphopeptide associated with the formation of the more slowly migrating phosphorylated form of connexin43 found in mitotic cells (Xie et al., 1997) is not phosphorylated directly by $p34^{cdc}$/cyclin B, suggesting that another kinase phosphorylating connexin43 may be acting downstream of the $p34^{cdc}$/cyclin B system.

Finally, a *Xenopus* oocyte expression system was used to show that the protein tyrosine kinase encoded by the viral oncogene v-src, $pp60^{v\text{-}src}$, induced tyrosine phosphorylation of connexin43 (Swenson et al., 1990). Swenson and co-workers went on to use site-directed mutagenesis to show that connexin43 was phosphorylated on tyrosine[265]. Consistent with their observations, $pp60^{v\text{-}src}$ induced tyrosine phosphorylation of connexin43 in fibroblasts (Crow et al., 1990; Goldberg and Lau, 1993) and in cells that were temperature sensitive for transformation (Crow et al., 1992). Later, Loo and associates (1995) used a recombinant baculovirus expressing connexin43 in SF-9 cells to show that $pp60^{v\text{-}src}$ directly phosphorylates connexin43 (Loo et al., 1995), and demonstrated a requirement for v-src SH2 and SH3 domains for the binding of a proline-rich region of connexin43 and for interaction with phosphorylated $tyr^{265}$ (Kanemitsu et al., 1997). More recent studies revealed that v-src–mediated gating did not require tyrosine phosphorylation but required SH3 binding domains and map kinase phosphorylation (Zhou et al., 1999). $pp60^{v\text{-}src}$, on the other hand, did not induce tyrosine phosphorylation of connexin32 (Swenson et al., 1990).

## C.  Role of Phosphatases

It is unclear what role phosphatases play in regulating the state of connexin phosphorylation during its life cycle. Two-dimensional tryptic maps suggest that the two major phosphorylated forms of connexin43 are differentially phosphorylated (Sáez, J.C. and Nairn, A.C., unpublished observation). In pulse-chase labeling experiments, the turnover rate of phosphate groups on combined connexin43-$P_1$ and connexin43-$P_2$ is not markedly faster then the half-life of the protein (Hertzberg et al., 1989; Laird et al., 1991). A slow turnover of phosphate groups is supported by the observation that several phosphatase inhibitors (except pervanadate) were ineffective in altering the phosphorylated state of connexin43 in fibroblasts (Husoy et al., 1993). Nevertheless, treatment for only 30 minutes with okadaic acid, an inhibitor of phosphoprotein phosphatases types 1 and 2A resulted in an increase in connexin43 phosphorylation (Sáez et al., 1993) and increased junctional conductance (Spray, D.C. and Sáez J.C., unpublished observation). Similarly when cells were treated with staurosporine, a protein kinase C and cyclic nucleotide–dependent protein kinase inhibitor, the state of connexin43-$P_2$ phosphorylation and intercellular communication were reduced (Sáez et al., 1993). In neonatal rat cardiac myocytes, TPA induced the recovery of cell coupling and the phosphorylation of connexin43 after staurosporine treatment. Consistent with the notion that differential phosphorylation of connexin43 leads to opposite functional consequences, the TPA-induced phosphorylation of connexin43 has been shown to occur at essentially the same sites detected under resting conditions (Sáez et al., 1997).

## D.  Functional Role of Phosphorylation

To date, all connexins examined in detail, except connexin26, have been shown to be phosphoproteins. Although connexin32 was the first gap junction protein shown to be phosphorylated, the functional role of this covalent modification remains unclear. It has been demonstrated that the phosphorylation of connexin32 can result in enhanced or decreased cell-cell communication. Agents that activate the cAMP-dependent protein kinase or protein kinase C increase the state of phosphorylation of connexin32 in adult hepatocytes within 15 minutes after treatment (Sáez et al., 1986; Takeda et al., 1987). The effect of the cAMP-dependent pathway has been found to correlate temporally with an increase in cell-cell communication in adult rat hepatocytes (Sáez et al., 1986), whereas other studies show that activation of the protein kinase C–dependent pathway induces inhibition of cell-cell communication between pancreatic acinar cells (Somogyi et al., 1989) but fails to reduce dye coupling in rat hepatocytes (Sáez et al., 1990). It has been reported that phosphorylation of connexin32 by protein kinase C prevents its proteolysis by two intracellular $Ca^{2+}$-dependent protease isoenzymes, calpains, which may be linked in some way to connexin turnover as discussed earlier (Elvira et al., 1993).

Frequently, enhanced connexin43 phosphorylation on serine/threonine residues correlates with the reduction in gap junction coupling (Brissette et al., 1991; Oh et al., 1991; Berthoud et al., 1992; Lau, et al., 1992; Reynhout et al., 1992; Berthoud et al., 1993). In the past, a close correlation between TPA-induced cell uncoupling and differential phosphorylation of connexin43 on serine residues was found in clone 9 cells (Berthoud et al., 1993). However, in other studies, phorbol ester activation of protein kinase C failed to inhibit junctional communication in neonatal rat cardiomyocytes and in human corpus cavernosum smooth muscle cells which also express connexin43 (Sáez et al., 1993; Moreno et al., 1993). Lampe (1994) developed this area further by demonstrating that TPA was acting by inhibiting the assembly of gap junctions.

Clearly, not all connexins have the same phosphorylation sites and it is not surprising to find that connexins are regulated differentially in response to phosphorylation. In many cells, rapid changes in cell-cell coupling in response to activators or inhibitors of protein kinases suggest that phosphorylation is responsible for channel gating. This was elegantly demonstrated by Atkinson and colleagues (1981), who demonstrated that the expression of the viral oncogene from Rous sarcoma virus pp60$^{v-src}$ led to a rapid inhibition of gap junction communication. Spray and associates (1992) and Takens-Kwak and Jongsma (1992) attributed the phosphorylation state of connexin43 to a wide range of gap junction channel conductances in cells stably transfected with connexin43 cDNA and in cardiac myocytes. Moreno and colleagues reported that large unitary conductances of 90 to 100 pS corresponded with unphosphorylated human connexin43 while smaller and more common unitary conductances of 60 to 70 pS were correlated with phosphorylated connexin43 (Moreno et al., 1994). Kwak and co-workers addressed this issue in further detail by examining the permeability and conductances of SKHep1 cells transfected with connexin43 and connexin45 under conditions where various protein kinases were activated or inhibited (Kwak et al., 1995a). These studies showed that connexins were differentially regulated even when cells were subjected to similar phosphorylating conditions (Kwak et al., 1995a). Furthermore, Kwak and colleagues showed in cardiac myocytes that TPA increased connexin43 conductance but decreased permeability (Kwak et al., 1995b). Together, these studies would strongly suggest that phosphorylation modulates gap junction channel activity.

Truncation and mutagenesis strategies are currently being used to examine the functional role of connexins. Interestingly, cells that express most mutated forms of connexin43 or connexin32 that cannot be extensively phosphorylated are still capable of forming functional cell-cell channels (Fishman et al., 1991; 1990; Werner et al., 1991; Dunham et al., 1992). Musil and Goodenough (1991) showed that the unphosphorylated form of connexin43 can indeed be transported to the cell surface of NRK cells and that the acquisition of Triton X-100 resistance (plaque formation) is associated with phosphorylation of connexin43 to the $P_2$ form. However, it has been shown that nonfunctional gap junction plaques do form in a communication-deficient mutant rat liver epithelial cell line that failed to phosphorylate connexin43 to the $P_2$ form (Oh et al., 1993) and other studies in rat mammary tumor cells suggest that extensive phosphorylation of connexin43 is not causal of gap junction plaque formation (Feldman et al., 1997). Extensive phosphorylation of connexin43 at the onset of mitosis in human vascular cells has been shown to be correlated with gap junction internalization, but it remains unclear whether this cell cycle specific phosphorylation event triggers the internalization process (Xie et al., 1997). The slow phosphorylation of connexin46 in comparison to connexin43 suggests that phosphorylation may play another undefined role (Jiang et al., 1993). Moreover, the identification of a possible endoplasmic reticulum–Golgi connexin43 phosphorylation event in some cell types leads one to speculate that there may be yet other roles for connexin43 phosphorylation (Laird et al., 1993; Puranam et al., 1993).

The state of connexin43 phosphorylation in adult rats varies from tissue to tissue (Kadle et al., 1991), further suggesting that *in vivo* phosphorylation or dephosphorylation, or both, are active processes, although, at least in the case of brain, dephosphorylation appears to be a postmortem artifact (Hossain et al., 1994). In support of this notion of close regulation of phosphorylation and dephosphorylation, it has been observed that during the ontogenesis of the pineal gland connexin43 expressed by astrocytes increases drastically at about the time the gland acquires its adult state of innervation, function, and protein kinase C activity (Berthoud and Sáez, 1993).

The connexin43 gene in several children with congenital heart diseases has been found to have substitutions of serine and threonine residues that are proposed to be phosphorylated in normal patients (Britz-Cunningham et al., 1995). In particular, five children had serines substituted with prolines at position 364 (Britz-Cunningham et al., 1995). These observations allude to the possibility that correct phosphorylation of connexin43 may be essential for regulated gap junction communication and for normal heart function.

In summary, while it is understood that the activation of a tyrosine kinase can lead to connexin43 phosphorylation and channel closure, no clear predictions can be made as to how connexins will respond to phosphorylation and dephosphorylation on serine residues in native tissues or cells.

## E. Other Modifications to Connexins

Apart from phosphorylation, the other post-transcriptional modifications associated with connexins are acylation and disulfide bridge formation. In 1989, Traub and co-workers were able to demonstrate that both tritiated palmitic and myristic acids could be incorporated into connexin32 (Traub et al., 1989). Similarly, the long chain fatty acid palmitate can be incorporated into connexin43 in leptomeningeal cells (Hertzberg et al., 1989). It is possible that the fatty acylation of connexins acts to regulate gap junctions in a yet undetermined manner. One can speculate that fatty acylation anchors the protein in the membrane in a manner that facilitates connexin oligomerization.

Several laboratories have now been able to demonstrate that connexin43 (El Aoumari et al., 1991; John and Revel, 1991) and connexin32 (Rahman et al., 1993) have at least one intramolecular disulfide linkage. The conservation of six cysteine residues (three in each extracellular loop) (Hoh et al., 1991) suggests that intramolecular disulfide bridges are common to all connexins. Dahl and colleagues (1991) used site-directed mutagenesis and the *Xenopus* oocyte expression system to show that the changing of anyone of the six cysteine residues resulted in the loss of gap junctional communication. This phenomena was even evident in asymmetrical conditions where one oocyte expressed the wild-type hemichannel (Dahl et al., 1991). In the liver, Rahman and co-workers suggested that the connexin32 interloop disulfide linkage(s) take place in the endoplasmic reticulum (Rahman et al., 1993). Formation of intramolecular disulfides during early connexin biogenesis may serve to prevent nonspecific associations between nascent polypeptides similarly to that proposed for two asialoglycoprotein subunits (Yilla et al., 1992).

Whether other post-translational modifications occur to connexins is poorly understood. Hertzberg(1995) and colleagues have used isoelectric focusing to show that connexin43, connexin32, and connexin26 have several pIs that are likely related to a covalent modification other than phosphorylation. It will be interesting to determine the nature of this modification and determine if it is directly linked to connexin43 assembly. There remains a general consensus that connexins are not glycoproteins. Earlier work has suggested that glycosylated proteins are not present in isolated gap junction plaques (Gilula, 1974; Hertzberg and Gilula, 1979). Experiments designed to determine whether connexin32 was sensitive to glycosidases also proved to be negative (Rahman et al., 1993), supporting the absence of connexin glycosylation (Wang et al., 1995).

## VI.    SUMMARY AND PROSPECTIVE FOR THE FUTURE

Soon after molecular probes for studying gap junction proteins were developed, insights into post-transcriptional modifications of connexins started to emerge. It is now known that the steady state levels of at least connexin26 and connexin32 are regulated post-transcriptionally. In addition, connexins are known to traffic through the Golgi apparatus and their oligomerization into hemichannels (connexons) has been reported to occur after the connexins leave the endoplasmic reticulum. Upon arrival at the cell surface, connexon pairing and plaque formation has been linked to the expression of cell adhesion molecules and to structural features of the extracellular domains of the gap junctional polypeptide subunits. Removal of gap junctions from the cell surface has been proposed to occur by the internalization of double membrane units that have been termed annular gap junctions. Subsequently, internalized gap junctions have been observed in electron dense membrane compartments suggestive of lysosomes. However, new insights suggest that proteasomes play a role in at least connexin43 degradation. It is also possible that there is more than one pathway for gap junction removal from the cell surface as evident from the fact that gap junctions have also been found in a disassembled state on the cell surface.

Connexins can be phosphorylated but, even though they enter the Golgi, they are not glycosylated. Phosphorylation of connexins likely has multiple roles as it has been shown to upregulate and downregulate cell-cell communication and to be correlated with gap junction assembly. The expression of chimeric connexins in exogenous expression systems is being actively used to obtain a better understanding of connexin phosphorylation. Many mutant forms of connexin32 and connexin43 whose C-terminals have been truncated, and presumably are not phosphorylated, are capable of forming functional channels at areas of cell-cell contact. However, it is possible that these connexin mutants have altered trafficking and assembly kinetics. Nevertheless, there is substantial evidence to suggest connexin phosphorylation is directly linked to channel gating.

We now know that gap junctions are highly dynamic structures. The half-life of connexin32 and connexin43 *in vitro* has been found to be approximately 1 to 3.5 hours, suggesting that the formation and removal of gap junctions play important roles in determining the degree of cell-cell communication. In the future, it will be important to establish whether molecular chaperones interact with these dynamic proteins. Two endoplasmic reticulum proteins, calnexin (Ahluwalia et al., 1992; Hochstenbach et al., 1992) and BiP (Pelham, 1989; Flynn et al., 1991) have been identified and are known to serve as molecular chaperones. These chaperones are long lived proteins that act by binding to nascent proteins until proper folding and assembly is complete. Co-immunoprecipitation studies suggest that calnexin does not act as a chaperone for connexin43 (Ou, Bergeron, and Laird, unpublished results). If chaperone interactions with connexins do exist in the endoplasmic reticulum, it is possible that such interactions are important, along with protein disulfide isomerase, in shielding or regulating cysteine interactions such that stable disulfide bonds are formed. It is believed that BiP may perform such a chaperone function during the folding of the influenza virus hemagglutinin (Segal et al., 1992). On the other hand, chaperones may be needed to prevent connexin oligomerization from occurring within the endoplasmic reticulum. The search for a connexin chaperone is currently in progress.

# ACKNOWLEDGMENTS

The authors would like to thank Elliot L. Hertzberg for reviewing this chapter and for his efforts in editing this series. Owing to the large volume of published reports and limits on space, the authors would like to apologize for any study that was not cited in this review. Work in the authors' laboratory was partially supported by grants from the Medical Research Council of Canada (MT-12241) to D.W. Laird, FONDECYT 1960559 to J.C. Sáez, and a National Institute of Health grant GM-30667 to E.L. Hertzberg and M.V.L. Bennett.

# REFERENCES

Aghion, J., Gueth-Hallonet, C., Antony, C., Gros, D., & Maro, B. (1994). Cell adhesion and gap junction formation in the early mouse embryo are induced prematurely by 6-DMAP in the absence of E-cadherin phosphorylation. J. Cell Sci. 107, 1369-1379.

Ahluwalia, N., Bergeron, J.J.M., Wada, I., Degan, E., & Williams, D.B. (1992). The p88 molecular chaperone is identical to the endoplasmic reticulum membrane protein, Calnexin. J. Biol. Chem. 267, 914-918.

Arneson, M.L., Cheng, H.L., & Louis, C.F. (1995). Characterization of the ovine-lens plasma membrane protein-kinase substrates. Eur. J. Biochem. 234, 670-679.

Atkinson, M.M., Lampe, P.D., Lin, H.H., Kollander, R., Li, X.-R., & Kiang, D.T. (1995). Cyclic AMP modifies the cellular distribution of connexin43 and induces a persistent increase in the junctional permeability of mouse mammary tumor cells. J. Cell Sci. 108, 3079-3090.

Atkinson, M.M., Menko, A.S., Johnson, R.G., Sheppard, J.R., & Sheridan, J.D. (1981). Rapid and reversible reduction of junctional permeability in cells infected with a temperature-sensitive mutant of avian sarcoma virus. J. Cell Biol. 91, 573-578.

Bachmair, A., Finley, D., & Varshavsky, A. (1986). In vivo half-life of a protein is a function of its amino terminal residue. Science 234, 179-186.

Berthoud, M., Beyer, E., Kurata, E., Lau, A.F., & Lampe, P.D. (1997). The gap-junction protein connexin 56 is phosphorylated in the intracellular loop and the carboxyl-terminal region. Eur. J. Biochem. 244, 89-97.

Berthoud, V.M., Ledbetter, M.L.S., Hertzberg, E.L., & Sáez, J.C. (1992). Connexin43 in MDCK cells: regulation by a tumor-promoting phorbol ester and $Ca^{2+}$. Eur. J. Cell Biol. 57, 40-50.

Berthoud, V.M., Rook, M., Hertzberg, E.L., & Sáez, J.C. (1993). On the mechanism of cell uncoupling induced by a tumor promoter phorbol ester in clone 9 cells, a rat liver epithelial cell line. Eur. J. Cell Biol. 62, 384-396.

Berthoud, V.M., & Sáez, J.C. (1993). Changes in connexin43, the gap junction protein of astrocytes, during development of the rat pineal gland. J. Pineal Res. 14, 67-72.

Beyer, E.C. (1990). Molecular cloning and developmental expression of two chick embryo gap junction proteins. J. Biol. Chem. 265, 14439-14443.

Beyer, E.C., Paul, D.L., & Goodenough, D.A. (1987). Connexin43: a protein from rat heart homologous to a gap junction protein from liver. J. Cell Biol. 105, 2621-2629.

Beyer, E.C., & Steinberg, T.H. (1991). Evidence that the gap junction protein connexin-43 is the ATP-induced pore of mouse macrophages. J. Biol. Chem. 266, 7971-7974.

Bjorkman, N. (1962). A study of the ultrastructure of the granulosa cells of the rat ovary. Acta Anat. 51, 125-147.

Brissette, J.L., Kumar, N.M., Gilula, N.B., & Dotto, G.P. (1991). The tumor promoter 12-O-tetradecanoylphorbol-13-acetate and the ras oncogene modulate expression and phosphorylation of gap junction proteins. Mol. Cell Biol. 11, 5364-5371.

Britz-Cunningham, S.H., Shah, M.M., Zuppan, C.W., & Fletcher, W.H. (1995). Mutations of the connexin43 gap junction gene in patients with heart malformations and defects of laterality. N. Engl. J. Med. 332, 1323-1329.

Broakman, I., Hoover-Litty, H., Wagner, K.R., & Hellenius, A. (1991). Folding of influenza hemagglutinin in the endoplasmic reticulum. J. Cell Biol. 114, 401-411.

Budunova, I.V., Williams, G.M., & Spray, D.C. (1993). Effect of tumor promoting stimuli on gap junction permeability and connexin43 expression in ARL 18 rat liver cell line. Arch. Toxicol. 67, 565-572.

Cascio, M., Kumar, N.M., Safarik, R., & Gilula, N.B. (1995). Physical characterization of gap junction membrane connexons (Hemichannels) isolated from rat liver. J. Biol. Chem. 270, 18643-18648.

Chen, M.A., Zhang, J.T., & Nicholson, B.J. (1990). Biosynthesis and membrane integration of connexin32 and connexin26 gap junction proteins in a reticulocyte cell free system. J. Cell Biol. 111, 80a.

Crow, D.S., Beyer, E.C., Paul, D.L., Kobe, S.S., & Lau, A.F. (1990). Phosphorylation of connexin43 gap junction protein in uninfected and Rous sarcoma virus-transformed mammalian fibroblasts. Mol. Cell Biol. 10, 1754-1763.

Crow, D.S., Kurata, W.E., & Lau, A.F. (1992). Phosphorylation of connexin43 in cells containing mutant src oncogenes. Oncogene 7, 999-1003.

Dahl, G., Levine, E., Rabadan-Diehl, C., & Werner, R. (1991). Cell/cell channel formation involves disulfide exchange. Eur. J. Biochem. 197, 141-144.

Dahl, G., Werner, R., Levine, E., & Rabadan-Diehl, C. (1992). Mutational analysis of gap junction formation. Biophys. J. 62, 187-195.

Dermietzel, R., Hertzberg, E.L., Kessler, J.A., & Spray, D.C. (1991). Gap junctions between cultured astrocytes immunohistochemical, molecular and electrophysiological analysis. J. Neurosci. 11, 1421-1432.

De Sousa, P.A., Valdimarsson, G., Nicholson, B.J., & Kidder, G.M. (1993). Connexin trafficking and the control of gap junction assembly in mouse preimplantation embryos. Development 117, 1355-1367.

DeVries, S.H., & Schwartz, E.A. (1992). Hemi-gap junction channels in solitary horizontal cells of the catfish retina. J. Physiol. (Lond.) 445, 201-230.

Díez, J.A., Elvira M., & Villalobo, A. (1995). Phosphorylation of connexin-32 by the epidermal growth factor receptor tyrosine kinase. Ann. N.Y. Acad. Sci. 766, 477-480.

Dunham, B., Liu, S., Taffet, S., Trabka-Janik, E., Delmar, M., Petryshyn, R., Zheng, S., Perzova, R., & Vallano, M.L. (1992). Immunolocalization and expression of functional and nonfunctional cell-to-cell channels from wild-type and mutant rat heart connexin43 cDNA. Circ. Res. 70, 1233-1243.

Ebihara, L., Berthoud, V.M., & Beyer, E.C. (1995). Distinct behavior of connexin56 and connexin46 gap junctional channels can be predicted from the behavior of their hemi-gap-junctional channels. Biophys. J. 68, 1796-1803.

El Aoumari, A., Dupont, E., Fromaget, C., Jarry, T., Briand, J.-P., Kreitman, B., & Gros. D. (1991). Immunolocalization of an extracellular domain of connexin43 in rat heart gap junctions. Eur. J. Cell Biol. 56, 391-400.

Elvira, M., Diez, J.A., Wang, K.K.W., & Villalobo, A. (1993). Phosphorylation of connexin-32 by protein kinase C prevents its proteolysis by $\mu$-calpain and $m$-calpain. J. Biol. Chem. 268, 14294-14300.

Epstein, M.L., Sheridan, J.D., & Johnson, R.G. (1977). Formation of low-resistance junctions *in vitro* in the absence of protein synthesis and ATP production. Exp. Cell Res. 104, 25-30.

Falk, M.M., Buehler, L.K., Kumar, N.M., & Gilula, N.B. (1997). Cell-free synthesis and assembly of connexins into functional gap junction membrane channels. EMBO J. 16, 2703-2716.

Falk, M.M., Kumar, N.M., & Gilula, N.B. (1994). Membrane insertion of gap junction connexins: polytopic channel forming membrane proteins. J. Cell Biol. 127, 343-355.

Fallon, J.F., & Goodenough, D.A. (1981). Five-hour half-life of mouse liver gap-junction protein. J. Cell Biol. 90, 521-526.

Feldman, P.A., Kim, J., & Laird, D.W. (1997). Loss of gap junction plaques and inhibition of intercellular communication in ilimaquinone-treated BICR-M1R$_k$ and NRK cells. J. Membr. Biol. 155, 275-287.

Fishman G.I., Spray D.C., & Leinwand L.A. (1990). Molecular characterization and functional expression of the human cardiac gap junction channel. J. Cell Biol. 111, 589-597.

Fishman, G.I., Spray, D.C., & Leinwand, L.A. (1991). Functional analysis of human cardiac gap junction channel mutants. Proc. Natl. Acad. Sci. U.S.A. 88, 3525-3529.

Flynn, G.C., Pohl, J., Flocco, M.T., & Rothman, J.E. (1991). Peptide-binding specificity of the molecular chaperon BiP. Nature 353, 726-730.

Foote, C.I., Zhu, X., Zhou, L., & Nicholson, B.J. (1998). The pattern of disulfide linkages in the extracellular loop regions of Cx32 suggest a model for the docking interface of gap junctions. J. Cell Biol. 140, 1187-1197.

Fujimoto, K., Nagafuchi, A., Tsukita, S., Kuraoka, A., Ohokuma, A., & Shibata, Y. (1997) Dynamics of connexins, E-cadherin and alpha-catenin on cell membranes during gap junction formation. J. Cell Sci. 110, 311-322.

Ghoshroy, S., Goodenough, D.A., & Sosinsky, G.E. (1995). Preparation, characterization and structure of half gap junctional layers split with urea and EGTA. J. Membr. Biol. 146, 15-28.

Gilula, N.B. (1974). Junctions between cells. In: Cell Communication (Coc, R.S., ed.), pp. 1-29. John Willey, New York.

Gimlich, R.L., Kumar, N.M., & Gilula, N.B. (1990). Differential regulation of the levels of three gap junction mRNAS in Xenopus embryos. J. Cell Biol. 110, 597-605.

Gingalewski C., Wang, K., Clemens, M.G., & De Maio A. (1996). Posttranscriptional regulation of connexin 32 expression in liver during acute inflammation. J. Cell. Physiol. 166, 461-467.

Ginzberg, R.D., & Gilula, N.B. (1979). Modulation of cell junctions during differentiation of the chicken otocyst sensory epithelium. Develop. Biol. 68, 110-129.

Goldberg, G.S., & Lau, A.F. (1993). Dynamics of connexin43 phosphorylation in pp60$^{v-src}$ -transformed cells. Biochem. J. 295, 735-742.

Guerrier, A., Fonlupt, P., Morand, I., Rabilloud, R., Audebet, C., Krutovskikh, V., Gros, D., Rousset, B., & Munari-Silem, Y. (1995). Gap junctions and cell polarity: connexin32 and connexin43 expressed in polarized thyroid epithelial cells assemble into separate gap junctions, which are located in distinct regions of the lateral plasma membrane domain. J. Cell Sci. 108, 2609-2617.

Gupta, V.K., Berthoud, V.M., Atal, N., Jarillo, J.A., Barrio, L.C., & Beyer, E.C. (1994). Bovine connexin44, a lens gap junction protein: molecular cloning, immunologic characterization, and functional expression. Invest. Opthalmol. Vis. Sci. 35, 3747-3758.

Hare, J.F. (1990). Mechanism of membrane protein turnover. Biochem. Biophys. Acta 1031, 71-90.

Haubrich, S., Schwarz, H.-J., Bukauskas, F., Lichtenberg-Frate, H., Traub, O., Weingart, R., & Willecke, K. (1996). Incompatibility of connexin40 and 43 hemichannels in gap junctions between mammalian cells is determined by intracellular domains. Mol. Biol. of the Cell, 7, 1995-2006.

Hendrix, E.M., Mao, S.J., Everson, W., & Larsen, W.J. (1992). Myometrial connexin 43 trafficking and gap junction assembly at term and in preterm labor. Mol. Reprod. Dev. 33, 27-38.

Hertig, C.M., Butz, S., Koch, S., Eppenberger-Eberhardt, M., Kemler, R., Eppenberger, H.M. (1996). N-cadherin in adult rat cardiomyocytes in culture. J. Cell Sci. 109, 11-20.

Hertzberg, E.L. (1984). A detergent-independent procedure for the isolation of gap junctions from rat liver. J. Biol. Chem. 259, 9936-9943.

Hertzberg, E.L. (1995). Isoelectric focusing of gap junction proteins indicates that covalent modification(s) other than phosphorylation alter pI and may be involved in gap junction assembly. Mol. Biol. Cell. 6, 189a.

Hertzberg, E.L., Corpina, R., Roy, C., Dougherty, M.J., & Kessler, J.A. (1989). Analysis of the 43 kDa heart gap junction protein in primary cultures of rat leptomeningeal cells. J. Cell Biol. 109, 47a.

Hertzberg, E.L., & Gilula, N.B. (1979). Isolation and characterization of gap junction from rat liver. J. Biol. Chem. 254, 2138-2147.

Hille, A., Waheed, A., & von Figura, K. (1990). Assembly of the ligand-binding conformation of the $M_r$ 46,000 mannose 6-phosphate specific receptor takes place before reaching the Golgi complex. J. Cell Biol. 110, 963-972.

Hochstenbach, F., David, V., Watkins, S., & Brenner, M.B. (1992). Endoplasmic reticulum resident protein of 90 kilodaltons associated with the T- and B-cell antigen receptors and major histocompatibility complex antigens during their assembly. Proc. Natl. Acad. Sci. U.S.A. 89, 4734-4738.

Hoh, J.H., John, S.A., & Revel, J.-P. (1991). Molecular cloning and characterization of a new member of gap junction gene family, connexin-31. J. Biol. Chem. 266, 6524-6531.

Hossain, M.Z., Murphy, L.J., Hertzberg, E.L., & Nagy, J.I. (1994). Phosphorylated forms of connexin43 predominate in rat brain: demonstration by rapid inactivation of brain metabolism. J. Neurochem. 62, 2394-2403.

Husoy, T., Mikalsen, S.O.,, & Sanner, T. (1993). Phosphatase inhibitors, gap junctional intercellular communication and $^{125}$I-EGF binding in hamster fibroblasts. Carcinogenesis 14, 2257-2265.

Imhof, B.A., Vollmers, P.H., Goodmand, S.L., & Birchmeir, W. (1983). Cell-cell interaction and polarity of epithelial cells: specific perturbation using a monoclonal antibody. Cell 35, 667-675.

Ito, S., & Loewenstein, W.R. (1969). Ionic communication between early embryonic cells. Dev. Biol. 19, 228-243.

Jiang, J.X., & Goodenough, D.A. (1996) Heteromeric connexons in lens gap junction channels. Proc. Natl. Acad. Sci. U.S.A. 93, 1287-1291.

Jiang, J.X., Paul, D.L., & Goodenough, D.A. (1993). Posttranslational phosphorylation of lens fiber connexin46: A slow occurrence. Invest. Ophthalmol. Vis. Sci. 34, 3558-3565.

John, S.A., & Revel, J.-P. (1991). Connexon integrity is maintained by non-covalent bonds: intramolecular disulfide bonds link the extracellular domains in rat connexin-43. Biochem. Biophys. Res. Com. 178, 1312-1318.

Johnson, R., Hammer, M., Sheridan, J., & Revel, J.P. (1974). Gap junction formation between reaggregated Novikoff hepatoma cells. Proc. Natl. Acad. Sci. U.S.A. 71, 4536-4540.

Jongen, W.M.F., Fitzgerald, D.J., Asamoto, M., Piccoli, C., Slaga, T.J., Gros, D., Takeichi, M., & Yamasaki, H. (1991). Regulation of connexin43-mediated gap junctional intercellular communication by calcium in mouse epidermal cells is controlled by E-cadherin. J. Cell Biol. 114, 545-555.

Kadle, R., Zhang, J.T., & Nicholson, B.J. (1991). Tissue-specific distribution of differentially phosphorylated forms of connexin43. Mol. Cell Biol. 11, 363-369.

Kanemitsu, M. Y., Jiang, W., & Eckhart, W. (1998). Cdc2-mediated phosphorylation of the gap junction protein, connexin43, during mitosis. Cell Growth Differ. 9, 13-21.

Kanemitsu, M.Y., & Lau, A.L. (1993). Epidermal growth factor stimulates the disruption of gap junctional communication and connexin43 phosphorylation independent of 12-O-tetradecanoylphorbol 13-acetate-sensitive protein kinase C: the possible involvement of mitogen-activated protein kinase. Mol. Biol Cell 4, 837-848.

Kanemitsu, M. Y., Loo, L. W., Simon, S., Lau, A. F., & Eckhart, W. (1997). Tyrosine phosphorylation of connexin 43 by v-Src is mediated by SH2 and SH3 domain interactions. J. Biol. Chem. 272, 22824-22831.

Katcher, B., & Reese, T. (1985). Rapid formation of gap-junction-like structures induced by glycerol. Anat. Rec. 213, 7-15.

Keane, R.W., Mehta, P.P., Rose, B., Honig, L.S., Loewenstein, W.R., & Rutishauser, U. (1988). Neural differentiation, N-CAM-mediated adhesion and gap junctional communication in neuroectoderm. A study *in vitro*. J. Cell Biol. 106, 1307-1319.

Kennelly, P.J., & Krebs, E.G. (1991). Consensus sequences as substrate-specificity determinants for protein kinases and protein phosphatases. J. Biol. Chem. 266, 15555-15558.

Kistler, J., Bond, J., Donaldson, P., & Engel, A. (1993). Two distinct levels of gap junction assembly *in vitro*. J. Struct. Biol. 110, 28-38.

Klausner, R.D., Rouault, T.A., & Harford, J.B. (1993). Regulating the fate of mRNA: the control of cellular iron metabolism. Cell 72, 19-28.

Kren, B.T., Kumar, N.M., Wang, S., Gilula, N.B., & Steer, C.J. (1993). Differential regulation of multiple gap junction transcripts and proteins during rat liver regeneration. J. Cell Biol. 123, 707-718.

Kumar, N.M., Friend, D.S., & Gilula, N.B. (1995). Synthesis and assembly of human beta-1 gap junctions in BHK cells by DNA transfection with the human beta-1 cDNA. J. Cell Sci. 108, 3725-3734.

Kwak, B.R., Hermans, M.M.P., De Jonge, H.R., Lohmann, S.M., Jongsma, H.J., & Chanson, M. (1995a). Differential regulation of distinct types of gap junction channels by similar phosphorylating conditions. Mol. Biol. Cell 6, 1707-1719.

Kwak, B.R., Van Veen, T.A.B., Analbers, L.J.S., & Jongsma, H. (1995b). TPA increases conductance by decreases permeability in neonatal rat cardiomyocyte gap junction channels. Exp. Cell Res. 220, 456-463.

Laing, J.G., & Beyer, E.C. (1995). The gap junction protein connexin43 is degraded via the ubiquitin proteasome pathway. J. Biol. Chem. 270, 26399-26403.

Laing, J. G., Tadros, P. N., Westphale, E. M., & Beyer, E. C. (1997). Degradation of connexin43 gap junctions involves both the proteasome and the lysosome. Exp. Cell Res. 236, 482-492.

Laing, J.G., Westphale, E.M., Engelmann, G.L., & Beyer, E.C. (1994). Characterization of the gap junction protein, connexin45. J. Membr. Biol. 139, 31-40.

Laird, D.W., Castillo, M., & Kasprzak, L. (1995) Gap junction turnover, intracellular trafficking, and phosphorylation of connexin43 in brefeldin A-treated rat mammary tumor cells. J. Cell Biol. 131, 1193-1203.

Laird, D.W., Puranam, K.L., & Revel, J.-P. (1991). Turnover and phosphorylation dynamics of connexin43 gap junction protein in cultured cardiac myocytes. Biochem. J. 272, 67-72.

Laird, D.W., Puranam, K.L., & Revel, J.-P. (1993). Identification of intermediate forms of connexin43 in rat cardiac myocytes. In: *Gap Junctions. Progress in Cell Research* (Hall, J.E., Zampighi, G.A., & Davis, R.M. eds.), vol. 3, pp. 263-268. Elsevier, Amsterdam.

Laird, D.W., & Revel, J.-P. (1990). Biochemical and immunochemical analysis of the arrangement of connexin43 in rat heart gap junction membranes. J. Cell Sci. 97, 109-117.

Lampe, P. (1994). Analyzing phorbol ester effects on gap junctional communication: a dramatic inhibition of assembly. J. Cell Biol. 127, 1895-1905.

Lampe, P., Kurata, W.E., & Lau, A.F. (1996). Phosphorylation of connexin43 in mitotic phase cells. Mol. Biol. of the Cell 7, 92a.

Lampe, P.D., Kurata, W.E., Warn-Cramer, B.J., & Lau, A.F. (1998). Formation of a distinct connexin43 phosphoisoform in mitotic cells is dependent upon p34$^{cdc2}$ kinase. J. Cell Sci. 111, 833-841.

Lane, N.J., & Swales, L.S. (1980). Dispersal of junctional particles, not internalization, during *in vitro* disappearance of gap junctions. Cell 19, 579-586.

Larsen, W.J. (1983). Biological implications of gap junction structure, distribution and composition: a review. Tissue Cell 15, 645-671.

Larsen, W.J., & Risinger, M.A. (1985). The dynamic life histories of intercellular membrane junctions. In: *Modern Cell Biology*, vol. 4, pp. 141-216. Alan R. Liss, New York.

Larsen, W.J., & Tung, H.N. (1978). Origin and fate of cytoplasmic gap junctional vesicles in rabbit granulosa cells. Tissue Cell 10, 585-598.

Larsen, W.J., Tung, H.N., Murray, S.A., & Swenson, C.A. (1979). Evidence for the participation of actin microfilaments and bristle coats in the internalization of gap junction membrane. J. Cell Biol. 83, 576-587.

Lau, A.F., Hatch-Pigott, V., & Crow, D.S. (1991). Evidence that heart connexin43 is a phosphoprotein. J. Mol. Cell Cardiol. 23, 659-663.

Lau, A.F., Kanemitsu, M.Y., Kurata, W.E., Danesh, S., & Boyton, A.L. (1992). Epidermal growth factor disrupt gap-junctional communication and induces phosphorylation of connexin43 on serine. Mol. Biol. Cell 3, 865-874.

Li, H., Liu, T.-F., Lazrak, A., Peracchia, C., Goldberg, G.S., Lampe, P.D., & Johnson, R.G. (1996) Properties and regulation of gap junctional hemichannels in the plasma membranes of cultured cells. J. Cell Biol. 134, 1019-1030.

Lippincott-Schwartz, J., Yuan, L.C., Bonifacio, J.S., & Klausner, R.D. (1989). Rapid redistribution of Golgi proteins into the ER in cells treated with brefeldin A: evidence for membrane recycling from the Golgi to ER. Cell 56, 801-813.

Loewenstein, W.R. (1967). On the genesis of cellular communication. Dev. Biol. 15, 503-520.

Loo, L.W., Berestecky, J.M., Kanemitsu, M.Y., & Lau, A.F. (1995). pp60$^{src}$-mediated phosphorylation of connexin43, a gap junction protein. J. Biol. Chem. 270, 12751-12761.

Makowski, L., Caspar, D.L.D., Phillips, W.C., & Goodenough, D.A. (1977). Gap junction structures. Analysis of the X-ray diffraction data. J. Cell Biol. 74, 629-645.

Maldonado, P.E., Rose, B., & Loewenstein, W.R. (1988). Growth factors modulate junctional cell-to-cell communication. J. Membr. Biol. 106, 203-210.

Manjunath, C.K., Goings, G.E., & Page, E. (1984). Cytoplasmic surface and intramembrane components of rat heart gap junctional proteins. Am. J. Physiol. 246, H856-H875.

Matesic, D.F., Rupp, H.L., Bonney, W.J., Ruch, R.J., & Trosko, J.E. (1994). Changes in gap junction permeability, phosphorylation, and number mediated by phorbol ester and non-phorbol ester tumor promoters in rat liver epithelial cells. Mol. Carcin. 10, 226-236.

Mazet, F., Wittenberg, B.A., & Spray D.C. (1985). Fate of intercellular junctions in isolated adult rat cardiac cells. Circ. Res. 56, 195-204.

Mehta, P.P., Lokeshwar, B.L., Schiller, P.C., Bendix, M.V., Ostenson, R.C., Howard, G.A., & Roos, B.A. (1996). Gap-junctional communication in normal and neoplastic prostate epithelial cells and its regulation by cAMP. Molec. Carcin. 15, 18-32.

Mehta, P.P., Yamamoto, M., & Rose, B. (1992). Transcription of the gene for the gap junctional protein connexin43 and expression of functional cell-to-cell channels are regulated by cAMP. Mol. Biol. Cell 3, 839-850.

Meyer, R.A., Laird, D.W., Revel, J.-P., & Johnson, R.G. (1992). Inhibition of gap junction and adherens junction assembly by connexin and A-CAM antibodies. J. Cell Biol. 119, 179-189.

Meyer, R.A., Lampe, P.D., Malewicz, B., Baumann, W., & Johnson, R.G. (1991). Enhanced gap junction formation with LDL and apolipoprotein B. Exp. Cell Res. 196, 72-81.

Montesano, R. (1980). Intramembrane events accompanying junction formation in a liver cell line. Anat. Rec. 198, 403-414.

Morello, D., Lavenu, A., & Babinet, C. (1990). Differential regulation and expression of jun, c-fos and c-myc proto-oncogenes during mouse liver regeneration and after inhibition of protein synthesis. Oncogene 5, 1511-1519.

Moreno, A.P., Campos de Carvalho, A.C., Christ, G., Melman, A., & Spray, D.C. (1993). Gap junctions between human corpus cavernosum smooth muscle cells: gating properties and unitary conductance. Am J. Physiol. 264, C80-C92.

Moreno, A.P., Sáez, J.C., Fishman, G.I., & Spray, D.C. (1994). Human connexin43 gap junction channels, regulation of unitary conductances by phosphorylation. Circ. Res. 74, 1050-1057.

Murray, S.A., Larsen, W.J., Trout, J., & Donta, S.T. (1981). Gap junction assembly and endocytosis correlated with patterns of growth in a cultured adrenocortical tumor cell (SW-13). Cancer Res 41, 4063-4069.

Musil, L.S., Beyer, E.C., & Goodenough, D.A. (1990a). Expression of the gap junction protein connexin43 in embryonic chick lens: molecular cloning, ultrastructural localization and post-translational phosphorylation. J. Membr. Biol. 116, 163-175.

Musil, L.S., Cunningham, B.A., Edelman, G.M., & Goodenough, D.A. (1990b). Differential phosphorylation of the gap junction protein connexin43 in junctional communication-competent and -deficient cell lines. J. Cell Biol. 111, 2077-2088.

Musil, L.S., & Goodenough, D.A. (1991). Biochemical analysis of connexin43 intracellular transport, phosphorylation, and assembly into gap junctional plaques. J. Cell Biol. 115, 1357-1374.

Musil, L.S., & Goodenough, D.A. (1993). Multisubunit assembly of an integral plasma membrane channel protein, gap junction connexin43, occurs after exit from the ER. Cell 74, 1065-1077.

Naus, C.C.G., Hearn, S., Zhu, D., Nicholson, B.J., & Shivers, R.S. (1993). Ultrastructural analysis of gap junctions in C6 glioma cells transfected with connexin43 cDNA. Exp. Cell Res. 206, 72-84.

Ne'eman, Z., Spira, M.E., & Bennett M.V.L. (1980). Formation of gap and tight junctions between reagreggated blastomeres of the killfish, *Fundulus*. Am. J. Anat. 158, 251-262.

Nishi, M., Kumar, N.M., & Gilula, N.B. (1991). Developmental regulation of gap junction gene expression during mouse embryonic development. Dev. Biol. 146, 117-130.

Oh, S.Y., Grupen, C.G., & Murray, A.W. (1991). Phorbol ester induces phosphorylation and down-regulation of connexin43 in WB cells. Biochem. Biophys. Acta 1094, 243-245.

Oh, S.Y., Dupont, E., Madhukar, B.V., Briand, J.-P., Chang, C.-C., Beyer, E., & Trosko, J.E. (1993). Characterization of gap junctional communication-deficient mutants of a rat liver epithelial cell line. Eur. J. Cell Biol. 60, 250-255.

Paul, D.L., Ebihara, L., Takemoto, L.J., Swenson, K.I., & Goodenough, D.L. (1991). Connexin46, a novel gap junction protein, induces voltage-gated currents in nonjunctional plasma membrane of *Xenopus* oocytes. J. Cell Biol. 115, 1077-1089.

Pelham, H.R.B. (1989). Control of protein exit from the endoplasmic reticulum. Annu. Rev. Cell Biol. 5, 1-23.

Pelletier, R.M. (1988). Cyclic modulation of Sertoli cell junctional complexes in a seasonal breeder: the mink *(Mustela vison)*. Am. J. Anat. 183, 68-102.

Porvaznik, M., Johnson, J.D., & Sheridan, J.D. (1979). Tight junction development between cultured hepatoma cells: possible stages in assembly and enhancement with dexamethasone. J. Supramol. Struct. 10, 13-30.

Puranam, K.L., Laird, D.W., & Revel, J.-P. (1993). Trapping an intermediate form of connexin43 in the Golgi. Exp. Cell Res. 206, 85-92.

Rahman, S., Carlile, G., & Evans, W.H. (1993). Assembly of hepatic gap junctions. J. Biol. Chem. 268, 1260-1265.

Reynhout J.K., Lampe P.D., & Johnson R.G. (1992). An activator of protein kinase C inhibits gap junction communication between cultured bovine lens cells. Exp. Cell Res. 198, 337-342.

Risley, M., Tan, I., Roy, C., & Sáez, J.C. (1992). Cell, age, and stage-dependent distribution of connexin43 gap junctions in testes. J. Cell Sci. 103, 81-96.

Rosemberg, E., Spray, D.C., & Reid, L.M. (1992). Transcriptional and posttranscriptional control of connexin mRNAs in periportal and pericentral rat hepatocytes. Eur. J. Cell Biol. 59, 21-26.

Sáez, J.C., Gregory, W.A., Watanabe, T., Dermieltzel, R., Hertzberg, E.L., Reid, L., Bennett, M.V.L., & Spray, D.C. (1989). cAMP delays disappearance of gap junctions between pairs of rat hepatocytes in primary cultures. Amer. J. Physiol. 257, C1-C11.

Sáez, J.C., Nairn, A.C., Czernik, A.J., Fishman, G.I., Spray, D.C., & Hertzberg, E.L. (1997). Phosphorylation of connexin43 and the regulation of neonatal rat cardiac myocyte gap junctions. J. Mol. Cell Cardiol. 29, 2131-2145.

Sáez, J.C., Nairn, A.C., Czernik, A.J., Spray, D.C., Hertzberg, E.L., Greengard, P., & Bennett, M.V.L. (1990). Phosphorylation of connexin 32, a hepatocyte gap-junction protein, by cAMP-dependent protein kinase, protein kinase C and $Ca^{2+}$/calmodulin-dependent protein kinase II. Eur. J. Biochem. 192, 263-273.

Sáez, J.C., Nairn, A.C., Spray, D.C., & Hertzberg, E.L. (1993). Rat connexin43: regulation by phosphorylation in heart. In: *Progress in Cell Research* (Hall, J.E., Zampighi, G.A., & Davis, R.M., eds.), pp. 275-281. Elsevier, Amsterdam.

Sáez, J.C., Spray, D.C., Nairn, A.C., Hertzberg, E.L., Greengard, P., & Bennett, M.V.L. (1986). cAMP increases junctional conductance and stimulates phosphorylation of the 27-kDa principal gap junction polypeptide. Proc. Natl. Acad. Sci. U.S.A. 83, 2473-2477.

Segal, M.S., Bye, J.M., Sambrook, J.F., & Geting, M.H. (1992). Disulfide bond formation during the folding of influenza virus hemaglutinin. J. Cell Biol. 118, 227-244.

Simon, A.M., Goodenough, D.A., Li, E., & Paul, D.L. (1997). Female infertility in mice lacking connexin37. Nature 385, 525-529.

Somogyi, R., Batzer, A., & Kolb, H.-A. (1989). Inhibition of electrical coupling in pairs of murine pancreatic acinar cells by OAG and isolated protein kinase C. J. Membr. Biol. 108, 273-282.

Sosinsky, G. (1995). Mixing of connexins in gap junction membrane channels. Proc. Natl. Acad. Sci. U.S.A. 92, 9210-9214.

Spray, D.C., Moreno, A.P., Eghbali, B., Chanson, M., & Fishman, G.I. (1992). Gating of gap junction channels as revealed in cells stably transfected with wild type and mutant connexin cDNAs. Biophys. J. 62, 48-50.

Stauffer, K.A. (1995). The gap junction proteins beta-1 connexin (connexin32) and beta-2 connexin (connexin26) can form heteromeric hemichannels. J. Biol. Chem. 270, 6768-6772.

Swenson, K.I., Pwnica-Worms, H., McNamee, H., & Paul, D.L. (1990). Tyrosine phosphorylation of the gap junction protein connexin43 is required for the pp60[v-src]-induced inhibition of communication. Cell Regulation 1, 989-1002.

Tadvalkar, G., & Pinto da Silva, G. (1983). In vitro, rapid assembly of gap junctions is induced by cytoskeleton disruptors. J. Cell Biol. 96, 1279-1287.

Takeda, A., Hashimoto, E., Yamamura, H., & Shimazu, T. (1987). Phosphorylation of rat liver gap junction protein by protein kinase C. FEBS Lett. 210, 169-172.

Takeda, A., Saheki, S., Shimazu, T., & Takeuchi, N. (1989). Phosphorylation of the 27-kDa gap junction protein by protein kinase C *in vitro* and in rat hepatocytes. J. Biochem (Tokyo) 106, 723-727.

Takens-Kwak, B.R., & Jongsma, H.J. (1992). Cardiac gap junctions: three distinct single channel conductances and their modulation by phosphorylating treatments. Pflugers Arch. 422, 198-200.

Takens-Kwak, B.R., Sáez, J.C., Wilders, R., Fishman, G.I., Hertzberg, E.L., Spray, D.C., & Jongsma, H.J. (1995). cGMP-dependent phosphorylation of connexin43: influence on gap junction channel conductance and kinetics. Pflugers Arch. 430, 770-778.

Traub, O., Look, J., Dermietzel, R., Brummer, F., Hülser, D., & Willecke, K. (1989). Comparative characterization of the 21 kD and 26 kD gap junction proteins in murine liver and cultured hepatocytes. J. Cell Biol. 108, 1039-1051.

Traub, O., Look, J., Paul, D., & Willecke, K. (1987). Cyclic adenosine monophosphate stimulates biosynthesis and phosphorylation of the 26 kDa gap junction protein in cultured mouse hepatocytes. Eur. J. Cell Biol. 43, 48-54.

Vaughan, D.K., & Lasater, E.M. (1992). Acid phosphatase localization in endocyted horizontal cell gap junctions. Visual Neurosc. 8, 77-81.

Voorter, C.E.M., & Kistler, J. (1989). cAMP-dependent protein kinase phosphorylates gap junction protein in lens cortex but not in lens nucleus. Biochem. Biophys. Acta 986, 8-10.

Wang, Y., Mehta, P.P., & Rose, B. (1995). Inhibition of glycosylation induces formation of open connexin43 cell-to-cell channels and phosphorylation and Triton X-100 insolubility of connexin43. J. Biol. Chem. 270, 26581-26585.

Wang, Y., & Rose, B. (1995). Clustering of connexin43 cell-to-cell channels into gap junction plaques: regulation by cAMP and microfilaments. J. Cell Sci. 108, 3501-3508.

Wang, Y., & Rose, B. (1997) An inhibition of gap-junctional communication by cadherins. J. Cell Sci. 110, 301-309.

Warn-Cramer, B.J., Cottrell, G.T., Burt, J.M.,, & Lau, A.F. (1998). Regulation of connexin-43 gap junctional intercellular communication by mitogen-activated protein kinase. J. Biol. Chem. 273, 9188-9196.

Warn-Cramer, B.J., Lampe, P.D., Kurata, W.E., Kanemitsu, M.Y., Loo, L.W.M., Eckhart, W., & Lau, A.F. (1996). Characterization of the mitogen-activated protein kinase phosphorylation sites on the connexin-43 gap junction protein. J. Biol. Chem. 271, 3779-3786.

Werner, R., Levine, E., Rabadan-Diehl, C., & Dahl G. (1991). Gating properties of connexin32 cell-cell channels and their mutants expressed in Xenopus oocytes. Proc. R. Soc. Lond. 243, 5-11.

Werner, R., Rabadan-Diehl, C., Levine, E., & Dahl, G. (1993). Affinity between connexins. In: Gap Junctions. Progress in Cell Research (Hall, J.E., Zampighi, G.A., & Davis, R.M., eds.), vol. 3, pp. 21-24. Elsevier, Amsterdam.

White, T.W., Bruzzone, R., Goodenough, D.A., & Paul, D.L. (1992). Mouse connexin50, a functional member of the connexin family of gap junction proteins, is the lens fibre protein MP70. Mol. Biol. Cell 3, 711-720.

Xie, H.Q., Laird, D.W., Chang, T.H., & Hu. V.W. (1997). A mitotis-specific phosphorylation of the gap junction protein connexin43 in human vascular cells: Biochemical characterization and localization. J. Cell Biol. 137, 203-210.

Yancey, S.B., Easter, D., & Revel, J.P. (1979). Cytological changes in gap junctions during liver regeneration. J. Ultrastr. Res. 67, 229-242.

Yancey, S.B., John, S.A., Lal, R., Austin, B.J., & Revel, J.-P. (1989). The 43-kD polypeptide of heart gap junction: immunolocalization, topology, and functional domains. J. Cell Biol. 108, 2241-2254.

Yancey, S.B., Nicholson, B.J., & Revel, J.-P. (1981). The dynamic state of liver gap junctions. J. Supramol. Struct. Cell Biochem. 16, 221-232.

Yee, A.G., & Revel, J.-P. (1978). Loss and appearance of gap junctions in regenerating liver. J. Cell Biol. 78, 554-564.

Yilla, M., Doyle, D., & Sawyer, J.T. (1992). Early disulfide bond formation prevents heterotypic aggregation of membrane proteins in a cell-free translation system. J. Cell Biol. 118, 245-252.

Zampighi, G., Kreman, M., Ramon, F., Moreno, A.L., & Simon, S.A. (1989). Structural characteristics of gap junctions. I. Channel number in coupled and uncoupled conditions. J. Cell Biol. 106, 1667-1678.

Zhang, J.-T., Chen, M., Foote, C.I., & Nicholson, B.J. (1996). Membrane integration of in vitro-translated gap junctional proteins: Co- and Post-translational mechanisms. Mol. Biol. Cell 7, 471-482.

Zhou, L., Kasperek, E.M., & Nicholson, B.J. (1999). Dissection of the molecular basis of pp60[v-src]-induced gating of connexin43 gap junctions. J. Cell Biol. 144, 1033-1045.

Zhu, D., Caveney, S., Kidder, G.M., & Naus, C.C.G. (1991). Transfection of C6 glioma cells with connexin43 cDNA: analysis of expression, intercellular coupling, and cell proliferation. Proc. Natl. Acad. Sci. U.S.A. 88, 1883-1887.

# GAP JUNCTION CHANNELS
## PERMEABILITY AND VOLTAGE GATING

Vytas K. Verselis and Richard Veenstra

I.   Introduction . . . . . . . . . . . . . . . . . . . . . . . . . . . . . . . . . . . . . . . . . . . . . . . . . . 130
II.  Quantifying Junctional Conductance and Permeability . . . . . . . . . . . . . . . . . . 131
     A.   Measurement of Macroscopic Junctional Conductance . . . . . . . . . . . . . . 132
     B.   Measurement of Single Channel Conductance . . . . . . . . . . . . . . . . . . . . . 135
     C.   Measurement of Junctional Permeability . . . . . . . . . . . . . . . . . . . . . . . . . 136
III. Permeability and Conductance . . . . . . . . . . . . . . . . . . . . . . . . . . . . . . . . . . . . 137
     A.   Permeability as Measured by Cell-to-Cell Diffusion of Tracers . . . . . . . . 137
     B.   Simple and Frictional Pore Models: Predictions for
          Unitary Currents, Selectivity, and Size . . . . . . . . . . . . . . . . . . . . . . . . . . 145
     C.   Studies of Unitary Conductance in Native Tissues . . . . . . . . . . . . . . . . . . 150
IV.  Voltage-Dependent Gating . . . . . . . . . . . . . . . . . . . . . . . . . . . . . . . . . . . . . . . 158
     A.   Multiplicity of Voltage Gating: Definitions and Descriptions. . . . . . . . . . 158
     B.   Slow $V_j$ Dependence . . . . . . . . . . . . . . . . . . . . . . . . . . . . . . . . . . . . . . . . . 160
     C.   Fast $V_j$ Dependence. . . . . . . . . . . . . . . . . . . . . . . . . . . . . . . . . . . . . . . . . . 169
     D.   $V_m$ or $V_{i\text{-}o}$ Dependence . . . . . . . . . . . . . . . . . . . . . . . . . . . . . . . . . . . . . . . 171
V.   Structure and Function . . . . . . . . . . . . . . . . . . . . . . . . . . . . . . . . . . . . . . . . . . 173
     A.   Molecular Studies of $V_j$ Dependence . . . . . . . . . . . . . . . . . . . . . . . . . . . . 173
     B.   Unitary Conductance of Connexin-Specific Channels . . . . . . . . . . . . . . . . 176
     C.   Where Is the Pore?. . . . . . . . . . . . . . . . . . . . . . . . . . . . . . . . . . . . . . . . . . . 181
     D.   Molecular Aspects of Connexin Compatibility. . . . . . . . . . . . . . . . . . . . . . 182

**Advances in Molecular and Cell Biology, Volume 30, pages 129-192.**
**Copyright © 2000 by JAI Press Inc.**
**All rights of reproduction in any form reserved.**
**ISBN: 0-7623-0599-1**

# I. INTRODUCTION

Gap junctions are specialized intercellular channels that are constructed as two halves, or hemichannels. The hemichannels are joined head to head as they extend into the intercellular gap between two closely apposed cells, thereby forming a continuous pore that links the cytoplasm of one cell to another. Each of the hemichannels, also termed *connexons* (Goodenough, 1975), consists of six sub-units, termed *connexins,* which in vertebrates comprise a multigene family with 15 distinct members thus far identified. More than 40 variants of connexins have been documented when species diversity is taken into consideration (see the chapter by Beyer and Willecke earlier in this volume). It is the intent of this chapter to survey the biophysical properties of gap junction channels, specifically conductance, permeability, and voltage-dependent gating.

Because of the large size of the gap junction channel pore, the scope of permeant molecules extends beyond the common intracellular inorganic ions such as $Na^+$, $K^+$, and $Ca^{2+}$ to include a variety of amino acids, nucleotides, small peptides, and sugars ranging in molecular weight up to approximately 1 kD in vertebrates (Bennett and Goodenough, 1978; Loewenstein, 1981; Spray et al., 1985). Over the years, studies of junctional permeability, predominantly using fluorescent tracers, have shown a great deal of tissue-to-tissue variation in cell-cell transfer rate and apparent size selectivity. Rudimentary estimates of the size of the channel pore based on diffusion cutoff range from 10 to 25 Å, depending on the tissue type or species examined (Weingart, 1974; Simpson et al., 1977; Brink and Dewey, 1978; Flagg Newton and Loewenstein, 1979; Zimmerman and Rose, 1985; Imanaga et al., 1987). Single channel studies similarly show considerable tissue-to-tissue variation in unitary conductance and, in many instances, multiple conductance states have been reported from single isolated cell pairs. These tissue-variable characteristics are now being addressed at the connexin level by exogenous expression of connexins in isolation. As it is turning out, the multiplicity of conductance states appears to be due, in part, to the expression of several connexins in a single cell, and to the presence of one or more subconductance states in a single connexin channel. Topics presented here include quantification of permeability to intercellular tracers in various tissues, measurements of unitary conductance in native tissues and in exogenously expressed systems, and examinations of models of permeation.

The second property that is the focus of this chapter is voltage-dependent gating. In vertebrates the conductance of gap junctions, $g_j$, is dependent on transjunctional voltage, $V_j$, the voltage between the coupled cells' interiors. In invertebrates, and in some vertebrate tissues, $g_j$ is sensitive to the absolute membrane voltage, $V_m$, or, perhaps more appropriately termed $V_{i-o}$, the voltage between the inside and outside of the cells and the channel interior (Verselis et al., 1991). In some tissues, $g_j$ is as sensitive to voltage as are ion channels in excitable membranes, and in other tissues $g_j$ is quite insensitive to voltage (Bennett and

Verselis, 1992). The degree of voltage dependence may reflect the physiological role that the particular gap junctions play. Heart cells and giant septate axons, which behave as through-conducting systems, possess junctions that are weakly sensitive (or insensitive) to voltage, whereas rectifying electrical junctions are strongly voltage dependent and fast acting, and provide a means of regulating impulse transmission (Furshpan and Potter, 1959; Auerbach and Bennett, 1969; Bennett, 1977). The finding that junctions between amphibian blastomeres are strongly $V_j$ dependent was unexpected (Harris et al., 1981; Spray et al., 1981). The sensitivity is great enough that differences in resting potentials could cause communication compartments to be established in the developing blastula (Harris et al., 1983) and suggests that other nonexcitable tissues may also utilize voltage as a means of controlling cell-cell communication. As for permeation and unitary conductance, connexins are being expressed exogenously in isolation to "sort out" the connexins' specific voltage gating properties. In this chapter, descriptions of the voltage-dependent gating characteristics of gap junctions are discussed at macroscopic and single channel levels along with structure-function studies that are just beginning to shed light onto the molecular basis of gap junction channel gating and permeation.

## II. QUANTIFYING JUNCTIONAL CONDUCTANCE AND PERMEABILITY

In intact tissues or organ preparations, the measurement of junctional conductance, $g_j$, and permeability, $p_j$, is made difficult by complex, three-dimensional cell-to-cell coupling topologies which produce multiple sites of contact with other cells and, in some instances, other cell types. In early studies, these problems were circumvented by utilizing preparations, usually invertebrate giant fiber systems, that contained homogeneous cells coupled end-to-end (e.g., Johnston and Ramon, 1981; Verselis and Brink, 1984). The advent of cell isolation provided an improved and more convenient means of simplifying cell-to-cell coupling topology by the generation of cell pairs. Concurrent with cell isolation was the advent of the dual voltage-clamp technique that ultimately permitted the direct measurement of junctional membrane currents (Spray, et al., 1981). The dual voltage-clamp technique, when exercised using patch-clamp amplifiers in the whole-cell configuration, extended the measurement of $g_j$ to the single channel level (Veenstra and DeHaan, 1986). The dual voltage clamp, applied at macroscopic and single channel levels, has been the workhorse in gap junction biophysics over the past 10 years. Techniques that attempt to obtain "attached-patches" directly onto junctional membranes (Brink and Fan, 1989) and onto plasma membranes of cells expressing functional connexin hemichannels (Trexler et al., 1996) represent newer directions for single channel analyses.

## A.    Measurement of Macroscopic Junctional Conductance

The equivalent circuit of a two cell preparation can be represented by a coupling conductance ($g_j$) connecting two lumped parameter input conductances for each cell, $g_1$ and $g_2$ (Figure 1). The minimal requirements for assessing $g_j$ include

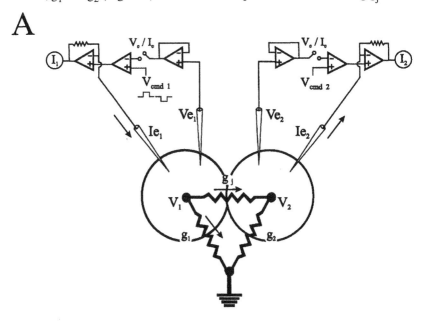

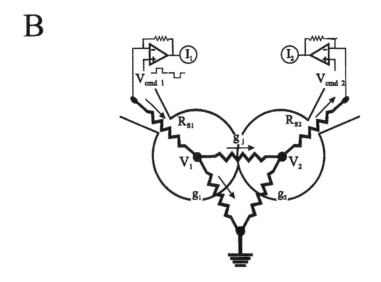

**Figure 1.** Techniques for the measurement of gap junctional conductance. Shown are two recording configurations, the double two-electrode current/voltage clamp (**A**) and the double whole-cell patch clamp (**B**). In both cases, an equivalent circuit of a cell pair coupled by gap junctions is represented by three principal conductances, the nonjunctional membrane conductances, $g_1$ and $g_2$, that connect the cytoplasms of the respective cells to the extracellular space, and the junctional conductance, $g_j$, that directly connects the cytoplasms of the two cells. **A.** In the double two-electrode configuration, each cell is impaled with a voltage-measuring ($V_e$) and current-passing ($I_e$) microelectrode. In current clamp, $g_1$, $g_2$, and $g_j$ can be determined by applying alternate current steps to each cell to measure input, $g_{11} = I_1/V_1$, $g_{22} = I_2/V_2$ and transfer, $g_{12} = I_1/V_2 = g_{21} = I_1/V_2$, conductances and applying a pi-tee transform (see Bennett, 1966). In voltage clamp, direct measurements of $g_1$, $g_2$, and $g_j$ are obtained by applying a voltage step to one cell (e.g., $\Delta V_1$) while keeping the voltage in the other cell ($V_2$) constant, and measuring the total change in current in the stepped cell, $\Delta I_1$, as well as the change in current in the unstepped cell, $\Delta I_2$. When stepping cell 1, $g_j$ and $g_1$ are obtained from $-\Delta I_2/\Delta V_1$ and $(\Delta I_1 + \Delta I_2)/\Delta V_1$, respectively; $g_j$ and $g_2$ are similarly obtained from $-\Delta I_1/\Delta V_2$ and $(\Delta I_1 + \Delta I_2)/\Delta V_2$ when stepping cell 2. **B.** In the double whole-cell recording configuration, access is gained to the interior of each cell by a patch electrode. $g_1$, $g_2$, and $g_j$ can be determined in current clamp and voltage clamp as described above in the double two-electrode configuration. However, significant errors in the measurement of $g_1$, $g_2$, and $g_j$ can occur due to the series electrode resistances, indicated as $R_{s1}$ and $R_{s2}$. Series resistances can be partially compensated electronically, but ideally cell pairs should be chosen in which $R_{s1}$ and $R_{s2}$ are low ($< 5\%$) relative to junctional resistance, $r_j = 1/g_j$, and in which nonjunctional resistances, $r_1$ and $r_2$, are high ($> 1$ Gohm) (see White et al., 1985; Veenstra and Brink, 1992).

---

the ability to measure the membrane voltages and input conductances of both cells. In large cells, this can be accomplished with four microelectrodes, one in each cell for measuring voltage and one for injecting current. With constant current application, this is known as the dual current clamp method. If constant current is injected into one cell (e.g., cell 1), and the membrane voltage deflections ($\Delta V_1$ and $\Delta V_2$) are measured in both cells, the ratio of $\Delta V_2/\Delta V_1$ provides a measure of the "coupling coefficient" in one direction, or $k_{12} = \Delta V_2/\Delta V_1 = 1/(1+g_2/g_j)$. The reciprocal injection of constant current into cell 2 provides a measure of the coupling coefficient in the reverse direction, $k_{21} = 1/(1+g_1/g_j)$. Although coupling coefficients have been used to estimate $g_j$, they do not provide an unambiguous measure of $g_j$ due to the influences of $g_1$ and $g_2$. However, $g_j$ can be assessed directly by applying a pi-tee transformation to the circuit in Figure 1 and measuring the "transfer conductance" ($g_{12}$ and $g_{21}$), which is defined as the ratio of the current injected into one cell to the voltage deflection in the other cell, or $g_{12} = g_{21}$ = $g_1 + g_2 + (g_1 g_2 / g_j)$, and the "total input conductance" ($g_{11}$ and $g_{22}$), which is defined as the ratio of current to voltage in the same cell (Bennett, 1966). The pi-tee transform provides $g_j$ in terms of the circuit constants $g_{11}$, $g_{22}$, and $g_{12} = g_{21}$ according to:

$$g_j = \frac{g_{11} g_{22} g_{12}}{(g_{12})^2 - g_{11} g_{22}} \tag{1}$$

The equality of the transfer conductance in either direction provides a means of testing for errors due to electrode placement or loss of voltage uniformity due to unfavorable geometry.

A direct means of assessing $g_j$ in cell pairs was made possible by utilizing a double voltage clamp (Spray et al., 1981). With this technique, each cell is voltage clamped independently using either a one- or two-electrode configuration. By stepping the voltage in one cell and keeping the other constant, junctional currents are observed directly as they flow into the unstepped cell. Thus, $g_j$ is obtained by dividing the junctional current by the applied voltage step (i.e., $g_j = I_j/V_j$). As in current clamp, the requirement of an accurately space clamped membrane does place certain limitations on the geometry of the preparation. In smaller cells, where impalement with two microelectrodes is not practical, the use of a single electrode in each cell in conjunction with switching amplifiers (Finkel and Redman, 1984) or patch-clamp amplifiers (Hamill et al., 1981) can be employed. Switching amplifiers, which alternate between brief periods of current injection and voltage recording (i.e., discontinuous single electrode voltage clamps), are inherently noisier than continuous voltage-clamp circuits and require fine tuning of the switching frequency relative to the membrane time constant to ensure that a steady state membrane potential is achieved. The patch-clamp technique, in the whole-cell configuration, offers the advantage of having low noise and refinements in the design of patch-clamp amplifiers and patch electrodes continue to improve the current resolution for single channel recordings.

When using the two-microelectrode or single-microelectrode switching configurations, problems associated with microelectrode series resistance are minimal. Series microelectrode resistance, however, imposes serious limitations when patch electrodes are used in the double whole-cell configuration. For the purpose of series resistance, it is useful to consider resistances rather than conductances. For the membrane potential ($V_m$) to equal the command potential ($V_c$) in any voltage-clamp circuit requires that the series resistance, $R_s$, be small relative to the input resistance ($R_{in}$) of the preparation, or $V_m = V_c - (R_s I)$. The appropriate equations for $V_j$ and $I_j$ have been derived elsewhere (Rook et al., 1988; Giaume, 1991; Veenstra and Brink, 1992) and are only briefly reviewed here. For a step command pulse applied to cell 1, $V_c = \Delta V_1$, $V_j = \Delta V_1 - \{(I_1 R_{s1}) - (I_2 R_{s2})\}$ and $I_j = \Delta I_2(1 + R_{s2}/R_{in2})$. Less obvious in these expressions is the influence of junctional resistance, $r_j = 1/g_j$, on the accuracy of these measurements, which is why alternative expressions for the deviations in $\Delta V_{m2} = V_2$, $\Delta V_{m1} = \Delta V_1$, and $I_j = -\Delta I_2$ as a function of $r_j$ have been derived elsewhere (Veenstra and Brink, 1992). Even under the ideal conditions when $R_s/R_{in} < 0.05$, the ratio of $r_j/(R_{s1} + R_{s2})$ must be considered since the $r_j$ estimate becomes less sensitive to $r_j$ as the value of $r_j$ approaches $(R_{s1} + R_{s2})$. This is due to the fact that low values of $r_j$ increase the magnitude of the whole cell currents, $(\Delta I_1$ and $\Delta I_2)$ when $V_j \rightarrow 0$, which leads to large series resistance errors. In other words, poor transjunctional voltage control can result from low $R_{in}$ or $r_j$ values. The double whole-cell clamp technique is most reliable when $r_j \approx R_{in} >> R_s$, which are the same conditions when the dual voltage-clamp procedures are least accurate ($k \rightarrow 0.5$). Conversely, dual current clamp procedures are most accurate when $R_{in} >> R \approx r_j$ ($k \rightarrow 1.0$), which are the very conditions when the double whole-cell recording technique is least accurate. However, it is possible to achieve a reliable $\delta V_{m1}$ using a voltage-clamp circuit as long as $R_{in} >> R_s$. Under these conditions, the coupling coefficient or transfer resistance can be determined by recording from the postjunctional cell while in the current clamp configuration to provide an accurate measure of $r_j$.

## B. Measurement of Single Channel Conductance

The double whole-cell patch recording technique also offers the capability of single channel recording. The best criteria for observing unitary currents are that $g_j < G_{in} << G_s$ and that there are approximately fewer than 5 to 10 operational gap junction channels present. When stepping $V_1$, unitary currents are recorded as discrete quantal changes in $I_2$, where $I_2 = I_j$, with equal and opposite quantal changes in $I_1$. The reverse would hold for stepping $V_2$. Discrete quantal current steps recorded by one voltage clamp that are not accompanied by equal and opposite current steps in the other clamp represent nonjunctional current events and are easily distinguished by these criteria. Since unitary currents can only be observed under conditions of low $g_j$ and low input conductance ($G_{in}$), series resistance contributions that plague macroscopic current measurements are negligible. The major limiting factor to the resolution of single gap junction channel currents are the variances of the background and open channel currents. As we discuss in this chapter, it is possible to derive additional information about single channel conductance, $g_j$, current variance, $s^2$, and open probability, $P_o$, from such junctional current records by deriving probability density functions (pdfs) from all points current amplitude histograms. These pdfs are based on the assumption that total junctional current is equal to the sum of the channel current amplitudes of all simultaneously open channels, all of which operate independently or:

$$I_j = g_j V_j = \sum_0^i (N^i P_o^i \gamma_j^i) \tag{2}$$

where $i$ is the number of independent populations of channels and N is the number of channels. Kinetic analysis of channel open and closed states is also possible, but measurements of single channel open and closed dwell times are best accomplished with one active channel and large numbers of events.

Most often, total $I_j$ between cells is too large to allow visualization of discrete unitary currents. This has necessitated the use of pharmacological agents to reduce the number of operational channels to a working level. In a limited number of cases, single channel records have been obtained from double whole-cell experiments without pharmacological intervention (Bukauskas and Weingart, 1993, 1994; Veenstra et al., 1994b; Oh et al., 1997). Single channel openings taken from multichannel records have also been used to obtain preliminary estimates of mean open time and $P_o$ (Chanson et al., 1993). However, the long-term stability of these recordings is generally not sufficient to permit quantitative analysis of channel open and closed dwell times (see, for example, Sigworth and Sine, 1987). The stability of stationary channel activity could be improved, in principal, by directly patching onto the exposed cytoplasmic surface of a junctional membrane. The organization of gap junction channels into plaques usually means that in a conventional (1 to 4 mm$^2$) membrane patch, either no channels will be present or considerably more than one channel will be present (Brink and Fan, 1989; Manivannan et al., 1992; Ramanan and Brink, 1993). The major drawbacks of the direct patch approach are technical; that is, the requirement for a clean exposed junctional membrane surface and the use of a cocktail of ion-selective channel blockers to isolate junctional channels from other single membrane ion channels (Brink and Fan, 1989; Veenstra and Brink, 1992). Despite these limitations, such direct patch recordings have been instrumental in developing new statistical techniques for estimating the conductance, subconductance state behavior, conductance shifts, open probabilities, open and closed channel variances, and cooperative or independent gating behavior of gap junction channels from single or multichannel records.

## C.  Measurement of Junctional Permeability

Junctional permeability, $p_j$, can be directly measured by injecting a traceable molecule in one cell and monitoring changes in concentration in the injected and recipient cells over time. As for $g_j$, measurement of $p_j$ is best obtained in isolated cell pairs or in preparations with a suitable geometry and coupling topology. Considering that the passage of molecules between cells is governed by laws derived from first principles of diffusion, $p_j$ can be defined as the ratio of transjunctional flux, $j_j$, to driving force. In the absence of a transjunctional voltage, $V_j$, $p_j = j_j/dC$, where $p_j$ is related to the diffusion coefficient, D, by $p_j = -D/dx$ and $dC/dx$ is the concentration gradient across the junctional membrane. Flux in the conventional sense, is defined as the number of moles of tracer that traverse a unit area of junctional membrane over time, and can be determined from the change in concentration of either cell over time according to:

$$j_j = -\frac{Vol_1 \, dC_1}{a \, dt} - \frac{Vol_2 \, dC_2}{a \, dt} \qquad (3)$$

where $a$ is junctional membrane area and $C_1$ and $C_2$ are the tracer concentrations in and $Vol_1$ and $Vol_2$ are the volumes of the injected and recipient cells, respectively. In the presence of a $V_j$, application of a modified form of the Goldman-Hodgkin-Katz constant field equation (see Brink and Dewey, 1978; Verselis et al., 1986) gives:

$$p_j = \frac{j_j 1 - e^{\frac{-zFV_j}{RT}}}{\frac{zFV_j}{RT} C_1 - C_2 e^{\frac{-zFV_j}{RT}}} \qquad (4)$$

where z is the valence of the tracer and F, R, and T have their usual meanings. With no leakage of tracer, $(Vol_1)(- dC_1) = (Vol_2)(dC_2)$. In practice, most fluorescent dyes have very low plasma membrane permeabilities so that leakage is not a problem. Nonetheless, $j_j$ is better measured as the appearance of tracer in the recipient cell due to ease of detection. If $j_j$ is sufficiently small such that the tracer concentration within each cell remains effectively uniform over a reasonable time interval, $p_j$ can be obtained simply by dividing the flux by the concentration difference between the cells, $j_j/\Delta C$.

Measurement of $p_j$ in intact tissues and cell monolayers requires application of multidimensional diffusion analysis (Crank, 1975). In such cases, diffusion is monitored over a distance from a point source of injected tracer and usually involves many cell boundaries. Given the potential tortuosity of the diffusion pathway within cells (Brink and Dewey, 1980) and the opportunity for leakage over such a large surface area, factors such as plasma membrane permeability and cytoplasmic diffusion can become significant contributors. Descriptions of quantitative analyses of axial and junctional $p_j$ in cell sheets and cell culture monolayers is provided in Safranyos et al., (1987), Brink and Ramanan (1985) and Ramanan and Brink (1990).

## III. PERMEABILITY AND CONDUCTANCE

### A. Permeability as Measured by Cell-to-Cell Diffusion of Tracers

#### *Quantification of Junctional Membrane Permeability*

Studies of gap junction channel permeability have taken advantage of the development of membrane-impermeant fluorescent dyes that could be injected into one cell and followed upon transfer to neighboring cells (see Loewenstein, 1981; Imanaga, et al., 1987). Many early studies attempted to assess channel size and selectivity by correlating the rate of dye transfer with the size and charge of the dye. However, the rate of dye transfer, which in essence examines junctional flux, does not provide an unequivocal measure of specific junctional membrane permeability. Excluding the binding of dyes to intracellular components (see Brink and Ramanan, 1985), intercellular flux obviously depends on the number of open gap junction channels and, perhaps less obviously, on cell volume. In tracer flux studies, cell volume is the analog to cell capacitance and for a brief injection

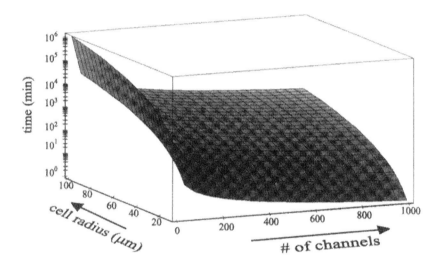

***Figure 2.*** Plot illustrating the effects of cell size and the strength of coupling on equilibration times for the diffusion of intercellular tracer from a tracer-injected cell to its coupled neighbor. The time to reach 90% equilibration, $t_{90}$, is plotted as a function of cell radius and the number of open channels. For a cell pair in which tracer is rapidly injected into one cell, the appearance of tracer in the recipient cell follows $C_0/2$ $(1 - \exp(- 3nP_c/2\pi a^3)$, where $C_0$ is the initial concentration of injected tracer, $n$ is the number of channels, $P_c$ is specific single channel permeability (in units of $cm^3/sec$) and a is the cell radius. Thus, $t_{90}$ follows $- (2\pi a^3(\ln(0.1))/(3nP_c)$. $P_c$ was fixed at $1.8 \times 10^{-14}$ $cm^3/sec$, using $P_c = (D\pi d^2)/(4L)$, where D is the tracer diffusion coefficient through the channel, and d and L are the channel diameter and length respectively. The following values were used: $D = 1 \times 10^{-6}$ $cm^2/sec$, $d = 15$ Å and $L = 100$ Å. The tracer was assumed to be freely diffusable, and impermeant to the nonjunctional membrane. The cells were assumed to be isopotential and of equal radii.

of tracer that establishes an initial concentration of $C_0$, the concentrations in the injected $(C_1)$ and recipient $(C_2)$ cells over time would follow:

$$C_2(t) = \frac{C_0}{2}\left(1 + e^{\frac{2ntp_c}{Vol}}\right) \tag{5a}$$

$$C_2(t) = \frac{C_0}{2}\left(1 - e^{\frac{2ntp_c}{Vol}}\right) \tag{5b}$$

where $p_c$ is the specific single channel permeability, n is the number of open channels and Vol is cell volume (assumed to be equal for both cells). The time required to reach 90% equilibration is plotted as a function of cell size and the number of open channels in Figure 2. Although specific channel permeability remains constant, the time to equilibration can vary significantly depending solely on cell size. Similarly, the time-to-equilibration is affected by the number of open channels, which depends on the level of expression, as well as on channel open probability. Thus, differences in flux due to differences in cell volume or open probability, or both, can be erroneously interpreted as differences in specific channel permeability or even impermeance if a low rate of transfer causes problems of detection. For fluorescent dyes, these problems can be exacerbated by optical interference, which can significantly diminish detection sensitivity.

As can be seen from Eqs. 3 and 4, a conventional measure of junctional membrane permeability, $p_j$, takes its analogy from classical membrane permeability theory in which flux through a unit area of membrane is measured and normalized to the driving force. In order to evaluate $p_j$ in this way, junctional membrane area must be assessed. Classically, junctional membrane area has been defined from freeze fracture replicas as the percentage of the total appositional membrane area that contains morphologically distinct plaques containing 60- 80-Å particles on one fracture face, usually the P-face, and corresponding pits on the complimentary E-face (Chalcroft and Bullivant, 1970; Goodenough and Revel, 1970; McNutt and Weinstein, 1970; Peracchia, 1980). In studies using isolated strips of cardiac tissue (Weidmann, 1966; Matter, 1973; Imanaga, 1974; Weingart, 1974; Tsien and Weingart, 1976) and the earthworm median giant septate axon (MGA) (Brink and Dewey, 1978), preparations in which the end-to-end coupling geometry of the cells made it possible to concentrate electron microscopic analyses to discrete regions of the membrane, junctional membrane area was found to occupy approximately 5% of the total appositional area. Quantifying $p_j$ using these calculated areas demonstrated specific junctional membrane permeability to be orders of magnitude larger than the specific nonjunctional membrane permeability. Channel diameter in each of these studies was estimated to be about 12 to 14 Å based on exclusion; $p_j$ was shown to be inversely proportional to molecular weight, and tracers with hydrated diameters >14 Å (e.g., halogenated fluorescein derivatives) were virtually impermeant. Hydration of the dyes inside the channel was assumed, a priori, because of the large size of gap junction channel pore. Supporting evidence that dyes diffuse through these MGA gap junction channels in a hydrated form came later from solvent substitution studies with deuterium oxide, $D_2O$ (Brink, 1983; Verselis and Brink, 1986). Substitution of $H_2O$ with $D_2O$, which magnifies the increase in solute hydration and hence effective solute radius with cooling, was shown to virtually block dye permeability at higher temperatures. These so-called solvent isotope effects were larger in dyes expected to be more hydrated consistent with their hydration inside the channel.

## Quantifying Total Permeability, $P_j$, and Conductance, $g_j$, Simultaneously—the $P_j/g_j$ Ratio

Quantitation of $p_j$ by normalization of flux to junctional membrane area is valid only if area is proportional to the number of open channels. This is problematic because morphological analyses are subject to errors, particularly in identifying particles as bona fide functional channels, and provide no information about channel open probabilities, which can vary considerably with physiological conditions. Another means of assessing $p_j$ is by normalizing the total permeability, which we will term $P_j$, by junctional conductance, $g_j$. $P_j$, with units of $cm^3/sec$, can be obtained by dividing the flux, $Vol_2\Delta C_2/\Delta t$, by the concentration difference between the cells, $\Delta C$, leaving out membrane area. Thus, $P_j$ is equivalent to conventional permeability, $p_j$, times membrane area and can be thought of as mean single channel permeability times the number of channels times the mean open probability. By analogy, $g_j$ is mean single channel conductance times the number of channels times the mean open probability. Differences in $P_j$ that result from differences in channel numbers or open probabilities, or both, will be reflected proportionally in $g_j$ (i.e., the $P_j/g_j$ ratio would be a constant). For the same permeant ion $A^\pm$, $g_j$ and $P_j$ are related by:

$$g_j = \frac{P_j z^2 F^2 [A^\pm]}{RT} \tag{6}$$

In practice, $P_j$ and $g_j$ are not interchangeable because they usually refer to different permeants. However, this difference in $P_j$ and $g_j$ provides a useful index, namely the $P_j/g_j$ ratio, which represents a measure of the relative difference in specific channel permeability to solutes that differ in size. When $g_j$ and $P_j$ are measured simultaneously in the same cell pair after injection of tracer, $g_j$ predominantly reflects the mean permeability of small inorganic ions (e.g., $K^+$ and $Cl^-$). Assuming independence of inorganic ion and tracer movement, the tracer, even if charged, would not appreciably contribute to $g_j$ because of the typically low concentrations of tracer used (<1 mM) and their relatively low mobilities.

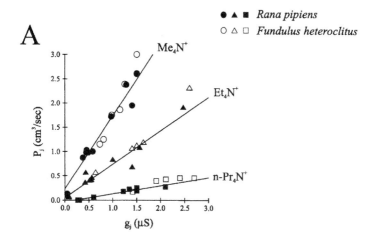

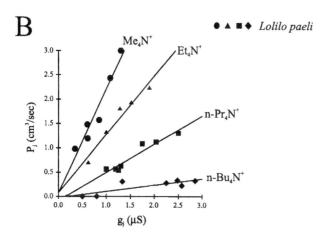

**Figure 3.** Plots of $P_j$ versus $g_j$ in cell pairs obtained from early embryos of *Rana pipiens*, *Fundulus heteroclitus*, and *Loligo paeli*. Each data point represents $P_j$ and $g_j$ obtained from a different cell pair. The data for *R. pipiens* (solid symbols) and *F. heteroclitus* (open symbols) are superimposable and are combined in **A**. The data for *L. paeli* are plotted separately in **B**. For each tetraalkylammonium ion, the relation between $P_j$ and $g_j$ was linear and inversely related to the size of the ion; that is, $Me_4N^+$ > $Et_4N^+$ > $Pr_4N^+$ > $Bu_4N^+$. The $P_j/g_j$ ratios were higher for *Loligo* than for *Rana* and *Fundulus*, suggesting that the squid channels are somewhat larger. The solid lines represent linear regressions with slopes of 1.50, 0.69, and 0.17 for $Me_4N^+$, $Et_4N^+$, and $Pr_4N^+$ in *Rana* and *Fundulus* and 2.23, 1.19, 0.58, and 0.13 for $Me_4N^+$, $Et_4N^+$, $Pr_4N^+$, and $Bu_4N^+$ in *Loligo*.

An example of the use of $P_j/g_j$ ratios to compare junctional permeabilities in different preparations is illustrated in Figure 3 using tetraalkylammonium ions, nonfluorescent tracers whose intracellular activities were monitored with ion-selective microelectrodes (see also Verselis, et al., 1986). $P_j$ values for teteramethylammonium (TMA), tetraethylammonium (TEA), tetrapropylammonium (TPA), and tetrabutylammonium (TBA) are plotted in Figure 3 as a function of $g_j$ in three different cell preparations obtained from embryos of frog *(Rana pipiens)*, fish *(Fundulus heteroclitus)*, and squid *(Loligo paeli)*. Each point represents $P_j$ for one of the tetraalkylammonium ions and the corresponding $g_j$ obtained from a single cell pair. $P_j$, although varying considerably among the cell pairs, remained proportional to $g_j$ for each of the tetraalkylammonium ions. The $P_j/g_j$ ratios for *Rana* and *Fundulus* were nearly the same, whereas in *Loligo* they were somewhat greater for each tetraalkylammonium ion. In *Fundulus,* as in *Rana,* permeability to TBA was undetectable, but was detectable in *Loligo* even with moderate coupling. The relatively low value of the $P_j/g_j$ ratio for TPA and the inability to detect transfer of TBA in *Rana* and *Fundulus* indicate that the effective gap junction channel diameter for cations in these embryos lies between that of these two ions. A somewhat larger channel size is indicated for *Loligo*. Also, the constancy of the $P_j/g_j$ ratio for each probe, regardless of its size, indicates that differences in $g_j$ reflect differences in the number of open channels rather than graded differences in specific channel permeability. The issue of different channel sizes among different tissues and organisms is examined in more detail next.

### Gap Junction Channel Size

Prior to the knowledge that connexins comprise a multigene family, estimates of channel size were obtained by assessing diffusion "cutoff" in preparations as distantly related as earthworm median giant axon (Brink and Dewey, 1978) and mammalian heart (Weingart, 1974). These early studies established a "canonical" gap junction channel diameter of approximately 12 to 14 Å. In subsequent studies, junctions between midge *(Chironomus)* salivary gland cells were shown to be permeable to considerably larger molecules, placing the channel diameter between 20 and 25 Å (Schwarzmann et al., 1981). Similarly, epidermal cells from beetles *(Tenebrio molitor)*, were also shown to pass rhodamine derivatives which spread poorly between mammalian cells in culture (Safranyos et al., 1987). On the other end of the spectrum, gap junction channels with smaller pores sizes were suspected from studies in which substantial electrical coupling was not accompanied by any detectable dye transfer to Lucifer yellow or 6-carboxyfluorescein. Examples of such preparations include insect epithelia at the borders of developmental compartments (Warner and Lawrence, 1982; Blennerhassett and Caveney, 1984), and vertebrate central nervous system (CNS) neurons (Dudek and Snow, 1985; Llinas, 1985).

Endowing different vertebrate tissues with gap junction channels of different pore diameters is plausible given the number of different connexins that have been identified and the expectation that their function is to communicate or restrict the communication of specific signaling molecules. However, corroborating evidence remains ambiguous. In cases where fluorescent dye transfer was monitored in whole tissues (e.g., CNS neurons), other complicating factors, such as intracellular binding, cell geometry, and tortuosity of the diffusion pathway, could slow intracellular diffusion sufficiently to give the appearance of reduced permeance or even impermeance. In a more recent study, electrically coupled neurons in the developing rat cortex did not exhibit dye coupling using the anionic dyes Lucifer yellow or 6-carboxyfluorescein, but did exhibit appreciable dye coupling using Neurobiotin, a cationic dye (Yuste et al., 1992; Peinado et al., 1993). Neurobiotin, although smaller than Lucifer yellow or 6-carboxyfluorescein, is positively charged and so these studies do not provide an unequivocal answer to the question of pore size; passage of Neurobiotin may have resulted from its smaller size or from a differential permeability for cations and anions, or both. It is also possible that cytoplasmic components in some neurons may differentially bind cationic and anionic dyes, further influencing the intercellular spread of dye independent of connexin properties.

With the advent of exogenous expression systems in which connexins can be expressed in isolation and examined in the same intracellular and extracellular environment, more rigorous examination of dye spread and its relation to conductance can be made. For example, injection of Lucifer yellow into UMR cells transfected with connexin45 (Cx45) or connexin43 (Cx43) showed spread to neighbors only in Cx43-transfected cells, even in those Cx45- transfected cells with comparable $g_j$ values to the Cx43-transfected cells (Steinberg et al., 1994). These data, although suggestive of a connexin-specific difference in channel size, remain inconclusive because of the possible influences of charge, as discussed earlier for Neurobiotin, and of channel gating. As an example of how channel gating can influence interpretations of data that compare conductance and permeability, single channel recordings of human connexin37 (Cx37) channels transfected in a rat Neuro2a cell line exhibit a main conductance state of approximately 300 pS and a long-lived substate of approximately 90pS (Veenstra et al., 1994b). The main state has an effective anion/cation permeability ratio of 0.43 and under conditions of moderate $V_j$, a low open probability. If the main open state was dye permeable and the substate was not, a large number of channels residing predominantly in the substate would provide for a high $g_j$ value, but limited dye transfer. Frequent residence in the main state would show concomitant dye transfer. These results indicate that macroscopic conductance and permeability are complexly related, depending not only on probe and pore structure, but also on gating characteristics.

How does channel pore size relate to unitary conductance? Insect gap junctions, which are presumed to have a larger pore size, are accompanied by large unitary

*Table 1.*  Gap Junction Channel Conductances in Native Tissues

| Preparation | Conductances (pS) | References |
| --- | --- | --- |
| Cardiac myocytes | | |
| Ventricular | | |
| embryonic chick | 166, 74 | Veenstra and DeHaan, 1986; Veenstra and DeHaan, 1988 |
| | 166, 58 | Veenstra, 1990 |
| | 240, 200, 158, 120, 81, 43 | Chen and DeHaan, 1992 |
| neonatal rat | 58 | Burt and Spray, 1988 |
| | 43, 18 | Rook et al., 1989; Rook et al., 1988 |
| | 20, 44, 70 | Takens Kwak and Jongsma, 1992 |
| neonatal hamster | 46, 33 | Wang et al., 1992 |
| adult guinea pig | 37 | Rudisuli and Weingart, 1989 |
| fibroblast-myocyte | 29 | Rook et al., 1989; Rook et al., 1992 |
| fibroblast-fibroblast | 22 | Rook et al., 1989; Rook et al., 1992 |
| Atrial | | |
| adult guinea pig | 36 | Lal and Arnsdorf, 1992 |
| Sinus node | | |
| adult rabbit | 92, 67, 52, 31 | Anumonwo et al., 1992 |
| Skeletal muscle | | |
| embryonic frog myotome | 100 | Chow and Young 1987 |
| Smooth muscle | | |
| vascular A7r5 cells | 89, 36 | Moore et al., 1991 |
| corpus cavernosum | 134, 93, 57, 38, 23 | Moreno et al., 1993 |
| Skin | | |
| human keratinocytes | 106, 78, 45 | Salomon et al., 1992 |
| Liver | | |
| white blood cells | 120, 87, 23 | Spray et al., 1991a |
| reconstituted membranes | 140–150 | Spray et al., 1986; Young et al., 1987 |
| Acinar cells | | |
| mouse pancreas | 130, 27 | Somogyi and Kolb, 1988 |
| | 96, 64, 31 | Somogyi et al., 1989 |
| rat lacrimal | 90–120 | Neyton and Trautmann, 1985; Neyton and Trautmann, 1986 |
| Ovary | | |
| Chinese hamster | 120, 70, 50 , 37, 22 | Somogyi and Kolb, 1988 |
| Brain | | |
| leptomeningeal | 80–90, 49 | Spray et al., 1991b |
| astrocytes | 50–60 | Dermietzel et al., 1991 |
| Retina, teleost | | |
| horizontal cells | 54 | McMahon et al., 1989 |
| Lens fiber cells | | |
| reconstituted membranes | 200 | Zampighi et al., 1985 |
| | 290, 90, 45 | Donaldson and Kistler, 1992 |
| Invertebrates | | |
| earthworm septate axon | 100 | Brink and Fan, 1989 |
| crayfish hepatopancreas | 250 | Chanson et al., 1994 |
| mosquito, C6/36 cell line | 375, 70 | Bukauskas and Weingart, 1993 |
| beetle epidermis | 197–347 | Churchill and Caveney, 1993 |

conductance values. Two different insect cell preparations, a cell line derived from the mosquito, *Aedes albopictus,* and cells isolated from epithelia of *T. molitor,* were reported to have unitary conductances of 365 and 345 pS, respectively (Bukauskas and Weingart, 1993; Churchill and Caveney, 1993; Weingart and Bukauskas, 1993). However, unitary conductances reported for mammalian junctional channels vary considerably (see Table 1), ranging from approximately 20 to 250 pS, and there appears to be little correlation between channel conductance and the ability to transfer dye. For example, gap junction channels between cardiac ventricular cells, with a unitary conductance of 50 to 60 pS, allow the passage of Lucifer yellow as well as gap junction channels between embryonic chick myocytes with a unitary conductance of 200 to 240 pS. Furthermore, no correlation was found between channel conductance and ionic or dye permeability among exogenously expressed connexins in Neuro2A cells (Veenstra et al., 1995; Veenstra, 1996). In single channel studies of a CMTX mutation in connexin32 (Cx32), S26L, unitary conductance using KCl in the patch pipettes was found to be indistinguishable from wild-type Cx32, yet measurement of channel size using polyethylene glycols (PEGs) indicated a significantly reduced diameter (Oh et al., 1997). These examples indicate that channel conductance, indeed, is not likely to be a useful index of channel pore size.

## B. Simple and Frictional Pore Models: Predictions for Unitary Currents, Selectivity, and Size

A simple pore model ideally assumes that channel dimensions alone determine the properties of conductance and permeability, which we will term $g_c$ and $p_c$, respectively. Such a channel would be nonselective except on the basis of size; that is, the selectivity sequence for monovalent cations would follow that in bulk solution with $Cs^+ > Rb^+ > K^+ > Na^+ > Li^+$. The relation between $g_c$ and $p_c$ in such a simple pore model would be:

$$g_c = \frac{Dpa^2}{L} * \frac{z^2 F^2 [A^+]}{RT} = p_c \left( \frac{z^2 F^2 [A^+]}{RT} \right) \tag{7}$$

where $p_c$ is defined by the first term, $Dpa^2/L$, in which D is the diffusion coefficient of the permeant ion in bulk solution and a and L are the channel radius and length, respectively. Although the predicted values for $g_c$ according to Eq. 7 are generally in the same range as those measured experimentally (see Table 1), a simple pore model fails to take into account a number of factors in nonideal conditions that are likely to exhibit considerable influence on ions and dyes as they traverse through the channel, namely, charge, friction, and access resistance.

## Influence of Charge

The influence of solute charge on permeability was evident even in the earliest quantitative fluorescent dye flux studies, when gap junction channels were generally regarded as nonselective aqueous pores. Using a series of halogenated derivatives of fluorescein, dyes of similar molecular weight, but increased negative charge, were found to have decreased junctional permeabilities (Brink and Dewey, 1980). The same conclusion was reached in somewhat less quantitative studies in which mammalian cell lines were systematically probed with a series of linear fluorescent amino acids and peptides that had similar abaxial widths (14 to 16 Å), but a series of gradually increasing backbone lengths (Flagg Newton and Loewenstein, 1979). Diminishing flux was evident by increasing the number of negative charges on the backbone. These data are consistent with the presence of a fixed negative charge within the gap junction channel pore and inconsistent with a simple nonselective pore. Slowed fluxes caused by the influence of charge would give artificially low estimates of channel size on the basis of exclusion. Although charge selectivity has not been examined in channels formed by all identified connexins, the effects of ion substitution on single channel conductance show a general bias towards cation selectivity and in some cases, substantial preference for cations (see the discussion that follows).

## Effects of Friction

Frictional forces encountered when solutes, charged or uncharged, diffuse through a restricted space can significantly reduce mobility. A relation describing reduced mobility in pores is given by:

$$ F = \frac{D_{pore}}{D} = C(1-\alpha)^2 \left( \frac{1 - 2.11\alpha + 2.09\alpha^3 - 1.71\alpha^5 + 0.73\alpha^6}{1 - 0.76\alpha^5} \right) \qquad (8) $$

where F is a reduction factor that represents the ratio of the diffusion coefficient in the pore, $D_{pore}$, to that in free solution, D, as it relates to $\alpha$, the ratio of ion radius to channel radius, a/r (see Dwyer et al., 1980). The first term, $C(1-\alpha)^2$, takes into account the statistical probability of solute entrance from free solution into the pore, where C is a constant used to calculate the area at the mouth of the pore, and the second term, $(1 - 2.11\alpha + 2.09\alpha^3 - 1.71\alpha^5 + 0.73\alpha^6)/(1 - 0.76\alpha^5)$, is the influence of viscous drag between water and the channel wall as approximated by the continuum expression describing the motion of spheres in a uniform cylinder (Renkin, 1955, Levitt, 1975, 1991). For a hydrated $K^+$ ion with a diameter of 5.5 Å, $D_{pore}$ would be reduced by 93% and $g_c$ to approximately 8.7 pS in a 12-Å channel and by 87% to approximately 25 pS in the 15-Å channel. Taking the dimension of Lucifer yellow as $\approx 11.5$ Å, it would have its mobility reduced by >99% even in channels as large as 15 Å.

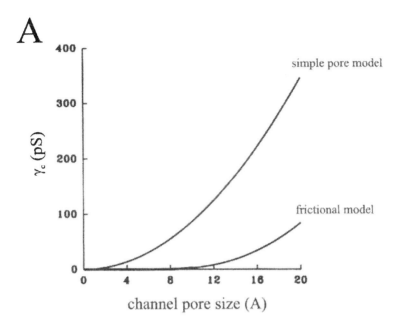

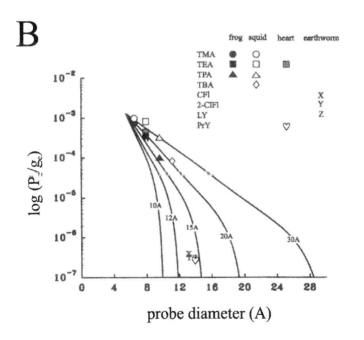

**Figure 4. A.** Theoretical predictions of single gap junction channel conductance ($\gamma_c$) as a function of pore size calculated for a simple pore and a pore in which frictional forces reduce solute (probe) mobility in the channel. For a simple pore bathed in symmetric KCl, $\gamma_c$ was calculated from $(D_K + D_{Cl})*(\pi a^2/L)*((z^2F^2[KCl])/(RT))$, where $D_K$ and $D_{Cl}$ are the bulk solution diffusion coefficients for $K^+$ and $Cl^-$, respectively at 25°C, a is channel radius, L is channel length, and z, F, R, and T have their usual meanings. For a frictional model, $D_K$ and $D_{Cl}$ were assumed to be reduced by a factor, $F(a,r) = (1 - \alpha^2)*(1 - 2.1\alpha + 2.09\alpha^3 - 0.95\alpha^5)$, where $\alpha = r/a$, the ratio of probe radius, r, to channel radius, a. The following values were used; $D_K = 1.96 \times 10^{-5}$ cm$^2$/sec, $D_{Cl} = 2.03 \times 10^{-5}$ cm$^2$/sec, [KCl] = 135 mM, L = 100 Å, $r_K = 3.05$ Å and $r_{Cl} = 3.01$ Å. $r_K$ and $r_{Cl}$ are estimated hydrated radii obtained from conductivity measurements at 25°C (Robinson and Stokes, 1965). **B.** Semilog plots of the ratio of $P_j$ to $g_j$ ($P_j/g_j$) as a function of probe size. The solid lines represent theoretically calculated $P_j/g_j$ values for five different fixed channel diameters of 10, 12, 15, 20, and 30 Å for probes ranging in size from that for $K^+$ (3.05 Å) up to the channel diameter. $g_j$ was calculated from $P_j z^2 F^2[KCl]/RT$ assuming that $K^+$ and $Cl^-$ were responsible for carrying all of the current and with $P_j$ assumed to follow $F(a,r)*D\pi a^2/L$. Superimposed on the theoretical plots are data obtained from various preparations in which $P_j$ and $g_j$ have been evaluated. The data for $Me_4N^+$, $Et_4N^+$, n − $Pr_4N^+$ and $nBu_4N^+$ in R. pipiens, F. heteroclitus, and L. paeli (shown in Figure 3) were obtained from simultaneous measurement of $P_j$ and $g_j$ in isolated blastomere pairs. $P_j/g_j$ values plotted for $Me_4N^+$ and $Et_4N^+$ in Aplysia buccal ganglion neurons (Bodmer et al., 1988), for 2–ClFl, 6CFl, and Lucifer Yellow in earthworm median giant septate axon (Brink and Dewey, 1978), for $Et_4N^+$ and Procion yellow in heart (Weingart, 1974) were calculated by multiplying the calculated permeabilities (in cm/sec) by the mean junctional membrane area (determined morphologically in separate studies), to give total permeability, $P_j$, in cm$^3$/sec and dividing by the mean $g_j$ for the corresponding cell preparation, also determined in separate studies.

---

Shown in Figure 4A are predicted values for $g_c$ as a function of channel size according to a simple pore model (Eq. 7) and a pore model that takes into account frictional forces (Eq. 8). In both cases, the exponential rise in $g_c$ reflects the dependence on the square of the channel radius. However, applying a frictional model considerably decreases the predicted conductance for a given channel size by a factor proportional to the ratio of solute radius to channel radius. Relating this to permeability, Figure 4B shows plots of theoretical $P_j/g_j$ ratios (solid lines) as a function of probe size for five different fixed channel sizes. $P_j$ and $g_j$ were computed according to Eq. 7 using values for the diffusion coefficient determined by multiplying D by F, the frictional mobility factor, according to Eq. 8. The theoretical values for $g_c$ in all cases were computed assuming symmetrical 135 mM KCl as the salt composition and thus, $K^+$ and $Cl^-$ as the principal current carrying ions. The theoretical $P_j/g_j$ ratios for all channel sizes converge as the probe radius approaches that of hydrated $K^+$ or $Cl^-$ and diverge strikingly with the increasing size of the probe. Superimposed on the theoretical plots are $P_j/g_j$ ratios obtained experimentally from a variety of preparations from which both $g_j$ and $P_j$ have been quantified, although not necessarily simultaneously. $P_j$ measurements were obtained for probes as small as TMA ($\sim$7 Å) and as large as Procion yellow ($\sim$17 Å). In no case do the ratios follow the theoretical predictions, typically deviating with increasing probe size for a given channel size. These deviations are possibly due to electrostatic interactions, which may be stronger with larger probes because charged moieties on these probes are more likely to interact with the channel wall. Nonetheless, rough estimates give predicted gap junction channel sizes in the 12- to 16-Å range for two vertebrate junctions, guinea pig heart (Weingart, 1974) and frog blastomeres (see also Figure 3), and one invertebrate junction, the earthworm median giant axon, MGA (Brink and Dewey, 1978). In squid blastomeres and insect epidermal cells, the predicted channel sizes are distinctly larger with diameters >20 Å. The tendency of invertebrate gap junctions to be larger was thought to be related to their morphological distinction; that is, upon freeze fracture the particles (assumed to represent channels) are larger in invertebrates (e.g., >150 Å in arthropods as compared to 100 to 120 Å in typical vertebrate junctions) and are retained on the E-face rather than on the P-face (Peracchia, 1980). These morphological distinctions have led to the classification of vertebrate junctions as A-type and invertebrate junctions as B-type. Interestingly, the earthworm MGA gap junctions, which, based on dye flux studies, are similar in size to vertebrate gap junctions are also morphologically similar; that is, they are classified as A-type (see Brink et al., 1981). It is now known that invertebrate gap junctions are composed of a gene family of proteins unrelated in primary sequence but with a predicted structure similar to that of the vertebrate connexins (Phelan et al., 1998).

### Effects of Series Access Resistance

As mentioned previously in section 1.A, series electrode resistance produces a voltage drop equal to $(I \times R_s)$ for each electrode which is minimized by keeping $R_s/R_{in} < 0.05$. Another series resistance component that contributes regardless of the recording configuration is the cytoplasmic access resistance. According to Eq. 7, the total resistance encountered upon traversing the channel depends on the resistivity, r, of the electrolyte solution and the radius ($a$) and length (L) of the path according to the relation $r = r/pa^2L$. When diffusing from bulk solution, solutes also encounter an access resistance of $r/2a$ that effectively increases the path length by a factor of $1.6a$ (see Hille, 1992). Wilders and Jongsma (1992) take the consideration of access resistance one step further in proposing that the 9.5-nm center-to-center packing of gap junction channels within a plaque, defined as $a_j$, produces a hemispherical shell of resistance equal to $r/2pa_j$. This model predicts that the applied $V_j$ for a given channel diminishes as its location nears the center of the gap junction plaque. This model would also predict that the apparent $g_j$ for a more centrally located channel would diminish as the number of open channels increased and that the gating of adjacent channels would be interactive.

The validity of the preceding access resistance model rests with the added assumption of flux coupling between uniformly spaced adjacent open channels, which greatly enhances the contribution of cytoplasmic access resistance to the overall measurement of $g_j$. The only evidence presented to support this model is the phenomenological reduction in $V_j$ sensitivity with increasing $g_j$ observed in a variety of cell pair preparations. High levels of connexin expression in *Xenopus* oocytes which, in turn, led to series access resistance problems is now believed to explain the lack of $V_j$ dependence originally described for Cx32 and Cx43 by (Swenson et al., 1989; Werner et al., 1989). In small cells, where $g_j$ is measured using the double whole-cell clamp technique, the effects of electrode series resistance must also be considered (Veenstra and Brink, 1992).

The general applicability of relating $g_j$ to the degree with which cytoplasmic series access resistance exerts an influence depends on the geometry of the junctional plaques. Numerous smaller plaques that confer the same macroscopic $g_j$ as a single large plaque will exhibit a reduced influence of series access resistance. Assembly of gap junctional channels into plaques introduces a type of cooperativity through the actions of series resistance; that is, as some channels close and reduce $g_j$, series access resistance diminishes which further influences channel closure. Observations at the single channel level indicate that while cooperative gating and conductance shifts in multichannel records do occur, some of the conductance shifts can readily be explained by substate behavior and fail to make any inference into mechanisms for cooperative gating behavior (Manivannan et al., 1992; Ramanan and Brink, 1993; Veenstra et al., 1994b). Theoretical studies by Ramanan and associates (1994) suggest that the ionic current flow around weakly ion selective channels is small and not likely to produce the magnitude of the effects modeled by Wilders and Jongsma (1992). An independent estimate of the contribution of cytoplasmic access resistances can be achieved by altering solvent viscosity with the addition of high molecular weight polymers that are impermeant to the channel (Bezrukov and Vodyanoy, 1993).

## C.  Studies of Unitary Conductance in Native Tissues

Unitary gap junction channel currents were initially recorded from rat lacrimal gland and embryonic chick heart cell pairs (Neyton and Trautmann, 1985; Veenstra and DeHaan, 1986). The 120- to 160-pS conductance values of these channels were definitively identified as gap junctional by the simultaneous occurrence of equal and opposite quantal current steps in both cells (Veenstra and DeHaan,

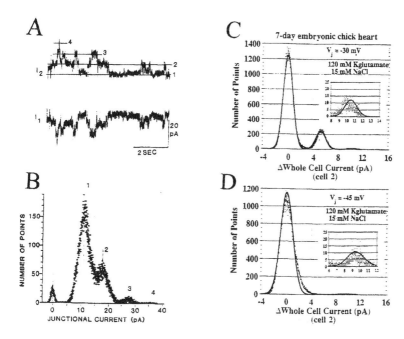

**Figure 5.** **A**. Equal amplitude and opposite polarity junctional whole cell currents ($I_1$ and $I_2$) from an 18-day embryonic chick heart cell pair clamped at a constant $V_j$ of 40 mV. Channel openings appear as upward deflections in $I_2$ (depicted by the thin solid lines) and downward deflections in $I_1$. Low-pass filter frequency was 200 Hz and the digital sample rate was 1 kHz. **B**. All points channel amplitude histogram of the 37-second $I_2$ record illustrated in panel **A**. $I_j = 0$ was determined by setting $V_j = 0$ mV before and after the 40-mV $V_j$ pulse. Four peaks are evident at 10.9, 17.6, 26.5, and 35.5 pA. Event counts were obtained by setting threshold detectors equal to the minimum (valley) between adjacent peaks. (**A** and **B** reproduced with permission from Veenstra, 1991). **C**. All points histogram obtained from the $\Delta I_2$ current trace (not shown) of a 7-day embryonic chick ventricular myocyte pair during a 2-minute, – 30-mV $V_j$ pulse. The probability density function (pdf) of the current distribution was determined assuming the presence of two simultaneously active 5.25-pA channels with a closed and an open channel variance of 0.55 pA. The inset illustrates the fit of the outermost peak. Event counts were 100 and 10 for the respective open current peaks. See text for further details about the equation used to generate the maximum likelihood estimate of channel parameters. **D**. Histogram produced by a 2-minute $V_j$ pulse to –45 mV in the same cell pair shown in panel **C**. Note the reduction in the relative open channel dwell time (area under the open current peak) to < 1%, down from 16% at –30 mV. There were 18 channel events and the open current variance was 1.1 pA. Acquisition parameters for **C** and **D** were analog filter = 100 Hz and digital sample rate = 2 kHz.

1986; Figure 5A). Channel conductances are typically determined from the statistical mean of Gaussian peaks apparent in the amplitude histograms compiled from junctional current recordings. Figure 5B is an example of an all-points current amplitude histogram from an 18-day embryonic chick heart cell pair. During the 37-second $V_j$ step of $-40$ mV, channel activity with two distinct amplitudes of 167 ($N = 73$ channel events) and 220 to 225 pS ($N = 7$ events) was observed. Although it is difficult to assess the basis for multiple conductance levels from similar multichannel recordings, the detection of direct transitions between different current levels is sufficient for determination of channel current amplitudes and, therefore, conductance. To ensure that the observed activity is representative of a given preparation, the histogram should be reproducible by longer repetitive recording periods in the same preparation. The 160-pS channel has been observed in the embryonic chick heart from 4 through 18 days of development despite differences in the $V_j$ sensitivity of macroscopic $g_j$ during this same time period (Veenstra, 1991).

Additional information can be obtained from amplitude histograms by fitting the Gaussian distributions with a probability density function (pdf) of the form:

$$f_i(M) = \sum_{n=0}^{M} B(M, n, p_i) f_b(I\ I_n, \sigma_b^2 + n\sigma_0^2) \tag{9}$$

where M is the number of channel types, n is the number of simultaneously open channels, $p_i$ is the open probability, $\sigma_b^2$ is the variance of the baseline noise, $\sigma_0^2$ is the variance of the open channel noise, and $I_n$ is the current of n open channels (Manivannan et al., 1992). Figure 5C is an all-points amplitude histogram compiled from a 2-minute junctional current recording from a 7-day chick heart cell pair held at a constant $V_j = -30$ mV. In this example, the current distribution (points) was fitted by a pdf (line) assuming two 5.25-pA (175-pS) channels with an identical $p_i = 0.09$. Figure 5D is an all-points histogram obtained from the same cell pair held at $V_j = -45$ mV for 2 minutes. At $-45$ mV only a single open channel is observed with an amplitude of 9.4 pA (208 pS) and $p_i = 0.01$. Taken together, this analysis illustrates the existence of multiple conductances and the $V_j$-dependent decline in open probability of embryonic chick heart gap junction channels. Kinetic analysis of gap junction channel gating is best obtained when a single channel is present and several hundred channel events are counted at a given voltage.

## Unitary Conductance and Connexin Expression

Using similar approaches, the channel conductances of native gap junctions have been determined from a variety of cell pair preparations and reconstituted membrane preparations, as summarized in Table 1. All of these experiments were performed with 120- to 150-mM electrolyte solutions and variations in channel conductances partially reflect the differences in electrolyte composition (e.g., $K^+$, $Cs^+$, Glutamate$^-$, $Cl^-$, etc.). The channel conductance effects of autonomic neurotransmitters and phosphorylating-dephosphorylating agents are not included in this listing and are only briefly discussed later in this chapter. Gap junction channel conductances vary from approximately 20 to 300 pS, indicative of the considerable species and tissue diversity of gap junction composition. Still, there are some similarities among the various preparations listed. Hepatocyte-derived cell lines, junctional membranes, and secretory acinar cells express gap junction channels with similar conductance properties. Likewise, mammalian cardiac myocytes, vascular smooth muscle cells, brain leptomeningeal cells, and astrocytes exhibit similar channel conductances.

The observation of similar gap junction channel conductances in functionally diverse tissues can readily be explained by the expression of the same connexins. Cx43 was detected in Western and Northern blot analyses of cultured astrocytes and leptomeningeal cells (Dermietzel et al., 1991; Spray et al., 1991b) and is known to be widely expressed in vascular smooth muscle (Moore et al., 1991), cardiac muscle (Beyer et al., 1987), endothelium (Larson et al., 1990), and lens epithelial tissues (Beyer, 1990). Cx32, cloned from a rat liver cDNA library, also exhibits a wide tissue distribution in Northern blot analysis (Paul, 1986).

Many tissues express multiple connexins as well, which contributes to the multiplicity of channel conductances found. Multiple connexin expression in a single tissue was first documented by the biochemical characterization and eventual cloning of a second connexin, connexin26 (Cx26), from mammalian liver (Nicholson et al., 1987; Zhang and Nicholson, 1989). Molecular cloning, Northern blot, and Western blot analyses have demonstrated the presence of connexin42 (Cx42), Cx43, and Cx45 in embryonic chick heart (Beyer, 1990; Musil et al., 1990a); connexin40 (Cx40), Cx43, and Cx45 in mammalian heart (Kanter et al., 1992), Cx37 and Cx43 in endothelium (Reed et al., 1993), and Cx40 and Cx43 in vascular smooth muscle cells (Beyer et al., 1992), to give a few examples. Indeed, multiple connexin expression within a given tissue is likely to be the rule rather than the exception. Based on sequence homologies, the connexin family of proteins can be subdivided into Groups I and II (Bennett et al., 1994). Typically, the Group I connexins (e.g., Cx26 and Cx32) are coexpressed in the same tissue while the Group II connexins (e.g., Cx37, Cx40, Cx42, Cx43, Cx45) are coexpressed in other tissues. Connexins from both Groups I and II (Cx26 and Cx43) were identified in Northern blots of isolated leptomeningeal cell RNA, although the observed biophysical channel properties were similar to those from Cx43-expressing tissues (Spray et al., 1991b). Multiple channel conductances may also result from modulation of channel conductance by second messenger pathways or the occurrence of channel subconductance states. Definitive proof of the molecular basis for multiple channel conductances in tissues requires comparison with the biophysical properties of the expressed connexin-specific channels. As we discuss later in this chapter, the functional expression of distinct connexins in communication-deficient cell lines is already providing evidence to support all three mechanisms (i.e., multiple connexin expression, modulation of channel conductance by second messenger pathways, and channel subconductance states).

## Selectivity

Experimental evidence of ion selectivity of native gap junctions at the single channel level is sparse. The relative gap junctional permeabilities of $Cl^-$ and $Na^+$ to $K^+$ in rat lacrimal gland cell pairs were estimated applying the Goldman-Hodgkin-Katz voltage equation under conditions in which asymmetrical electrode solutions were used (Neyton and Trautmann, 1985). The $P_{Na}/P_K$ ratio of 0.81 can be explained by the different mobilities of the two ions in bulk solution. However, the $P_{Cl}/P_K$ ratio of 0.69 is unexpectedly low given the similar mobilities of $Cl^-$ and $K^+$. Direct patch experiments on the 100-pS channel of the earthworm MGA septal membrane produced similar relative conductance ($G_X/G_K$) ratios of 1.0, 0.84, 0.64, and 0.52, and 0.2 for $Cs^+$, $Na^+$, $TMA^+$, $Cl^-$, and $TEA^+$, respectively (Brink and Fan, 1989). Again, the relative cation conductances follow the predictions for an aqueous pore while the relative chloride conductance is slightly lower than predicted for a completely nonselective aqueous ion channel. Channel conductances of A7r5 vascular smooth muscle and rat liver-derived WB cell pairs also increased upon switching from Kglutamate to CsCl- or KCl-containing electrode solutions (Moore et al., 1991; Spray et al., 1991a). All of these experiments demonstrate that both cations and anions can permeate gap junction channels in mammalian tissues despite slight differences in the cation/anion permeability or conductance ratios. A summary of the relative permeability or conductance ratios of gap junction channels from rat lacrimal gland, earthworm septate axon, and cardiac myocyte intercalated disc demonstrates that the permeability data from a wide variety of gap junctions, plotted as a function of the unhydrated solute diameter, closely follow a common second-order polynomial relation (Brink, 1991). These results were interpreted as consistent with a large aqueous pore containing a slight fixed negative electrostatic charge. Conductance-mobility plots produced by equimolar ion substitutions for exogenously expressed Cx26, Cx37, Cx40, Cx43, and Cx45 channels reveal a selectivity for cations ranging from < 2:1 for Cx43 to 10:1 for chicken Cx45 (Veenstra et al., 1995; Veenstra, 1996; Wang and Veenstra, 1997), again consistent with a negative electrostatic charge associated with the connexin pore. Thus far, only Cx32 appears to be slightly anion selective.

More recently, permeability ratios obtained from reversal potential measurements in salt gradients for Cx46 hemichannels (Trexler et al., 1996) and Cx43 and Cx40 channels (Beblo and Veenstra, 1997; Wang and Veenstra, 1997) demonstrate $P_K/P_{Cl}$ to be higher than that calculated from conductance ratios or conductance-mobility plots. These studies demonstrate that the conductance of a connexin channel or hemichannel is not necessarily a good predictor of permeability and that ionic fluxes through these large aqueous pores need not obey the laws of simple diffusion as predicted by the independent movement of ions. Fits of monovalent ion permeablity ratios for Cx40, Cx43, and Cx46 to the "hydrodynamic equation" (Eq. 8) estimate limiting pore diameters in the range of 10 to 13 Å (Beblo and Veenstra, 1997; Wang and Veenstra, 1997). Thus, given that the hydrated diameter of a $K^+$ or $Cl^-$ ion is less that 50% of the estimated connexin pore diameter, stronger interactions between ions and the channel wall may be occurring than previously expected. Further investigation of ionic selectivities and the mechanisms of ion permeation are needed to determine the relationship between gap junction channel conductance and ionic selectivity.

### Connexin Hemichannels

The often rapid formation of coupling between cells placed in contact has suggested that assembled hemichannels (or connexons) are present in the plasma membrane or situated close by where they could be rapidly recruited for cell-cell channel formation. Rate of channel formation as a function of injected RNA concentration exhibits properties of a bimolecular reaction, providing indirect support for the existence of membrane pools of hemichannels (Dahl et al., 1991). There are data supporting oligomerization of Cx43 protein into multimers, perhaps hemichannels, prior to membrane insertion (Musil and Goodenough, 1993). However, most connexins appear to be nonfunctional as unapposed hemichannels, suggesting that they closed by some mechanism in order to maintain the integrity of the cell.

In a few cases, unapposed hemichannels have been reported to be functional, at least under certain experimental conditions. In mouse macrophages and the J774 macrophagelike cell line, the application of $> 100$ μM extracellular $ATP^{4-}$ dramatically increased the membrane conductance and permeability to large hydrophilic molecules (e.g., fura-2, $M_r = 831$). These adenosine triphosphate (ATP)-permeabilized cells were shown to express Cx43 by Northern and Western blot analysis (Beyer and Steinberg, 1991). In contrast, Alves and associates (1996) reported that the ATP-induced uptake of Lucifer yellow did not increase and decrease with treatments expected increase and decrease hemichannel opening, respectively. Also, there is no direct electrophysiological evidence that Cx43 forms functional hemichannels in these cells or that extracellular ATP (not intracellular MgATP, Sugiura et al., 1990) modulates hemichannels.

Permeabilization of cell membranes, which ultimately resulted in cell death, was observed in *Xenopus* oocytes following injection of Cx46 mRNA (Paul et al., 1991; Ebihara and Steiner, 1993). Cx46 mRNA injections induced the expression of a large time- and voltage-dependent current, $I_h$, which was suppressed by extracellular acidification and increased extracellular $Ca^{2+}$ or trivalent cations $La^{3+}$ and $Gd^{3+}$. Similar currents were not observed in oocytes injected with mRNAs for

Cx32, suggesting that the induction of $I_h$ was owing to the expression of Cx46. The preceding findings were explained by the formation of functional Cx46 hemichannels in the oocyte membrane prior to gap junction formation initiated by cell-cell contact. Single (hemi)channel recordings from Cx46 expressing oocytes (Figure 6) exhibit a large (> 300 pS) conductance with properties consistent with those observed macroscopically (Trexler et al., 1996). These hemi(channels) display a substantial preference for cations over anions, yet have a relatively large pore as inferred from permeability to TEA (~8.5 Å). At hyperpolarized potentials, Cx46 hemichannels are fully closed and activate steeply with moderate depolarization. The polarity of activation and kinetics of this gating indicate that it is the mechanism by which these hemichannels open and close in the cell surface membrane when unapposed by another hemichannel (Trexler et al., 1996). Thus, Cx46 hemichannels can activate at physiological voltages and could induce substantial cation fluxes in cells expressing Cx46, such as lens (Goodenough, 1992) and proliferating Schwann cells (Chandross et al., 1995).

Several reports have indicated that Cx43, chick Cx56 (a likely ortholog to rat Cx46), and XenCx38 also have the capacity to form functional hemichannels when exposed to low external $Ca^{2+}$ concentrations (Ebihara et al., 1995; Ebihara, 1996; Li et al., 1996). Thus, it is possible that most, if not all connexins under suitable conditions may function as unapposed hemichannels.

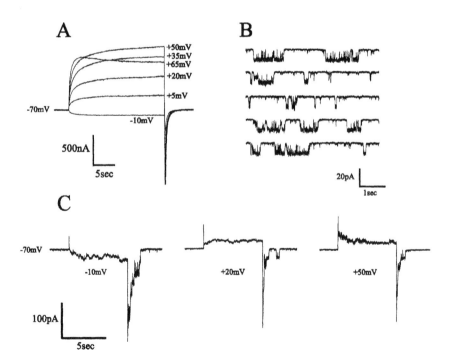

***Figure 6.*** Examples of macroscopic and single channel recordings from Xenopus oocytes expressing Cx46. **A**. Macroscopic currents elicited in a two-electrode voltage clamp experiment from a single *Xenopus* oocyte injected with Cx46 mRNA together with DNA antisense to the endogenous *Xen*Cx38. From a holding potential of $-70$ mV, voltage steps from $-10$ to $+65$ mV in increments of $+15$ mV were applied to this cell. Currents activated with depolarization, and reversed near 0 mV. At large inside positive voltages, conductance decreased as indicated by a reduced incremental change in current between $+35$ mV and $+50$ mV, and a pronounced current relaxation followed a peak at $+65$ mV. Repolarization to $-70$ mV produces inward tail currents. Time constants of the tail currents are strongly voltage dependent (not shown). The bath solution contained (in mM) 88 KCl, 1 NaCl, 2 $MgCl_2$, 5 glucose, 10 HEPES, pH $= 7.6$. The microelectrodes contained 1M KCl. **B**. Cell-attached patch recording of a single Cx46 hemichannel. The currents shown were elicited at a net transmembrane voltage, $V = -35$ mV ($V = V_{rp} - V_{pipette}$). $V_{rp}$ is the resting potential of the cell as measured with a separate intracellular microelectrode (for this cell, $V_{rp} = -20$ mV). The patch pipette solution contained (in mM) 100 KCl, 1 $CaCl_2$, 2 $MgCl_2$, 5 EGTA, 10 HEPES, pH $= 7.6$. Data were filtered at 1 kHz with a 4-pole lowpass Bessel filter and digitized at 5 kHz. **C**. Excised (inside-out) patch recording containing several Cx46 hemichannels from the same oocyte as in **B**. From a holding potential of $-70$ mV, steps to $-10$ mV and $+20$ mV elicited slowly activating inward and outward currents, respectively. Stepping to $+50$ mV elicited a faster rising outward current (not resolvable at this time scale) that slowly decayed to a lower value. Single channel currents were not resolvable during any of the three depolarizing steps, but were clearly visible in the tail currents upon repolarization back to $-70$ mV. Ensemble averaging of the tail currents and extrapolation to the peak indicated that there were 7 hemichannels in the patch that were activated with depolarization. The bath solution contained (in µM) 88 NaCl, 1 KCl, 2 $MgCl_2$, 5 glucose, 10 HEPES, pH $= 7.6$. Nominal $Ca^{2+}$ was 8 to 10 (M. The patch pipette solution, filtering and acquisition were the same as in **B**. (From Trexler et al., 1996.)

Evidence for the existence of functional hemichannels in vivo comes from horizontal cells of the teleost retina. Gap junctions are known to mediate the lateral inhibition of excitatory stimuli in the horizontal cell layer of the outer retina, although the expressed connexin has yet to be identified. Dopamine, acting through a cyclic adenosine monophosphate (cAMP)-dependent protein kinase pathway, reduces $g_j$ by lowering the apparent open probability of a 50- 60-pS junctional channel (McMahon et al., 1989). In an independent study, DeVries and Schwartz (1992) demonstrated that horizontal cells exhibit a $Ca^{2+}$ and dopamine-sensitive ionic current ($I_g$) in the plasma membrane carried predominantly by monovalent cations (i.e., $Na^+$, $K^+$, $Cs^+$), although $I_g$ exhibits some finite anion permeability as well. $I_g$ is activated by lowering extracellular $Ca^{2+}$ from 1 mM to 10 μM and suppressed by intracellular acidification or activation of the adenylate cyclase or guanylate cyclase second messenger cascade pathways. $I_g$ also undergoes a time- and voltage-dependent decay when the membrane potential is hyperpolarized, but not when the membrane potential is depolarized to values greater than 0 mV. All of these regulatory properties are consistent with the expected behavior of a gap junction hemichannel. Supporting evidence that $I_g$ is produced by a hemichannel is the observation that isolated horizontal cells are permeable to Lucifer yellow under conditions that activate $I_g$ and are rendered impermeable to Lucifer yellow in the presence of dopamine or high $Ca^{2+}$. In cell-attached or excised patches, the activity of a population of 145-pS channels was observed. The activity of these 145-pS channels in the cell-attached patch-clamp configuration was similarly suppressed by the extracellular application of 1 mM dopamine. It is of interest to note that the conductance of the presumed hemichannel of the catfish horizontal cells is more than double the 50- to 60-pS conductance of the intact gap junction channels from the perch horizontal cells. The physiological suppression of $g_j$ by dopamine in both species suggests that the same connexin may be present in retinal horizontal cells of all teleosts, yet the 145-pS conductance of the hemichannel (in 70 mM Cs-D-aspartate and 25 mM TEA-D-aspartate) is 2.4 to 2.9 times the conductance of the intact gap junction channel (in 72 mM Kgluconate, 48 mM KF, and 4 mM KCl; McMahon et al., 1989). The only additional evidence that could address this hypothesis further would be the identification of the connexin expressed in horizontal cells and transmission electron micrographs of the junctional membranes of paired and isolated cells, possibly performed in conjunction with immunolabeling of the connexin, to confirm the structure and composition of the intact and isolated gap junctions. Evidence for an in vivo functional role for these hemichannels remains to be established.

## IV.  VOLTAGE-DEPENDENT GATING

### A.  Multiplicity of Voltage Gating: Definitions and Descriptions

There is now incontrovertible evidence that all vertebrate gap junctions are sensitive to voltage to some extent. Conflicting reports of voltage sensitivity or a lack thereof have now been shown to be in error due to the effects of uncompensated series electrode resistance (see section II.A) or cytoplasmic series access resistance (see section III.B), or both. Voltage dependence in gap junctions has been described to be of two fundamental types, transjunctional, $V_j$, and transmembrane, $V_m$, or inside-outside, $V_{i-o}$. This distinction is based on the sensitivity of gap junctions to two different types of voltage stimuli, one that imposes a potential difference between the cells ($V_j$) and one that imposes a

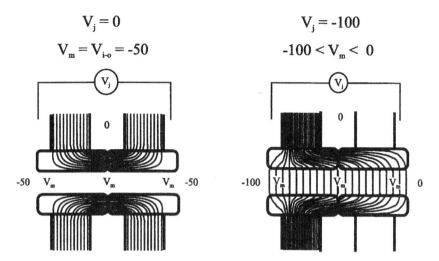

**Figure 7.** Schematic representation of a gap junction channel with presumed isopotential lines when the cells are held at the same voltage (i.e., transjunctional voltage, $V_j$, is absent) and when the cells are held at different voltages (i.e., $V_j$ is present). $V_j$ is defined as the voltage difference between the cytoplasms of the two cells. $V_m$ is defined as the voltage difference between the inside of the cells and the channel lumen relative to the outside extracellular space (also termed $V_{i-o}$). With both cells at, for example, $-50$ mV, the voltage in the channel lumen ideally can be assumed to be isopotential with the cytoplasms of both cells. This condition establishes a large voltage drop across the channel wall (i.e., between the channel lumen and the extracellular gap between the two cells) that is essentially uniform along the channel length. Gap junction channels that respond to the magnitude of this voltage drop are termed $V_m$ dependent and could conceivably involve voltage-sensing residues placed in the channel wall at or near the intercellular gap. In the presence of $V_j$ (e.g., with one cell at $-100$ mV) and the other at $0$ mV, ideally $V_j$ is constant along the channel length, whereas $V_m$ changes along the channel length from $-100$ to $0$ mV. Gap junction channels that respond only to the voltage difference between the two cells, regardless of the magnitude of $V_m$, are termed $V_j$ dependent and are likely to involve voltage-sensing residues that respond to the local field in the lumen. Some gap junction channels can respond to both $V_j$ and $V_m$.

voltage difference across the cell membrane ($V_m$) or, perhaps more appropriately, between the inside of each cell and the channel pore relative to the extracellular space. Shown in Figure 7 are schematic diagrams of presumed isopotential lines and associated fields in gap junction channels exposed to $V_m$

and $V_j$ stimuli. Considering the channel pore as perfectly insulated from the extracellular space, under conditions in which both cells are at the same voltage (Figure 7A), the channel pore would be isopotential with the cytoplasm of both cells. This condition generates a uniform voltage drop across each cell membrane and the channel wall as it traverses the intercellular gap. This voltage drop defines $V_m$ or $V_{i-o}$. Under conditions in which $V_j$ is imposed (i.e., both cells at different voltages; see Figure 7B), $V_m$ again also is established, but is different in the two cell membranes and is nonuniform across the channel wall along its length. Only the voltage difference between the cytoplasm of one cell relative to the other defines $V_j$.

$V_j$ dependence was first described in electrical synapses in nervous tissue (Furshpan and Potter, 1959; Auerbach and Bennett, 1969). These junctions were characterized by steep rectification in which $g_j$ increased and decreased when the presynaptic side was made relatively positive and negative, respectively. The changes in $g_j$ with $V_j$ displayed extraordinarily ($< 1$ msec) fast kinetics (Margiotta and Walcott, 1983; Jaslove and Brink, 1986; Giaume et al., 1987). A slow form of voltage dependence was subsequently described in amphibian blastomeres and was characterized by reductions in $g_j$ for either polarity of $V_j$; $g_j$ decreased symmetrically about $V_j = 0$ with kinetics that took hundreds of milliseconds to seconds to reach steady state (Harris et al., 1981; Spray et al., 1981). More recent studies using exogenously expressed connexins in various pairing combinations have shown that junctions can exhibit both fast and slow forms of symmetrical and asymmetrical $V_j$ dependence. In homotypic junctions, in which the pairing of identical hemichannels gives rise to structural symmetry, symmetry in $V_j$ dependence about $V_j = 0$ is usually found. In heterotypic junctions, in which cells expressing different connexins are paired, structural asymmetry is introduced and asymmetry in $V_j$ dependence usually follows. The differential sensitivity and kinetics of fast versus slow $V_j$ dependence suggests that the distinction between these forms of $V_j$ dependence is mechanistic. It is uncertain that the mechanism underlying the fast $V_j$ dependence in exogenously expressed heterotypic junctions is the same as that present in rectifying electrical synapses; in the latter, $V_j$ dependence is considerably steeper. Discussions of fast and slow $V_j$ dependence are considered separately in the next sections.

## B. Slow Vj Dependence

### Homotypic Junctions

In homotypic junctions, the pairing of two identical hemichannels in a head-to-head fashion makes for twofold symmetry along the channel axis with respect to a field imposed by $V_j$. If the constituent hemichannels could each gate separately in response to $V_j$, a given polarity of $V_j$ would reduce $g_j$ by closing only one of the hemichannels. The opposite polarity of $V_j$ would reverse the configuration of open/closed hemichannels, thereby achieving symmetrical reductions in $g_j$ about $V_j = 0$. This type of mechanism was originally proposed by Harris and associates (1981) to explain the $V_j$ dependence in amphibian blastomeres. Steady state $g_j$-$V_j$ relations obtained from frog blastomere pairs are shown in Figure 8A with examples of current relaxations in response to applied $V_j$s of $\pm20$ and $\pm40$ mV. Steady state $g_j$ decreases symmetrically about $V_j = 0$, with time constants in the hundreds of milliseconds to seconds range. The current relaxations can be fit by single exponentials suggestive of a two-state open-closed equilibria for each hemichannel according to:

$$\text{Closed}_1 \underset{\beta_1}{\overset{\alpha_1}{\rightleftharpoons}} \text{Open}_1 \qquad \text{Closed}_2 \underset{\beta_2}{\overset{\alpha_2}{\rightleftharpoons}} \text{Open}_2$$

**A**

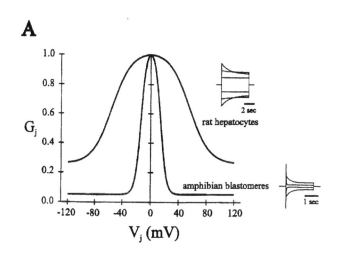

**B**

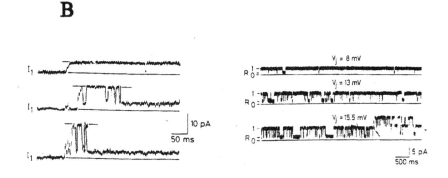

***Figure 8.*** Vj dependence of junctional conductance. **A**. Shown are examples of macroscopic steady state $G_j$-$V_j$ relations and corresponding junctional currents obtained from isolated pairs of rat hepatocytes and amphibian blastomeres using the double two-electrode voltage clamp technique. $G_j$ is junctional conductance, $g_{j}$, normalized to the value at $V_j = 0$. Characteristic features of $V_j$ dependence include symmetrical reductions in $G_j$ about $V_j = 0$ that plateau to a residual conductance, $G_{min}$, and slow kinetics to steady state. These features, however, can vary in magnitude among different cell types. When comparing the properties of amphibian blastomeres to those of rat hepatocytes, the blastomeres display a considerably higher sensitivity to $V_j$, with a half-maximal conductance occurring at ~14 mV compared to ~55 mV for hepatocytes, faster kinetics, reaching steady state in tens of milliseconds compared to seconds for hepatocytes, and exhibiting a lower $G_{min}$, 5% compared to 25% for hepatocytes. Such differences among cell types is known to represent connexin specific differences; rat hepatocytes express Cx32 and Cx26 (Zhang and Nicholson, 1989) and amphibian blastomeres are likely to express Cx38 (Ebihara et al., 1989). **B**. Examples of $V_j$ dependence at the single channel level obtained from insect (*Aedes albopictus*) cell pairs using the double whole-cell patch-clamp technique. Shown on the left are examples of first openings of newly formed gap junction channels, followed by $V_j$ gating. In each example, two cells were individually whole-cell voltage clamped and brought into contact. Records shown are with standing $V_j$ gradients of –15 mV (*top*), –28 mV (*middle*), and -44 mV (*bottom*). $I_1$ represents the junctional current. Several minutes after contact, $I_1$ rises from baseline (bottom solid lines) with an apparent slow and sometimes multistep fashion to reach the fully open state, termed $g_j$ (main state) (upper solid lines). Subsequent to the first opening, the channel gates between $g_j$ (main state) and a subconducting state, termed $g_j$ (residual), that is about one-fifth that of $g_j$ (main state); transitions to the fully closed state are no longer evident. Residence times in $g_j$ (residual) increase with increasing $V_j$s and the transitions are rapid and typical of channel gating. The initial, apparently slow transitions from baseline are believed to represent "docking" currents as two hemichannels link to form a functional cell-cell channel. The latter rapid transitions between $g_j$ (main state) and $g_j$ (residual) are believed to represent $V_j$ gating. Thus, closure of the $V_j$ gate in one hemichannel would leave the cell-cell channel in $g_j$ (residual) and explains the macroscopically observed residual conductance plateaus of $G_j$-$V_j$ relations typical of gap junctions (see panel **A**). (From Bukauskas and Weingart, 1994). **C**. Example of $V_j$-dependence at the single channel level. Shown is a single active channel between two *A. albopictus* cells. Gating is between $g_j$ (main state) and $g_j$ (residual). Residence times of the cell-cell channel in $g_j$ (residual) increase and in $g_j$ (main state) decrease with increasing $V_j$, consistent with a dependence of channel open probability on $V_j$. (From Bukauskas and Weingart, 1994).

where $Open_1/Closed_1$ and $Open_2/Closed_2$ represent the open and closed states of the individual component hemichannels and $\alpha$ and $\beta$ are the corresponding opening and closing rate constants, respectively. Accordingly, the steady state $g_j$-$V_j$ relation can be considered as resulting from two simple open/closed equilibria, one for each hemichannel. Each open/closed equilibria could be fit to a Boltzmann equation of the form:

$$g_j = [(g_{jmax} - g_{jmin})/(1 + e^{A(V_j - V_0)})] + g_{jmin} \qquad (10)$$

where $g_{jmax}$ and $g_{jmin}$ are the maximum and minimum (residual) conductances and $V_0$ is the voltage at which $g_j$ is halfway between the maximum and minimum values (see Spray et al., 1981). The constant A is a measure of voltage sensitivity in terms of nq/RT, where n is the number of elementary charges, q, that move through the entire $V_j$ field. The value of n is commonly referred to as the gating charge and in a two-state system directly correlates with the steepness of the $g_j$-$V_j$ relation; the larger the value of n, the steeper the $g_j$-$V_j$ relation. Residual $g_j$ is a distinctive property of $V_j$ dependence in which $g_j$, rather than declining to zero at large $V_j$s, declines and asymptotes to a substantial, nonzero value. Residual $g_j$ resembles a voltage insensitive conductance that can be large (e.g., ~40% of the maximum $g_j$ in Cx43), and is connexin specific.

Each of the thirteen distinct mammalian connexins listed in Table 2 have been expressed in *Xenopus* oocytes or one or more mammalian cell lines, or both. Two of these, connexin31.1 (Cx31.1) and connexin33 (Cx33), have thus far failed to produce functional coupling, suggesting that some connexins may not have the capacity to form functional channels (Hennemann et al., 1992a; Chang et al., 1996). All connexins that produce functional homotypic channels are $V_j$ dependent and are characterized by slow, symmetrical reductions in $g_j$ about $V_j = 0$. The exceptions are Cx26 and Cx43 which have slight asymmetries in their $g_j$-$V_j$ relations about $V_j = 0$ when expressed in *Xenopus* oocytes due to a small degree of $V_m$ ($V_{i-o}$) dependence (Barrio et al., 1991; Rubin et al., 1992b; Bruzzone et al., 1994). $V_j$ sensitivities among the homotypic junctions range from those that are comparable to ion channels in excitable membranes; that is, displaying e-fold changes in $g_j$ every 2 to 3 mV at large $V_j$s (e.g., Cx37, Cx40, and Cx50), to those that are 10 fold less sensitive (e.g., Cx30.3, Cx32, Cx43, and Cx46). Similarly, the kinetics range considerably, exhibiting time constants from tens of milliseconds to seconds. The relaxations of the currents are generally multiexponential and indicative of multistate rather than simple two-state open-closed equilibria as originally proposed in amphibian blastomeres. Although a multistate kinetic scheme makes explicit application of the Boltzmann relation inappropriate, calculated Boltzmann parameters from fits to the $g_j$-$V_j$ relations according to Eq. 10 provide useful indexes for comparison and are summarized in Table 3.

Studies in which cell pairs displaying inherently low levels of coupling (Chanson et al., 1993) and in which individual cells were placed together to generate

**Table 2.** Mammalian Connexin Compatibility Determined Using Exogenous Expression in *Xenopus* Oocytes and Communication-Deficient Cell Lines

| 26 | 30 | 30.3 | 31 | 31.1 | 32 | 33 | 37 | 40 | 43 | 45 | 46 | 50 | Cx |
|---|---|---|---|---|---|---|---|---|---|---|---|---|---|
| C/C | C/C | | □/N | | C/C | | □/N | N/N | N/N | N/N | C/□ | C/□ | 26 |
| | C/( | | | | C/( | | | | | | | | 30 |
| | | C/C | | | | | | | | | | | 30.3 |
| | | | C/C | | □/N | | □/N | □/N | □/N | | | | 31 |
| | | | | N/N | | | | N/□ | | | | | 31.1 |
| | | | | | C/C | | N/♦ | N/N | N/♦ | N/□ | C/□ | C/□ | 32 |
| | | | | | | N/□ | N/□ | | N/□ | | | | 33 |
| | | | | | | | C/C | C/C | C/C | □/C | | | 37 |
| | | | | | | | | C/C | N/N | □/C | N/□ | N/□ | 40 |
| | | | | | | | | | C/C | C/C | C/□ | N/□ | 43 |
| | | | | | | | | | | C/C | N/□ | | 45 |
| | | | | | | | | | | | C/□ | C/□ | 46 |
| | | | | | | | | | | | | C/□ | 50 |

*Notes:* The first symbol indicates data from the *Xenopus* oocyte expression system and the second symbol indicates data from transfected communication deficient cell lines. The absence of any symbols (blank) indicates that there is no reported data using either *Xenopus* oocytes or transfected cell lines.

C indicates coupling.

N indicates no coupling.

□ indicates no reported data.

♦ indicates conflicting data. Cx32/Cx37 and Cx32/Cx43 heterotypic junctions were reported as non-functional in Elfgang et al. (1995) using a dye diffusion assay in transfected HeLa cells; these same heterotypic junctions in HeLa cells were reported to be electrically coupled by Bukauskas and Weingart (Abstract International 1995 Gap Junction Meeeting, L'Ile de Embiez, France) using the dual whole-cell voltage-clamp technique.

Electrophysiological data on the newest members of the connexin family, Cx36 and Cx57, have not yet been reported and are not included in the table.

low levels of coupling (Bukauskas and Weingart, 1993, 1994) have provided a glimpse into the operation of $V_j$ dependence at the single channel level. In the latter technique, individual cells are placed into contact after establishing whole-cell recordings to promote and record de novo channel formation. Using a clonal cell line derived from larvae of the Asian mosquito *(Aedes albopictus)*, Bukauskas and Weingart (1993, 1994) observed that the first channel usually appeared 10 to 15 minutes after placing the cells into contact (Figure 8B). The appearance of the second channel generally took an additional several minutes allowing sufficient time for analysis of single channel activity. The first opening transition was always slow, taking 20 to 30 msec to open fully. The channel opened to a conductance of 375 pS, termed $g_j$(main state). After the first slow opening transition, subsequent opening and closing transitions were rapid, resembling gating typical of ion channels. These subsequent transitions occurred between $g_j$(main state), and a residual conductance state of 70 pS, termed $g_j$(residual). Transitions into a nonconducting or closed state, termed $g_j$(closed), were rare. The initial slow transition was postulated to represent the de novo opening of a newly formed channel after hemichannel alignment and docking. The subsequent fast transitions were

***Table* 3.** $V_j$ Sensitivities of Connexins Expressed in *Xenopus* Oocytes

| Connexin | Sensitivity to $V_j$ | References |
|---|---|---|
| Cx26 | $n = 4, V_0 = 85$ | Barrio et al., 1991; Rubin et al., 1992b |
| Cx31 | N.D. | Hoh et al., 1991 |
| Cx31.1 | No coupling | Hennemann et al., 1992a |
| Cx30.3 | $n = 2; V_0 = 72$ | Hennemann et al., 1992a |
| Cx32 | $n = 2, V_0 = 55$ | Barrio et al., 1991; Rubin et al., 1992b |
| Cx33 | No coupling | Chang et al., 1996 |
| Cx37 | $n = 11, V_0 = 16$ | Willecke et al., 1991 |
| Cx38 | N.D. | Ebihara et al., 1989 |
| Cx40 | $n = 8, V_0 = 35$ | Willecke et al., 1991 |
| Cx43 | $n = 2, V_0 = 60$ | White et al., 1994b |
| Cx45 | N.D. | Moreno et al., 1995 |
| Cx46 | $n = 2, V_0 = 67$ | White et al., 1994b |
| Cx50 | $n = 8.5, V_0 = 18$ | White et al., 1994b |

***Note:*** N.D., no data.

ascribed to $V_j$ gating of the stably formed channel. Transitions between an open state and a long-lived residual subconducting state have also been reported in single channel studies of vertebrate connexins, such as those expressed in Leydig cells from mouse testis (Perez Armendariz et al., 1994) and in exogenously expressed Cx43 (Moreno et al., 1994) and Cx37 (Veenstra et al., 1994b). In contrast to *Aedes* insect cells, more frequent transitions into a closed (nonconducting) state were observed in the vertebrate connexins. (Moreno et al., 1994) suggested that the residual conductance state represents a voltage-insensitive substate. (Bukauskas and Weingart, 1994) suggested that the residual state represents the extent of channel closure by $V_j$. In any case the residual conductance state is believed to be responsible for the macroscopically observed residual $g_j$ at high $V_j$s. In support of this view, macroscopic residual $g_j$ appears to be absent in cultured Schwann cells and no residual substate was observed at the single channel level (Chanson et al., 1993). The connexin present in these cultured Schwann cells has not been identified.

As demonstrated in *Aedes* cell pairs with only a single active gap junction channel, moderate increases in $V_j$ (10 to 20 mV), decrease open probability by reducing dwell times in $g_j$(main state) and increasing dwell times in $g_j$(residual) (Bukauskas and Weingart, 1994). Similar conclusions were reached in cultured rat Schwann cells (Chanson et al., 1993), although changes in open probability were inferred from cell pairs containing multiple (~17) active channels. These data indicate $V_j$ dependence governs rate constants between discrete high-, low-, and nonconducting states. The multiplicity of states is consistent with the observed multiexponential behavior of the macroscopic current relaxations.

## Heterotypic Junctions

Among the 15 distinct connexin genes (or paralogues), there are 106 unique heterotypic junctions that can be formed. Thirty-six heterotypic junctions have been tested to date, 16 of which have been found to be functional (see Table 2). Connexins of Groups I and II will form both intra- and intergroup heterotypic junctions. Some connexins are less discriminating (e.g., Cx46 forms functional channels in four out of six tested combinations) whereas others are very discriminating (e.g., Cx40 only forms functional channels in one out of seven tested combinations).

Heterotypic junctions typically show asymmetrical $g_j$-$V_j$ relations about $V_j = 0$. However, there are two general types of asymmetry; the type in which each half of the $g_j$-$V_j$ relation resembles the characteristics of the constituent hemichannels as inferred from the corresponding homotypic junctions (e.g., Cx37/Cx40; Hennemann et al., 1992b), and the type in which the $g_j$-$V_j$ relation bears no obvious resemblance to the corresponding homotypic junctions (e.g. Cx32/Cx26 and Cx46/Cx32l; Barrio et al., 1991; White et al., 1994a). In heterotypic junctions that retain the characteristics of the component hemichannels, $V_j$ gating would appear to be an intrinsic hemichannel property, with each hemichannel possessing its own $V_j$ sensor ("separate-sensor" model; see Harris et al., 1981; Verselis et al., 1994). In such junctions, hemichannel interactions either play no significant role in determining $V_j$ dependence or play equivalent roles in the corresponding homotypic and heterotypic pairing combinations. Conversely, in heterotypic junctions in which the $g_j$-$V_j$ relations bear no obvious resemblance to either of the component hemichannels, $V_j$ dependence would not appear to be an intrinsic hemichannel property, with $V_j$ sensors formed, and perhaps shared, by the two hemichannels ("shared-sensor" model; see White et al., 1994a).

Recent mutagenic studies of homotypic and heterotypic junctions composed of Cx32 and Cx26 appear to resolve the unpredicted asymmetry in slow $V_j$ dependence of Cx32/Cx26 and, perhaps, some other heterotypic junctions. As indicated previously, Cx32/Cx26 junctions do not resemble either Cx32 or Cx26 homotypic junctions, most notably displaying slow $V_j$ dependent closure for only a single polarity of $V_j$ and a novel fast $V_j$-dependent rectification reminiscent of fast rectifying electrical synapses (Figure 9A). By plotting the ratio of steady state to initial $g_j$, termed G (Figure 9A, plot on right), the slow changes in $g_j$ can be viewed without the fast changes, thereby revealing the single polarity of closure; G decreases only when the Cx26 side is made relatively positive on its cytoplasmic side (or Cx32 is made relatively negative). Mutational analyses of Cx32 and Cx26 showed that this asymmetry in slow $V_j$ dependence resulted from the opposite gating polarities of the component Cx32 and Cx26 hemichannels with Cx26 hemichannels closing on relative positivity on their cytoplasmic side and Cx32 on relative negativity (Verselis et al., 1994). The consequences of pairing hemichannels having the same and opposite polarities are illustrated diagrammatically in

Figure 9B. Shown are two homotypic junctions composed of connexins, termed Cx(+) and Cx(−), that close for opposite polarities of $V_j$. Both Cx(+) and Cx(−) homotypic junctions display symmetrical $g_j$-$V_j$ relations whose opposite gating polarities are obscured by symmetry. Although symmetrical in both cases, $V_j$s that make the hemichannel on the right relatively positive would close the hemichannel on that same side in Cx(+) junctions, but on the opposite side in Cx(−) junctions. The configuration of open/closed hemichannels would reverse for the opposite polarity of $V_j$. In heterotypic Cx(−)/Cx(+) junctions, relative positivity on the Cx(+) side would tend to close both Cx(+) and Cx(−) hemichannels, whereas neither hemichannel would tend to close for $V_j$s of the opposite polarity. The resultant $g_j$-$V_j$ relation, although bearing no obvious relation to the corresponding homotypic junctions, is the expected result when the component hemichannels have opposite gating polarities.

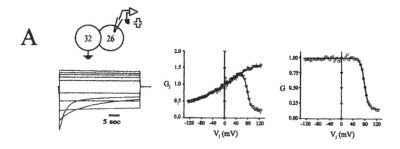

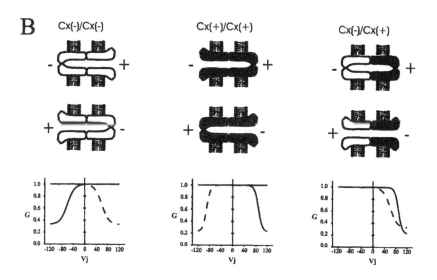

*Figure 9.* **A**. Example of the asymmetrical behavior of the Cx32/Cx26 heterotypic junction. Shown in the left panel are currents elicited by (20, 40, 60, 80, and 100 mV voltage steps applied to the Cx26-expressing cell from a holding voltage of –30 mV. The voltage in the Cx32-expressing cell was held constant at the holding voltage of – 30 mV. In the middle panel, the steady state (open triangles) and instantaneous (filled triangles) $G_j$s are plotted as a function of $V_j$. Solid line represent fits to the data using a Boltzmann relation (see Eq. 10 in text). $G_j$ is $g_j$ normalized to its value at $V_j = 0$. Because steady state value of $G_j$ in these junctions depends on both fast and slow processes, the data and fits are also plotted (right panel) as the ratio of steady state $G_j$ to initial $G_j$. The ratio is termed G and is used only to illustrate changes in the slow process to which these data on polarity reversal apply (shown in **B**). Independence of fast and slow processes is not implied. **B**. Expected $G_j$-$V_j$ relations and diagrammatic representations of gap junction hemichannel open/closed configurations at large $V_j$s of either sign for homotypic junctions in which the component hemichannels gate on negativity, Cx(–), and positivity Cx(+), relative to their cytoplasmic ends and a heterotypic junction composed of Cx(–) and Cx(+) hemichannels. The solid and dashed lines represent Boltzmann relations describing the gating of hemichannels on the right and left side of the junction, respectively. Homotypic junctions maintain symmetry about $V_j = 0$, but opposite hemichannels close for a given polarity of $V_j$. The $G_j$-$V_j$ relation of the heterotypic Cx(–)/Cx(+) junction is asymmetrical because both hemichannels close for $V_j$ that is both relatively positive to Cx(+) and relatively negative to Cx(–). Neither hemichannel closes for the opposite polarity of $V_j$. (From Verselis et al., 1994).

---

Although opposite gating polarities can explain the behavior of Cx32/Cx26, as well as Cx32/Cx46 heterotypic junctions, the characteristics of some heterotypic junctions remain unexplained. One example is the Cx46/Cx50 junction, which exhibits nearly symmetrical slow $V_j$-dependent closure for both polarities of $V_j$ (White et al., 1994b). Although the characteristics of Cx46 and Cx50 homotypic junctions are quite different, both in sensitivity and kinetics, the characteristics of Cx46/Cx50 junctions resemble neither that of Cx46 nor Cx50 for either polarity of $V_j$. In another example, Cx43/Cx38 junctions show an asymmetry consistent with closure on relative positivity for both hemichannels whereas Cx43/Cx45 junctions show an asymmetry consistent with closure on relative negativity for both hemichannels (Moreno et al., 1995; Verselis and Bargiello, unpublished observations). Thus, the polarity of Cx43 closure would appear to depend on the connexin to which it is paired. Results of this kind led to the suggestion that $V_j$ gating is not an intrinsic property of the hemichannel, but rather a property of the channel as a whole (White et al., 1994a). Alternatively, it is possible that the major features of the conformational change of $V_j$ gating are intrinsic to the hemichannel (Verselis et al., 1994; see also Ebihara et al., 1995), and differences in hemichannel properties in different pairing combinations may result from "hemichannel interactions" that are, in essence, modulatory much like the interaction of ancillary subunits with the subunits of $Na^+$, $K^+$, and $Ca^{2+}$ channels. Single channel studies of heterotypic junctions will be critical in determining the role of hemichannel interactions in the expression of $V_j$ dependence.

## C.   Fast $V_j$ Dependence

In addition to displaying an asymmetrical slow $V_j$ dependence, Cx32/Cx26 junctions possess an asymmetrical $V_j$-dependent rectification reminiscent of rectifying electrical synapses (see Figure 9A). Relative positivity on the Cx26 side results in an increase in initial $g_j$, while relative negativity results in a decrease. The kinetics, as for rectifying electrical synapses, are fast; the time course of the changes in $g_j$ could not be resolved within the settling time of the voltage clamp.

Furshpan and Potter (1959) likened rectification at the crayfish electrical synapse to a diode. A diode, in a strict sense, would imply that rectification occurs by a mechanism analogous to that which occurs in a P-N semiconductor junction, except that fixed charges in the membrane (or channel) are responsible for separating charge. Mauro (1962) and Finkelstein (1963) have shown that ionic systems containing fixed charge regions of opposite sign are endowed with properties of capacitance that can give rise to sufficiently steep rectifying I-V relations which could account for the behavior of rectifying synapses. Alternatively, rectification can occur by fast gating. Upon cooling the crayfish lateral giant-motor electrical synapse, a kinetic component of fast $V_j$ dependence was resolvable consistent with gating, although an unresolvable component still remained (Jaslove and Brink, 1986). Bukauskas and associates (1995b) showed that the fast $V_j$ dependence observed macroscopically in Cx32/Cx26 junctions correlated with rectification of the conductance of the single channel. These data suggest that permeation, rather than gating, underlies fast rectification in Cx32/Cx26 junctions. However, the degree of rectification in Cx32/Cx26 junctions is considerably less than that of rectifying electrical synapses so that permeation cannot fully explain the steep rectification observed at electrical synapses. It is possible that the kinetically unresolvable component of rectification in the crayfish rectifying synapse is due to permeation, analogous to that in Cx32/Cx26 junctions, leaving the resolvable kinetic component with cooling, perhaps, due to fast $V_j$ gating.

It should be noted that rat Cx37 homotypic junctions expressed in *Xenopus* oocytes also display fast $V_j$ dependence (Willecke et al., 1991). Unlike Cx32/Cx26 junctions, fast $V_j$ dependence in Cx37 homotypic junctions is symmetrical about $V_j = 0$. The sensitivity and extent of closure by fast $V_j$ dependence is typically less that for slow $V_j$ dependence with maximal changes in $g_j$ of approximately 50% over 100 mV. Single channel studies have not rigorously explored whether changes in single channel conductance correlate with fast $V_j$ dependence in these homotypic junctions. Existing views of ion channel permeation provide plausible mechanisms that can account for nonlinearities in single channel I-V relations of homotypic and heterotypic junctions, without invoking P-N theory. For example, ions traversing gap junction channels can be viewed as binding to sites within the pore and traversing energy barriers in order to "hop" from one binding site to another (see Hille, 1992, for discussion of Eyring rate theory). The barrier profile would determine the single channel I-V relation, which could substantially rectify if a single rate-limiting barrier were located at one end of the channel. Linking identical hemichannels in series, to form a homotypic channel, would create a symmetrical barrier profile about the midline of the channel and would produce a symmetrical, although not necessarily linear, single channel I-V relation. Linking different hemichannels in series, to form a heterotypic channel, would create an asymmetrical barrier profile that would give rise to rectification. Studies of fast $V_j$ dependence are in their early stages and will require the rigorous testing of P-N, permeation, and gating models.

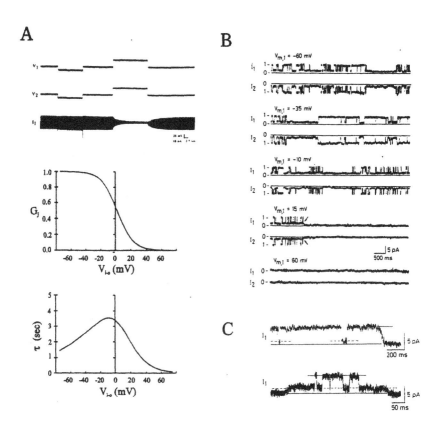

**Figure 10.**   **A**. Example of $V_m$ or $V_{i-o}$ dependence of $g_j$ in *Drosophila melenogaster* salivary glands. The three traces shown (*top*) represent the voltage applied to cell 1 and cell 2 ($V_1$ and $V_2$) and the current recorded in cell 1 ($I_1$) obtained from a double two-electrode voltage-clamp configuration. Small, brief voltage steps were alternately applied to each cell to measure $g_j$. The upward defection in $I_1$ represents junctional currents. Superimposed on the small alternating steps are two long-duration voltage steps applied simultaneously to both cells. Simultaneous hyperpolarization produced a small increase in $g_j$, whereas depolarization produced a marked decrease. Calibration bar, 25 mV, 10 nA, 2.5 seconds. A typical steady state $G_j$-$V_m$ relation for *D. melanogaster* salivary gland cell pair is shown below the voltage and current traces; $V_m$ is indictated as $V_{i-o}$ on the x-axis). $G_j$ is normalized to its maximum at hyperpolarized voltages. $G_j$ reaches a maximum plateau at hyperpolarized voltages and declines with depolarization. Unlike $V_j$ dependence in which steady state $g_j$ declines to residual conductance at large $V_j$s of either polarity, $g_j$ declines to zero at sufficiently positive $V_m$s. **B**. Example of $V_m$ dependence of $g_j$ at the single channel level. Records were obtained from cell pairs from an insect cell line, *Aedes albopictus*, using the double whole cell patch-clamp configuration. Shown is single channel activity at $V_m$s of –60, –35, –10, 15 and 60 mV. $V_m$ was imposed by applying equal and simultaneous polarizations to both cells. $V_j$ was held constant throughout at –15 mV. At $V_{i-o}$s of –60, –35 and –10 mV, the channel gates between $\gamma$(main state), indicated by level 1, and $\gamma$(residual). Depolarized from –60 to –10 decreased the relative time spent in $\gamma$(main state). Further depolarization to +15 mV caused the channel to enter into $\gamma$(closed), indicated by level 0. At +60 mV, the channel resided only in $\gamma$(closed). The transition to $\gamma$(closed) had an apparently slow time course and is ascribed to $V_m$ gating. Transitions between $\gamma$(main state) and $\gamma$(residual) were rapid and are ascribed to $V_j$ gating. No $V_j$ gating is observed when the channel resides in $\gamma$(closed). (From Bukauskas and Weingart, 1994). **C**. Examples of fast versus slow transitions associated with $V_j$ and $V_m$ gating, respectively. The top trace shows two transitions between $\gamma$(main state) and $\gamma$(residual) followed by a slow transition to $\gamma$(closed). $V_1$ was held at –60 mV and $V_2$ at -75 mV. The bottom trace shows a slow transition from $\gamma$(closed) to $\gamma$(residual) followed first by rapid gating between $\gamma$(residual) and $\gamma$(main state), and then by a slow transition back to $\gamma$(closed), which terminated the fast gating activity. $V_1$ was held at –61 mV and $V_2$ at –75 mV. (From Bukauskas and Weingart, 1994).

---

## D.   $V_m$ or $V_{i-o}$ Dependence

Sensitivity to $V_j$ is widespread among gap junctions in vertebrates and inverte-brates, whereas sensitivity to $V_m$ is less frequent and, until recently, was described only in invertebrates. As it is turning out, there are a number of gap junctions that express $V_j$ dependence and not $V_m$ dependence, whereas there are no examples to date in which the converse is true.

$V_m$ dependence was first described in *Chironomus* salivary glands (Obaid et al., 1983). Equal and simultaneous hyperpolarization of both cells caused $g_j$ to increase to a maximum plateau, whereas depolarization caused $g_j$ to decrease to virtually zero. Subsequent quantitative analyses of junctions in salivary glands of the fruit fly *Drosophila melanogaster* (Verselis et al., 1991) showed a sigmoidal $g_j$-$V_m$ relation that exhibited an e-fold change in $g_j$ for $\pm 12$ mV and slow kinetics; time constants ranged from hundreds of milliseconds to seconds depending on membrane voltage (Figure 10A). Unlike in *Chironomus*, gap junctions in *D. melanogaster* show considerable $V_j$ dependence in addition to $V_m$ dependence.

Evidence suggesting that $V_m$ dependence differs mechanistically from $V_j$ dependence comes from single channel studies in *Aedes* cells (Bukauskas and Weingart, 1994). Shown in Figure 10B are records from an *Aedes* cell pair in which there appeared to be only one active channel. With an imposed $V_j$, the channel gates between the open state, $g_j$(main state), and the residual substate, $g_j$ (residual). Transitions to a fully closed state are rare. However, if the same $V_j$ is maintained with the cells held at more positive membrane voltages, transitions to a fully closed state, $g_j$(closed), increase in frequency. Residence times in $g_j$(closed) are prolonged, at the same $V_j$, with further depolarization of both cells indicating that the transitions associated with $g_j$(closed) are due to $V_m$ ($V_{i-o}$). The transitions to and from $g_j$(residual) and $g_j$ (closed) are unusual in that the kinetics appear to be slow, unlike the fast transitions associated with $V_j$ gating and typically with other ion channels. This behavior is very similar to that observed in single Cx46 hemichannels at moderate inside negative voltages (Trexler et al., 1996). Transitions between the fully open state, termed $\gamma_{open}$, and the fully closed state, termed $\gamma_{closed}$, are slow and usually involve several transient subconductance states. In some cases, these transitions appear to be continuous, lasting 20 to 30 msec and may be attributable to rapid gating among the transient substates. The resemblance of these "slow" or multistep current transitions to the "formation" currents observed during de novo channel opening (Bukauskas and Weingart, 1994) suggest that these gating mechanisms may be related and may represent conformational changes associated with the extracellular loops of the connexin subunits. Recent studies using cysteine replacement mutagenesis in Cx46 hemichannels suggest that a voltage gate, possibly equivalent to the "loop" gate proposed by Trexler and co-workers (1996), is located extracellular to amino acid residue 35 in the first transmembrane domain (Pfahnal Dahl, 1998).

The differences in the steady state and kinetic properties of gating by $V_m$ and $V_j$ suggest that these stimuli induce gap junction channels to adopt different conformational states, consistent with their being different gating mechanisms. However, there appears to be an interaction between the two gating mechanisms, because the dwell times in $g_j$(main state) and $g_j$(residual) at the same $V_j$, change with $V_m$. The data are consistent with results obtained in macroscopic studies in

*Drosophila,* which showed an increase in the efficacy of $V_j$ dependence with depolarizing $V_m$ stimuli (Verselis et al., 1991).

# V.  STRUCTURE AND FUNCTION

Structure-function studies of gap junctions are relatively few even though exogenous expression systems have been established for some time both in *Xenopus* oocytes and in various communication-deficient cell lines. This relative paucity of structure-function studies stems, in part, from the complex construction of gap junction channels and from a lack of sequence homology with any ion channel proteins, most notably lacking homologs to the voltage-sensing S4 and pore-forming P domains in $Na^+$, $K^+$, and $Ca^{2+}$ channels (Hartmann et al., 1991; Liman et al., 1991). Next we describe studies that, using a combination of point mutational and chimeric approaches, have provided first glimpses into the molecular basis of gap junction channel gating, permeation, and connexin (hemichannel) compatibility.

## A.  Molecular Studies of $V_j$ Dependence

### *Identification of a Putative $V_j$ Sensor*

Molecular studies of voltage gating have sought to identify charged amino acid residues that respond to changes in voltage, thereby coupling charge movement or shifts in dipole to changes in channel conformation. For $V_j$ dependence, in which the only relevant voltage difference is that established between the cells, it has been proposed that the $V_j$ sensor resides in the pore where $V_j$ can be sensed independently of $V_m$ (Harris et al., 1981). Although placement of the $V_j$ sensor inside the pore is logical, its placement outside the pore is also feasible with the condition that the sensor be restricted to moving along an electrical field generated by $V_j$, and not $V_m$. Early attempts to elucidate the molecular basis of $V_j$ dependence utilized domain substitution between Cx32 and Cx26 (Rubin et al., 1992a). Prior to the knowledge that Cx32 and Cx26 had opposite gating polarities, these studies sought to identify domains that accounted for the differences in voltage dependence observed for homotypic Cx32 and Cx26 junctions. Substitution into Cx32 of Cx26 sequence that included either or both extracellular loops (E1, E2, E1 + E2), the third transmembrane domain (TM3) and the cytoplasmic loop (CL) failed to significantly alter $g_j$-$V_j$ relations of mutant homotypic junctions as compared to wild-type Cx32 (Rubin et al., 1992a, 1992b).

Utilizing the opposite $V_j$ gating polarities of Cx32 and Cx26, Verselis and associates (1994) successfully used a chimeric approach to localize the molecular basis for this difference in gating polarity (Figure 11A). The surprising result was that the opposite gating polarities of Cx32 and Cx26 resulted from a charge differ-

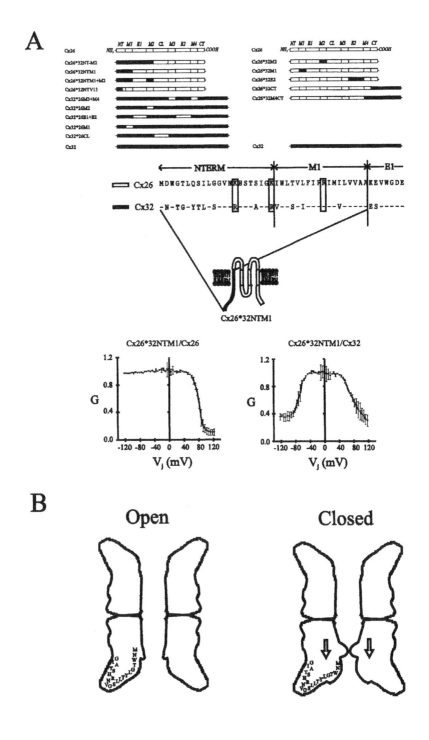

*Figure 11.* **A**. Reversal of Cx26 gating polarity is evident in chimeras in which the proximal N-terminal segment of Cx26 is replaced by Cx32 sequence. Shown schematically (*top*) is a series of Cx32 and Cx26 chimeras segmented according to the accepted membrane topology of connexins. Cx26 sequence is indicated by open boxes. Chimeras on the left close on relative negativity on their cytoplasmic ends; those on the right on relative positivity. Also shown are the primary sequences of Cx26 and Cx32 from the N-terminus through the proximal portion of E1, with a schematic of the putative connexin membrane topology and G-$V_j$ relations of chimeric Cx26*32NTM1/ Cx26 and Cx26*32NTM1/Cx32 heterotypic junctions. Reductions in G (the ratio $g_j$ [steady state]/$g_j$ [initial]) for a single polarity in Cx26*32NTM1/Cx26 junctions and for both polarities in Cx26*32NTM1/Cx32 junctions indicates that component hemichannels composed of Cx26*32NTM1 close on relative negativity as do hemichannels composed of Cx32. The steady state properties for Cx26*32NTM1 (right side of Cx26*32NTM1/Cx32 G-$V_j$ relation) is less steep than for Cx32 and does not reach a plateau by 120 mV. Error bars are standard deviations of means obtained for three cell pairs of each heterotypic junction. (From Verselis, et al., 1994). **B**. Diagrammatic representation of a gap junction channel in which the N-terminus of Cx32 is looped back toward the membrane to form the vestibule of the channel pore at the cytoplasmic end. The amino acid at the N-terminus forms part of the $V_j$ sensor. The arrows indicate the direction of the movement of the voltage sensor that leads to channel closure.

---

ence on a single amino acid residue located at the second position of the N-termi- nus. Other charge substitutions at the TM1/E1 border were also found to influence gating polarity and to suppress the effects of charge substitutions in the N-termi- nus. Strong evidence that these residues constitute part of the $V_j$ sensor comes from the correlation between the gating polarity and the net charge of these resi- dues; net negative and positive charges correlate with closure on relative positiv- ity and negativity, respectively, on the cytoplasmic side of the hemichannel. The simplest picture that emerges from these studies (Figure 11B) is that the $V_j$ sensor in these connexins is composed of a distributed charge complex whose net charge can be of either sign, and whose movement toward the cytoplasm is associated with hemichannel closure. A significant structural implication of this picture is that the proximal portion of the N-terminus, presumed to be in the cytoplasm, must be in the membrane or, perhaps, looped back to form the vestibule of the pore in order to sense the $V_j$-induced field. Chou-Fasman secondary structure pre- dictions, indeed, place a reverse turn in the N-terminus at G12, V13, N14, and R15 in these two connexins (Purnick et al., unpublished results). It is interesting to note that three mutations in Cx32, G12S, V13L, and R15Q, have been found in kindred with CMTX, a peripheral neuropathy that is linked to loss of Cx32 function (Bergoffen et al., 1993; Fairweather et al., 1994). These N-terminal CMTX mutations indicate that the N-terminus is critical to channel function. Also, the structural changes affecting the pore of the gap junction with gating, as deduced by Unwin and Ennis (1984), are greatest at the cytoplasmic membrane surface and, perhaps, are indicative of structural alterations involving the N-terminus. Mutations of other members of the connexin family will determine whether a common structure motif governs gating in all gap junction channels. Group II connexins lack the GVN motif at positions 12 to 14 and suggest that the structure of the N-terminus of Group II connexins may be different.

### Voltage Transduction

Noting a proline, P87, in the second transmembrane domain (TM2) conserved among all connexins, Suchyna and associates (1993) mutated this position in Cx26 and found profound effects on $V_j$ dependence. Heterotypic pairings of P87G and P87L mutants with wild-type Cx32 and Cx26 produced novel voltage-gating properties, which were interpreted as due to the alteration of the structural conformation of TM2 imposed by P87. The structural conformation of TM2 imposed by P87 was proposed to play a role in voltage gating by acting as a trans-ducing element that coupled the movement of a $V_j$ sensor to that of a gate. The effects of the P87 mutations were specific to $V_j$ gating as no effects were observed on $H^+$-induced gating. Interestingly, the P87G substitution in Cx32 failed to affect $V_j$ dependence when paired wild-type Cx32 or Cx26 and had no effect on the ability of charge substitutions in the N-terminus to reverse gating polarity (Verselis et al., 1994). Thus, if P87 acts as a transduction element in $V_j$ gating, the differential effects of P87G substitutions in Cx32 and Cx26 indicate that there are differences in the transmembrane domain interactions within these two connexins. These interactions may involve functional properties of the P87 analogous to those pos-tulated for prolines in signal transduction in G-protein–coupled receptors (Ri et al., 1999).

### B.   Unitary Conductance of Connexin-Specific Channels

The functional expression of connexins in communication-deficient cell lines compliments the *Xenopus* oocyte expression system by providing data at the single channel level. Inherently more time consuming than RNA injection into oocytes, stable expression of connexin cDNAs does not lend itself as well to site-directed mutagenesis experiments. Hence, most transfection studies so far have focused on the determination of connexin-specific unitary channel currents. Reported single channel conductances are listed in Table 4. The channel conduc-tances of Cx32, Cx43, and Cx42, for example, reassuringly correlate closely with the observed channel conductances in the tissues where each connexin is known to be expressed.

***Table 4.*** Channel Conductances of Transfected Connexins

| Preparation | Conductances (pS) | References |
|---|---|---|
| Cx26: mouse | 140, 25* | Bukauskas et al., 1995b |
| Cx32: mouse | 45, 10* | Bukauskas et al., 1995b |
| rat | 120–150 | Eghbali et al., 1990 |
| | 123 | Moreno, et al., 1991 |
| Cx37, human | 219, 165, 123. 53 | Reed et al., 1993 |
| | 295, 71 | Veenstra, et al., 1994b |
| Cx40: mouse | 198, 36* | Bukauskas et al., 1995a |
| rat | 158 | Beblo et al., 1995 |
| chick Cx42 | 236, 201, 158, 121, 86 | Veenstra et al., 1992 |
| Cx43: human | 55-60 | Fishman et al., 1990 |
| human | 90, 60 | Moreno et al., 1992 |
| rat | 57 | Veenstra et al., 1995 |
| chick | 67, 44, 28 | Veenstra et al., 1992 |
| chick | 166 | Veenstra et al., 1995 |
| Cx45: chick | 29 | Veenstra et al., 1992 |
| chick | 19, 26 | Veenstra et al., 1994a |
| human | 32 | Moreno et al., 1995 |
| Cx56: chick | 96 | Rup et al., 1993 |

**Note:** *residual subconductance state

The correlation between the channel conductances in exogenously expressed and in native tissues is best illustrated by the independent observations of multiple channel conductances in 7-day embryonic chick heart (Chen and DeHaan, 1992) and the functional expression of chick Cx42, Cx43, and Cx45 (Veenstra et al., 1992). As illustrated previously in Figure 5, channel events with conductances of 160 to 170 pS are frequently observed in the embryonic chick heart preparation. Channel activity recorded from paired neuro2A cells transfected with chick Cx42 also exhibit 160-pS channel activity (Figure 12A and B). The other chick heart connexins form smaller conductance channels of 40 to 50 pS for Cx43 (92% homologous to rat Cx43) and 20 to 30 pS for Cx45 (Figure 12C through F; Veenstra, et al., 1992).

Northern blots, immunocytochemical staining, and electrophysiological data on the developmental changes in $V_j$ dependence of embryonic chick heart all support the interpretation that Cx42 is the predominant connexin in the avian heart and Cx45 is expressed primarily early in development (Beyer, 1990; Veenstra et al., 1992). To correlate the observed gap junction channel behavior of 7-day embryonic chick heart with those of the three functionally expressed connexins, event amplitude histograms were constructed from the channel activity observed in the embryonic chick heart (Chen and DeHaan, 1992) and connexin-transfected Neuro2A cell lines (Veenstra et al., 1992). As illustrated in Figure 13, there is a striking similarity in the amplitudes of the channel activity observed in the native and functionally expressed avian heart gap junctions. Although the relative proportions of the event counts for the six major peaks vary slightly between the two

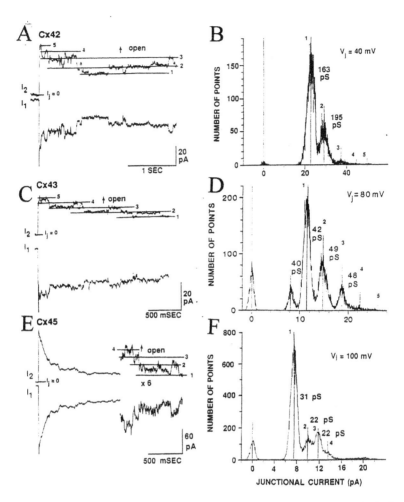

***Figure 12.*** Junctional channel currents of functionally expressed chick heart connexins. Panels **A**, **C**, and **E** illustrate the paired whole cell currents (I₁ and I₂) obtained from each pair of Cx42-, Cx43-, and Cx45-transfected Neuro2A (N2A) cells during a transjunctional voltage (V$_j$) pulse to the indicated value. The corresponding current amplitude histograms are shown in panels **B**, **D**, and **F** for each connexin. Pulse duration was 20 seconds and all current tracings were low-pass filtered at 125 Hz and digitized at 1 kHz. Corresponding histogram peaks and current levels evident in each recording are labeled accordingly. All three connexins form channels with distinct conductances which are summarized in Table 3. Event counts were 44, 38, and 45 for Cx42, Cx43, and Cx45, respectively. (Reproduced from Veenstra et al., 1992 with permission.)

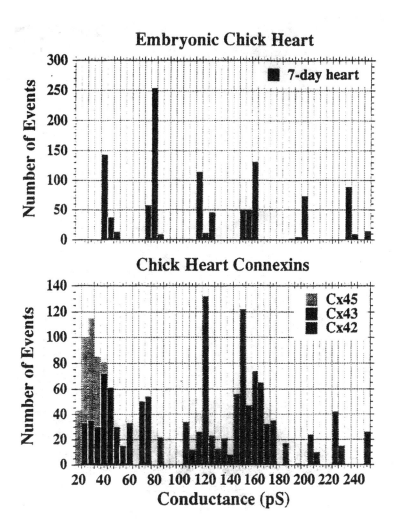

**Figure 13.** Channel event amplitude histograms from 7-day embryonic chick heart and the functional expression of chick Cx42, Cx43, and Cx45 in (neuro2A) cells. The conductances of channel events recorded in paired chick ventricular myocytes or connexin-transfected N2A cells as originally reported in Chen and DeHaan, 1992) and (Veenstra, et al., 1992) were binned in 5 pS intervals from 20 to 260 pS. Mean values of the major conductance peaks in each amplitude histogram are listed in Tables 1 and 3. The 120-, 160-, 200-, and 240-pS channel conductances are all attributable to the activity of Cx42 channels, whereas the 80- and 40-pS channels could result from the presence of more than one type of connexin channel.

preparations, the peak conductances are nearly identical. These data confirm that the basis for the occurrence of multiple channel conductances in embryonic chick heart results from multiple conductance levels of Cx42 channels and the expression of multiple connexins.

The observation of multiple conductance states is not unique to chick Cx42. Chick Cx43 exhibited multiple levels of 67, 44 and 28 pS. These values are nearly identical to the values of 70, 44, and 20 pS reported for rat ventricular (Cx43) gap junction channels (see Table 1; Takens Kwak and Jongsma, 1992). The higher channel conductances of 90 and 60 pS channels reported for human Cx43 expressed in SKHep1 cells can be largely attributed to differences in chloride content of the electrolyte solutions, as Moreno and colleagues (1992) used 135-mM CsCl in their patch pipettes compared to 130-mM Kgluconate and 10-mM KCl for the rat heart studies. The chick connexin channel conductances were determined using a 120-mM Kglutamate, 15-mM NaCl pipette solution. The 30-pS channel events in the Cx43-transfected SKHep1 cells have been attributed to an endogenous gap junction channel (Moreno et al., 1991), but would be indistinguishable from a 20- to 30-pS event associated with Cx43 expression as seen for rat ventricular cells. The endogenous gap junction channel was identified to be formed of Cx45, consistent with previous accounts of Cx45 channel conductance and voltage sensitivity (see Table 4; Moreno et al., 1995).

One obvious issue to be addressed is how the multiple conductances of gap junction channels are modulated. Voltage, cytoplasmic $H^+$ and $Ca^{2+}$, and phosphorylation are all known to modulate junctional conductance in various tissues. In the mammalian myocardium, β-adrenergic stimulation increases junctional conductance via the cAMP-dependent protein kinase (PKA) second messenger pathway whereas muscarinic receptor activation decreases junctional conductance by acting through the cyclic guanosine monophosphate (cGMP)-dependent protein kinase cascade (Burt and Spray, 1988; Takens Kwak and Jongsma, 1992). Despite their opposite actions on macroscopic junctional conductance in rat myocardium, activation of either of these second messenger pathways appears to reduce single channel conductance. Dephosphorylating treatments (e.g., alkaline phosphatase or protein kinase inhibitors) in both cases increased single channel conductance. Cx43 is capable of existing in any of at least three different forms which, based on their electrophoretic mobility, are presumed to correlate with an unphosphorylated and two phosphorylated forms, termed $P_1$ and $P_2$ (Musil et al., 1990b; Kadle et al., 1991). Taken together with the biophysical studies, it would appear that the unphosphorylated Cx43 channel corresponds to the largest, 70- to 90-pS conductance state, the $P_1$ form to the intermediate, 44- to 60-pS state, and the $P_2$ form to the lowest, 20- to 30-pS conductance state. It is not known if the $P_2$ form of Cx43 produced by cAMP-dependent or cGMP-dependent protein kinases are identical, but preliminary electrophysiological evidence suggests that they are functionally identical. This similar action of two different protein kinases on junctional channel conductance also raises new and interesting questions about how the macroscopic junctional conductance is regulated in conjunction with channel gating and conductance properties of the connexins.

## C.    Where Is the Pore?

TM3 was originally implicated as the most likely pore-lining domain because of its amphipathic nature (see Bennett et al., 1991). TM3 also contains a series of small polar residues flanked by large hydrophobic residues that align in a ridge parallel to the channel axis (Unwin, 1989). With the helices more or less inclined, the large hydrophobic residues could move in and out of the pore and thus provide a physical correlate for gating (Unwin, 1989; Bennett et al., 1991). Some early supporting evidence came from site-directed mutagenesis of a highly conserved residue in TM3, S158 (Spray et al., 1992). The point mutation S158K in Cx43 was found to increase single channel conductance and concomitantly reduce Lucifer yellow transfer; the latter suggests an effect on channel selectivity, although the sign of the mutation is opposite to that expected to reduce anion permeability.

Truncation of the carboxy-terminus (CT) of human Cx43 by 80 or 138 amino acids altered the single channel conductance without affecting $V_j$ sensitivity (Fishman et al., 1991). The 80-amino acid deletion, which produces a Cx43 protein lacking the serine-rich region containing several consensus phosphorylation sites, led to the formation of 160-pS channels. The 138-amino acid deletion produced a 50-pS channel. Although these data suggest that CT of Cx43 plays a role in determining single channel conductance, it does not directly implicate CT as a pore-forming domain. Clearly, CT cannot form the pore itself, as its removal should prohibit the formation of a functional channel. Synthetic peptides corresponding to Cx32 sequences were tested for their ability to inhibit cell-cell channel formation (Dahl et al., 1994). Although inhibition implicated portions of both extracellular loops, E1 and E2, the properties of one 12-mer, corresponding to positions 156 to 166, did not inhibit formation of coupling, but rather induced channel activity both in *Xenopus* oocyte membranes and artificial bilayers. Complex behavior was observed, with unitary conductances ranging from 20 to 160 pS. I-V curves showed strong voltage dependence with depolarization inducing slowly activating outward currents. Although Dahl and associates (1994) suggest that this E2 peptide may form part of the Cx32 hemichannel pore analogous to the P-region of $K^+$ channels (see Durell and Guy, 1992), the ability of isolated peptides to form pores in membranes is well documented and need not be indicative of its role as the pore-forming domain in the native channel (Montal, 1990).

More recent studies in Cx46 hemichannels and hemichannels formed of a chimera in which E1 of Cx32 was substituted with Cx43 sequence, Cx32E143, suggest a role for TM1 as contributing to the pore. Cysteine substitution of a subset of TM1 and TM3 residues and subsequent assessment of accessibility using the thiol reagent maleimido-butyryl-biocytin (MBB) revealed two residues at equivalent positions in TM1, I34 and L35 in Cx46 and I33 and M34 in Cx32E143, as accessible and contributing to the lining of the pore (Zhou et al., 1997). Studies also suggesting TM1 as contributing to the pore include biophysical characterization of a missense mutation in TM1, S26L, that causes CMTX (Oh et al., 1997). This mutation produces little structural perturbation as evidenced by a nearly wild-type $g_j$-$V_j$ relation, but reduces the effective pore radius of Cx32 from approximately 7 Å to less than 3 Å. These results are most easily explained if the substitution of Ser by the more bulky Leu causes a local constriction in the channel near the cytoplasmic aspect of TM1. The lack of an effect of the thiol reagent MBB on S26C in Cx32E143 indicates either that there may be steric hindrance, as MBB is a large 537 D reagent, or that S26L causes a conformational change that causes another residue to move into the pore and partially obstruct it. Cysteine scanning of TM3 in Cx32E143 hemichannels suggest S138, E146, and M150 are also reactive residues although the level of inhibition by MBB at these positions was considerably smaller than in TM1.

As discussed previously in this chapter, molecular studies of Cx32 and Cx26 identified amino acid residues located in the N-terminus and at the E1/TM1 border as components of the $V_j$ sensor (Verselis et al., 1994). The requirement that $V_j$-sensing residues be located in the transjunctional field where they can sense $V_j$ and not $V_m$ (see Harris et al., 1981; Verselis et al., 1994) suggests that the N-terminus loops back to place the components of the $V_j$ sensor in or near the channel lumen, where the $V_j$ field remains constant at all values of $V_m$. It also places residues at the E1/TM1 in or near the pore. These studies, together with those of Zhou and associates (1997) and Oh and co-workers (1997) paint a picture of the pore as formed by a number of domains that are distributed (i.e., not collinear with the primary sequence). Given the current present data with regard to channel lining sequences, an open mind to the interpretation of these results is merited.

## D.  Molecular Aspects of Connexin Compatibility

Not all heterotypic combinations of connexins are functional. The best studied of these nonfunctional heterotypic junctions is Cx43/Cx40 (Bruzzone et al., 1993). Purkinje cells, which are part of the conductive myocardium, express Cx40 and cells of the working myocardium express Cx43 (Bastide et al., 1993; Gros et al., 1994). The absence of coupling between these two systems assures a high speed of excitation spread in the branches of the bundle of His. In search of a molecular basis for connexin compatibility, the extracellular loops, E1 and E2, are natural candidates as they must mediate hemichannel docking. All connexins conserve three cysteines with a $CX_6CX_3C$ motif in E1 and a $CX_4CX_5C$ motif in E2. A change in any of these cysteines to serine results in a loss of functional expression of coupling in pairs of *Xenopus* oocytes (Dahl et al., 1992). These data are in agreement with the effects of thiol-specific reagents on formation of coupling and prompted a hypothesis for functional channel formation (i.e., hemichannel docking or opening, or both) that involves disulfide exchange reactions involving the cysteines in E1 and E2 (Dahl et al., 1991). Supporting the view that the extracellular loops are involved in channel formation, a number of synthetic peptides representing portions of Cx32 E1 and E2 sequences were shown to inhibit cell-cell channel formation (Dahl et al., 1994). Analysis of mutants and chimeras similarly implicates the extracellular loops. Although Cx38/Cx32 junctions do not form, Cx38/Cx43 junctions form readily. The double mutant Cx32I52R+ K167T, in which Cx43 residues are substituted at the same positions in Cx32, was shown to be capable of forming functional heterotypic junctions with Cx38 (Dahl et al., 1992).

Studies, based on the observation that Cx43 can form heterotypic junctions with Cx46 and not with Cx50, showed that replacement of E2 in Cx46 with the corresponding sequence of Cx50 abolished the ability Cx46 to recognize Cx43 (White et al., 1995). The ability of the reciprocal chimera, in which E2 in Cx50 was replaced with the corresponding sequence of Cx46, to confer recognition of Cx50 with Cx43 suggests that the interactions between connexins that determines incompatibility are localized in E2. However, these studies used functional electrical coupling alone as an assay for compatibility and thus do not distinguish channels that do not form or dock, or both, from those that form, but do not conduct. Haubrich and associates (1996) found that the extracellular loops, E1 and E2, were important in forming functional channels. In studies using Cx40 and Cx43, (Haubrich et al., 1996) showed using antibody staining that plaques are absent between Cx40/Cx43 HeLa cell pairs. Molecular analyses found that the extracellular loops, E1 and E2, were important in forming functional channels, in agreement with White and colleagues (1995). These data suggest that the incompatibilty between Cx40 and Cx43 channels results from an inability of the component hemichannels to dock or form cell-cell channels and that E1 and E2 play a role. In the peptide inhibition assay, however, recognition sites involved in connexin compatibility were complex, involving fractions of both extracellular loops (Dahl et al., 1994). As can be seen from Table 2, the pattern of heterotypic junction formation does not always follow a simple transitive rule; for example, Cx43 couples to Cx46 and Cx46 couples to Cx32, but Cx43 does not couple to Cx32, suggesting that connexin recognition may not simply involve matching extracellular loops. Examination of the primary sequences of connexins indicates a lack of correlation between the overall degree of sequence similarity in the conserved regions (including E1 and E2) and compatibility. This may implicate local sequence or structure may be a more useful index of compatibility.

## REFERENCES

Alves, L.A., Coutinho-Silva, R., Persechini, P.M., Spray, D.C., Savino, W., & Campos de Carvalho, A.C. (1996). Are there functional gap junctions or junctional hemichannels in macrophages? Blood 88, 328-334.

Anumonwo, J.M., Wang, H.Z., Trabka-Janik, E., Dunham, B., Veenstra, R.D., Delmar, M., & Jalife, J. (1992). Gap junctional channels in adult mammalian sinus nodal cells. Immunolocalization and electrophysiology. Circ. Res. 71, 229-239.

Auerbach, A.A., & Bennett, M.V.L. (1969). A rectifying electrotonic synapse in the central nervous system of a vertebrate. J. Gen. Physiol. 53, 211-237.

Barrio, L.C., Suchyna, T., Bargiello, T., Xu, L.X., Roginski, R.S., Bennett, M.V., & Nicholson, B.J. (1991). Gap junctions formed by connexins 26 and 32 alone and in combination are differently affected by applied voltage. Proc. Natl. Acad. Sci. U.S.A. 88, 8410-8414.

Bastide, B., Neyses, L., Ganten, D., Paul, M., Willecke, K., & Traub, O. (1993). Gap junction protein connexin40 is preferentially expressed in vascular endothelium and conductive bundles of rat myocardium and is increased under hypertensive conditions. Circ. Res. 73, 1138-1149.

Beblo, D.A. & Veenstra, R.D. (1997). Monovalent cation permeation through the connexin40 gap junction channel. Cs, Rb, K, Na, Li, TEA, TMA, TBA, and effects of anions Br, Cl, F, acetate, aspartate, glutamate, and $NO_3$. J. Gen. Physiol. 109, 509-522.

Beblo, D.A., Wang, H.Z., Beyer, E.C., Westphale, E.M., & Veenstra, R.D. (1995). Unique conductance, gating, and selective permeability properties of gap junction channels formed by connexin40. Circ. Res. 77, 813-822.

Bennett, M.V., Barrio, L.C., Bargiello, T.A., Spray, D.C., Hertzberg, E., & Saez, J.C. (1991). Gap junctions: new tools, new answers, new questions. Neuron 6, 305-320.

Bennett, M.V., & Goodenough, D.A. (1978). Gap junctions, electrotonic coupling, and intercellular communication. Neurosci. Res. Program. Bull. 16, 1-486.

Bennett, M.V., & Verselis, V.K. (1992). Biophysics of gap junctions. Semin. Cell Biol. 3, 29-47.

Bennett, M.V., Zheng, X., & Sogin, M.L. (1994). The connexins and their family tree [review]. Soc. Gen. Physiol. Ser. 49, 223-233.

Bennett, M.V.L. (1966). Physiology of electrotonic junctions. Ann. N.Y. Acad. Sci. 37, 509-539.

Bennett, M.V.L. (1977). Electrical transmission: a functional analysis and comparison to chemical transmission. In: *The Handbook of Physiology* (Kandel, E., ed.), pp. 357-416. American Physiological Society, Washington, D.C.

Bergoffen, J., Scherer, S.S., Wang, S., Scott, M.O., Bone, L.J., Paul, D.L., Chen, K., Lensch, M.W., Chance, P.F., & Fischbeck, K.H. (1993). Connexin mutations in X-linked Charcot-Marie-Tooth disease. Science 262, 2039-2042.

Beyer, E.C. (1990). Molecular cloning and developmental expression of two chick embryo gap junction proteins. J. Biol. Chem. 265, 14439-14443.

Beyer, E.C., Paul, D.L., & Goodenough, D.A. (1987). Connexin43: a protein from rat heart homologous to a gap junction protein from liver. J. Cell Biol. 105, 2621-2629.

Beyer, E.C., Reed, K.E., Westphale, E.M., Kanter, H.L., & Larson, D.M. (1992). Molecular cloning and expression of rat connexin40, a gap junction protein expressed in vascular smooth muscle. J. Membr. Biol. 127, 69-76.

Beyer, E.C., & Steinberg, T.H. (1991). Evidence that the gap junction protein connexin-43 is the ATP-induced pore of mouse macrophages. J. Biol. Chem. 266, 7971-7974.

Bezrukov, S.M., & Vodyanoy, I. (1993). Probing alamethicin channels with water-soluble polymers. Effect on conductance of channel states. Biophys. J. 64, 16-25.

Blennerhassett, M.G., & Caveney, S. (1984). Separation of developmental compartments by a cell type with reduced junctional permeability. Nature 309, 361-364.

Brink, P.R. (1983). Effect of deuterium oxide on junctional membrane channel permeability. J. Membr. Biol. 71, 1-9.

Brink, P.R. (1991). Gap junction channels and cell-to-cell messengers in myocardium. J. Cardiovasc. Electrophys. 2, 360-366.

Brink, P.R., & Dewey, M.M. (1978). Nexal membrane permeability to anions. J. Gen. Physiol. 72, 62-86.

Brink, P.R., & Dewey, M.M. (1980). Evidence for fixed charge in the nexus. Nature 285, 101-102.

Brink, P.R., Dewey, M.M., Colflesh, D.E., & Kensler, R.W. (1981). Polymorphic nexuses in the earthworm Lumbricus terrestris. J. Ultrastruct. Res. 77, 233-240.

Brink, P.R., & Fan, S.F. (1989). Patch clamp recordings from membranes which contain gap junction channels. Biophys. J. 56, 579-593.

Brink, P.R. & Ramanan, S.W. (1985). A model for the diffusion of fluorescent probes in the septate giant axon of earthworm. Axoplasmic diffusion and junctional membrane permeability. Biophys. J. 48, 299-309.

Bruzzone, R., Haefliger, J.A., Gimlich, R.L., & Paul, D.L. (1993). Connexin40, a component of gap junctions in vascular endothelium, is restricted in its ability to interact with other connexins. Mol. Biol. Cell. 4, 7-20.

Bruzzone, R., White, T.W., & Paul, D.L. (1994). Expression of chimeric connexins reveals new properties of the formation and gating behavior of gap junction channels. J. Cell Sci. 107, 955-967.

Bukauskas, F.F., Elfgang, C., Willecke, K., & Weingart, R. (1995a). Biophysical properties of gap junction channels formed by mouse connexin40 in induced pairs of transfected human HeLa cells. Biophys. J. 68, 2289-2298.

Bukauskas, F.F., Elfgang, C., Willecke, K., & Weingart, R. (1995b). Heterotypic gap junction channels (connexin26-connexin32) violate the paradigm of unitary conductance. Pflugers Arch. 429, 870-872.

Bukauskas, F.F., & Weingart, R. (1993). Multiple conductance states of newly formed single gap junction channels between insect cells. Pflugers Arch. 423, 152-154.

Bukauskas, F.F., & Weingart, R. (1994). Voltage-dependent gating of single gap junction channels in an insect cell line. Biophys. J. 67, 613-625.

Burt, J.M., & Spray, D.C. (1988). Single-channel events and gating behavior of the cardiac gap junction channel. Proc. Natl. Acad. Sci. U.S.A. 85, 3431-3434.

Chandross, K.J., Chanson, M., Spray, D.C., & Kessler, J.A. (1995). Transforming growth factor-beta 1 and forskolin modulate gap junctional communication and cellular phenotype of cultured Schwann cells. J. Neurosci. 15, 262-273.

Chang, M., Werner, R., & Dahl, G. (1996). A role for an inhibitory connexin in testis? Dev. Biol. 175, 50-56.

Chanson, M., Chandross, K.J., Rook, M.B., Kessler, J.A., & Spray, D.C. (1993). Gating characteristics of a steeply voltage-dependent gap junction channel in rat Schwann cells. J. Gen. Physiol. 102, 925-946.

Chanson, M., Roy, C., & Spray, D.C. (1994). Voltage-dependent gap junctional conductance in hepatopancreatic cells of Procambarus clarkii. Amer. J. Physiol. 266, C569-C577.

Chen, Y.H., & DeHaan, R.L. (1992). Multiple-channel conductance states and voltage regulation of embryonic chick cardiac gap junctions. J. Membr. Biol. 127, 95-111.

Chow, I., & Young, S.H. (1987). Opening of single gap junction channels during formation of electrical coupling between embryonic muscle cells. Dev. Biol. 122, 332-337.

Churchill, D., & Caveney, S. (1993). Double whole-cell patch-clamp characterization of gap junctional channels in isolated insect epidermal cell pairs. J. Membr. Biol. 135, 165-180.

Dahl, G., Levine, E., Rabadan Diehl, C., & Werner, R. (1991). Cell/cell channel formation involves disulfide exchange. Eur. J. Biochem. 197, 141-144.

Dahl, G., Nonner, W., & Werner, R. (1994). Attempts to define functional domains of gap junction proteins with synthetic peptides. Biophys. J. 67, 1816-1822.

Dahl, G., Werner, R., Levine, E., & Rabadan Diehl, C. (1992). Mutational analysis of gap junction formation. Biophys. J. 62, 172-180.

Dermietzel, R., Hertzberg, E.L., Kessler, J.A., & Spray, D.C. (1991). Gap junctions between cultured astrocytes: immunocytochemical, molecular, and electrophysiological analysis. J. Neurosci. 11, 1421-1432.

DeVries, S.H., & Schwartz, E.A. (1992). Hemi-gap-junction channels in solitary horizontal cells of the catfish retina. J. Physiol. 445, 201-230.

Donaldson, P., & Kistler, J. (1992). Reconstitution of channels from preparations enriched in lens gap junction protein MP70. J. Membr. Biol. 129, 155-165.

Dudek, F.E., & Snow, R.W. (1985). Electrical interactions and synchronization of cortical neurons: electrotonic coupling and field effects. In: *Gap Junctions* (Bennett, M.V.L., & Spray, D.C., Eds.), pp. 325-336. Cold Spring Harbor Laboratory, Cold Spring Harbor, NY.

Durell, S.R., & Guy, H.R. (1992). Atomic scale structure and functional models of voltage-gated potassium channels. Biophys. J. 62, 238-247; discussion 247-250.

Dwyer, T.M., Adams, D.J., & Hille, B. (1980). The permeability of the endplate channel to organic cations in frog muscle. J. Gen. Physiol. 75, 469-492.

Ebihara, L. (1996). Xenopus connexin38 forms hemi-gap-junctional channels in the nonjunctional plasma membrane of Xenopus oocytes. Biophys. J. 71, 742-748.

Ebihara, L., Berthoud, V.M., & Beyer, E.C. (1995). Distinct behavior of connexin56 and connexin46 gap junctional channels can be predicted from the behavior of their hemi-gap- junctional channels. Biophys. J. 68, 1796-1803.

Ebihara, L., Beyer, E.C., Swenson, K.I., Paul, D.L., & Goodenough, D.A. (1989). Cloning and expression of a Xenopus embryonic gap junction protein. Science 243, 1194-1195.

Ebihara, L., & Steiner, E. (1993). Properties of a nonjunctional current expressed from a rat connexin46 cDNA in Xenopus oocytes. J. Gen. Physiol. 102, 59-74.

Eghbali, B., Kessler, J.A., & Spray, D.C. (1990). Expression of gap junction channels in communication-incompetent cells after stable transfection with cDNA encoding connexin 32. Proc. Natl. Acad. Sci. U.S.A. 87, 1328-1331.

Elfgang, C., Eckert, R., Lichtenberg-Frate, H., Butterweck, A., Traub, O., Klein, R.A., Hulser, D.F., & Willecke, K. (1995). Specific permeability and selective formation of gap junction channels in connexin-transfected HeLa cells. J.Cell biol. 29, 805-817.

Fairweather, N., Bell, C., Cochrane, S., Chelly, J., Wang, S., Mostacciuolo, M.L., Monaco, A.P., & Haites, N.E. (1994). Mutations in the connexin 32 gene in X-linked dominant Charcot-Marie-Tooth disease (CMTX1). Hum. Mol. Genet. 3, 29-34.

Finkel, A.S., & Redman, S. (1984). Theory and operation of a single microelectrode voltage clamp. J. Neurosci. Methods 11, 101-127.

Finkelstein, A. (1963). The role of time variant resistance and electromotive force in ionic systems related to cell membranes: The excitability properties of frog skin and toad bladder. PhD Thesis, The Rockefeller University, New York.

Fishman, G.I., Moreno, A.P., Spray, D.C., & Leinwand, L.A. (1991). Functional analysis of human cardiac gap junction channel mutants. Proc. Natl. Acad. Sci. U.S.A. 88, 3525-3529.

Fishman, G.I., Spray, D.C., & Leinwand, L.A. (1990). Molecular characterization and functional expression of the human cardiac gap junction channel. J. Cell Biol. 111, 589-598.

Flagg Newton, J., & Loewenstein, W.R. (1979). Experimental depression of junctional membrane permeability in mammalian cell culture. A study with tracer molecules in the 300 to 800 Dalton range. J. Membr. Biol. 50, 65-100.

Furshpan, E.J., & Potter, D.D. (1959). Transmission at the giant motor synapses of the crayfish. J. Physiol. (Lond.) 145, 289-325.

Giaume, C. (1991). Application of the patch clamp technique to the study of junctional conductance. In: Biophysics of Gap Junction Channels. (Peracchia, C., Ed.), pp. 175-190. CRC Press, Boca Raton, Florida.

Giaume, C., Kado, R.T., & Korn, H. (1987). Voltage-clamp analysis of a crayfish rectifying synapse. J. Physiol. (Lond.) 386, 91-112.

Goodenough, D.A. (1975). The structure of cell membranes involved in intercellular communication. Am. J. Clin. Pathol. 63, 636-645.

Goodenough, D.A. (1992). The crystalline lens. A system networked by gap junctional intercellular communication. Semin. Cell Biol. 3, 49-58.

Goodenough, D.A., & Revel, J.P. (1970). A fine structural analysis of intercellular junctions in the mouse liver. J. Cell Biol. 45:272-290.

Gros, D., Jarry-Guichard, T., Ten Velde, I., de Maziere, A., van Kempen, M.J., Davoust, J., Briand, J.P., Moorman, A.F., & Jongsma, H.J. (1994). Restricted distribution of connexin40, a gap junctional protein, in mammalian heart. Circ. Res. 74, 839-851.

Hamill, O.P., Marty, A., Neher, E., Sakmann, B., & Sigworth, F.J. (1981). Improved patch-clamp techniques for high-resolution current recording from cells and cell-free membrane patches. Pflugers Arch., Eur. J. Physiol. 391, 85-100.

Harris, A.L., Spray, D.C., & Bennett, M.V. (1981). Kinetic properties of a voltage-dependent junctional conductance. J. Gen. Physiol. 77, 95-117.

Harris, A.L., Spray, D.C., & Bennett, M.V. (1983). Control of intercellular communication by voltage dependence of gap junctional conductance. J. Neurosci. 3, 79-100.

Hartmann, H.A., Kirsch, G.E., Drewe, J.A., Taglialatela, M., Joho, R.H., & Brown, A.M. (1991). Exchange of conduction pathways between two related K+ channels. Science 251, 942-944.

Haubrich, S., Schwarz, H.J., Bukauskas, F., Lichtenberg-Frate, H., Traub, O., Weingart, R., & Willecke, K. (1996). Incompatibility of connexin 40 and 43 hemichannels in gap junctions between mammalian cells is determined by intracellular domains. Mol. Biol. Cell. 7, 1995-2006.

Hennemann, H., Dahl, E., White, J.B., Schwarz, H.J., Lalley, P.A., Chang, S., Nicholson, B.J., & Willecke, K. (1992a). Two gap junction genes, connexin 31.1 and 30.3, are closely linked on mouse chromosome 4 and preferentially expressed in skin. J. Biol. Chem. 267, 17225-17233.

Hennemann, H., Suchyna, T., Lichtenberg Frate, H., Jungbluth, S., Dahl, E., Schwarz, J., Nicholson, B.J., & Willecke, K. (1992b). Molecular cloning and functional expression of mouse connexin40, a second gap junction gene preferentially expressed in lung. J. Cell Biol. 117, 1299-1310.

Hille, B. (1992). Ionic Channels of Excitable Membranes. Sinauer Associates, Inc., Sunderland, Massachusetts.

Hoh, J.H., John, S.A., & Revel, J.P. (1991). Molecular cloning and characterization of a new member of the gap junction gene family, connexin-31. J. Biol. Chem. 266, 6524-6531.

Imanaga, I., Kameyama, M., & Irisawa, H. (1987). Cell-to-cell diffusion of fluorescent dyes in paired ventricular cells. Am. J. Physiol. 252, H223-H232.

Jaslove, S.W., & Brink, P.R. (1986). The mechanism of rectification at the electrotonic motor giant synapse of the crayfish. Nature 323, 63-65.

Johnston, M.F., & Ramon, F. (1981). Electrotonic coupling in internally perfused crayfish segmented axons. J. Physiol. (Lond.) 317, 509-518.

Kadle, R., Zhang, J.T., & Nicholson, B.J. (1991). Tissue-specific distribution of differentially phosphorylated forms of Cx43. Mol. Cell Biol. 11, 363-369.

Kanter, H.L., Saffitz, J.E., & Beyer, E.C. (1992). Cardiac myocytes express multiple gap junction proteins. Circ. Res. 70, 438 444.

Lal, R., & Arnsdorf, M.F. (1992). Voltage-dependent gating and single-channel conductance of adult mammalian atrial gap junctions. Circ. Res. 71, 737-743.

Larson, D.M., Haudenschild, C.C., & Beyer, E.C. (1990). Gap junction messenger RNA expression by vascular wall cells. Circ. Res. 66, 1074-1080.

Levitt, D.G. (1975). General continuum analysis of transport through pores. I. Proof of Onsager's reciprocity postulate for uniform pore. Biophys. J. 15, 533-551.

Levitt, D.G. (1991). General continuum theory for multiion channel. II. Application to acetylcholine channel. Biophys. J. 59, 278 288.

Li, H., Liu, T.F., Lazrak, A., Peracchia, C., Goldberg, G.S., Lampe, P.D., & Johnson, R.G. (1996). Properties and regulation of gap junctional hemichannels in the plasma membranes of cultured cells. J. Cell Biol. 134, 1019-1030.

Liman, E.R., Hess, P., Weaver, F., & Koren, G. (1991). Voltage-sensing residues in the S4 region of a mammalian K+ channel. Nature 353, 752-756.

Llinas, R. (1985). Electrotonic transmission in the mammalian central nervous system. In: Gap Junctions. (Bennett, M.V.L. & Spray, D.C. Eds.), pp. 337-353. Cold Spring Harbor Laboratory, Cold Spring Harbor, N.Y.

Loewenstein, W.R. (1981). Junctional intercellular communication: the cell-to-cell membrane channel. Physiol. Rev. 61, 829-913.

Manivannan, K., Ramanan, S.V., Mathias, R.T., & Brink, P.R. (1992). Multichannel recordings from membranes which contain gap junctions. Biophys. J. 61, 216-227.

Margiotta, J.F., & Walcott, B. (1983). Conductance and dye permeability of a rectifying electrical synapse. Nature 305, 52-55.

Matter, A. (1973). A morphometric study on the nexus of rat cardiac muscle. J. Cell Biol. 56, 690-696.

Mauro, A. (1962). Space charge regions in fixed membranes and the associated property of capacitance. Biophys. J. 2, 179-198.

McMahon, D.G., Knapp, A.G., & Dowling, J.E. (1989). Horizontal cell gap junctions: single-channel conductance and modulation by dopamine. Proc. Natl. Acad. Sci. U.S.A. 86, 7639-7643.

McNutt. N., & Weinstein, R.S. (1970). The ultrastructure of the nexus. A correlated thin section and freeze-cleave study. J. Cell Biol. 47, 666-688.

Montal, M. (1990). Channel protein engineering. An approach to the identification of molecular determinants of function in voltage-gated and ligand-regulated channel proteins. Ion Channels. 2, 1-31.

Moore, L.K., Beyer, E.C., & Burt, J.M. (1991). Characterization of gap junction channels in A7r5 vascular smooth muscle cells. Am. J. Physiol. 260, C975-C981.

Moreno, A.P., Campos de Carvalho, A.C., Christ, G., Melman, A., & Spray, D.C. (1993). Gap junctions between human corpus cavernosum smooth muscle cells: gating properties and unitary conductance. Am. J. Physiol. 264, C80-C92.

Moreno, A.P., Eghbali, B., & Spray, D.C. (1991). Connexin32 gap junction channels in stably transfected cells: unitary conductance. Biophys. J. 60, 1254-1266.

Moreno, A.P., Fishman, G.I., & Spray, D.C. (1992). Phosphorylation shifts unitary conductance and modifies voltage dependent kinetics of human connexin43 gap junction channels. Biophys. J. 62, 51-53.

Moreno, A.P., Laing, J.G., Beyer, E.C., & Spray, D.C. (1995). Properties of gap junction channels formed of connexin 45 endogenously expressed in human hepatoma (SKHep1) cells. Am. J. Physiol. 268, C356-C365.

Moreno, A.P., Rook, M.B., Fishman, G.I., & Spray, D.C. (1994). Gap junction channels: distinct voltage-sensitive and -insensitive conductance states. Biophys. J. 67, 113-119.

Musil, L.S., Beyer, E.C., & Goodenough, D.A. (1990a). Expression of the gap junction protein connexin43 in embryonic chick lens: molecular cloning, ultrastructural localization, and post-translational phosphorylation. J. Membr. Biol. 116, 163-175.

Musil, L.S., Cunningham, B.A., Edelman, G.M., & Goodenough. D.A. (1990b). Differential phosphorylation of the gap junction protein connexin43 in junctional communication-competent and -deficient cell lines. J. Cell Biol. 111, 2077-2088.

Musil, L.S., & Goodenough, D.A. (1993). Multisubunit assembly of an integral plasma membrane channel protein, gap junction connexin43, occurs after exit from the ER. Cell 74, 1065-1077.

Neyton, J., & Trautmann, A. (1985). Single-channel currents of an intercellular junction. Nature 317, 331-335.

Neyton, J., & Trautmann, A. (1986). Physiological modulation of gap junction permeability. J. Exp. Biol. 124, 993-114.

Nicholson, B., Dermietzel, R., Teplow, D., Traub, O., Willecke, K., & Revel. J.P. (1987). Two homologous protein components of hepatic gap junctions. Nature 329, 732-734.

Obaid, A.L., Socolar, S.J., & Rose, B. (1983). Cell-to-cell channels with two independent regulated gates in series: analysis of junctional channel modulation by membrane potential, calcium and pH. J. Membr. Biol. 73, 69-89.

Oh, S., Ri, Y., Bennett, M.V., Trexler, E.B., Verselis, V.K., & Bargiello, T.A. (1997). Changes in permeability caused by connexin 32 mutations underlie X-linked Charcot-Marie-Tooth disease. Neuron 19, 927-938.

Paul, D.L. (1986). Molecular cloning of cDNA for rat liver gap junction protein. J. Cell Biol. 103, 123-134.

Paul, D.L., Ebihara, L., Takemoto, L.J., Swenson, K.I., & Goodenough, D.A. (1991). Connexin46, a novel lens gap junction protein, induces voltage-gated currents in nonjunctional plasma membrane of Xenopus oocytes. J. Cell Biol. 115, 1077-1089.

Peinado, A., Yuste, R., & Katz, L.C. (1993). Extensive dye coupling between rat neocortical neurons during the period of circuit formation. Neuron 10, 103-114.

Peracchia, C. (1980). Structural correlates of gap junction permeation. Int. Rev. Cytol. 66, 81-146.

Perez Armendariz, E.M., Romano, M.C., Luna, J., Miranda, C., Bennett, M.V.L., & Moreno, A.P. (1994). Characterization of gap junctions between pairs of Leydig cells from mouse testis. Am. J. Physiol. 267, C570-C580.

Pfahnl, A. & Dahl, G. (1998). Localization of a voltage gate in connexin46 gap junction hemichannels. Biophys. J. 75, 2323-2331.

Phelan, P., Stebbings, L.A., Baines, R.A., Bacon, J.P., Davies, J.A., & Ford, C. (1998). Drosophila Shaking-B protein forms gap junctions in paired Xenopus oocytes. Nature 391, 181-184.

Ramanan, S.V., & Brink, P.R. (1990). Exact solution of a model of diffusion in an infinite chain or monolayer of cells coupled by gap junctions. Biophys. J. 58, 631-639.

Ramanan, S.V., & Brink, P.R. (1993). Multichannel recordings from membranes which contain gap junctions. II. Substates and conductance shifts. Biophys. J. 65, 1387-1395.

Ramanan, S.V., Mesimeris, V., & Brink, P.R. (1994). Ion flow in the bath and flux interactions between channels. Biophys. J. 66, 989-995.

Reed, K.E., Westphale, E.M., Larson, D.M., Wang, H.Z., Veenstra, R.D., & Beyer, E.C. (1993). Molecular cloning and functional expression of human connexin37, an endothelial cell gap junction protein. J. Clin. Invest. 91, 997-1004.

Renkin, E.M. (1955). Filtration, diffusion and molecular sieving through porous cellulose membranes. J. Gen. Physiol. 38, 225-243.

Ri, Y., Ballesteros, J.A., Abrams, C.K., Oh, S., Verselis, V.K., Weinstein, H., & Bargiello, T.A. (1999). The role of a conserved proline residue in mediating conformational changes associated with voltage gatine of Cx32 gap junctions. Biophys. J. 76, 2887-2898.

Rook, M.B., Jongsma, H.J., & de Jonge, B. (1989). Single channel currents of homo- and heterologous gap junctions between cardiac fibroblasts and myocytes. Pflugers Arch. 414, 95-98.

Rook, M.B., Jongsma, H.J., & van Ginneken, A.C. (1988). Properties of single gap junctional channels between isolated neonatal rat heart cells. Am. J. Physiol. 255, H770-H782.

Rook, M.B., van Ginneken, A.C., de Jonge, B., el Aoumari, A., Gros, D., & Jongsma, H.J. (1992). Differences in gap junction channels between cardiac myocytes, fibroblasts, and heterologous pairs. Am. J. Physiol. 263, C959-C977.

Rubin, J.B., Verselis, V.K., Bennett, M.V., & Bargiello, T.A. (1992a). A domain substitution procedure and its use to analyze voltage dependence of homotypic gap junctions formed by connexins 26 and 32. Proc. Natl. Acad. Sci. U.S.A. 89, 3820-3824.

Rubin, J.B., Verselis, V.K., Bennett, M.V., & Bargiello, T.A. (1992b). Molecular analysis of voltage dependence of heterotypic gap junctions formed by connexins 26 and 32. Biophys. J. 62, 183-193.

Rudisuli, A., & Weingart, R. (1989). Electrical properties of gap junction channels in guinea-pig ventricular cell pairs revealed by exposure to heptanol. Pflugers Arch. 415, 12-21.

Rup, D.M., Veenstra, R.D., Wang, H.Z., Brink, P.R., & Beyer, E.C. (1993). Chick connexin-56, a novel lens gap junction protein. Molecular cloning and functional expression. J. Biol. Chem. 268, 706-712.

Safranyos, R.G., Caveney, S., Miller, J.G., & Petersen, N.O. (1987). Relative roles of gap junction channels and cytoplasm in cell-to-cell diffusion of fluorescent tracers. Proc. Natl. Acad. Sci. U.S.A. 84, 2272-2276.

Salomon, D., Chanson, M., Vischer, S., Masgrau, E., Vozzi, C., Saurat, J.H., Spray, D.C., & Meda, P. (1992). Gap junctional communication of primary human keratinocytes: characterization by dual voltage clamp and dye transfer. Exp. Cell Res. 201, 452-461.

Schwarzmann, G., Wiegandt, H., Rose, B., Zimmerman, A., Ben Haim, D., & Loewenstein, W.R. (1981). Diameter of the cell-to-cell junctional membrane channels as probed with neutral molecules. Science 213, 551-553.

Sigworth, F.J., & Sine, S.M. (1987). Data transformations for improved display and fitting of single-channel dwell time histograms. Biophys. J. 52, 1047-1054.

Simpson, I., Rose, B., & Loewenstein, W.R. (1977). Size limit of molecules permeating the junctional membrane channels. Science 195, 294-296.

Somogyi, R., Batzer, A., & Kolb, H.A. (1989). Inhibition of electrical coupling in pairs of murine pancreatic acinar cells by OAG and isolated protein kinase C. J. Membr. Biol. 108, 273-282.

Somogyi, R., & Kolb, H.A. (1988). Cell-to-cell channel conductance during loss of gap junctional coupling in pairs of pancreatic acinar and Chinese hamster ovary cells. Pflugers Arch. 412, 54-65.

Spray, D.C., Chanson, M., Moreno, A.P., Dermietzel, R., & Meda, P. (1991a). Distinctive gap junction channel types connect WB cells, a clonal cell line derived from rat liver. Am. J. Physiol. 260, C513-C527.

Spray, D.C., Harris, A.L., & Bennett, M.V. (1981). Equilibrium properties of a voltage-dependent junctional conductance. J. Gen. Physiol. 77, 77-93.

Spray, D.C., Moreno, A.P., Eghbali, B., Chanson, M., & Fishman, G.I. (1992). Gating of gap junction channels as revealed in cells stably transfected with wild type and mutant connexin cDNAs. Biophys. J. 62, 48-50.

Spray, D.C., Moreno, A.P., Kessler, J.A., & Dermietzel, R. (1991b). Characterization of gap junctions between cultured leptomeningeal cells. Brain Res. 568, 1-14.

Spray, D.C., Saez, J.C., Brosius, D., Bennett, M.V., & Hertzberg, E.L. (1986). Isolated liver gap junctions: gating of transjunctional currents is similar to that in intact pairs of rat hepatocytes. Proc. Natl. Acad. Sci. U.S.A. 83, 5494-5497.

Spray, D.C., White, R.L., Mazet, F., & Bennett, M.V. (1985). Regulation of gap junctional conductance. Am. J. Physiol. 248, H753-H764.

Steinberg, T.H., Civitelli, R., Geist, S.T., Robertson, A.J., Hick, E., Veenstra, R.D., Wang, H.Z., Warlow, P.M., Westphale, E.M., Laing, J.G., et al. (1994) Connexin43 and connexin45 form gap junctions with different molecular permeabilities in osteoblastic cells. EMBO J. 13, 744-750.

Suchyna, T.M., Xu, L.X., Gao, F., Fourtner, C.R., & Nicholson, B.J. (1993). Identification of a proline residue as a transduction element involved in voltage gating of gap junctions. Nature 365, 847-849.

Sugiura, H., Toyama, J., Tsuboi, N., Kamiya, K., & Kodama, I. (1990). ATP directly affects junctional conductance between paired ventricular myocytes isolated from guinea pig heart. Circ. Res. 66, 1095-1102.

Swenson, K.I., Jordan, J.R., Beyer E.C., & Paul, D.L. (1989). Formation of gap junctions by expression of connexins in Xenopus oocyte pairs. Cell. 57, 145-155.

Takens Kwak, B.R., & Jongsma, H.J.. (1992). Cardiac gap junctions: three distinct single channel conductances and their modulation by phosphorylating treatments. Pflugers Arch. 422, 198-200.

Trexler, E.B., Bennett, M.V., Bargiello, T.A., & Verselis, V.K. (1996). Voltage gating and permeation in a gap junction hemichannel. Proc. Natl. Acad. Sci. U.S.A. 93, 5836-5841.

Tsien, R., & Weingart, R. (1976). Inotropic effect of cyclic AMP in calf ventricular muscle studied by a cut-end method. J. Physiol. (Lond.) 260, 117-141.

Unwin, N. (1989). The structure of ion channels in membranes of excitable cells. Neuron 3, 665-676.

Unwin, P.N., & Ennis, P.D. (1984). Two configurations of a channel-forming membrane protein. Nature 307, 609-613.

Veenstra, R.D. (1990). Voltage-dependent gating of gap junction channels in embryonic chick ventricular cell pairs. Am. J. Physiol. 258, C662-C672.

Veenstra, R.D. (1991). Developmental changes in regulation of embryonic chick heart gap junctions. J. Membr. Biol. 119, 253-265.

Veenstra, R.D. (1996). Size and selectivity of gap junction channels formed from different connexins. J. Bioenerg. Biomembr. 28, 327-337.

Veenstra, R.D. & Brink, P.R. (1992). Patch clamp analysis of gap junctional currents. In: Cell–Cell Interactions: A Practical Approach. (Paul, D.L. & Gallin, Eds.), pp. 167-201. IRL Press, Oxford, England.

Veenstra, R.D., & DeHaan, R.L. (1986). Measurement of single channel currents from cardiac gap junctions. Science 233, 972-974.

Veenstra, R.D., & DeHaan, R.L. (1988). Cardiac gap junction channel activity in embryonic chick ventricle cells. Am. J. Physiol. 254, H170-H180.

Veenstra, R.D., Wang, H.Z., Beblo, D.A., Chilton, M.G., Harris, A.L., Beyer, E.C., & Brink, P.R. (1995). Selectivity of connexin-specific gap junctions does not correlate with channel conductance. Circ. Res. 77, 1156-1165.

Veenstra, R.D., Wang, H.Z., Beyer, E.C., & Brink, P.R. (1994a). Selective dye and ionic permeability of gap junction channels formed by connexin45. Circ. Res. 75, 483-490.

Veenstra, R.D., Wang, H.Z., Beyer, E.C., Ramanan, S.V., & Brink, P.R. (1994b). Connexin37 forms high conductance gap junction channels with subconductance state activity and selective dye and ionic permeabilities. Biophys. J. 66, 1915-1928.

Veenstra, R.D., Wang, H.Z., Westphale, E.M., & Beyer, E.C. (1992). Multiple connexins confer distinct regulatory and conductance properties of gap junctions in developing heart. Circ. Res. 71, 1277-1283.

Verselis, V., & Brink, P.R. (1986). The gap junction channel. Its aqueous nature as indicated by deuterium oxide effects. Biophys. J. 50, 1003-1007.

Verselis, V., White, R.L., Spray D.C., & Bennett, M.V. (1986). Gap junctional conductance and permeability are linearly related. Science 234, 461-464.

Verselis, V.K., Bennett, M.V., & Bargiello, T.A. (1991). A voltage-dependent gap junction in Drosophila melanogaster. Biophys. J. 59, 114-126.

Verselis, V.K., Ginter, G.S., & Bargiello, T.A. (1994). Opposite voltage gating polarities of two closely related connexins. Nature 368, 348-351.

Wang, H.Z., Li, J., Lemanski, L.F., & Veenstra, R.D. (1992). Gating of mammalian cardiac gap junction channels by transjunctional voltage. Biophys. J. 63, 139-151.

Wang, H.Z., & Veenstra, R.D. (1997). Monovalent ion selectivity sequences of the rat connexin43 gap junction channel. J. Gen. Physiol. 109, 491-507.

Warner, A.E. & Lawrence, P.A. (1982). Permeability of gap junctions at the segmental border in insect epidermis. Cell 28, 243-252.

Weidmann, S. (1966). The diffusion of radiopotassium across intercalated discs of mammalian cardiac muscle. J. Physiol. (Lond.) 187, 323-342.

Weingart, R. (1974). The permeability to tetraethylammonium ions of the surface membrane and intercalated disks of sheep and calf myocardium. J. Physiol. (Lond.) 240, 741-762.

Weingart, R., & Bukauskas, F.F. (1993). Gap junction channels of insects exhibit a residual conductance. Pflugers Arch. 424, 192-194.

Werner, R., Levine, E., Rabadan Diehl, C., & Dahl, G. (1989). Formation of hybrid cell-cell channels. Proc. Natl. Acad. Sci. U.S.A. 86, 5380-5384.

White, R.L., Spray, D.C., Campos de, C.A.C., Wittenberg, B.A., & Bennett, M.V. (1985). Some electrical and pharmacological properties of gap junctions between adult ventricular myocytes. Am. J. Physiol. 249, C447-C455.

White, T.W., Bruzzone, R., Goodenough, D.A., & Paul, D.L. (1994a). Voltage gating of connexins. Nature 371, 208-209.

White, T.W., Bruzzone, R., Wolfram, S., Paul, D.L., & Goodenough, D.A. (1994b). Selective interactions among the multiple connexin proteins expressed in the vertebrate lens: the second extracellular domain is a determinant of compatibility between connexins. J. Cell Biol. 125, 879-892.

White, T.W., Paul, D.L., Goodenough, D.A., & Bruzzone, R. (1995). Functional analysis of selective interactions among rodent connexins. Mol. Biol. Cell. 6, 459-470.

Wilders, R., & Jongsma, H.J. (1992). Limitations of the dual voltage clamp method in assaying conductance and kinetics of gap junction channels. Biophys. J. 63, 942-953.

Willecke, K., Heynkes, R., Dahl, E., Stutenkemper, R., Hennemann, H., Jungbluth, S., Suchyna, T., & Nicholson, B.J. (1991). Mouse connexin37: cloning and functional expression of a gap junction gene highly expressed in lung. J. Cell Biol. 114, 1049-1057.

Young, J.D., Cohn, Z.A., & Gilula, N.B. (1987). Functional assembly of gap junction conductance in lipid bilayers: demonstration that the major 27 kd protein forms the junctional channel. Cell. 48, 733-743.

Yuste, R., Peinado, A., & Katz, L.C. (1992). Neuronal domains in developing neocortex. Science. 257, 665-669.

Zampighi, G.A., Hall, J.E., & Kreman, M. (1985). Purified lens junctional protein forms channels in planar lipid films. Proc. Natl. Acad. Sci. U.S.A. 82, 8468-8472.

Zhang, J.T., & Nicholson, B.J. (1989). Sequence and tissue distribution of a second protein of hepatic gap junctions, Cx26, as deduced from its cDNA. J. Cell Biol. 109, 3391-3401.

Zhou, X.W., Pfahnl, A., Werner, R., Hudder, A., Llanes, A., Luebke, A., & Dahl, G. (1997). Identification of a pore lining segment in gap junction hemichannels. Biophys. J. 72, 1946-1953.

Zimmerman, A.L., & Rose, B. (1985). Permeability properties of cell-to-cell channels: kinetics of fluorescent tracer diffusion through a cell junction. J. Membr. Biol. 84, 269-283.

# GAP JUNCTIONS IN DEVELOPMENT

Cecilia W. Lo and Norton B. Gilula

|  |  |  |
|---|---|---|
| I. | Gap Junctions and Development | 193 |
| II. | Gap Junctional Communication in Embryogenesis and Development | 194 |
|  | A. Communication Restrictions and Compartments in the Insect Epidermis | 197 |
|  | B. Gap Junctional Communication in the Frog Embryo | 200 |
|  | C. Gap Junctional Communication in Mouse and Chick Embryos | 200 |
|  | D. Gap Junctional Communication in Molluscan Embryos | 203 |
|  | E. Gap Junctional Communication in Nematode Embryos | 204 |
|  | F. Mechanisms Underlying Restrictions in Gap Junctional Communication | 205 |
| III. | Gap Junction Perturbation and Disruptions in Development | 208 |
|  | A. Blockage of Coupling with Antibody Reagents | 208 |
|  | B. Transgenic Mouse Models | 209 |
|  | C. Visceroatrial Heterotaxia and Connexin43 Mutations | 210 |
|  | D. Other Connexin Knockout Mouse Models | 210 |
|  | E. Gene Knockouts and Phenotypes | 211 |
|  | F. Dominant Negative Inhibition of Gap Junctions | 212 |
| IV. | Gap Junctions and Cell-Cell Signaling in Development | 213 |

## I. GAP JUNCTIONS AND DEVELOPMENT

A role for gap junctions in development has been proposed since the early studies showing the presence of low resistance junctions in nonexcitable cells and tissues

**Advances in Molecular and Cell Biology, Volume 30, pages 193-219.**
**Copyright © 2000 by JAI Press Inc.**
**All rights of reproduction in any form reserved.**
**ISBN: 0-7623-0599-1**

(see, for example, Furshpan and Potter, 1968; Sheridan 1966, 1968; Ito and Loewenstein, 1969; Bennett and Trinkaus, 1970). These studies demonstrated that cells of various adult tissues and those of the early embryo generally exhibited low resistance pathways through which ions and electrical current can pass from cell to cell (ionic coupling). Thus, microelectrode impalements into squid embryos revealed extensive coupling between the yolk sac and other cells throughout the embryo (Potter et al., 1966; Furshpan and Potter, 1968). This was observed even after cells in the embryo had acquired a different developmental fate. Subsequently, it was shown that these low resistance pathways were likely due to the presence of gap junctions (Gilula et al., 1972).

Gap junctions are, in fact, found in embryos from a diverse array of organisms, including rodents, chick, frog, newt, nematode, teleost, echinoderms, ascidians, and molluscs. It has been suggested that cell-cell interaction mediated by gap junctions may play a role in various physiological functions, such as in maintaining tissue homeostasis or in growth regulation. It has also been suggested that gap junctions may play a role in tissue morphogenesis and cell differentiation. One intriguing possibility is that gap junctions may help organize patterning events by mediating the formation of chemical gradients that specify positional information in a developmental field (Wolpert, 1978).

Two major avenues of investigation have been pursued by various laboratories to examine the possible role of gap junctions in development. One approach entails characterizing changes in the pattern of gap junctional communication in the embryo as development progresses. Such studies revealed that gap junction–mediated cell-cell communication is often spatially restricted in a developmentally significant manner, often in a pattern that is suggestive of a role in the underlying patterning and developmental events. A second line of investigation has entailed examining the developmental effects of disrupting or perturbing gap junctional communication. Such studies revealed striking developmental perturbations. In this chapter, we summarize some of the results obtained from such studies, and close with some speculations as to the possible role of gap junctions in development.

## II.   GAP JUNCTIONAL COMMUNICATION IN EMBRYOGENESIS AND DEVELOPMENT

Gap junction–mediated coupling is established in most embryos within the first few cleavages, and results in the entire embryo becoming interconnected as a syncytium, at least for passive diffusion of low molecular weight hydrophilic molecules (Ito and Hori, 1966; Ito and Loewenstein, 1969; Bennett and Trinkaus, 1970; Lo and Gilula, 1979; Guthrie, 1984; Serras and van den Bigelaar, 1987). This can be demonstrated by using microelectrode impalements to monitor the cell-to-cell movement of ions (ionic coupling). Ionic coupling, once established,

frequently persists even between cells of different differentiated states or with differing developmental potentials. For example, in the chick embryo, presumptive
notochord cells are ionically coupled to cells of the neural plate and to paraxial
mesodermal cells (Sheridan, 1968), or in axolotl embryos, ionic coupling is
observed between neural and lateral ectoderm, even after neural induction has
been completed (Warner, 1973). However, it is interesting to note that in some
instances, ionic coupling is apparently lost as development progresses. Thus in
several developmental systems, ionic coupling is not detected between the
embryo proper and cells of the extraembryonic tissue. This was observed for
squid and *Fundulus* embryos; in both instances, the embryo proper became
uncoupled from the yolk cell (Potter et al., 1966; Kimmel et al., 1984). Similarly,
in the early postimplantation mouse embryo, ionic coupling is lost between the
embryo proper and the trophoectoderm-derived placental precursor tissue (see the
discussion that follows; Lo and Gilula, 1979b; Kalimi and Lo, 1988). Given that
gap junctions can mediate the cell-to-cell movement of ions and low molecular
weight metabolites, including second messenger molecules, the loss of coupling
between the extraembryonic tissues and the embryo proper may serve to isolate
the developing embryo from metabolite pools of the extraembryonic tissues. In
mammalian embryos where the placenta of the developing embryo makes intimate cell-cell contact with the maternal tissues, this isolation may be of added
importance.

Further analysis of gap junctional communication in embryos with the injection
of fluorescein or Lucifer yellow has revealed dynamic changes in coupling that
were not evident with the ionic coupling measurements. In many instances, the
pattern of fluorescent dye spread (dye coupling) delineated boundaries defining
restrictions in gap junctional communication. Such restrictions effectively segregated the developing embryo or tissue into a number of "communication compartment" domains. Thus, cells lying within a communication compartment are well
coupled, exhibiting both ionic and dye coupling, while there is little or no coupling between cells situated across a compartment border. In some cases, ionic
coupling persists across a compartment boundary in the absence of dye coupling,
while in other instances, both ionic and dye coupling persist across a compartment
boundary but at a reduced level. Segregation of cells into communication com
partments was first observed in mouse embryos (Lo and Gilula, 1979b), but now
has been described in many developmental systems, both in vertebrates and invertebrates. In the following discussion, we summarize observations from a number
of organisms to highlight the fact that the restriction of gap junctional communication and the segregation of cells into communication compartment domains are
almost always associated with embryogenesis and development. Moreover, in
many instances, communication compartments appear to be established in a
developmentally regulated manner, and segregate the embryo or developing tissue into developmentally significant domains. Such observations strongly suggest

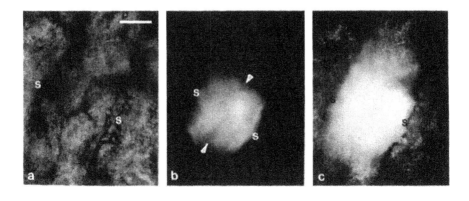

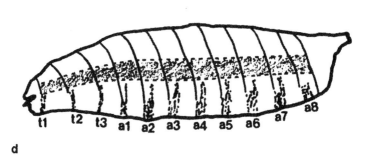

**Figure 1.**    Restriction of dye coupling in the *Drosophila* larval epidermis. **a–c**. Three *Drosophila* larval epidermal segments from a third instar larva can be seen in the phase contrast image in (a). The segmental borders are indicated as (s). A microelectrode was impaled into an epidermal cell in the middle segment and Lucifer yellow was iontophoretically injected intracellularly. The fluorescent dye spread (**b, c**) indicate dye movement is restricted along the segment border (s), and along two boundaries that lie orthogonal to the segmental border (see white arrowheads in **b**). **d**. A *Drosophila* larva with the approximate position of the mediolateral communication compartment denoted with dotted lines (filled in region). Each thoracic (t) and abdominal (a) segment is subdivided into four communication compartments, two smaller mediolateral compartments, and two larger dorsal/ventral compartments. The orthogonally positioned communication restriction boundaries delineating the mediolateral communication compartment is situated slightly dorsal of the denticle belts (black dots at the ventral aspect of each thoracic and abdominal segment).

a functional role for gap junction–mediated cell-cell communication in various aspects of patterning and development.

## A. Communication Restrictions and Compartments in the Insect Epidermis

The insect integument has served as a paradigm for studying pattern formation. Most recently, the analysis of *Drosophila* developmental mutants (see, for example, Nusslein-Volhard and Wieschaus, 1980; for review, see McGinnis and Krumlauf, 1992) have provided a wealth of insights into the cell and molecular mechanisms underlying patterning of the insect segmental body plan. Moreover, lineage studies in the insect epidermis have previously suggested that early in development, interactions between groups of cells (i.e., polyclones; Crick and Lawrence, 1975) are required for the formation of patterning units known as "developmental compartments." Such compartments do not define specific cell differentiation fates; rather, they define domains within which differentiation is spatially organized to reflect subdivisions in body parts, such as anterior versus posterior, or dorsal versus ventral aspects of the wing, leg, and so on.

The examination of gap junctional communication in the insect epidermis has revealed the presence of both ionic and dye coupling, with ionic coupling observed across each segment border, whereas little or no dye coupling is detected between cells of adjacent body segments (Warner and Lawrence, 1982; Blennerhasset and Caveney 1984; Ruangvoravat and Lo, 1992). Hence, each insect epidermal segment corresponds to a communication compartment (for example, see Figure 1). This has been demonstrated for a number of different insects including *Calliphora*, *Oncopeltus*, and *Drosophila*. That these communication compartment domains may be of developmental significance is suggested by the fact that each body segment is a unit within which patterning is regulated (Lawrence, 1966; Stumpf, 1966; for review, see, Locke, 1967), with the boundary of each segment coinciding with a lineage compartment border. In addition, each segment appears to serve as a domain within which pattern regulating genes of the Bithorax (Lewis, 1978) and Antennapedia (Kaufman and Abbot, 1984) complexes are differentially expressed.

However, molecular analyses of segmentation in the *Drosophila* epidermis have shown that the initial specification of the segmental body plan in the embryo proceeds in a "parasegmental" rather than segmental fashion (Martinez-Arias and Lawrence, 1985). In addition, many of the early events involved in segmentation in the *Drosophila* embryo occur either before cellularization or during the period when cytoplasmic bridges continue to link all the cells of the blastoderm. Thus cell-cell interactions mediated by gap junctions are not likely to be involved in the initial segmentation process. Nevertheless, beyond the blastoderm stage of development, the next tier of pattern regulating genes involved in insect patterning are the segment polarity genes, and these are, in fact, predominantly involved in

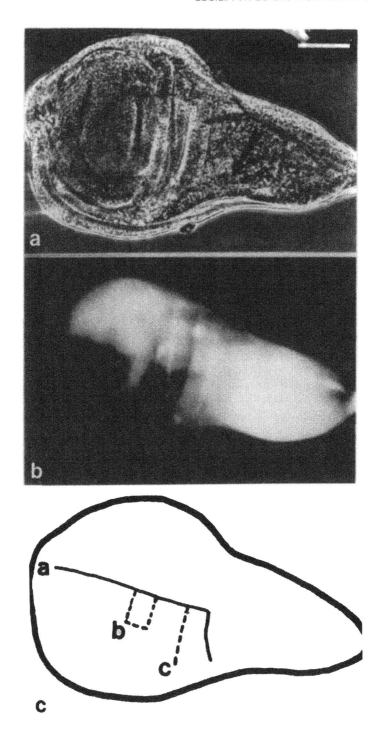

**Figure 2.** Restriction of dye coupling in the *Drosophila* wing imaginal disk. **a**. Phase contrast image of a third instar wing disk showing the site of microelectrode impalement and dye injection at the presumptive notal region (denoted by arrow). **b**. Darkfield fluorescence image at 52 minutes after the start of dye injection (Lucifer yellow). Dye has spread in a highly asymmetrical pattern, with sharp boundaries delineated as outlined in **c**. Note that boundary (a) approximates the position of the anterior/posterior lineage compartment border. Also note the partial nature of this restriction, as some dye has spread across boundary (a). (Reprinted with permission from Weir and Lo, 1982.)

cell-cell interactions (Ingham and Martinez Arias, 1992). For example, the segmental polarity gene *wingless* encodes a short-range diffusible cell signaling molecule, or the gene *armadillo* encodes a cell adhesion molecule (Peifer and Bejsovec, 1992). It is perhaps in these later events that cell-cell interactions mediated by gap junctions may play a role, either in maintaining the segmental pattern established earlier, or in facilitating patterning within each body segment. With regard to these possibilities, it is interesting to note that dye injection studies in the *Drosophila* larval epidermis revealed that each body segment in the thorax and abdomen is subdivided into four compartments (see Figure 1; Ruangvoravat and Lo, 1992), with the two lateral compartments likely overlapping or colocalizing with the two lateral bands of *wingless* expression.

An examination of dye coupling in the epidermal cells that give rise to the *Drosophila* adult integument, that is the imaginal disk epithelium, provided a similar finding. Thus, cells in the wing disk epithelium are segregated into communication compartment domains. Of particular interest is the finding that the communication compartment boundaries coincided with those of lineage compartment borders (Weir and Lo, 1982, 1984). For example, dye coupling is restricted along a boundary that bisects the disk along its long axis, a position identical to that of the anterior/posterior (A/P) lineage compartment border (Figure 2; Weir and Lo, 1982, 1984). Interestingly, in the wing disk of *engrailed* mutants, this A/P boundary is maintained (Weir and Lo, 1985), even though no A/P lineage restriction border is observed. This finding is consistent with the notion that gap junctional communication and communication compartments may play a role in specifying positional information in the wing disk epithelium, as positional information in *engrailed* mutants appears to be unaffected by the loss of *engrailed* function (as cell fate is entirely position appropriate but transformed in a mirror symmetric fashion).

The finding that gap junctional communication is restricted at lineage compartment borders in the *Drosophila* epidermis is also of interest, as it further suggests a role for gap junctions in growth and size regulation, processes that are essential to appropriate patterning and morphogenesis. Thus, cells within the A/P or dorsal/

ventral lineage compartments do not proliferate beyond their respective lineage compartment borders, even when given a growth advantage (see Morata and Lawrence, 1978). In addition, in *Drosophila* imaginal disk overgrowth mutants, excessive cell proliferation in the imaginal disk is accompanied by a significant reduction in gap junctional communication (Jursnich et al., 1990). It should be noted that although the role of gap junctions in growth regulation and carcinogenesis has been under investigation for some time (for review, see Loewenstein and Rose, 1992), there is as yet little consideration given to the possible role of gap junctions in regulating growth in embryogenesis and development.

## B. Gap Junctional Communication in the Frog Embryo

Examination of gap junctional communication in the *Xenopus* frog embryo has revealed that cell-cell communication as monitored by dye coupling is first detected at the 16-cell stage (Guthrie, 1984; Cardellini et al., 1988; Guthrie et al., 1988). Dye coupling was found at a high level along the presumptive dorsal half of the embryo, while only low levels of dye coupling were observed ventrally. At the 32-cell stage, dye coupling continued to be maintained at a high level dorsally, and at a low level ventrally except at the primary cleavage axis (which bisects the embryo into right and left halves), where no dye coupling was observed. As the earliest patterning events in frog embryogenesis entails the specification of the dorsal/ventral axis, perhaps this dorsal/ventral difference in coupling efficiency may play a role in establishing the dorsal/ventral axis. Consistent with this possibility is the fact that treatments that dorsalize (lithium; Kao et al., 1986) or ventralize (UV irradiation; Scharf and Gerhart, 1983) *Xenopus* embryos also resulted in a corresponding increase or decrease in dye coupling within the ventral or dorsal side of the treated embryo, respectively (Nagajski et al., 1989). Moreover, dye coupling in the ventral aspect of the *Xenopus* embryo is elevated upon ectopic expression of *wnt-1* (vertebrate homolog of *wingless*; Olson et al., 1991, 1992), a treatment that results in dorsal anterior transformation, accompanied by axial duplication (Olson et al., 1991, 1992). These observations and those discussed earlier with the blockage of coupling using gap junction antibodies suggest that gap junctional communication is likely to play a role in axial patterning in the frog embryo. This may include mediating interactions encoded by diffusible factors such as those of the *wingless* or *wnt* gene family.

## C. Gap Junctional Communication in Mouse and Chick Embryos

Examination of gap junctional communication in mammalian development has focused mainly on the mouse embryo. Mouse embryogenesis appears to differ greatly from that of frogs and insects, as maternal gene products play little or no role in organizing the initial body plan of the mammalian embryo. One indication of this differing developmental strategy is evidenced by the fact that zygotic gene expression is activated much earlier in the mouse embryo, that is from the two-cell stage, whereas in frog and fly embryos, zygotic gene expression is not turned on until midblastula transition or the cellular blastoderm stage, respectively. In the mouse embryo, gap junctional communication is initiated at the eight-cell stage, resulting in all cells becoming linked as a syncytium via gap junctions (Lo and Gilula, 1979a). As this is also the time that the first cell determination event occurs, with outside cells giving rise to trophectoderm and inside cells to the yolk sac and embryo proper (inner cell mass), gap junctional coupling may play a role in this early lineage segregation event (Lo and Gilula, 1979a). Consistent with this possibility is the result of antibody blocking studies. Also consistent with a role for gap junctions in mouse preimplantation development is the results of coupling studies in embryos derived from DDK mouse eggs fertilized with sperm of other mouse strains such as C3H (Buehr et al., 1987; Becker et al., 1992). Normally such intercrossed embryos exhibit preimplantation lethality, with a failure to form blastocysts. This embryonic lethality is accompanied by a reduced level of dye coupling. However, when such embryos were treated with cyclic adenosine monophosphate (cAMP) or methylamine (a weak base that would raise the intra-cellular pH), dye coupling increased and, concomitantly, there was significant developmental rescue of such embryos (to the blastocyst stage). In contrast, treatment of normal embryos with butyrate, a weak acid, lowered intracellular pH and resulted in developmental arrest.

Further examination of ionic and dye coupling of the egg cylinder stage postimplantation mouse embryo showed that as development progresses, the embryo becomes segregated into a number of communication compartment domains. Particularly interesting is the finding that the embryo proper (inner cell mass–derived) and the placental precursor cells (trophectoderm derived) are each segregated into separate communication compartment domains (Lo and Gilula, 1979b; Kalimi and Lo, 1988, 1989). As a result, the egg cylinder embryo is said to be subdivided into two global communication compartments. The isolation of the embryo from the placental precursor cells may be of functional importance to the development of the embryo proper, or some aspect of placentation, or both (Kalimi and Lo, 1988, 1989).

Within the egg cylinder embryo, each of the two global communication compartments is further subdivided into smaller communication compartment domains. Such communication compartments are usually characterized by the absence of dye coupling, but nevertheless a low level of ionic coupling is retained across the compartment border. Thus, each of the embryonic germ layer in the embryo proper corresponds to a separate dye fill compartment, with ionic coupling continuing to link cells of adjacent germ layers. Interestingly, each germ layer is further subdivided into smaller communication compartments. Most striking was the finding of dye-filled regions that described boxlike domains in the ectoderm and mesoderm (Figure 3; Kalimi and Lo, 1988). Results obtained from multiple impalements into the same embryo suggested that tandem array of such

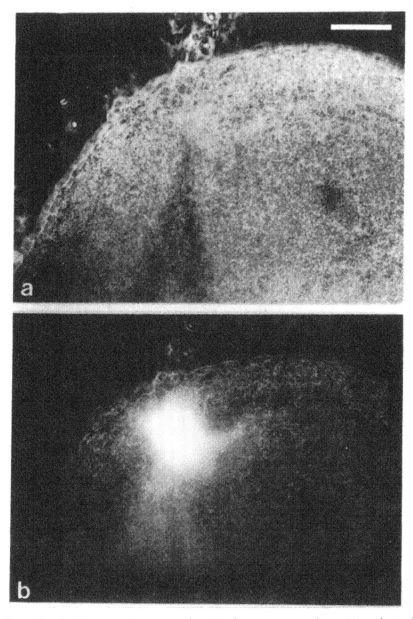

***Figure 3.*** Boxlike compartments in the gastrulating mouse embryo. Microelectrode impalement and the iontophoretic injection of carboxyfluorescein into the embryo proper of a 7.5-dg mouse embryo revealed dye spread which delineated a boxlike domain (**b**). Outline of the microelectrode can be seen in the phase contrast image in **a**.

boxlike compartments may exist in the egg cylinder stage embryo. As this was observed at gestational day 7.5—that is, at least 12 to 24 hours before somites are first observed and 36 to 48 hours before neuromeres are formed—we suggested that such boxlike compartments may be the anlage of somites or neuromeres, or both.

Regarding this possibility, it is interesting to note that grafting experiments in chick embryos have suggested that multicellular units representing precursors to somites may exist in the unsegmented paraxial mesoderm (for review, see Keynes and Stern, 1988). Moreover, morphological analyses of chick and mouse embryos have indicated that the paraxial mesoderm is segregated into segmental units prior to the visible subdivision of cells into somites (Meier, 1979; Meier and Tam, 1982; Tam et al., 1982). That somites and neuromeres indeed correspond to communication compartment domains has been demonstrated by dye injection experiments into chick embryos (Martinez et al., 1992; Bagnall et al., 1992). In neuromeres, dye coupling was observed to be restricted at interneuromeric boundaries. Such compartments are only partially restrictive in gap junctional communication, as ionic coupling was maintained across the interneuromeric boundaries (Martinez et al., 1992). Dye injection studies of cells in the segmental plate also revealed that each somite constitutes a separate communication compartment (Bagnall et al., 1992). In addition, within each somite, cells in the outer epithelium are not dye coupled to cells in the somitocoele. As cells in the somite disperse, the dermatome cells continued to exhibit dye coupling, while the sclerotomal derivatives were further subdivided into rostral and caudal communication compartments, with the intrasclerotomal fissure delineated by a group of well-coupled cells (Bagnall et al., 1992). These results are consistent with earlier studies in amphibian embryos, which showed the absence of ionic coupling between the dermatome and myotomal layers and the loss of ionic coupling between the segmental plate and forming somites (Blackshaw and Warner, 1976). Overall, these studies suggest that the specification of the segmental body plan in verterbate embryos is accompanied by the segregation of cells into communication compartments that reflect the underlying segmental units. This is not unlike the results obtained in the insect system.

## D. Gap Junctional Communication in Molluscan Embryos

Development of molluscan embryos is highly mosaic and uniquely characterized by the specification of mesodermal cell lineages via the asymmetrical segregation of cytoplasmic determinants from the egg. Given this striking "mosacisim," it is of interest to examine whether gap junction–mediated cell-cell interactions may nevertheless be of relevance to patterning in the molluscan embryo. Dye injection studies, in fact, showed that in embryos of *Patella vulgata* and *Lymnaea stagnalis*, embryonic development is accompanied by the progressive subdivision of cells into communication compartment domains (Serras and van den Bigelaar, 1987; Serras, et al., 1989). Initially the entire embryo is dye coupled, but by the 48- *(Lymnaea)* or 72-cell stage *(Patella)*, restrictions in dye spread are observed. Thus, dye injections into the ectoderm revealed that the embryos are first segregated into a pretrochal versus post-trochal communication compartment. This is achieved through the loss of dye coupling between the pre- and post-trochal cells and the prototroch, a ring of ciliated cells that divides the trochophore into an anterior (pretrochal) versus posterior (post-trochal) domain. As the embryo further develops, the ectodermal cells in the pre- and post-trochal domains further segregate into smaller communication compartments, and these appear to coincide with developmentally significant domains. For example, in the pretrochal region of the *Patella* embryo, no dye coupling is observed between the apical organ or the head anlage, even though efficient dye coupling is detected between cells within each structure (Serras et al., 1989). In the post-trochal region, a dorsal versus ventral compartment is observed (which later becomes further subdivided into smaller compartments)—dorsally these compartments encompass the mantle epithelium versus outer edge of the mantle epithelium, and ventrally, they encompass the foot anlage versus ectoderm around the foot and mantle epithelium (Serras et al., 1989). Similarly, in *Lymnaea* embryos (Serras and van den Bigelaar, 1987), the two cephalic plates in the pretrochocal domain each constitute a separate communication compartment, with no dye coupling observed with surrounding cells of larval fates (apical plate and head vesicle). In the post-trochal domain, restrictions in dye spread between regions of differing developmental fates was also observed. As has been observed in the insect epi- dermis and in mouse embryos, some of these dye coupling restrictions are of a partial nature. For example, in *Lymnaea* embryos, dye spread was observed between the shell field and the remainder of the post-trochal ectoderm. In *Patella* embryos, some experiments revealed dye spread between the foot anlage and the communication compartment constituting the ectoderm around the foot. More- over, ionic coupling was observed between regions of the *Patella* embryo where dye coupling was not observed, that is between the dorsal versus ventral com- partments.

## E.  Gap Junctional Communication in Nematode Embryos

The development of nematode embryos is highly mosaic as well, but neverthe- less, recent studies suggest that cell-cell interactions also play an important role in various aspects of cell lineage choice and cell fate specification. Dye injection studies in *Caenorhabditis elegans* embryos have revealed the segregation of cells into communication compartment domains (Bossinger and Schierenberg, 1992). Dye coupling is observed between all cells of the embryo from the four-cell stage onwards, but starting at the 24-cell stage, dye coupling to the germ cell (P4) pro- genitor disappears. Moreover, the daughters of P4 are also found to be uncoupled from the somatic cells. In contrast, all of the somatic cells are dye coupled to one another, thus making the soma a single communication compartment. The loss of dye coupling between the germ cell progenitor with the somatic cell populations perhaps may play a role in the segregation and specification of the germ cell lineage. Interestingly, as development progresses, the single somatic cell compartment becomes subdivided into smaller communication compartment domains. This was demonstrated with dye injections into the alimentary tract, which showed segregation of the intestinal versus pharyngeal primordium into separate communication compartments. These results are very much reminiscent of the observations in insects, mammals, and frogs, and suggest that despite apparent differences, gap junctions may participate in the development of evolutionarily diverse organisms.

## F. Mechanism Underlying Restrictions in Gap Junctional Communication

Given the likely developmental importance of restricting gap junctional communication and subdividing cells into communication compartment domains, the outstanding question is how are such restrictions in coupling established. Elucidating the mechanism whereby communication restrictions are established may provide further insights into the role of gap junctions in embryogenesis and development. Recent advances in cloning and characterization of different members of the gap junction gene family provide some clue as to possible mechanisms. Gap junctions are encoded by a multigene family comprising at least a dozen members (Beyer et al., 1990; Willecke, et al., 1991; also see the chapter by Yeager and Nicholson earlier in this volume). As gap junctional channels encoded by each connexin isotype exhibit unique conductance and gating properties (for review, see Bennett et al., 1991), restrictions in coupling may result merely from changes in channel permeability elicited by various gating mechanisms. Moreover, as each connexin gene exhibits different spatially restricted patterns of expression in the embryo and adult tissues, segregation of cells into communication compartments may also arise from the differential expression of different members of the gap junction gene family. Particularly interesting is the finding that connexin40 (Cx40) is restricted in its ability to interact with other connexin isotypes (Hennemann et al., 1992; Bruzzone et al., 1993). Also of relevance is the observation that gap junction channels formed by some connexin isotypes exhibit selective dye and ion permeability. For example, gap junctions formed by connexin45 (Cx45) provided dye coupling for dichorofluorescein but not carboxyfluorescein (Veenstra et al., 1994). Hence, the differential expression of gap junction genes may allow communication restrictions to be established such that ionic coupling may persist in the absence of dye coupling. In addition, as some heterotypic gap junction combinations exhibit asymmetrical coupling properties (Barrio et al., 1991), this could further ccount for the finding that cells situated at a communication

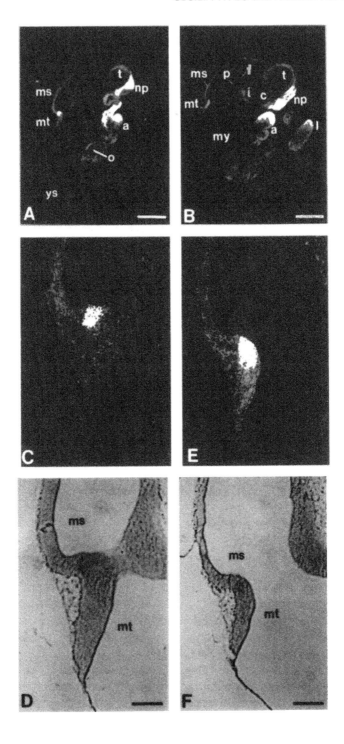

**Figure 4.** In situ hybridization analysis showing Cx43 transcript expression in the midbrain and hindbrain in a pattern similar to that of *wnt-1*. **A, B**. Stereomicroscope image showing Cx43 transcript distribution in sagittal sections of a 10.5-day mouse embryo. (C-E). Higher magnification images of midbrain/hindbrain region in (**A**) and (**B**). Note band of high level transcript expression in the midbrain/hindbrain junction (**A, C**). Along the dorsal midline, only a sharp anterior boundary is observed (**B, D**). This pattern of Cx43 transcript expression is identical to that of *wnt-1*. **a**, branchial arch; **c**, optic chiasma; **f**, pontine flexure; **i**, infundibulum; **l**, limb; **ms**, mesencephalon; **mt**, metencephalon; **my**, myelencephalon; **np**, nasal placode; **o**, otic vesicle; p, pons; **t**, telencephalon; **ys**, yolk sac. (Reprinted with permission from Ruangvoravat and Lo, 1992.)

---

restriction boundary may be coupled to cells within the communication compartment, but not to adjacent cells situated across the compartment boundary.

The evaluation of these possibilities will first require a detail characterization of connexin gene expression pattern in the embryo. In the early postimplantation mouse embryo where gap junctional communication is restricted in two global compartments, Cx43 is expressed exclusively in the ICM-derived embryonic compartment, and not in the trophectoderm-derived extraembryonic compartment (Ruangvoravat and Lo, 1992; Reuss et al., 1997). In contrast, connexin26 (Cx26) expression is not found in the embryo proper, but is expressed in the extraembryonic region, within a subset of cells in the developing placenta (Pauken and Lo, 1995). Subsequent to implantation, connexin31 (Cx31) appears to be expressed only in the ectoplacental cone and extraembryonic ectoderm (Reuss et al., 1997). These results suggest that the differential expression of gap junction genes may indeed play a role in restricting gap junctional communication in the mouse embryo. Consistent with this localization of Cx26 is a role for gap junctions consisting in transfer of metabolites from the mother to the fetus. A decrease in transplacental movement of glucose has been found in Cx26 knockout mice, perhaps accounting for the early embryonic lethality in this genotype (Gabriel et al., 1998). It is interesting to note that the pattern of Cx26 expression in the mouse placenta also appears to coincide with the segregation of cells in the developing placenta into separate communication compartment domains (Kalimi and Lo, 1989).

Other studies examining gap junction gene expression in mouse and chick embryos have further shown that gap junction gene expression is regionalized in developmentally significant domains (Minkoff et al., 1991; Ruangvoravat and Lo, 1992; Risek et al., 1992; Yancey et al., 1992; Risek et al., 1994; Pauken and Lo, 1995; Minkoff et al., 1997). Particularly striking is the finding that connexin43 (Cx43) in the embryonic brain exhibits expression domains that are similar to those of various members of the *wnt* gene family (Figures 4 and 5). As *wnt* genes encode short-range diffusible factors that regulate patterning of the central ner-

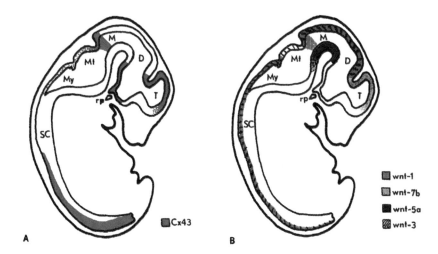

**Figure 5.** Cx43 and *wnt* gene expression in the central nervous system. Cx43 and *wnt* transcripts are expressed in a similar pattern, with either overlapping or complementary expression domains. Note that expression of Cx43 and *wnt-1* extends dorsolaterally as a ring in the region just anterior to junction of the midbrain and hindbrain. Along the spinal cord, Cx43 expression at the rostral end is localizes dorsally, mainly in the ventricular ependymal cells, while caudally, expression extends through out. (Wnt expression pattern according to Wilkinson et al., 1987; Gavin et al., 1990; McMahon et al., 1992; Roelink and Nusse, 1991.) **T**, telencephalon; **D**, diencephalon; **M**, mesencephalon; **Mt**, metencephalon; **My**, myelencephalon; **rp**, RAthke's pocket; **SC**, spinal cord.

vous system, it is tantalizing to speculate that gap junctions may play a role in brain development by helping to relay signal transduction cascades triggered by wnt and other short-range signaling molecules. The validity of these ideas must await the characterization of gap junctional communication in the early embryonic nervous system.

## III. GAP JUNCTION PERTURBATION AND DISRUPTIONS IN DEVELOPMENT

### A. Blockage of Coupling with Antibody Reagents

The inhibition of gap junctional communication in a number of different developmental systems using antibodies to gap junction proteins has also suggested a role for gap junctions in development. In the *Xenopus* embryo, gap junction anti-body injections into the eight-cell embryo resulted in tadpoles exhibiting head defects (Warner et al., 1984). These were characterized by underdevelopment of the brain or eye, or both. Similar injections of gap junction antibodies into mouse embryos resulted in the blockage of coupling and the extrusion of the antibody-injected cells (Lee et al., 1987). The examination of chick limb bud digit formation after loading of gap junction antibody into zone of polarizing activity (ZPA) cells and cells in the limb bud anterior mesenchyme further indicated a role for gap junctions in limb patterning (Allen et al., 1990), with fibroblast growth factor 4 subsequently shown to induce an increase in gap junctions in limb bud mesen-chyme (Makarenkova et al., 1997). Similarly, the loading of gap junction antibod-ies in regenerating hydra suggests that gap junctional communication may play a role in establishing the head inhibition gradient required for patterning in the hydra (Fraser et al., 1987). All of these studies suggest a role for gap junctions in the development of diverse organisms.

## B.  Transgenic Mouse Models

More recently, gene knockouts have been carried out using the embryonic stem cell method of gene disruption to examine the function of various gap junction genes. Null mutant mouse embryos containing a disrupted gene encoding Cx43 (Gjα1) develop to term, but die shortly after birth (Reaume et al., 1995). Such mutants exhibit conotruncal heart malformations, with death likely resulting from a failure to establish normal pulmonary circulation as a result of severe outflow tract obstructions (Reaume et al., 1995; Ya et al., 1998). Studies of homozygous (Cx43 knockout) preimplantation embryos demonstrated significantly reduced cell coupling and selectivity for different dyes, suggesting that other connexins, perhaps including Cx45, may adequately fulfill potential roles for gap junctional communication at these early stages (De Sousa et al., 1997). Interestingly, studies from our lab also showed conotruncal heart defects in animals that are hemizy-gous for the Cx43 null allele, suggesting that the level of gap junction gene expression may be of importance in conotruncal heart development (Huang et al., 1998). As with most knockout animal models, the precise role of Cx43 gap junc-tions in conotruncal heart development is not evident from the knockout heart phenotype. However, some insights into this question have arisen from recent studies of transgenic mouse models exhibiting overexpression of Cx43 targeted by the cytomegaloviral promoter (CMV43 mice; Ewart et al., 1997). In CMV43 transgenic mice, conotruncal heart defects are found, and of significance is the fact that in these mice, the transgene provides for Cx43 overexpression restricted to subpopulations of neural crest cells and neural crest progenitors situated in the dorsal neural tube. It should be noted that in these transgenic mice, there is no Cx43 transgene expression in the myocardium. Based on these and other findings, it was proposed that the conotruncal heart malformations may arise from the perturbation of neural crest cells that normally migrate to and participate in outflow tract morphogenesis (Ewart et al., 1997), a possibility further indicated by recent studies of a dominant negative loss of function transgenic animal model (see following discussion; Sullivan et al., 1998). It should be noted that neural crest cells have in fact been found to express Cx43 gap junctions, and they are also gap junction communication competent as indicated by dye coupling studies (Lo et al., 1997).

## C. Visceroatrial Heterotaxia and Connexin43 Mutations

Mutations in the Cx43 gene have been found in pediatric cardiac transplant patients with visceroatrial heterotaxia (VAH) (Britz-Cunningham et al., 1995). This syndrome involves fundamental perturbations in left/right patterning and is characterized by complex heart malformations in addition to visceral organ defects. Given the wide ranging variability of VAH phenotypes, this syndrome is likely to have multigenic origins. Of significance is the fact that five of the six VAH patients showed defects associated with the pulmonary outflow tract (pulmonic atresia or stenosis), a region of the heart malformed in the Cx43 knockout and CMV43 transgenic mice (and mice expressing a dominant negative Cx43 gap junction fusion protein, Sullivan et al., 1998). These findings suggest that there may be a special subset of VAH patients who harbor Cx43 mutations.

In all six VAH patients analyzed, point mutations were found in the Cx43 gene. These are composed of the substitution of one or more of the phosphorylatable serine or threonine residues in the cytoplasmic tail of the Cx43 polypeptide. Expression of these connexin polypeptides in tissue culture cells showed that they can generate functional gap junctions, but with aberrant regulation by protein kinases. These observations further indicate a role for Cx43 gap junctions in conotruncal heart development, and together with the CMV43 transgenic and Cx43 knockout mouse studies, suggest that the precise level of Cx43-mediated coupling is critically important to conotruncal heart development. The finding of Cx43 mutations in VAH patients also suggests a role for gap junctions in left/right patterning (see, too, Debrus et al., 1997). Most significantly, recent studies showed that the expression of the VAH-associated mutant Cx43 proteins in *Xenopus* embryos resulted in heterotaxia (Levin and Mercola, 1998). This provides further evidence of a role for Cx43 gap junctions in left/right patterning. With regards to the latter possibility, it is intriguing to note that expression of Cx43 has been found to be left/right asymmetrical and of reverse sidedness in the fore versus hindlimb buds of mouse embryos (Meyer et al., 1997).

## D. Other Connexin Knockout Mouse Models

In addition to the Cx43 knockout mice, null mutant mice recently have also been generated with disruption of the connexin genes encoding connexin32

(Cx32) (Gjβ1; Nelles et al., 1996), connexin37 (Cx37) (Gjα4; Simon et al., 1997), connexin46 (Cx46) (Gjα3; Gong et al., 1997), and Cx26 (Gjβ2; Gabriel et al., 1998). Except for the Cx26 knockout, which dies early in embryogenesis, the other knockout mice are homozygote viable, but show various phenotypes restricted to specific tissues that correspond to a subset of the domains where each of these genes are expressed. Null mutants for Cx32 exhibit perturbation of inner-vation-dependent glycogen metabolism in the liver (Nelles et al., 1996) as well as an increased propensity towards formation of tumors in liver (Temme et al., 1997), a tissue that normally expresses Cx32 in abundance. This may contrast with the results in humans where null mutations in the Cx32 gene (X-linked Char-cot-Marie-Tooth syndrome) showed no evidence of liver dysfunction, but rather is associated with peripheral nerve demyelineation (Bergohoffen et al, 1993; Bruzzone et al., 1994). In the Cx37 knockout mice, homozygous animals show female sterility. Analysis of the ovarian tissue revealed defects in oocyte matura-tion related to the loss of functional coupling between oocytes and granulosa cells (Simon et al., 1997). The oocytes in these knockout mice failed to achieve meiotic competence. There is also evidence of abnormal corpora luteal development. Recent studies of the Cx46 knockout mice revealed the presence of cataracts in young adult homozygous animals (Gong et al., 1997). As lens tissue is the major site of Cx46 expression, these results clearly indicate a role for Cx46 gap junc-tions in the maintenance of lens tissue homeostasis.

## E.  Gene Knockouts and Phenotypes

It should be noted that given the large number of connexin genes in the verte-brate genome, developmental perturbations in null mutants may be restricted or masked by functional redundancies mediated by other members of the gap junc-tion gene family. This may account for the finding of rather restricted develop-mental perturbations associated with each connexin knockout mutant. That the phenotype in human versus mouse Cx32 null mutants differs is somewhat surpris-ing and perhaps may reflect anatomical and physiological differences in periph-eral nerve conduction in mouse versus human. It is also possible that differences in the expression pattern of gap junction genes may account for such discrepan-cies. Thus, a comparison of the expression pattern of connexin genes in various species shows differences in tissue distribution. This has been reported for the expression of pattern of Cx26 and Cx43 in the mouse versus rat placenta (Pauken and Lo, 1995). The cumulative studies from various laboratories further showed that the distribution of Cx32 and Cx26 in the liver differs among mouse, rat, and guinea pig (Dermietzel et al., 1984; Nicholson et al., 1987; Traub et al., 1989; Kuraoka et al., 1993). Disparity in the pattern of Cx43 was also observed in the dog versus rat heart (Van Kemper et al, 1991; Gourdie et al., 1992; Kanter et al., 1993), and between the mouse and chick limb (Green et al., 1994). In light of these observations, the validity of extrapolating phenotypes and function from observations obtained in one species to that of another must be carefully examined. These observations also suggest the possibility that functional redundancy in the gap junction gene family may have relaxed the requirement for the conservation of gene expression pattern.

## F.  Dominant Negative Inhibition of Gap Junctions

Another approach used to explore the role of gap junctions in development is the making of recombinant gap junction gene constructs that act in a dominant fashion in disrupting gap junction function. The feasibility of this approach is based on the fact that gap junctions are oligomeric in nature, being comprised of a hexameric array of connexin polylpeptides. Hence, the expression of mutant connexin polypeptides may allow the formation of mixed oligomers that would interfere with the assembly, trafficking, or gating of gap junctions. This may provide the means to disrupt multiple connexin isotypes simultaneously, and thus allow the detection of phenotypes that otherwise may be masked by functional redundancies in the gap junction gene family. Two such dominant negative mutations have been generated, one comprised of a chimeric construct containing portions of Cx32 and Cx43 (Paul et al., 1995), and another consisting of full-length Cx43 fused with bacterial β-galactosidase (Cx43/lacZ; Sullivan and Lo, 1995). The injection of the chimeric construct in the anterodorsal blastomere of the *Xenopus* embryo resulted in extrusion and loss of cells derived from the antibody-injected blastomere (Paul et al., 1995). This observation is intriguing in that it is somewhat reminiscent of the mouse embryo studies where injection of gap junction antibodies also resulted in cell extrusion. Significantly, transgenic mice generated expressing the Cx43/lacZ fusion protein exhibited conotruncal heart defects (Sullivan et al., 1998). This finding was associated with fusion protein expression and the inhibition of gap junctional communication in presumptive neural crest cells. These results further indicate a role for Cx43-mediated gap junctional communication in cardiac neural crest cells that participate in conotruncal heart development. In addition to the finding of developmental perturbation in *Xenopus* and mouse models exhibiting the dominant negative inhibition of gap junctional communication, recent studies of human populations suggest that there may be naturally existing dominant negative gap junction mutations that underlie various human diseases. Thus, a linkage has been established between Cx26 mutations and hereditary nonsyndromic sensorineural deafness (Denoyelle et al., 1997; Kelsell et al., 1997; Zelante et al., 1997). It is also possible that the dominant negative inhibition of gap junctions may play a role in Charcot-Marie-Tooth disease. This is indicated based on the findings using the paired *Xenopus* oocyte system, that some of the Cx32 mutations in Charcot-Marie-Tooth patients can inhibit Cx26 gap junctions.

# IV. GAP JUNCTION AND CELL-CELL SIGNALING IN DEVELOPMENT

The cumulative evidence from the more descriptive studies to those in which gap junction function is perturbed would indicate that gap junctional communication may play a role in various aspects of patterning and development. In light of the permeability properties of gap junctions, this likely entails providing an efficient intracellular pathway for the transmission of second messengers and other cell signaling molecules. In this manner, gap junctions may facilitate inductive interactions, or otherwise play a role in the specification of positional information in a developmental field. As gap junctional communication in most embryos is organized in a compartmental fashion, communication compartment domains may provide the context within which cell signaling is relayed, amplified, and restricted. With regard to these possibilities, it is interesting to consider the comment of Furshpan and Potter (1968) that gap junction–mediated cell-cell interaction "would presumably not aid in understanding how differences between cells arise in the first place; it can only exploit existing differences. . . . the most obvious consequence of intercellular exchange is to make cells more alike." Although longterm gap junctional communication is indeed likely to "make cells more alike," differences in local environment and cell lineage may elicit biochemical differences that may allow chemical gradients to be generated within the context of communication compartment domains. Hence, communication compartments may serve as the functional units within which positional signaling is established. It should be noted that the feasibility of generating gradients via gap junctional pathways has been demonstrated by studies in model tissue culture system (Michalke, 1977). The finding that communication compartments are only partially restricted in coupling (as indicated either by the maintenance of ionic coupling or by a low level of dye coupling) may reflect the cross talk needed for coordinating developmental events in different regions of the embryo. Ultimately, the verification of these ideas will require the use of novel approaches for identifying cell signaling molecules that traverse gap junctions, and also experimental approaches that can demonstrate the participation of these cell signaling molecules in embryogenesis and development.

## REFERENCES

Allen F., Tickle, C., & Warner, A. (1990). The role of gap junctions in patterning of the chick limb bud. Development 108, 623-634.

Bagnall, K.M., Sanders, E.J., & Berdan, R.C. (1992). Communication compartments in the axial mesoderm of the chick embryo. Anat. Embryol. 186, 195-204.

Barrio, L.C., Suchyna, T., Bargiello, T., Xu, L.X., Roginski, R.S., Bennett, M.V.L., Nicholson, B.J. (1992). Gap junctions formed by connexins 26 and 32 alone and in combination are differently affected by applied voltage. Proc. Natl. Acad. Sci. 88, 8410-8414.

Becker, D.L., Leclerc-David, C., & Warner, A. (1992). The relationship of gap junctions and compaction in the preimplantation mouse embryo. Develop. Suppl. 113-118.

Bennett, M.V.L., Barrio, LC., Bargiello, T.A., Spray, D.C., Hertzberg, E.L., & Sáez, J.C. (1991). Gap junctions: new tools, new answers, new questions. Neuron 6, 305-320.

Bennett, M.V.L., & Trinkaus, J.P. (1970). Electrical coupling between embryonic cells by way of extracellular space and specialized junctions. J. Cell Bio. 44, 592-610.

Bergohoffen, J., Scherer, S.S., Wang, S., Oronzi Scott, M., Bone, L.J., Paul, D.L., Chen, K., Lensch, M.W., Chance, P.F., & Fischbeck, K.H. (1993). Connexin mutations in X-linked Charcot Marie Tooth disease. Science 262, 2039-2042.

Beyer, E.C., Paul, D.L., & Goodenough, E.A. (1990). Connexin family of gap junction proteins. J. Membr. Biol. 116, 187-194.

Blackshaw, S.E., & Warner, A.E. (1976). Low resistance junctions between mesoderm cells during development of trunk muscles. J. Physiol. 255, 209-230.

Blennerhassett, M., & Caveney, S. (1984). Separation of developmental compartments by a cell type with reduced junctional permeability. Nature 24, 361-364.

Bossinger, O. & Schierenberg, E. (1992). Cell–cell communication in the embryo of Caenorhabditis elegans. Develop. Biol. 151, 401-409.

Britz-Cunningham, S.H., Shah, M.M., Zuppan, C.W., & Fletcher, W.H. (1995). Mutations of the connexin43 gap junction gene in patients with heart malformations and defects of laterality. New Engl. J. Med. 332, 1323-1329.

Bruzzone, R., Haefliger, J., Gimlich, R.L., & Paul, D.L. (1993). Connexin41, a component of gap junctions in vascular endothelium, is restricted in its ability to interact with other connexins. Mol. Biol. Cell 4, 7-20.

Bruzzone, R., White, T.W., Scherer, S.S., Fischbeck, K.H., & Paul, D.L. (1994). Null mutations of connexin32 in patients with X-linked Charcot-Marie-Tooth disease. Neuron 13, 1253-1260.

Buehr, M., Lee, S., McLaren, A., & Warner, A. (1987). Reduced gap junctional communication is associated with the lethal condition characteristic of DDK mouse eggs fertilized by foreign sperm. Development 101, 449-459.

Cardellini, P., Rasotto, M.B., Tertoolen, G.J., & Durston, A.J. (1988). Intercellular communication in the eight-cell stage of Xenopus laevis development: a study using dye coupling. Develop. Bio. 129, 265-269.

Crick, F.C.H., & Lawrence, P.A. (1975). Compartments and polyclones in insect development. Science 189, 341-347.

De Sousa, P.A., Juneja, S.C., Caveney, S., Houghton, F.D., Davies, T.C., Reaume, A.G., Rossant, J., & Kidder, G.M. (1997). Normal development of preimplantation mouse embryos deficient in gap junctional coupling. J. Cell Sci. 110, 1751-17588.

Debrus, S., Sauer, U., Gilgenkrantz, S., Jost, W., Jesberger, H.J., & Bouvagnet, P. (1997). Autosomal recessive lateralization and midline defects: blastogenesis recessive 1. Am. J. Med. Genet. 68, 401-4044.

Denoyelle, F., Weil, D., Maw, M.A., Wilcox, S.A., Lench, N.J., Allen-Powell, D.R., Osborn, A.H., Dahl, H.H., Middleton, A., Houseman, M.J., Dode, C., Marlin, S., Boulila-ElGaied, A., Grati, M., Ayadi, H., BenArab, S., Bitoun, P., Lina-Granade, G., Godet, J., Mustapha, M., Loiselet, J., El-Zir, E., Aubois, A., Joannard, A., Petit, C., et al. (1997). Prelingual deafness: high prevalence of a 30delG mutation in the connexin 26 gene. Hum. Mol. Genet. 6, 2173-21777.

Dermietzel, R., Leibstein, A., Frixen, U., Janssen-Timmen, U., Traub, O., & Willecke, K. (1984). Gap junctions in several tissues share antigenic determinants with liver gap junctions. EMBO J. 3, 2261-2270.

Ewart, J.L., Cohen, M.F., Wessels, A., Gourdie, R.G. , Chin, A.J., Park, S.M.J., Lazatin, B.O., Villabon, S., & Lo, C.W. (1997). Heart and neural tube defects in transgenic mice overexpressing the Cx43 gap junction gene. Development 124, 1281-1292.

Fraser, S.E., Green, C.R., Bode, H.R., & Gilula, N.B. (1987). Selective disruption of gap junctional communication interferes with a patterning process in hydra. Science 237, 49-55.

Furshpan, E.J., & Potter, D.D. (1968). Low-resistance junctions between cells in embryos and tissue culture. Curr. Top. Dev. Biol. 3, 95-126.

Gabriel, H.D., Jung, D., Butzler, C., Temme, A., Traub, O., Winterhager, E., & Willecke, K. (1998). Transplacental uptake of glucose is decreased in embryonic lethal connexin26-deficient mice. J. Cell Biol. 140, 1453-1461.

Gavin, B.J., McMahon, J.A., & McMahon, A.P. (1990). Expression of multiple novel wnt-1/int-1 related genes during fetal and adult mouse development. Genes Develop. 4, 2319-2332.

Gilula, N.B., Reeves, O.R., & Steinbach, A. (1972). Metabolic coupling, ionic coupling and cell contacts. Nature 235, 262-265.

Gong, X., Li, E., Klier, G., Huang, Q., Wu, Y., Lei, H., Kumar, N.M., Horwitz, J., and Gilula, N.B. (1997). Disruption of alpha3 connexin gene leads to proteolysis and cataractogenesis in mice. Cell 91, 833-843.

Gourdie, R.G., Green, C.R., Severs, N.J., & Thompson, R.P. (1992). Immunolabeling patterns of gap junction connexins in the developing and mature rat heart. Anat. Embryol. 185, 363-378.

Green, C.R., Bowles, L., Crawley, A., & Tickle, C. (1994). Expression of the connexin 43 gap junctional protein in tissues at the tip of the chick limb bud is related to the epithelial mesenchymal interactions that mediate morphogenesis. Dev. Biol. 161, 12-21.

Guthrie, S., Turin, L., & Warner, A. (1988). Patterns of junctional communication during development of the early amphibian embryo. Development 103, 769-783.

Guthrie, S. (1984). Patterns of junctional permeability in the early amphibian embryo. Nature 311, 149-141.

Hennemann, H., Suchyna, T., Lichtenberg-Frate, H., Jungbluth, S., Dahl, E., Schwartz, J., Nicholson, B.J., & Willecke, K. (1992). Molecular cloning and functional expression of mouse connexin40, a second gap junction gene preferentially expressed in lung. J. Cell Biol. 117, 1299-1310.

Huang, G.Y., Wessels, A., Smith, B.R., Linask, K.K., Ewart, J.L., & Lo, C.W. (1998). Alteration in connexin 43 gap junction gene dosage impairs conotruncal heart development. Dev. Biol. 198, 32-44.

Ingham, P.W., & Martinez Arias, A. (1992). Boundaries and fields in early embryos. Cell 68, 221-235.

Ito, S., & Hori, N. (1966). Electrical characteristics of Triturs egg cells during cleavage. J. Gen. Physiol. 19, 1019-1027.

Ito, S., & Loewenstein, W.R. (1969). Ionic communication between early embryonic cells. Dev. Biol. 19, 228-243.

Jursnich, V.A., Fraser, S.E., Held, L.I., Ryerse, J., & Bryant, P.J. (1990). Defective gap junctional communication associated with imaginal disc overgrowth and degeneration caused by mutants of the dco Gene in Drosophila. Dev. Biol. 140, 413-429.

Kalimi, G., & Lo, C.W. (1988). Communication compartments in the gastrulating mouse embryo. J. Cell Biol. 107, 241-255.

Kalimi, G., & Lo, C.W., (1989). Gap junctional communication in the extraembryonic tissues of the gastrulating mouse embryo. J. Cell Biol. 109, 3015-3026.

Kanter, H.L., Laing, J.G., Beau, S.L., Beyer, E.C., & Saffitz, J.E. (1993). Distinct patterns of connexin expression in canine Purkinje fibers and ventricular muscle. Cir. Res. 73, 1124-1131.

Kao, K.R., Masui, Y., & Elison, R.P. (1986). Lithium-induced respecification of pattern in Xenopus laevis embryos. Nature 322, 371-373.

Kaufman, T., & Abbot, M. (1984). Homeotic genes and the specification of segmental identity in the embryo and adult thorax of Drosophila melanogaster. In: *Molecular Aspects of Early Development* (Malacinski, G.M., & Klein, W., eds.), pp. 189-218. Plenum Press, New York.

Kelsell, D.P., Dunlop, J., Stevens, H.P., Lench, N.J., Liang, J.N., Parry, G., Mueller, R.F., & Leigh, I.M. (1997). Connexin 26 mutations in hereditary non-syndromic sensorineural deafness. Nature 387, 80-83.

Keynes, R.J., & Stern, C.D. (1988). Mechanisms of vertebrate segmentation. Development 103, 413-429.

Kimmel, C.B., Spray, D.C., & Bennett, M.V.L. (1984). Developmental uncoupling between blasto-derm and yolk cell in the embryo of the teleost Fundulus. Dev. Biol. 102, 483-487.

Kuraoka, A., Iida, H., Hatae, T. Shibata, Y., Itoh, M., Kurita, T. (1993). Localization of gap junction proteins connexins 32 and 26, in rat and guinea pig liver as revealed by quick-freeze, deep-etch immunoelectron microscopy. J. Histochem. Cytochem. 41, 971-80.

Lawrence, P.A. (1966). Gradients in the insect segment: the orientation of hairs in the milkweed bug Oncopeltus fasciatus. J. Exp. Biol. 44, 607-620.

Lee, S., Gilula, N.B., & Warner, A.E. (1987). Gap junctional communication and compaction during preimplantation stages of mouse development. Cell 51, 851-860.

Levin, M. & Mercola, M. (1998). Gap junctions are involved in the early generation of left–right asymmetry. Dev. Biol. 203, 90-105.

Lewis, E.B. (1978). A gene complex controlling segmentation in Drosophila. Nature 276, 565-570.

Lo, C.W., Cohen, M.F. , Ewart, J.L., Lazatin, B.O., Patel, N., Sullivan, R., Pauken, C., & Park, S.M.J. (1997). Cx43 gap junction gene expression and gap junctional communication in mouse neural crest cells. Dev. Gen. 20, 119-132.

Lo, C.W., & Gilula, N.B. (1979a). Gap junctional communication in the preimplantation mouse embryo. Cell 18, 399-409.

Lo, C.W., & Gilula, N.B. (1979b). Gap junctional communication in the postimplantation mouse embryo. Cell 18, 411-422.

Locke, M. (1967). The development of patterns in the integument of insects. Adv. Morphogenesis 6, 33-88.

Loewenstein, W.R., & Rose, B. (1992). The cell-cell channel in the control of growth. Sem.Cell Biol. 3, 59-79.

Makarenkova, H., Becker, D.L., Tickle, C., & Warner, A.E. (1997). Fibroblast growth factor 4 directs gap junction expression in the mesenchyme of the vertebrate limb bud. J. Cell Biol. 138, 1125-1137.

Martinez-Arias, A., & Lawrence, P.A. (1985). Parasegments and compartments in the Drosophila embryo. Nature 313, 639-642.

Martinez, S., Giejo, E., Sachez-Vives, M.V., Puelles, L., & Gallego, R. (1992). Reduced junctional permeability at interrhombomeric boundaries. Development 116, 1069-1076.

McGinnis, W., & Krumlauf, R. (1992). Homeobox genes and axial patterning. Cell 68, 283-302.

McMahon, A.P., Joyner, A.L., Bradley, A., & McMahon, J.A. (1992). The midbrain-hindbrain pheno-type of wnt-1–/wntp1– mice results from stepwise deletion of engrailed-expressing cells by915 days postcoitum. Cell 69, 581-595.

Meier, S. (1979). Development of the chick embryo mesoblast. Formation of the embryonic axis and establishment of the metameric pattern. Dev. Biol. 73, 24.

Meier, S., & Tam, P.P.L. (1982). Metameric pattern development in the embryonic axis of the mouse. I. Differentiation of the cranial segments. Differentiation 21, 95-108.

Meyer, R.A., Cohen, M.F., Recalde, S., Zakany, J., Bell, S., Scott, W.J., & Lo, C.W. (1997). Develop-mental regulation and asymmetric expression of the Cx43 gap junction gene in the mouse limb bud. Dev. Genet. 21, 290-300.

Michalke, W. (1977). A gradient of difusible substance in a monolayer of cultured cells. J. Membr. Biol. 33, 1-20.

Minkoff, R., Parker, S.B., & Hertzberg, E.L. (1991). Analysis of distribution patterns of gap junctions during development of embryonic chick facial primordia and brain. Development 111, 509-522.

Minkoff, R., Parker, S.B., Rundus, V.R., & Hertzberg, E.L. (1997). Expression patterns of connexin43 protein during facial development in the chick embryo: associates with outgrowth, attachment, and closure of the midfacial primordia. Anat. Rec. 248, 279-290.

Morata, G., & Lawrence, P.A. (1978). Cell lineage and homeotic mutants in the development of imaginal discs of Dorsophila. In: *The Clonal Basis of Development* (Subtelny, S., & Sussex, I.M., eds.), pp. 45-60. Academic Press, New York.

Nagajski. D.K., Guthrie, S.C., Ford, C.C., & Warner, A.E. (1989). The correlation between patterns of dye transfer through gap junctions and future developmental fate in Xenopus; the consequences of UV irradiation and lithium treatment. Development 105, 747-757.

Nelles, E., Butzler, C., Jung, D., Temme, A., Gabriel, H.-D., Dahl, U., Traub, O., Stumpel, F., Jungermann, K., Zielasek, J., Toyka, K.V., Dermietzel, R., & Willecke, K. (1996). Defective propagation of signals generated by sympathetic nerve stimulation in the liver of connexin32-deficient mice. Proc. Natl. Acad. Sci. U.S.A. 93, 9565-9570.

Nicholson, B.J., Dermietzel, R., Teplow, D., Traub, O., Willecke, K., & Revel, J. (1987). Two homologuous protein components of hepatic gap junctions. Nature 329, 732-734.

Nusslein-Volhard, C., & Wieschaus, E. (1980). Mutations affecting segment number and polarity in *Drosophila*. Nature 287, 795-801.

Olson, D.J., Christian, J.L., & Moon, R.T. (1991). Effect of wnt-1 and related proteins on gap junctional communication in Xenopus embryos. Science 252, 1173-1176.

Pauken, C.M., & Lo, C.W. (1995). Nonoverlapping expression of Cx43 and Cx26 in the mouse placenta and decidua: a pattern of gap junction gene expression differing from that in the rat. Mol. Reprod. Devel. 41, 195-203.

Paul, D.L., Yu, K., Bruzzone, R., Gimlich, R.L., & Goodneough, D.A. (1995). Expression of a dominant negative inhibitor of intercellular communication in the early Xenopus embryo causes delamination and extrusion of cells. Development 121, 371-381.

Peifer, M., & Bejsovec, A. (1992). Knowing your neighbors: cell interactions determine intrasegmental patterning in Drosophila. Trends Genet. 8, 243-249.

Potter, D.D., Furshpan, E.J., & Lennox, E.S. (1966). Connections between cells of the developing squid as revealed by electrophysiological methods. Proc. Natl. Acad. Sci. U.S.A. 55, 328-336.

Reaume, A.G., de Sousa, P.A., Kulkarni, S., Langille, B.L., Zhu, D., Davies, T.C., Junija, S.C., Kidder, G.M., & Rossant, J. (1995). Cardiac malformation in neonatal mice lacking connexin 43. Science 267, 1831-1834.

Reuss, B., Hellmann, P., Traub, O., Butterweck, A., & Winterhager, E. (1997). Expression of connexin31 and connexin43 genes in early rat embryos. Dev. Genet. 21, 82-90.

Risek, B., Klier, F.G., & Gilula, N.B. (1992). Multiple gap junction genes are utilized during rat skin and hair development. Development 116, 639-651.

Risek, B., Klier, F.G., & Gilula, N.B. (1994). Developmental regulation and structural organization of connexins in epidermal gap junctions. Dev. Biol. 164, 183-196.

Roelink, H., & Nusse, R. (1991). Expression of two members of the *wnt* family during mouse development restricted temporal and spatial patterns in developing neural tube. Genes Devel. 5, 381-388.

Ruangvoravat, C.P., & Lo, C.W. (1992). Restrictions in gap junctional communication in the Drosophila larval epidermis. Devel. Dynamics 193, 70-82.

Scharf, S.R.,& Gerhart, J.C. (1983). Axis determination in eggs of Xenopus laevis: a critical period before first cleavage, identified by the common effects of cold, pressure, and ultraviolet irradiation. Dev. Biol. 99, 75-87.

Serras, F., Damen, P., Dictus, W.J.A.G., Notenboom, R.G.E., & Van den Biggelaar, J.A.M. (1989). Communication compartments in the ectoderm of embryos of Patella vulgata. Roux. Arch. Devel. Biol. 198, 191-200.

Serras, , F., & van den Biggelaar, J.A.M. (1987). Is a mosaic embryo also a mosaic of communication compartments? Dev. Biol. 120, 132-138.

Sheridan, J.D. (1966). Electrophysiological study of special connections between cells in the early chick embryo. J. Cell Biol. 31, C1-C5.

Sheridan, J.D. (1968). Electrophysiological evidence for low resistance intercellular junctions in the early chick embryo. J. Cell Biol. 37, 650-659.

Simon, A.M., Goodenough, D.A., Li, E., & Paul, D.L. (1997). Female infertility in mice lacking connexin 37. Nature 385, 525-529.

Stumpf, H. (1966). Mechanism by which cells estimate their location within the body. Nature 212, 430-431.

Sullivan, R., & Lo, C.W. (1995). Expression of a connexin43/b-galactosidase fusion protein inhibits gap junctional communication in NIH3T3 cells. J. Cell Bio. 130, 419-429.

Sullivan, R., Meyer, R., Huang, G.Y., Cohen, MF. Wessels, A., Linask, K.K., & Lo, C. (1998). Heart malformations in transgenic mice exhibiting dominant negative inhibition of gap junctional communication. Dev. Biol. 204, 224-234.

Tam, P.P.L., Meier, S., & Jacobson, A.G. (1982). Differentiation of the metameric pattern in the embryonic axis of the mouse. II. Somitomeric organization of the presomitic mesoderm. Differentiation 21, 109-122.

Temme, A., Buchmann, A., Gabriel, H. D., Nelles, E., Schwarz, M., & Willecke, K. (1997). High incidence of spontaneous and chemically induced liver tumors in mice deficient for connexin32. Curr. Biol. 7, 713-716.

Traub, O., Look, J., Dermietzel, R., Brummer, F., Hulser, D., & Willecke, K. (1989). Comparative characterization of a 21-kD and 26-kD gap junction protein in murine liver an cultured hepatocytes. J. Cell Biol. 108, 1039-1051.

Van Kemper, M.J.A., Fromagel, C., Gros, D., Moorman, A.F.M., & Lamers, W.H. (1991). Spatial distribution of connexin 43, the major cardiac gap junction protein, in the developing and adult rat heart. Circ. Res. 68, 1638-1651.

Veenstra, R.D., Wang, H.-Z., Beyer, E.C., & Brink, P.R. (1994). Selective dye and ion permeability of gap junction channels formed by connexin45. Circ. Res. 75, 483-490.

Warner, A.E. (1973). The electrical properties of the ectoderm in the amphibian embryo during induction and early development of the nervous system. J. Physiol. 235, 267-286.

Warner, A.E., Guthrie, S.C., & Gilula, N.B. (1984). Antibodies to gap-junctional protein selectively disrupt junctional communication in the early amphibian embryo. Nature 311, 127-131.

Warner, A.E., & Lawrence, P.A. (1982). Permeability of gap junctions at the segmental border in insect epidermis. Cell 28, 243-252.

Weir, M.P., & Lo, C.W. (1982). Gap junctional communication compartments in the Drosophila wing disk. Proc. Natl. Acad. Sci. U.S.A. 79, 3232-3235.

Weir, M.P., & Lo, C.W. (1984). Gap junctional communication compartments in the Drosophila wing imaginal disk. Dev. Biol. 102, 130-146.

Weir, M.P., & Lo, C.W. (1985). An anterior/posterior communication compartment border in engrailed wing discs: possible implications for Drosophila pattern formation. Dev. Biol. 110, 84-90.

Wilkinson, D.G., Bailes, J.A., & McMaho, A.P. (1987). Expression of the proto-oncogene int-1 is restricted to specific neural cells int he developing mouse embryo. Cell 50, 79-88.

Willecke, K., Hennemann, H., Dahl, E., Jungbluth, S., & Heynkes, R. (1991). The diversity of connexin genes encoding gap junctional proteins. European J. Cell Bio. 56, 1-7.

Wolpert, L. (1978). Gap junctions: Channels for communication in development. In: Intercellular Junctions and Synapses. (Feldman, J., Gilula, N.B., & Pitts, J.D., Eds.), pp. 83-94. Chapman and Hall, London, England.

Ya, J., Erdtsieck-Ernste, E.B., de Boer, P.A., van Kempen, M.J., Jongsma, H., Gros, D., Moorman, A.F., & Lamers, W.H. (1998). Heart defects in connexin43-deficient mice. Circ. Res. 82, 360-366.

Yancey, S.B., Biswal, S., & Revel, J.P. (1992). Spatial and temporal patterns of distribution of the gap junction protein connexin43 during mouse gastrulation and organogenesis. Development 114, 203-212.

Zelante, L., Gasparini, P., Estivill, X., Melchionda, S., D'Agruma, L., Govea, N., Mila, M., Monica, M. D., Lutfi, J., Shohat, M., Mansfield, E., Delgrosso, K., Rappaport, E., Surrey, S., & Fortina, P. (1997). Connexin26 mutations associated with the most common form of non-syndromic neurosensory autosomal recessive deafness (DFNB1) in Mediterraneans. Hum. Mol. Genet. 6, 1605-1609.

# GAP JUNCTIONS DURING NEOPLASTIC TRANSFORMATION

Mark J. Neveu and John Bertram

I. Introduction . . . . . . . . . . . . . . . . . . . . . . . . . . . . . . . . . . . . . . . . . . . . . . . . . . . 222
II. The Carcinogenic Process . . . . . . . . . . . . . . . . . . . . . . . . . . . . . . . . . . . . . . . 222
III. Functional Modulation of Gap Junctional Communication in
Development and in Adult Tissues—Tissues Differ in Their
Expression of Gap Junctions . . . . . . . . . . . . . . . . . . . . . . . . . . . . . . . . . . . . . 225
IV. Proliferation-Associated Changes in Connexin Expression in
Nontransformed Cell Types . . . . . . . . . . . . . . . . . . . . . . . . . . . . . . . . . . . . . . . 227
V. Association of Altered Gap Junctional Communication with
Oncogensis—Studies of Gap Junctional Communication during
Oncogenesis in Animal Models . . . . . . . . . . . . . . . . . . . . . . . . . . . . . . . . . . . 228
   A. Liver Tumor Induction . . . . . . . . . . . . . . . . . . . . . . . . . . . . . . . . . . . . . . 229
   B. Skin Tumor Induction . . . . . . . . . . . . . . . . . . . . . . . . . . . . . . . . . . . . . . . 230
   C. Other Tumor Types . . . . . . . . . . . . . . . . . . . . . . . . . . . . . . . . . . . . . . . . . 232
VI. *In Vitro* Studies of Gap Junctional Intercellular Communication
During Oncogensis . . . . . . . . . . . . . . . . . . . . . . . . . . . . . . . . . . . . . . . . . . . . . . 233
   A. Carcinogen-Induced Transformants . . . . . . . . . . . . . . . . . . . . . . . . . . . . 233
   B. Oncogene-Induced Alterations in Gap Junctional Communication . . . . . . 234
   C. Modulation for Gap Junctional Communication by Tumor Promoters . . . . 234

**Advances in Molecular and Cell Biology, Volume 30, pages 221-262.**
**Copyright © 2000 by JAI Press Inc.**
**All rights of reproduction in any form reserved.**
**ISBN: 0-7623-0599-1**

D.   Mechanisms for Tumor Promoter-Inhibition of
     Gap Junctional Communication .................................. 237
E.   Evidence for Altered Gap Junctional Communication in
     Metastatic Tumors. ........................................... 239
VII.   Mechanisms Responsible for Altered Gap Junctional Communication
       during Oncogenesis—Oncogene Products ........................... 240
VIII.  Compounds that Inhibit the Carcinogenic Process, Increase Connexin
       Expression or Gap Junctional Intercellular Communication, or Do Both ...... 241
       A.   Retinoids. ............................................. 241
       B.   Carotenoids. ........................................... 242
IX.    Restoration of Gap Junctional Communication Inhibits Oncogenesis ........ 243
X.     Growth Regulatory Signal Hypothesis ............................. 244
XI.    Future Directions ............................................. 245
       A.   Chemoprevention: As an Intermediate Marker of Response ........... 245
       B.   Toxicology: As a Predictive Assay for Tumor Promoters ............. 246
       C.   Chemotherapy: The "Bystander Effect" ......................... 246
       D.   Signal Transduction ...................................... 248

# I. INTRODUCTION

Studies by Stoker in 1967 found that gap junction–permeable molecules can act to inhibit the proliferation of polyoma transformed cells. Within the same year, Loewenstein observed that the abundance of gap junctions and the level of gap junction–mediated intercellular communication was diminished in liver neoplasms (Loewenstein et al., 1966). Following these novel observations made over three decades ago, many attempts to further define the role of gap junctional communication during neoplasia have provoked consternation rather than insight. No consensus has emerged. At least two factors account for the ambiguity. Firstly, cloning studies show that gap junctions are constructed of a family of proteins termed connexins (Beyer et al., 1990). Prior to such molecular analysis, all gap junctions were erroneously thought to be composed of the same protein. Secondly, the majority of investigations of gap junctional communication during carcinogenesis have been performed *in vitro* (Trosko et al., 1990b; Yamasaki, 1990). As both the proliferation of neoplastic cells and gap junctional communication are controlled by the cellular microenvironment, the behavior of connexin expression in cell cultures may not duplicate neoplasia *in vivo* (Larsen, 1983; Tomomura et al., 1990). This chapter examines the role of gap junctional communication during normal cell proliferation, describes the cellular machinery that controls gap junctional communication, and describes the many mechanisms that can modify gap junctional communication during neoplastic development. We discuss the positive and negative evidence for an obligatory role of alterations in gap junctional communication during oncogenic transformation.

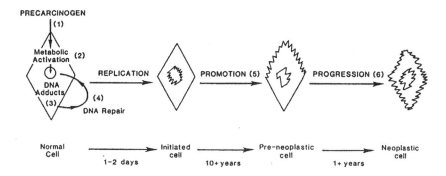

***Figure 1.*** The stepwise development of neoplasia: proposed modulation by gap junctional communication. In steps 1-3, DNA damage to a crucial gene occurs. For most chemicals, this requires metabolic activation to a reactive intermediate. Physical carcinogens (e.g., X-rays or Ultraviolet) or free radicals generated from oxidative stress can damage DNA directly. If DNA replication occurs prior to the repair of damaged DNA, a stable mutation can be introduced resulting in the production of an initiated cell with a proliferative advantage (due to aberrant growth– or death–control). Over a period of many years, additional mutations in oncogenes and tumor suppressor genes (Weinberg, 1989) accumulate in the progeny of this initiated cell during a process termed promotion (Step # 5). This process requires additional DNA damage and proliferation. Compounds called tumor promoters, such as TPA, accelerate this process, while cancer preventive agents, such as retinoids, inhibit promotion. The former compounds inhibit gap junctional communication whereas the latter enhance gap junctional communication. It is proposed that one component of the action of promoters (preventives) is to interfere with junctionally mediated growth regulatory signals, thus augmenting (inhibiting) proliferation required for the growth of cell clones bearing additional genetic damage (Bertram, 1990). Most neoplastic cells are deficient in the ability to undergo heterologous gap junctional communication with normal cells; this may be due to selection for cells capable of autonomous growth. Restoration of heterologous gap junctional communication by pharmacological means (Nishino et al., 1984), or the overexpression of connexin genes (Mehta et al., 1986) can partially restore the normal phenotype and suppress tumorigenicity. It is proposed that gap junctional communication, by its effects on proliferation, can influence the production of mutations on oncogenes and tumor suppressor genes and the expression of these mutations as disregulated proliferation. (Taken from Bertram et al., 1987, with permission.)

## II.  THE CARCINOGENIC PROCESS

It is now well established that the natural development of most cancers is a multi-stage process encompassing both genotoxic and nongenotoxic mechanisms (Bertram et al., 1987; Ames, 1989; Knudson, 1989; Loeb, 1989; Weinberg, 1989). For the purposes of this chapter, we shall limit the discussion to solid tumors, principally the carcinomas and sarcomas, in which gap junctional communication can occur. Mouse skin has been the most thoroughly investigated experimental model; here the process can be divided into (1) initiation, a rapid process involving the interaction of a reactive chemical species with DNA to form DNA adducts and the subsequent conversion of these adducts during DNA replication to a stable mutation in a gene responsible for the control of growth, a process that results in initiated cells, and (2) promotion, where proliferation of initiated cells results in a population of sufficient size in which subsequent genetic damage can occur to other critical genes. The summation of this sequential damage to genes involved in growth control is the expression of the genotoxic damage as neoplasia (Figure 1) (Bertram et al., 1987). In humans the situation appears analogous. Elegant dissection of the process of carcinogenesis has been recently achieved in colorectal cancer. Here the conversion of normal colonic epithelium to minimally transformed cells, to polyps, to frank carcinoma has been followed at the molecular level. In this disease it appears that at least five genetic alterations are required to complete the conversion from normal to malignant (Powell et al., 1992). These changes in general involve the activation, or inappropriate expression, of oncogenes (e.g., ras) and the inactivation of tumor suppressor genes (e.g., p53) (Weinberg, 1989).

There is little to indicate that the induction of mutations at genetic loci involved in carcinogenesis occurs at higher frequencies than at other loci, such as HGPRT, also mutated by carcinogens; here frequencies of about $10^{-6}$ have been reported after high dose treatment with mutagens (Ames, 1989). In view of the large number of normal cells at risk of being initiated by a carcinogen (it has been estimated that humans contain some $10^{14}$ cells, of which not all are capable of initiation) and the unavoidable exposure to genotoxins, adults probably contain many initiated cells. Yet most individuals will not suffer from this disease. In order for the necessary second and subsequent mutagenic events to occur with any degree of probability, it seems necessary that the initiated population both proliferate and increase its size (Cohen et al., 1991; Pitot et al., 1991). These two factors are not identical; for example, normal stem cells proliferate but in adults do not increase their population size. Proliferation is required for conversion of DNA adducts to mutations, while an increase in population size makes probable the occurrence of additional mutations (Ames et al., 1993).

One action of tumor promoters may be to allow initiated cells to achieve this goal by escaping growth control mechanisms. It is clear that initiated cells are primed to do this because a common feature of oncogenes and tumor suppressor genes is their involvement in molecular pathways concerned with the control of cell proliferation or cell death (terminal differentiation). Unfortunately, many of the epigenetic alterations responsible for tumor promotion remain an enigma. Clues to the mechanisms of tumor promotion may reside in a deeper understanding of the cellular controls that inhibit the growth of initiated cells despite the presence of mutations in one or more growth regulatory genes. This question is complex as tumor promoters induce a cascade of phenotypic changes; might a subset of these changes be necessary for rendering initiated cells less susceptible to local growth controls? Changes in cell-to-cell and cell-to-substratum interactions are likely candidates (Yamasaki, 1990). *Given that gap junctional intercellular communication is the most intimate aspect of a cell's microenvironment, alterations in this process could represent a necessary alteration during oncogenic transformation* (Loewenstein, 1979; Bertram, 1990; Trosko et al., 1990a).

## III. FUNCTIONAL MODULATION OF GAP JUNCTIONAL COMMUNICATION IN DEVELOPMENT AND IN ADULT TISSUES—TISSUES DIFFER IN THEIR EXPRESSION OF GAP JUNCTIONS

Cancer is best characterized as a disease of abnormal differentiation in which immature cells retain their proliferative capacity and fail to mature, or mature cells fail to undergo normal terminal differentiation and accumulate. In view of the necessity of most organs to constantly monitor and modify rates of cellular replenishment, it seems apparent that extensive intercellular communication must occur. In this section, we review the evidence that gap junctional communication plays an important role in this process.

Consistent with subtle differences in the morphological appearance of gap junctions in different tissues and species (Larsen, 1977), subsequent studies have identified a family of proteins that make up these structures (Beyer et al., 1990; Willecke et al., 1991; see also the chapter by Beyer and Willecke earlier in this volume). A nomenclature system has been established in which the members of this family are distinguished by their predicted molecular mass calculated from cDNA sequences (Tomomura et al., 1990). In addition, another classification system has been developed in which gap junction proteins are divided into $\alpha$-, $\beta$-, and, more recently, $\gamma$-groupings (Gimlich et al., 1990). The former naming system will be used throughout this review to avoid confusion. In rodents, at least 15 different connexins have been identified by various cloning strategies (Beyer et al., 1990; Willecke et al., 1991; Bennett, 1994; see also Beyer and Willecke in this volume). The amino acid homology between connexins ranges from 42% to 70%. The length of the C-terminal tail, which may be responsible for controlling the gating of the connexon intercellular pore, accounts for the most of the difference in molecular mass.

Spatial and temporal changes in the expression of connexin family members are a common feature during cellular morphogenesis and differentiation (Dermietzel et al., 1989; Beyer et al., 1990; Laird et al., 1992; Winterhager et al., 1993). Although the mechanisms are not completely understood, defined areas of gap junctional communication, within which communication is restricted within boundaries, have been observed in all multicellular organisms examined, including both animals and plants (Pitts et al., 1987; Revel, 1988; Meiners et al., 1991; see, too, the chapter by Lo and Gilula in this volume). These regions have been termed communication compartments. These compartments have been mapped in several organisms and were found to correspond to lineage compartments (Serras and van-den, 1987; Lo, 1989; Warner, 1992). These observations suggest that the boundaries may be important in the establishment of gradients for pattern formation and localized organization (Wolpert, 1978). The question arises as to whether different connexin proteins can form gated heterotypic connexon channels that allow passage of vital tracers. Although heterotypic connexons formed by injecting oocytes with different connexin proteins exhibit electrical coupling (Swenson et al., 1989; see, too, the chapters by Yeager and Nicholson, and by Verselis and Veenstra earlier in this volume), their response to applied voltage is different than homotypic connexins (Barrio et al., 1991). Furthermore, heterotypic dye coupling between human mammary tumor cells transfected with either connexin43 (Cx43) or connexin26 (Cx26) does not occur despite robust homotypic dye coupling (Tomasetto et al., 1993). The development of boundaries seems to represent a common phenotypic change during oncogenesis (see section VI).

The spatiotemporal expression of connexins suggests that gap junctions may provide a mechanism for the transmission of timed growth and positional signals. Several lines of evidence support this conclusion: Inhibition of gap junctional communication during early amphibian embryogenesis by connexin antibodies or antisense mRNAs results in lethal abnormalities in both invertebrates and vertebrates (Warner et al., 1984; Fraser et al., 1987; Lee et al., 1987; Bevilacqua et al., 1989). Mutations of the lethal discs overgrown locus (*dco* gene) in *Drosophila* results in defective gap junctional communication and hyperplastic overgrowth of imaginal disks (Jursnich et al., 1990). A lethal mouse mutant, repeated epilation, associated with overgrowth of epidermis, is associated with defective gap junctional communication in this tissue (Kam et al., 1989). Loss-of-function mutations in the human connexin32 (Cx32) gene result in a peripheral neuropathy, termed X-linked Charcot-Marie-Tooth disease (Bergoffen et al., 1993; Bennett, 1994; Fairweather et al., 1994). The surprising lack of obvious developmental defects and no associated increase in tumor incidence in humans suggests that other connexins may substitute for Cx32 function in these individuals. In contrast, Cx32 knockout mice have been found to have elevated levels of spontaneous and inducible hepatic cancers (Temme et al., 1997). Studies of humans with other connexin mutations as well as transgenic animals will continue to be instrumental in addressing the role of different connexins during normal and neoplastic development.

In fully differentiated tissues, gap junctions are thought to be important for tissue integration and homeostasis. Each tissue so far examined has been found to express multiple connexins (Beyer et al., 1990); indeed, some tissues express five different connexin mRNAs (Hennemann et al., 1992) and heterotypic communication has been observed (Werner et al., 1989; Barrio et al., 1991). In hepatocytes both Cx32 and Cx26 are expressed and they appear to colocalize within the same junctional plaque (Zhang and Nicholson, 1989). The physiological requirement for two hepatocyte connexins is unclear; it may allow hepatocytes to have multiple mechanisms of controlling gap junctional communication during different physiological conditions. In support of this concept, different connexin family members can be differentially gated by post-translational modifications such as phosphorylation (Arellano et al., 1990; Crow et al., 1990; Filson et al., 1990; Stagg and Fletcher, 1990; Swenson et al., 1990; Brissette et al., 1991; Moreno et al., 1992; Takens-Kwak and Jongsma, 1992; see, too, the chapter by Laird and Sáez earlier in this volume).

Within a developed tissue, gap junctional communication integrates cellular activity necessary for many glandular functions. Of interest to the control of growth during the carcinogenic process, the meiotic arrest of oocytes prior to ovulation is modulated by the transfer of $Ca^{2+}$ between the membrana granulosa and cumulus oophorus cells (Sandberg et al., 1990). Changes in the level of luteinizing and follicle-stimulating hormones results in the inhibition of heterologous transfer of $Ca^{2+}$ ions, thereby resulting in ovulation (Larsen et al., 1986; Wert and Larsen, 1990). Activity in glial cells results in focal ion transients that are dissipated via a recovery process mediated by gap junctions (Charles et al., 1991), a function that may have relevance to Charcot-Marie-Tooth disease, referenced earlier.

Although these observations reinforce the importance of gap junctional communication during development, they do not identify the important morphogenic or growth regulatory molecules that traverse gap junctions. However it can be demonstrated that certain second messengers such as $Ca^{2+}$ ions, inositol trisphosphates, and cyclic nucleotides pass through gap junctions and modify cellular behavior (Fletcher et al., 1987a; Sáez et al., 1989; Boitano et al., 1992; Charles et al., 1992). Hence, sharing of ions and second messengers via gap junctions can allow for various physiological activities that require an integrated cellular response.

## IV. PROLIFERATION-ASSOCIATED CHANGES IN CONNEXIN EXPRESSION IN NONTRANSFORMED CELL TYPES

It is known that neoplastic cells differ from normal cells in their cell cycle control. Furthermore the most frequently involved tumor suppressor genes, p53 (Levine et al., 1991; Kuerbitz et al., 1992; Yin et al., 1992) and Rb (DeCaprio et al., 1989; Mittnacht and Weinberg, 1991), are cell cycle regulatory proteins. It therefore seems appropriate to ask if connexins are differentially regulated during the cell cycle, and thus whether the reported changes in connexin expression in transformed cells are merely epiphenomena. Unfortunately, although morphological and physiological studies show that the abundance or function, or both, of gap junctions are indeed modified during the cell cycle, the results have not been consistent between the different cell types examined. A question that remains is whether different connexin proteins exert unique control of cell cycle regulation.

Proliferation-associated changes in connexin expression have been most extensively studied in regenerating rat liver. Compensatory liver hyperplasia following a 70% partial hepatectomy (PH) provides a reproducible model to study cell proliferation *in vivo*, because both hepatocytes and other liver nonparenchymal cells undergo near synchronous waves of DNA synthesis followed by mitosis (Michalopoulos, 1990). Following partial liver resection, the extent of gap junctional communication, the abundance of morphologically identifiable gap junctions, and the expression of Cx32 and Cx26 are downregulated during DNA synthesis (Yancey et al., 1979; Dermietzel et al., 1987; Hendrix et al., 1992; Neveu et al., 1995). The decrease in Cx32 and Cx26 expression post-hepatectomy results from post-transcriptional alterations in mRNA stability (Kren et al., 1993). Alterations in gap junctional communication in hepatocytes appear insufficient to induce proliferation because bile duct ligation and phenobarbital treatment cause a diminution of hepatocyte gap junctions without increasing proliferation (Robenek et al., 1981; Neveu et al., 1990).

In contrast to hepatocytes, proliferating keratinocytes (in "wounded" skin) and regenerating tracheal epithelium exhibit a maximum level of gap junctions during the S-phase of the cell cycle (Gordon et al., 1982; Kam et al., 1987). In a similar manner, increased levels of Cx43 mRNA, protein, and gap junctional communication have been observed in wounded monolayers of bovine capillary endothelial cells (Larson and Handenschild, 1988; Pepper et al., 1989). In mouse fibroblasts, increased cell proliferation induced by transforming growth factor-$\beta$ (TGF-$\beta$) in defined conditions, is accompanied within 6 hours of treatment by upregulated Cx43 mRNA expression followed by increased gap junctional communication (Gibson et al., 1994). In synchronized human mammary cell cultures, Cx26 and Cx43 mRNAs display different patterns of expression during the cell cycle: while Cx43 mRNA was constitutively expressed throughout the cell cycle, the abundance of Cx26 transcripts was enhanced in S-phase cells (Lee et al., 1992). Taken together, these results indicate that different tissues have unique mechanisms for the regulation of gap junctional communication during the cell cycle and that generalizations cannot be made in the comparison of normal and tumor cells.

# V. ASSOCIATION OF ALTERED GAP JUNCTIONAL COMMUNICATION WITH ONCOGENESIS—STUDIES OF GAP JUNCTIONAL COMMUNICATION DURING ONCOGENESIS IN ANIMAL MODELS

The early studies of altered gap junctional communication during the process of oncogenesis were conducted prior to the molecular cloning of connexins genes and identified alterations in gap junctional communication using ultrastructural or physiological techniques, or both. These methods revealed an association between altered gap junctional communication and neoplasia. In addition, several investigators found that gap junctional communication between normal and various transformed cell types was associated with growth arrest of the tumor cells, suggesting the transfer of growth regulatory molecules. Such observations led to the hypothesis that gap junction–permeable factors present in normal cells can act to repress tumor cell growth (Stoker, 1967; Loewenstein, 1979; Bertram and Faletto, 1985). In support of this theory, a diminution in the abundance of morphologically identifiable gap junctions was found in many primary tumors and tumorigenic cell lines.

To determine at what stage(s) alterations in gap junctional communication occur during neoplastic development, investigators have used various models of multistage carcinogenesis. Using two-stage models of carcinogenesis, it has been shown that changes in connexin expression and gap junctional communication occur early during the process of oncogenesis. Briefly, the first stage involves treatment of the animal with a genotoxic agent followed by administration of a nongenotoxic tumor promoter. This protocol results in the clonal expansion of the initiated cells. Unlike the irreversible changes that occur during initiation, withdrawal of the promoter prior to onset of neoplasia causes the majority of preneoplastic cells to undergo apoptosis or remodeling, or both. The most studied *in vivo* multistage models are the induction in rodents of altered hepatic foci (AHF) (Farber, 1984), and the induction of skin papillomas (Yuspa and Poirier, 1988).

## A. Liver Tumor Induction

Early morphological studies of primary and transplanted liver tumors from rats, mice, and humans demonstrated that these neoplasms have a quantitative decrease in the number or area of gap junctions, or both, as compared to normal liver (Weinstein et al., 1976). The alterations may represent an early change, perhaps reflecting toxicity and regenerative proliferation because reduced abundance of gap junctions are also observed in non-neoplastic pathologies in human liver (i.e., cholestasis and cirrhosis) (Robenek et al., 1981)

Examination of *in vivo* models of rat and mouse multistage hepatocarcinogenesis, demonstrate quantitative alterations in the expression of Cx32 protein in preneoplastic AHF and carcinomas (Beer et al., 1988; Fitzgerald et al., 1989; Neveu et al., 1990; Neveu et al., 1994b). Prior to administration of the tumor promoter, diethylnitrosamine-initiated cells display typical gap junctions with surrounding hepatocytes; however, in the presence of the tumor promoter phenobarbital (PB), cells in most AHF exhibit reduced Cx32 protein expression (Mesnil et al., 1988a; Neveu et al., 1990). Unlike other markers of AHF, these alterations are observed during several different chemical carcinogenesis regimens utilizing different promoting agents and complete carcinogens. RNA *in situ* hybridization demonstrated that most AHF generated by DEN initiation and PB promotion downregulate Cx32 expression independent of changes in mRNA abundance (Neveu et al., 1994b). This decreased expression of connexins appears to be functionally linked to decreased gap junctional communication, as reduced Lucifer yellow dye transfer has also been shown in AHF using liver slices (Krutovskikh et al., 1991). Although cells within some AHF still retained the ability to communicate, deficient heterologous coupling between cells of the AHF and those of normal liver was observed. During induction of heptocarcinogenesis in the rat by nitrosamine, a progressive decrease in Cx32 expression was found to correlate inversely with BrdU labeling used as a measure of proliferation. No detectable Cx32 expression in was found in lung metastases (Tsuda et al., 1995).

The mechanisms responsible for selective communication of AHF are presently unknown: differential expression of cell adhesion molecules (Hixson et al., 1985), of connexin proteins (Sakamoto et al., 1992), or increased internalization of gap junctions (Neveu et al., 1994b) have all been reported. Reduction in Cx32 expression by PB is not dependent on prior carcinogen treatment; PB treatment of noninitiated rats also reduces gap junctions (Sugie et al., 1987; Tateno et al., 1994). Examination of tissue sections revealed that PB reversibly decreases Cx32 expression within centrilobular hepatocytes of the hepatic acinus; however, this reduction occurred independently of changes in cell proliferation (Neveu et al., 1990). Thus, alterations in gap junctional communication are not sufficient to stimulate the growth of noninitiated hepatocytes. In rodent hepatomas downregulation of connexin levels is apparently achieved by diverse mechanisms. Although some hepatomas downregulate Cx32 expression by decreased mRNA abundance, many lesions utilize post-translational mechanisms (Neveu et al., 1994b). Increased Cx43 mRNA levels are present in many hepatic neoplasms, but immunoreactivity is restricted to nonparenchymal cell types (Oyamada et al., 1990; Wilgenbus et al., 1992; Neveu et al., 1994b). Taken together, transformed hepatocytes fail to use a common mechanism to modify connexin expression.

## B. Skin Tumor Induction

Human skin is extensively coupled as revealed by dye transfer experiments. The most widely coupled cells are dermal fibroblasts followed by suprabasal keratinocytes. No direct coupling was observed between dermal and epidermal cells (Pitts et al., 1987; Saloman et al., 1988). Morphological examination of mouse skin 2 to 72 hours after a single application of TPA demonstrated widening of extracellular spaces (10 hours) and a reversible decrease in the abundance of morphologically recognizable gap junctions in basal and suprabasal layers (10 to 42 hours). The decrease in plasma membrane gap junctions was paralleled by an increase in internalized gap junctions. An unexplained finding was that a single TPA treatment results in extensive epidermal-dermal coupling (Pitts et al., 1987), whereas as in human skin, dye transfer between dermal and epidermal cells is normally infrequent. The situation in mouse, and probably human skin, is complex: four distinct tumor promoters, TPA, okadaic acid, chrysarobin, and benzoyl peroxide all transiently increased expression of Cx26 and Cx43 while inhibiting expression of connexin30.1 (Cx30.1) (Budunova et al., 1996a). This suggests that measurements of coupling frequency between different cell types is very dynamic and difficult to interpret.

The complete carcinogen dimethylbenzanthracene (DMBA) also decreases the abundance of gap junctions in squamous cell carcinoma and in surrounding skin (Tachikawa et al., 1984). In CD-1, mice no change in expression of Cx26 or Cx43 were observed in papillomas induced by DMBA/TPA; however, in tumors that had progressed to squamous cell carcinoma (SCC), clear reductions in expression of these genes occurred (Kamibayashi et al., 1995). Application of fluocinolone acetonide, an inhibitor of stage I and stage II tumor promotion, prevented the TPA-induced decrease in communication (Kalimi and Sirsata, 1984) as well as reduction of expression of Cx26 in a manner consistent with downregulation of glucocorticoid receptors in these cells during skin carcinogenesis (Budunova et al., 1997). Retinoic acid, also a potent inhibitor of experimental two-stage carcinogenesis and a clinically active cancer preventive agent in skin, caused the upregulated expression of Cx43 in human epidermis (Guo et al., 1992), and of morphologically identified gap junctions in basal cell carcinoma (Elias et al., 1980). Consistent with these *in vivo* observations, cell lines derived from different stages of SENCAR mouse skin carcinogenesis, initiated with DMBA and promoted with TPA, exhibit a reduced capability for gap junctional communication with increasing tumorigenicity (Klann et al., 1989; Budunova et al., 1996b) as well as a decrease in expression of Cx26 (Stern et al., 1998). However, in other studies, gap junctional communication in cultured cells was reported not to be decreased as cells progressed from normal to papilloma to SCC, but was inhibited on the further progression to spindle cells, a defect linked to deficient expression of E-cadherin (Holden et al., 1997). These data suggest that gross defects in connexin expression are a late event in mouse skin carcinogenesis. Given the difficulties in measuring gap junctional communication itself in this organ, defects in function may occur at a much earlier phase as is suggested by the effects of the cancer preventive agents discussed earlier. As may be expected, gap junctions are still expressed in skin tumors, being recognized morphologically (Prutkin, 1975), by electric coupling (Flaxman and Cavoto, 1973), and by Cx26 and Cx43 immunoreactivity (Wilgenbus et al., 1992; Rundhaug et al., 1997). Further molecular and physiological studies are warranted to determine if preneoplastic and neoplastic epidermal cells exhibit common alterations in connexin expression or function.

## C.  Other Tumor Types

As in liver and skin, a diminution in gap junctions has been shown to be associated with preneoplastic lesions in human cervix. Whereas a reduced incidence of gap junctions were found in preinvasive cervical carcinomas (severe dysplasia and carcinoma *in situ*), benign epithelium (metaplasia and mild dysplasia) failed to exhibit alterations in gap junction morphology (McNutt, 1969; McNutt et al., 1971). Most cervical cancer is a consequence of infection with human papilloma virus (HPV), and it is of interest that transfection of the E5 open reading frame of the high-risk HPV 16 virus into human keratinocytes causes a dramatic reduction in gap junctional communication. This was accompanied by dephosphorylation of Cx43 (Oelze et al., 1995). In stomach (Kanno and Matsui, 1968) and thyroid neoplasms (Jamakosmanovic and Loewenstein, 1968) decreased electrical coupling has been reported, while in the prostate, immunohistochemical analysis revealed a progressive decrease in Cx43 expression with disease severity (Van Lieshout et al., 1996). In culture, human mammary carcinoma (Lee et al., 1992; Hirschi et al., 1996) and primary human colon carcinoma cell lines (Friedman and Steinberg, 1982) exhibit diminished gap junctional communication.

Although gap junctions appear to be absent in many neoplasms, some tumors continue to display morphologically identifiable gap junctions (Weinstein et al., 1976; Larsen, 1983). In contrast to a quantitative decrease in gap junction abundance, morphological analysis of cultured rat bladder cells demonstrated that normal and tumor cells display different sized gap junctions (Pauli and Weinstein, 1981). Differential expression of connexins in neoplasms has also been reported; this may allow tumor cells to communicate but not allow heterologous communication with surrounding normal cells (i.e., selective communication) (Mesnil and Yamasaki 1988b; Yamasaki and Fitzgerald, 1988). Molecular examination of Cx26, Cx32, and Cx43 expression by immunostaining in a limited set of malignant human tissues demonstrated that connexin expression varies widely: some tumors exhibited a loss of immunoreactivity, whereas others displayed increased expression of a connexin not observed in normal tissue (i.e., metaplasia) (Wilgenbus et al., 1992; Neveu et al. 1994b). Furthermore, not all connexins expressed in neoplasms are associated with the plasma membrane; in rat hepatomas Cx32 and Cx26 immunoreactivity was found to be abnormally diffuse or cytoplasmic (Neveu et al., 1994b). Consistent with these observations, increased levels of internalized gap junctions have been observed in renal adenocarcinomas (Letourneau et al., 1975) and in sarcoma cell lines lacking expression of cell-cell adhesion molecules (Musil et al., 1990). Punctate immunostaining should not be automatically considered to represent plasma membrane gap junctions, because internalized gap junctions are circular and may also generate punctate immunostaining. In addition, intracellular pools of connexin protein exist during the formation of gap junctions (Puranam et al., 1993). Additional studies are needed to determine if tumors that display connexin immunoreactivity are indeed electrically or dye coupled.

# VI.   *IN VITRO* STUDIES OF GAP JUNCTIONAL INTERCELLULAR COMMUNICATION DURING ONCOGENESIS

## A.   Carcinogen-Induced Transformants

Detailed studies have been conducted in the mouse Balb/3T3 and the C3H/ 10T1/2 cell culture systems. As was originally reported by Borek and associates (1969), transformed cells are deficient in their ability to undergo heterologous communication with their normal counterparts. Indeed a low level of gap junctional communication may be a requirement for the success of an *in vitro* transformation system: in a comparison of two Balb/3T3 cell lines, the transformable line was found to downregulate gap junctional communication at confluence, whereas in the nontransformable line communication remained high. Similarly in 10T1/2 cells, gap junctional communication, and Cx43 mRNA and protein expression, is dramatically decreased as cells reach confluence (Hossain and Bertram, 1994). One explanation is that maintenance of a high level of gap junctional communication in confluent cultures would suppress the ability of initiated cells to proliferate and undergo transformation; furthermore, should transformed cells exhibit a high capacity to undergo heterologous junctions with surrounding normal cells, they would be unable to express their transformed phenotype. Support for this explanation comes from several sources: Herschman reported that ultraviolet (UV) irradiation of C3H10T1/2 cells led to the production of cells that would transform only when effectively removed from 10T1/2 cells or when treated with TPA (Herschman and Brankow, 1986, 1987); also working with 10T1/2 cells, Bertram's group showed that carcinogen-initiated cells would only transform when effectively separated from surrounding normal cells by the formation of large colonies of initiated cells (Mordan et al., 1983). Accordingly, in both the mouse C3H/ 10T1/2 and the BALB/3T3 cell culture systems, transformants produced by chemical carcinogens communicate poorly if at all with their nontransformed counterparts (Mehta et al., 1986; Yamasaki, 1990). As will be discussed later, the opposite situation produces the expected response: when gap junctional communication is upregulated, transformation is inhibited.

Thus, it would seem that in the process of selecting a cell line for transformation studies, only those lines having low gap junctional communication at confluence, when transformed foci appear, are found suitable. A similar situation of suppression of expression of the transformed phenotype by normal cells was encountered in early studies using the NIH/3T3 cell line as recipient of transfected oncogenes. Here, in a line selected for transfection efficiency not necessarily for transformation efficiency, several oncogene-induced transformants were found to be latent unless the nontransfected cells were removed (Weinberg, 1989). By analogy, the same process may be considered to occur spontaneously *in vivo*; only those cells capable of escaping growth-controlling signals, whether junctional or not, will be detectable and available for study.

## B.  Oncogene-Induced Alterations in Gap Junctional Communication

Proto-oncogenes and their mutated counterparts have also been found to modify gap junctional communication in a cell-specific manner. The oncogenic proteins examined include c-src, v-src (Azarnia et al., 1989), activated c-Ha-ras (Dotto et al., 1989; Vanhamme et al., 1989; Brissette et al., 1991; Hayashi et al., 1998), c-myc, v-raf, c-fos, v-fms, v-fgr, v-mos, polyoma virus middle and large T antigens, and SV40 large T antigen (Atkinson et al., 1988; Bignami et al., 1988a; Martin et al., 1991; Brownell et al., 1997; Hayashi et al., 1998). Only v-src, c-src, v-ras, and polyoma middle and large T antigen have been found to decrease homologous gap junctional communication, and interactions among these oncogenes appear to potentiate these effects (Hayashi et al., 1998). Loewenstein and his colleagues, as well as other investigators, have made an extensive study of the effects of v-src on gap junctional communication. Here, inhibition of gap junctional communication is a rapid event upon upregulating expression of this gene and requires tyrosine kinase activity (Azarnia and Loewenstein, 1984b; Azarnia et al., 1988; Swenson et al., 1990); indeed, Cx43 is a substrate for the src kinase (Filson et al., 1990; Loo et al., 1995). The mechanisms responsible for such changes are discussed in section VII. Dual-transfection of NIH-3T3 cells with both c-src and adenovirus E1A genes resulted in transformation and a pronounced reduction in gap junctional communication. In contrast, transfection of either gene alone only slightly reduced gap junctional communication, and did not induce growth in soft agar (Azarnia et al., 1989). Many oncogenes that lack the ability to diminish homologous communication selectively modulate heterologous communication (Bignami et al., 1988a; Kalimi et al., 1992). These results indicate that some oncogenes modulate genes other than connexins, which in turn regulate heterologous gap junctional communication. Possible targets include modifications in cell-cell adhesion or cytoskeletal proteins (see section IV).

## C.  Modulation of Gap Junctional Communication by Tumor Promoters

The association of alterations in gap junctional communication with early stages of neoplasia were driven by the co-discovery in the laboratories of Trosko (Yotti et al., 1979) and Murray (Murray and Fitzgerald, 1979) that the mouse skin tumor promoter TPA rapidly inhibits metabolic coupling. Shortly thereafter,

Enomoto and colleagues (1981) demonstrated that TPA treatment also inhibits electric coupling. However, the association between tumor promoters and their ability to reduce gap junctional communication is far from perfect. In view of the diverse nature of promotion and promoters, for example, a persistent proliferative stimulus may act as promoter in some circumstances (Fürstenberger et al., 1989). and our poor understanding of the mechanism of action of promoters, some degree of confusion is to be expected. Furthermore, it may be unreasonable to expect that *all* promoters should inhibit gap junctional communication; some classes could inhibit the synthesis or reception of the putative junctionally transmitted signal. This activity would not be detected in most of the assays that have been utilized in investigations of promoters.

The following discussion reviews our interpretation of how various factors that modulate gap junctional communication should be considered when interpreting these sometimes conflicting results, and when designing future *in vitro* studies to test the hypothesis that tumor promoting agents inhibit gap junctional communication and stimulate proliferation.

### Communication Assay

Some negative or contradictory findings can be explained by differences in the assays used to measure gap junctional communication. Several methods have been developed to quantitate gap junctional communication, as follows: (1) electric coupling measures ionic movement; (2) metabolic coupling assesses transfer of low molecular weight molecules (radiolabeled or toxic metabolites); and (3) dye coupling measures the movement of tagged-fluorochromes (up to 1000 daltons [D]). Because each assay quantitates the movement of different size molecules, partial inhibition of gap junctional communication may not be observable by electric or metabolic coupling. For example, scrape loading dye transfer is reported to be more sensitive than metabolic cooperation in detecting dieldrin's inhibitory effects on gap junctional communication (Loch-Caruso et al., 1990); the metabolic coupling "kiss of death " assay may not be able to detect temporal changes in gap junctional communication because resumption of communication after a transient inhibition would lead to the killing of 6-thioguanine resistant cells (Trosko et al., 1987) As neither the critical molecular size nor the period of time necessary to transmit growth regulatory signals through gap junctions and achieve growth inhibition is not known, we recommend testing dye coupling (most sensitive) and electric coupling.

### Culture Conditions

Interlaboratory reproducibility of gap junctional communication assays can be strongly influenced by culture conditions. For example, the pH and serum concentration of the culture medium influences the inhibitory effects of TPA on gap junctional communication (Ruch et al., 1990). Furthermore, changes in the composition of culture media (i.e., serum, hormones, etc.) can alter connexin expression. For example Cx43 expression is modulated in appropriate cells by retinoic acid (Rogers et al., 1990) and transcriptionally by estrogen through the possession of an estrogen response element in the promoter region (Musil and Goodenough, 1993; Yu et al., 1994). Hepatocytes can dramatically switch from Cx32 and Cx26 to Cx43 depending on culture conditions (Stutenkemper et al., 1992; Neveu et al., 1994c). Many growth factors that promote oncogenesis can also modulate gap junctional communication. Because growth factors are usually present in culture media, either added in serum or as part of a defined supplement, their influence cannot be ignored; moreover, effects are often cell type specific. For example, epidermal growth factor, TGF-β, or bovine pituitary extract diminished gap junctional communication in human keratinocytes (Madhukar et al., 1989), whereas TGF-β upregulated Cx43 expression and gap junctional communication in fibroblastic C3H10T1/2 cells (Gibson et al., 1994). Similarly, platelet-derived growth factor decreased gap junctional communication in normal rat kidney and 3T3 cells but did not decrease communication in rat liver epithelial cells (Maldonado et al., 1988; Pelletier et al., 1994). The ability of growth factors to modulate gap junctional communication complicates the interpretation of assay results for tumor promoters because assays frequently utilize diverse culture media or cell lines, or both.

### Dose and Duration

There is only a narrow time window in which some chemicals effect gap junctional communication and this is cell type specific. For example, maximal downregulation of gap junctional communication in V79 cells is seen at 5 hours, whereas Swiss 3T3 cells respond at 30 hours (Zeilmaker and Yamasaki, 1986). In primary hepatocytes, PB inhibited gap junctional communication within 8 hours, but not until 96 hours in V79 cells. To complicate the issue further, PB stimulated gap junctional communication in V79 cells at 5 hours (Noda et al., 1981; Zeilmaker and Yamasaki, 1986). In addition to temporal changes, the concentration of test chemical may be critical. Retinoic acid, for example, has a biphasic effect on gap junctional communication: in C3H10T1/2 cells, low concentrations ($10^{-10}$ M) inhibit gap junctional communication and, as would be predicted, enhance transformation; higher concentrations ($10^{-8}$ to $10^{-6}$ M) strongly upregulate gap junctional communication and inhibit transformation, whereas still higher toxic levels again inhibit communication (Hossain et al., 1989).

### Cell Specificity

Inhibition of gap junctional communication by certain promoters is cell type specific (Bombick, 1990), activity being dependent on the connexin type and presumably on cell-specific intercellular signal transduction machinery. For example, Cx32 and Cx43 are known to be differentially modulated by phosphorylation; cyclic adenosine monophosphate (cAMP) increases gap junctional communication in rat hepatocytes (Klaunig and Ruch, 1987b) (which express Cx32 and Cx26), but inhibits gap junctional communication in cardiac myocytes (De Mello, 1984) (which express Cx43). Because an important class of promoters, the phorbol esters, owe their activity to activation of C-kinase, whereas growth factors activate other kinases, it is to be expected that different connexins will differ in their response to agents affecting protein kinases.

Care must also be taken in comparing responses of normal cells and transformed cells to promoters, as transformation of some cell lines alters their response. Human liver cells, for example, after transformation with SV-40 increase their sensitivity to inhibition of gap junctional communication by polychlorinated biphenyls (Swierenga et al., 1990). In another example, transformation of primary human bronchial cells with an adenovirus-12/SV40 hybrid virus altered their response to TGF-$\beta$: in normal bronchial cells TGF-$\beta$ inhibited dye transfer, but in transformed cells gap junctional communication was stimulated (Albright et al., 1991). These results show that the cell type chosen to examine tumor promoters *in vitro* should closely approximate the susceptible cell type *in vivo*.

### Proximate or Ultimate Effects

Some compounds that may have *in vivo* tumor-promoting activities may act only after metabolic activation or through secondary (i.e., indirect) effects that are not properly replicated in standard cell culture systems. Thus, while several bile acids that act as promoters in rat liver decrease gap junctional communication in V79 cells, cholic and deoxycholic acid do not. It is thought that these bile acids are metabolized *in vivo* during enterohepatic circulation to ultimate forms that act as tumor promoters (Noda et al., 1981). The importance of metabolism with regards to inhibition of gap junctional communication has also been shown for carbon tetrachloride (Sáez et al., 1987), sodium cyclamate, and phenol (Malcolm and Mills, 1987).

Because of a need for metabolism, the use of primary hepatocytes has been suggested as an alternative assay to detect chemicals that may require metabolic activation to become tumor promoters (Tong and Williams, 1987). However, carbon tetrachloride or chloroform, which decrease the abundance of Cx32 protein in hepatocytes *in vivo* (Miyashita et al., 1991), failed to inhibit gap junctional communication in primary hepatocytes in culture (Ruch and Klaunig 1986). As discussed earlier, the dose, duration, or culture media used may have resulted in these negative findings.

## D. Mechanisms for Tumor Promoter-Inhibition of
## Gap Junctional Communication

Tumor promoters appear to possess multiple mechanisms for inhibition of gap junctional communication; TPA has been studied most extensively. TPA inhibits dye transfer 5 to 10 minutes after addition to mouse fibroblasts, and this inhibition was fully reversed 100 minutes after removing TPA (Fitzgerald et al., 1983). Subsequent studies show that TPA activates protein kinase C (PKC), which phosphorylates serine and threonine residues of Cx43 (Berthoud et al., 1992) and Cx32 (Takeda et al., 1987; Sáez et al., 1990). Following TPA-induced phosphorylation in MDCK cells, Cx43 immunoreactivity is rapidly internalized from the plasma membrane (Berthoud et al., 1992). In the rat liver cell line, ARL18 (Budunova et al., 1993), and Syrian hamster embryo cells, post-translational downregulation of gap junctional communication by TPA was observed (Rivedal et al., 1994). However, in primary keratinocytes, treatment with TPA reduced Cx43 mRNA levels after 4 to 8 hours by an unknown mechanism (Brissette et al., 1991). Diacylglycerol, the endogenous ligand of PKC, mimics TPA's inhibitory effects on gap junctional communication in mouse epidermal cells (Muir and Murray, 1986) and Balb/C 3T3 cells (Enomoto and Yamasaki, 1985). Studies in normal and transformed mouse keratinocytes have implicated both TPA and benzoyl peroxide as causing decreases in E-cadherin levels in the plasma membrane and contributing to the loss of gap junctional communication (Jongen et al., 1991; Jansen et al., 1996). This adhesion molecule has in other studies been shown capable of modulating gap junctional communication and, thus, changes in expression could in part contribute to the actions of diverse tumor promoters on gap junctional communication (Wang and Rose, 1997).

Kinetic studies of PB, a liver promoter, on hepatocyte gap junctional communication demonstrated inhibition of dye transfer within 30 to 60 minutes, and recovery by 24 hours. Rapid recovery of gap junctional communication occurs within 15 minutes of removal of PB (Ruch and Klaunig, 1988b). Treatment of rats with PB decreases the morphological abundance of gap junctions (Sugie et al., 1987). Examination of Cx32 and Cx26 mRNA and protein levels following PB treatment indicated that these changes occur through post-translational mechanisms (Neveu et al., 1994b). Suppression of gap junctional communication may be vital to PB promotion, as the percent inhibition of gap junctional communication by PB in livers of different strains of rodents correlated with its *in vivo* promotion activity (Klaunig and Ruch, 1987a). Activity may require metabolism since the inhibitory effects of PB in primary hepatocytes can be prevented by the cytochrome P450 inhibitor SKF-525A. In contrast, lindane and DDT treatment of hepatocyte cultures induced effects on gap junctional communication that were not inhibited by SKF-525A (Klaunig and Ruch, 1987a, 1990). A role for metabolism of PB is again implied by the coincident centrolobular decrease in Cx32 and induction of cytochrome P450 (Neveu et al., 1994a).

Of interest to much current research linking oxidative damage to carcinogenesis (Trush and Kensler, 1991), free radicals inhibit gap junctional communication. Noncytotoxic concentrations of paraquat-generated free radicals inhibited dye-coupling in primary mouse hepatocytes. The effects could be inhibited by antioxidants (vitamin E or diphenyl-p-phenylenediamine), or potentiated by the glutathione depleter (diethylmaleate) or a catalase inhibitor (Ruch and Klaunig, 1988a). Similarly, hydrogen peroxide has been shown to inhibit gap junctional communication under conditions of glutathione deficiency (Upham et al., 1997). Vitamin E also increased gap junctional communication in C3H10T1/2 cells without altering levels of Cx43 mRNA or protein, suggesting a direct effect on the gap junction (Zhang et al., 1991). Arachidonic acid and $CHCl_3$ also block gap junctional communication in hepatocytes; that this action of arachidonic acid is blocked by cyclooxygenase/lipooxygenase inhibitors suggests free radical involvement (Sáez et al., 1987). Free radical generation may inhibit gap junctional communication by causing oxidization of sulfhydryl groups on connexins that are thought important for modulating channel assembly and gating (Dahl et al., 1991). Additionally, cholesterol is known be subject to oxidative damage, and its oxidation products inhibit gap junctional communication (Chang et al., 1988a., 1988b).

## E. Evidence for Altered Gap Junctional Communication in Metastatic Tumors

Cells capable of autonomous growth may have very different requirements for altered gap junctional communication at foreign sites distant from the primary tumor. Here, the need to undergo metabolic cooperation may be a major factor in survival. An additional observation, of as yet unknown relevance *in vivo*, is that junctionally communicating cell spheroids are resistant to the cytotoxic effects of lymphotoxins (Fletcher et al., 1987b).

The capability of normal mammary cells to heterologously communicate with fibroblasts is inhibited in various metastatic cell lines derived *in vivo* (Hamada et al., 1988; Bräuner and Hülser, 1990; Nicolson et al., 1988); however, this correlation was not observed in ras transfected cells (Nicolson et al., 1990). This difference may be related to selection pressure *in vivo* for weakly communicating cells. In contrast, there is evidence that heterologous communication with vascular endothelium may be required for metastasis. In an *in vitro* model for invasion and metastasis, mammary tumor and glioma cells that communicated with chick embryo heart fragments were able to invade the tissue, whereas gap junctional communication deficient tumor cells were unable invade but formed a cellular capsule around the tissue (Brauner et al., 1990). Accordingly, highly metastatic murine melanoma and rat mammary adenocarcinoma cell lines communicate with vascular endothelium (El-Sabban and Pauli, 1991). The apparent differences between minimally altered preneoplastic cells, discussed earlier, and metastatic cells may be explained by differences in the needs of these two cell types; preneo-plastic cells may require a decrease in gap junctional communication to escape the growth inhibitory effects of surrounding cells, whereas metastatic cells, which no longer require clonal expansion in order to evolve, may need to depend on meta-bolic communication for their growth prior to the onset of angiogenesis.

## VII.   MECHANISMS RESPONSIBLE FOR ALTERED GAP JUNCTIONAL COMMUNICATION DURING ONCOGENESIS—ONCOGENE PRODUCTS

The mechanisms responsible for oncogene induced inhibition of gap junctional communication are just beginning to be explored. The role of the v-src oncogene in the tyrosine-phosphorylation of Cx43 and the subsequent inhibition of gap junctional communication has been most extensively studied (Crow et al., 1990; Filson et al., 1990; Swenson et al., 1990; Loo et al., 1995). The central role of this phosphorylation is indicated by the demonstration that inhibition of gap junc-tional communication by temperature sensitive mutants of v-src occurs rapidly (5 to 10 minutes) upon switching to the permissive temperature and is not dependent on changes in morphology (Azarnia and Loewenstein, 1984a, 1984b). Site-directed mutagenesis of Cx43 suggests that phosphorylation of Tyr 265 is necessary for v-src inhibition of gap junctional communication (Swenson et al., 1990). The causal role of inhibition of gap junctional communication in src-medi-ated loss of growth control is also indicated by the association between the ability of several different src mutations to inhibit gap junctional communication and deregulate growth control in transformed mammalian and avian cells (Loewen-stein and Rau, 1992). In contrast to src, inhibition of gap junctional communica-tion by PDGF in C3H/10T1/2 cells (Pelletier and Boynton, 1994) or by EGF in T51B rat liver cells (Lau et al., 1992), is mediated by phosphorylation of serine residues of Cx43. In the case of EGF, phosphorylation of serines 255, 279, and 282 appears the direct result of mitogen-activaged protein (MAP) kinase activa-tion (Warn-Cramer et al., 1996).

The mechanism whereby ras decreases gap junctional communication is not clear. In rat liver epithelial cells the (Mehta et al., 1989), abundance of activated ras oncoprotein levels is inversely related to gap junctional communication (De Feijter et al., 1990, 1992); however, inhibition of gap junctional communication by ras in these cells and in mouse primary keratinocytes is independent of changes in Cx43 mRNA levels, although the level of serine phosphorylation is changed (Brissette et al., 1991). Such changes could be significant in view of the known effects of TPA (a C-kinase activator) and cAMP (an A-kinase activator) on gap junctional communication. Interestingly, inhibition of ras farnesylation, and pre-sumably plasma membrane localization, by lovastatin specifically enhanced gap junctional communication and reduced the tumorigenicity of rat liver epithelial cells transfected with ras, but not src, neu, or raf/myc (Ruch et al., 1993). Other oncogenes may indirectly modulate gap junctional communication by regulating genes associated with connexin assembly or turnover. For example, transfection of gap junctional communication-deficient but connexin-expressing tumor cells (S180) with E-cadherin cDNA resulted in serine phosphorylation of Cx43, punctate Cx43 immunoreactivity, and dye coupling (Musil et al., 1990; Jongen et al., 1991).

## VIII.  COMPOUNDS THAT INHIBIT THE CARCINOGENIC PROCESS, INCREASE CONNEXIN EXPRESSION OR GAP JUNCTIONAL INTERCELLULAR COMMUNICATION, OR DO BOTH

### A.  Retinoids

The prevention of cancer by drugs that inhibit the promotional phase of carcinogenesis is now a clinical reality. The most widely tested such agents are the retinoids, derivatives of vitamin A. In a clinical study, 13-cis retinoic acid was shown to dramatically reduce the frequency of second primary malignancies in the head and neck region in patients presenting with a primary, smoking-related malignancy (Hong et al., 1990). The airway epithelium of such patients is considered to contain many premalignant lesions, "field cancerization," as a result of extensive exposure to tobacco carcinogens. The primary cancer is thus just the first of a series of cancers separated by location and time. The inability of retinoids to influence the growth of the primary tumor, but to strongly suppress the development of second primaries which had not yet progressed to cancers at the time of initial treatment, clearly demonstrates the preventive action of the retinoids. Chemoprevention by retinoids has also been demonstrated in other epithelia: in human cervix, bladder, and skin, retinoids have been shown to reduce incidence rates in high-risk patients (Kelloff et al., 1992).

The chemopreventive ability of the retinoids was predicted by numerous successful demonstrations of inhibition of experimental carcinogenesis in animal and in cell culture systems. In the mouse C3H/10T1/2 cell culture system, a widely used assay system for carcinogens and inhibitors of carcinogenesis (Bertram et al., 1989), retinoids are potent inhibitors of carcinogen-induced neoplasia. Here, this activity has been demonstrated to be highly correlated with the ability of retinoids to upregulate gap junctional communication (Hossain et al., 1989), and the expression of Cx43, the only connexin known to be expressed in these cells. Increased expression of Cx43 was seen at the mRNA and protein level (Rogers et al., 1990) and was accompanied by increased growth control, as measured by a decreased saturation density and proliferation rate at confluence (Mehta et al., 1989). Retinoic acid has since been shown to increase stability of Cx43 mRNA in a manner involving the 3' untranslated region of the Cx43 gene (Clairmont and Sies, 1997).

Effects on growth and cell morphology were seen only in retinoid-treated confluent cells, not in cells seeded sparsely without the ability to form cell-cell contacts (Hossain and Bertram, 1994). Furthermore, the inhibitory effects of the tumor promoter TPA on gap junctional communication were antagonized by retinoids in these 10T1/2 cells (Hossain et al., 1989). In the Balb/C cell transformation system, inhibition of transformation by retinoic acid, glucocorticoids, and cAMP agonists has also been associated with the ability of all three agents to upregulate gap junctional communication (Yamasaki and Katoh, 1988).

Although these studies have not been extended to the human airway epithelium to test the relevance of these findings to the clinical head and neck cancer study discussed earlier, in hamster tracheal cells retinol was shown to counteract the inhibition of dye coupling caused by cigarette condensate (Rutten et al., 1988; Van der Zandt et al., 1990). Thus, as with TPA, the promoting activity of cigarette smoke is associated with decreased gap junctional communication and retinoids can counter this action. Studies conducted in human skin have confirmed the ability of retinoic acid to upregulate expression of connexin 43 in human tissues (Guo et al., 1992). These molecular studies extend earlier morphological descriptions of increased junctions in the retinoic acid–treated skin tumors, basal cell carcinoma, and keratocanthoma (Prutkin, 1975; Elias et al., 1980).

Retinoic acid is a potent inhibitor of TPA-mediated promotion in rodent skin (Verma, 1987), and it is tempting to speculate that its action in this organ is due to its ability to upregulate gap junctional communication. However, retinoic acid has many other functions as a locally produced hormone in skin and alters the expression of many differentiation-specific genes (Kopan et al., 1987). It remains to be determined which, if any, of these additional activities is controlled via gap junctional communication.

## B.  Carotenoids

An extensive epidemiological literature supports the concept that carotenoids have cancer preventive properties (reviewed in Bertram et al., 1987), and the conversion of some dietary carotenoids to retinoids suggests overlapping actions on Cx43 expressing cells. All the major dietary carotenoids have been shown to increase gap junctional communication in 10T1/2 cells, and this action correlates with their ability to inhibit neoplastic transformation in these cells (Bertram et al., 1991; Zhang et al., 1991). As with retinoids, these effects on gap junctional communication were accompanied by increases in Cx43 message and protein. A surprising observation was that carotenoid activity was not dependent on the potential conversion of carotenoids to retinoids. Thus, canthaxanthin, a nonprovitamin A carotenoid was able to upregulate Cx43 expression in 10T1/2 cells and in F9 cells but did not upregulate expression of a known retinoid-responsive gene

RAR-β. In contrast, a retinoid was capable of upregulating both genes in these cells (Zhang et al., 1992). This action of carotenoids on Cx43 expression suggests a mechanistic basis for their protection against risk of cancer in humans.

The antipromoting bioflavonoid quercetin, prevents TPA's inhibitory effects on gap junctional communication in V79 cells (Flodstrom et al., 1988); however, several other antipromoters do not antagonize TPA's effects on gap junctional communication in these cells. The agents tested include anti-inflammatory sterols, protease inhibitors, calcium agonists, retinoids, and antioxidants (Radner and Kennedy, 1984; Hartman and Rosen, 1986). The significance of these negative reports is unclear. Although V79 cells respond to many classes of promoter by decreasing gap junctional communication, and have been proposed as a test system for promoters (Ljungquist and Warngard, 1990), the effects of antipromoters may be expected to be cell type specific. Other negative reports may also be related to the test system or concentration of test chemical. Thus noncytotoxic concentrations of the mouse skin antitumor promoter glycyrrhetinic acid, were reported to inhibit metabolic communication between V79 cells and to act synergistically with TPA to inhibit gap junctional communication (Tsuda and Okamoto, 1986). Retinoic acid has also been reported to enhance the inhibitory effect of TPA on V79 cells and rat liver epithelial cells (Radner and Kennedy, 1984; Hartman and Rosen, 1986). In C3H10T1/2 cells, retinoic acid will also inhibit gap junctional communication; however, inhibition only occurs at low (about $10^{-10}$ M), or at toxic concentrations ($>10^{-6}$ M) (Mehta et al., 1989). Thus, some of these observations may be dose dependent. The reader is referred to section VII for a discussion of other potential problems associated with these studies.

## IX. RESTORATION OF GAP JUNCTIONAL COMMUNICATION INHIBITS ONCOGENESIS

A positive association between restoration of gap junctional communication and decreased cell proliferation have been described by several investigators (Mehta et al., 1986, 1989; Bignami et al., 1988b). To determine if alterations in gap junctional communication are necessary during oncogenesis, several laboratories have transfected connexins into various rodent and human tumor cells to examine whether restoration of gap junctional communication normalizes growth or tumorigenicity, or both (Eghbali et al., 1991; Mehta et al., 1991; Zhu et al., 1991, 1992; Rose et al., 1993). The results have supported the hypothesis that gap junctions can transmit growth-repressing signals. After transfection of human hepatoma cells with Cx32 (Eghbali et al., 1991; Rae et al., 1998; Ruch et al., 1998), C6 glioma cells with Cx43 (Zhu et al., 1992), and 10T1/2 fibroblasts with Cx43 (Mehta et al., 1991), all tranfectants showed at least a partial normalization of aberrant growth control *in vitro* or *in vivo*, or both. In the two tumors that arose in animals injected with transformed 10T1/2 cells transfected with Cx43, molecular analysis revealed that both had lost the transfected Cx43 gene but retained the G418 selectable marker (Rose et al., 1993). These studies suggest that transfection had restored the ability of host cells to suppress the growth of tumor cells. After transfection of rat C6 glioma cells with Cx43, alterations in insulinlike growth factor binding proteins and insulinlike growth factors were noted (Bradshaw et al., 1993), suggesting that restoration of homologous gap junctional communication can modify gene expression.

Subtractive hybridization of mRNA from normal and tumorigenic human mammary cells revealed loss of expression of connexin26 in the tumor cells, and identified it as a potential tumor suppressor gene. Southern blot analysis indicated no gross structural alteration in Cx43 or Cx26 genes in these neoplasms (Lee et al., 1991). Restoration of gap junctional function in a human mammary carcinoma cell line by stable transfection of either Cx26 or Cx43 resulted in cells with reduced growth rates, decreased capacity to form tumors in nude mice, and a partial restoration of their capacity to differentiate (Hirschi et al., 1996). Studies in HeLa cells have suggested the additional complexity that not all connexins can function to restore growth control: whereas both Cx26 and Cx43 restored gap junctional communication, only cells expressing Cx26 were found to decrease growth rates and tumorigenicity (Mesnil et al., 1995). It is interesting to speculate that, at least in HeLa cells, these differences reflect differences in permeability of connexons produced by Cx26 versus Cx43 which act to restrict passage of growth regulatory molecules in the latter case.

Reasons for the inability of tumor cells to engage in heterologous communication with surrounding normal cells are unclear. However, if it can be accepted that heterologous communication results in growth control, then it follows that only those cells losing this ability will have the capacity to proliferate and be recognized. It should be again pointed out that even communication-competent cells may have lost the ability to respond to the junctionally transmitted signal(s). Moreover, other genes may prevent restoration of gap junctional communication when connexin mRNAs are overexpressed or phenotypic or genotypic alterations in highly metastatic tumor cells may be dominant over restoration of gap junctional communication. Some clues as to reasons for lack of gap junctional communication in connexin-expressing cells come from research showing that inhibition of glycosylation allowed heterologous coupling (Wang and Mehta, 1995). Expression of cadherins allowed communication to occur between mouse hepatoma cells, but inhibited it in mouse L cells. This inhibition could be relieved by blocking glycosylation (Wang and Rose, 1997). These studies suggest that some tumor cells either fail to express appropriate adhesion molecules (cadherins) or that access to these molecules is blocked by inappropriate glycosylation. These studies strongly imply that targeted upregulation of connexin expression may be an appropriate strategy for cancer therapy but that many tumors will present additional difficulties in restoring gap junctional communication.

## X.   GROWTH REGULATORY SIGNAL HYPOTHESIS

First elegantly proposed by Loewenstein (1979), the hypothesis that growth regulatory signals traverse gap junctions and control cell population density has recently received strong support from the molecular studies described earlier, in which overexpression of connexin genes restored many aspects of the normal phenotype of transformed cells. In this model of growth control, a communicating cell population would be established containing discrete signal sources in which the extent of growth inhibition would be determined by the number of sources and the rate of signal transfer. The former could be determined by cell type (in the extreme, neoplastic cells forming fewer sources than normal cells), the latter by the number of open junctions. If the number of sources is proportional to the total number of cells, then an increase in cell number would increase the number of sources and thus inhibit additional proliferation. In principal, the production of discrete signaling sources in an originally homogeneous population is not dissimilar to the concept of pattern formation during morphogenesis (Lo, 1989). Here all cells are initially equal sources, with each source suppressing signaling in adjacent cells. Owing to random fluctuations in signaling, one cell becomes dominant within a discrete population and suppresses the signaling of other cells within that population. Within the concept of growth control, loss of signal sources as a result of injury or differentiation would lead to proliferation, whereas an increase in junctional transfer of signals would restrict the proliferation of still more cells. This is the response observed in the model C3H10T1/2 cells after increasing junctional communication by molecular means in transfection experiments (Mehta et al., 1991), or by pharmacological means using retinoids in the homologous situation (Mehta et al., 1989), and cAMP in the heterologous (normal/transformed cell) situation (Mehta et al., 1986).

Studies in C3H10T1/2 cells demonstrate that in this system, at least chemically, certain oncogene-transformed cells are capable of both generating and responding to the putative growth regulatory signal(s). They do, however, appear less capable of transmitting such signals (owing to reduced gap junctional communication) and less capable of responding to such signals (even restored gap junctional communication does not fully restore a normal saturation density). Clearly, the ability to generate and respond need not always be present in transformed cells. In view of the recent detection in human tumors of multiple aberrations in genes controlling growth, it is to be expected that not all transformed cells will have retained the ability to respond to junctionally transmitted signals. Thus, an observation of restoration of gap junctional communication without restoration of growth control should not be construed as a fatal flaw in this hypothesis.

# XI. FUTURE DIRECTIONS

## A. Chemoprevention: As an Intermediate Marker of Response

A major requirement for the rational choice of chemopreventive agent to be tested in clinical trials, and for the monitoring of such trials, is the development of intermediate markers of response. Without intermediate endpoints most clinical trials require many years before success or failure can be judged. Success may hinge as much on drug protocol as on drug choice. For two important classes of chemopreventive agents, the retinoids and the carotenoids, chemopreventive action in model C3H10T1/2 cells is tightly correlated with upregulated expression of Cx43 and gap junctional communication (Hossain et al., 1989; Zhang et al., 1991); furthermore, human skin retinoids also upregulate CX43 expression (Guo et al., 1992). We would suggest that Cx43 expression may be a useful intermediate marker of response for those chemopreventive agents that alter expression of this gene. Other members of the connexin family may have a similar utility in cancers that show consistent alterations in gap junctional communication.

## B. Toxicology: As a Predictive Assay for Tumor Promoters

The development of a test for tumor promoters has been hampered because these agents can induce cell proliferation, differentiation, or diverse phenotypic effects depending on the cell type examined. Several alternative assay systems to the classical mouse skin system have been proposed as a short-term tests for tumor promoters. Inhibition of gap junctional communication has been extensively investigated by techniques such as inhibition of metabolic cooperation or of transfer of fluorescent dye. Over 200 chemicals have been examined for their ability to interfere with metabolic communication in V79 cells. The predictive accuracy as a marker for other toxicities was as follows: teratogenicity or reproductive toxicity, 76% of 103 chemicals positive in both tests; carcinogenicity, 69% of 85 chemicals; mutagenicity, 56% of 100 chemicals; and neurotoxicity, 73% of 80 chemicals (Bohrman et al., 1988). Future studies using a panel of lines expressing different connexin proteins may be useful in determining how the cell and tissue specificity of tumor promoters correlates with their inhibition of gap junctional communication.

## C. Chemotherapy: The "Bystander Effect"

Striking success in retroviral-mediated treatment of tumors has recently been described. The procedure utilizes ganciclovir and cells producing replication-deficient murine retrovirus particles encoding herpes simplex-thymidine kinase (HStk) (Hard et al., 1990; Jyonouchi et al., 1991; Eichner et al., 1992). Ganciclovir is modified into a cytotoxic nucleotidelike precursor by HStk, thus destroying cells that produce the enzyme but not mammalian cells that do not metabolize this drug to a toxic nucleotide (Rheinwald and Beckett, 1981; Simeone et al., 1990; Beato, 1991). When HStk retrovirus-producing cells are placed in brain tumors *in vivo*, complete tumor regression can be seen in over 83% of the experimental animals when exposed to ganciclovir . However, only 10% to 70% of the tumor cells are actually transduced by the HStk gene even in the event of complete tumor regression. Furthermore, cells transfected with the HStk gene without (retroviral) infectious capability are also able to cause tumor regression

## HYPOTHESIS: SELECTIVE LOSS OF GJIC LEADS TO INCREASED PROLIFERATION AND MUTATION

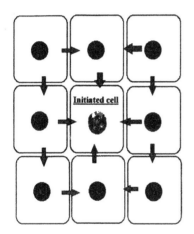 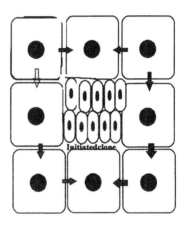

GJIC High: Proliferation inhibited in normal cells and in central initiated cell. Probability of additional mutation low.

GJIC Low: Proliferation of initiated cells leads to increased clone size. High probability of additional mutations and selection of more highly neoplastic cells

*Figure 2.* Clonal expansion of initiated cells as a consequence of decreased gap junctional communication. *Left panel:* gap junctional communication, indicated by arrows, restricts proliferation of an initiated cell. This low level of proliferation reduces the probability of DNA adducts being "fixed" as mutations, and reduces the population size of cells in which mutations can occur. In this situation, second and subsequent mutations are rare and progression to neoplasia is prevented. By stimulating gap junctional communication, the retinoids and carotenoids can act as a cancer preventive agents. *Right panel:* Decreased gap junctional communication has allowed the clonal expansion of the initiated cell population. When the clone size approaches $10^6$ cells, a second mutation resulting in oncogene activation or tumor suppressor gene inactivation can occur with high probability. This doubly mutated cell must then expand its clone size in order for subsequent mutation to occur. For additional discussion, see the text.

(Hard et al., 1990; Jyonouchi et al., 1991). This has been termed the "bystander effect" (Hard et al., 1990; Jyonouchi et al., 1991; Kaneko et al., 1991; Eichner et al., 1992). To explain how the toxic nucleotide is transferred to nontransduced cells, it was proposed and demonstrated, in at most cases, that gap junctional communication is a likely mechanism by which HStk producing cells transfer the lethal metabolite to neighboring cells (Goldberg and Bertram, 1994; Mesnil et al., 1996; Shinoura et al., 1996; Dilber et al., 1997; Vrionis et al., 1997). This phenomenon, previously termed the "kiss of death," has been noted in cell culture experiments involving mixed populations of HGPRT+ and mutant HGPRT– cells exposed to 8-azaguanine (Hooper, 1982). It may be notable that, in one case, the presence of gap junctions actually protected infected cells from death, presumably by effectively diluting the toxic nucleotide product to coupled cells, a process referred to as the "Good Samaritan effect" (Wygoda et al., 1997).

Although the bulk of evidence indicates a role for gap junctional communication in mediating the bystander effect for the HStk/ganciclovir system, other such pharmacological interventions also based upon the production of toxic nucleotide analogs *in situ* do not appear to be mediated via gap junctions (Denning and Pitts, 1997; Lawrence et al., 1998). Insofar as gap junctional communication is responsible for this phenomenon in model systems extended to clinical gene transfer studies, it may exploited to result in enhanced selective killing of tumor cells. For example, it may be possible to utilize HStk producer cells that are able to communicate with target tumor cells but not surrounding normal tissue. In addition, agents that specifically increase homologous communication between target tumor cells (Mehta et al., 1989) would be expected to augment effects of this procedure, as has recently been demonstrated *in vitro* and *in vivo* for retinoic acid (Park et al., 1997).

## D. Signal Transduction

Identification of the putative junctionally transmitted signal(s) responsible for growth control is an exciting goal with many implications for basic and clinical research. Such signals would be expected to be charged, so as to be restricted to junctional transfer, small, below about 1000 D, so as to pass the junction, and capable of being rapidly produced and inactivated. The second messengers, cAMP, inositol phosphate, and $Ca^{2+}$, which have all been demonstrated to pass the junction, satisfy these requirements (Fletcher et al., 1987a; Sáez et al., 1989; Boitano et al., 1992; Charles et al., 1992).

## REFERENCES

Albright, C.D., Grimley, P.M. , Jones, R.T., Fontana, J.A., Keenan, K.P. & Resau, J.H. (1991). Cell-to-cell communication: A differential response to TGF-β in normal and transformed (BEAS-2B) human bronchial epithelial cells. Carcinogen 12, 1993-1999.

Ames, B.N. (1989). Mutagenesis and carcinogenesis: endogenous and exogenous factors. Envir. Mol. Mutagen 14(suppl. 16), 66-77.

Ames, B.N., Shigenaga, M.K., & Gold, L.S. (1993). DNA lesions, inducible DNA repair, and cell division: three key factors in mutagenesis and carcinogenesis. Envir. Health Perspec. 101(suppl. 5), 35-44.

Arellano, R.O., Rivera, A., & Ramón, F. (1990). Protein phosphorylation and hydrogen ions modulate calcium-induced closure of gap junction channels. Biophys. J. 57, 363-367.

Atkinson, M.M., & Sheridan, J.D. (1988). Altered junctional permeability between cells transformed by v-ras, v-mos, or v-src. Am. J. Physiol. 255, C674-C683.

Azarnia, R., & Loewenstein, W.R. (1984a). Intercellular communication and the control of growth:X. Alteration of junctional permeability by the src gene. A study with temperature-sensitive mutant rous sarcoma virus. J. Memb. Biol. 82, 191-205.

Azarnia, R., & Loewenstein, W.R. (1984b). Intercellular communication and the control of growth: XI. Alteration of junctional permeability by the src gene in a revertant cell with normal cytoskeleton. J. Memb. Biol. 82, 207-212.

Azarnia, R., Mitcho, M., Shalloway, D., & Loewenstein, W.R. (1989). Junctional intercellular communication is cooperatively inhibited by oncogenes in transformation. Oncogene 4, 1161-1168.

Azarnia, R., Reddy, S., Kmiecik, T.E., Shalloway, D., & Loewenstein, W.R. (1988). The cellular src gene product regulates junctional cell-to-cell communication. Science 239, 398-401.

Barrio, L.C., Suchyna, T., Bargiello, T., Xu, L.X., Roginski, R.S., Bennett, M.V.L., & B.J. Nicholson, B.J. (1991). Gap junctions formed by connexins 26 and 32 alone and in combination are differently affected by applied voltage. Proc. Natl. Acad. Sci. U.S.A. 88, 8410-8414.

Beato, M. (1991). Transcriptional control by nuclear receptors. FASEB J. 5, 2044-2051.

Beer, D.G., Neveu, M.J., Paul, D.L., Rapp, U.R., & Pitot, H.C. (1988). Expression of the c-raf proto-oncogene, γ-glutamyltranspeptidase, and gap junction protein in rat liver neoplasms. Canc. Res. 48, 1610-1617.

Bennett, M.V.L. (1994). Connexins in disease. Nature 368, 18-19.

Bergoffen, J., Scherer, S.S., Wang, S., Oronzi Scott, M., Bone, L.J., Paul, D.L., Chen, K., Lensch, M.W., Chance, P.F., & Fischbeck, K.H. (1993). Connexin mutations in X-linked Charcot-Marie-Tooth Disease. Science 262, 2039-2042.

Berthoud, V.M., Ledbetter, M.L.S., Hertzberg, E.L., & Sáez, J.C. (1992). Connexin43 in MDCK cells: Regulation by a tumor-promoting phorbol ester and $Ca^{2+}$. Eur. J. Cell Biol. 57, 40-50.

Bertram, J.S. (1990). Role of gap junctional cell/cell communication in the control of proliferation and neoplastic transformation. Radiat. Res. 123, 252-256.

Bertram, J.S., & Faletto, M.B. (1985). Requirements for and kinetics of growth arrest of neoplastic cells by confluent 10T1/2 fibroblasts induced by a specific inhibitor of cyclic adenosine 3':5'-phosphodiesterase. Cancer Res. 45, 1946.

Bertram, J.S., Hossain, M.Z., Pung, A., & Rundhaug, J.E. (1989). Development of *in vitro* systems for chemoprevention research. Prev. Med. 18, 562-575.

Bertram, J.S., Kolonel, L.N., & Meyskens, F.L. (1987). Rationale and strategies for chemoprevention of cancer in humans. Cancer Res. 47, 3012-3031.

Bertram, J.S., Pung, A., Churley, M., Kappock, T.J.I., Wilkins, L.R., & Cooney, R.V. (1991). Diverse carotenoids protect against chemically induced neoplastic transformation. Carcinogenesis 12, 671-678.

Bevilacqua, A., Loch-Caruso, R., & Erickson, R.P. (1989). Abnormal development and dye coupling produced by antisense RNA to gap junction protein in mouse preimplantation embryos. Proc. Natl. Acad. Sci. U.S.A. 86, 5444-5448.

Beyer, E.C., Paul, D.L., & Goodenough, D.A. (1990). Connexin family of gap junction proteins. J. Membr. Biol. 116, 187-194.

Bignami, M., Rosa, S., Falcone, G., Tato, F., Katoh, F., & Yamasaki, H. (1988a). Specific viral oncogenes cause differential effects on cell-to-cell communication, relevant to the suppression of
the transformed phenotype by normal cells. Mol. Carcinogen 1, 67-75.

Bignami, M.,. Rosa, S, La Rocca, S.A., Falcone, G., & Tatò, F. (1988b). Differential influence of adjacent normal cells on the proliferation of mammalian cells transformed by the viral oncogenes
myc, ras and src. Oncogene 2, 509-514.

Bohrman, J.S., Burg, J.R., Elmore, E., Gulati, D.K., Barkfnecht, T.R., Niemeier, R.W., Dames, B.L.,
Toraason, M., & Langenbach, R. (1988). Interlaboratory studies with the Chinese hamster V79
cell metabolic cooperation assay to detect tumor-promoting agents. Environ. Mol. Mutagen
12, 33-51.

Boitano, S., Dirksen, E.R., & Sanderson, M.J. (1992). Intercellular propagation of calcium waves
mediated by inositol trisphosphate. Science 258, 292-295.

Bombick, D.W. (1990). Gap junctional communication in various cell types after chemical exposure.
In Vitro Toxicol. 3, 27-39.

Borek, C., Higashino, S., & Loewenstein, W.R. (1969). Intercellular communication and tissue
growth. IV. Conductance of membrane junctions of normal and cancerous cells in culture. J.
Membr. Biol.1, 274-293.

Bradshaw, S.L., Naus, C.C., Zhu, D., Kidder, G.M., D'Ercole, A.J., & Han, V.K. (1993). Alterations
in the synthesis of insulin-like growth factor binding proteins and insulin-like growth facors in
rat C6 glioma cells transfected with a gap junction connexin43 cDNA. Reg. Pept. 48, 99-112.

Bräuner, T. & Hülser, D.F. (1990). Tumor cell invasion and gap junctional communication. II. Normal
and malignant cells confronted in multicell spheroids. Invasion Metastasis 10, 31-48.

Brauner, T., Schmid, A., & Hulser, D.F. (1990). Tumor cell invasion and gap junctional communication. I. Normal and malignant cells confronted in monolayer cultures. Invasion Metastasis 10,
18-30.

Brissette, J.L., Kumar, N.M., Gilula, N.B., & Dotto, G.P. (1991). The tumor promoter 12-O-tetrade-
canoylphorbol-13-acetate and the ras oncogene modulate expression and phosphorylation of
gap junction proteins. Mol. Cell Biol. 11, 5364-5371.

Brownell, H.L., Whitfield, J.F., & Raptis, L. (1997). Elimination of intercellular junctional communication requires lower Ras(leu61) levels than stimulation of anchorage-independent proliferation. Cancer Detect. Prev. 21, 289-294.

Budunova, I. V., Carbajal, S., Kang, H., Viaje, A., & Slaga, T. J. (1997). Altered glucocorticoid receptor expression and function during mouse skin carcinogenesis. Mol. Carcinog. 18, 177-85.

Budunova, I.V., Carbajal. S., & Slaga, T.J (1996a). Effect of diverse tumor promoters on the expression of gap-junctional proteins connexin (Cx) 26, Cx31.1, and Cx43 in SENCAR mouse epidermis. Mol. Carcinog. 15, 202-214.

Budunova, I.V., Carbajal, S., Viaje, A., & Slaga, T.J. (1996b). Connexin expression in epidermal cell
lines from SENCAR mouse skin tumors. Mol. Carcinog. 15, 190-201.

Budunova, I.V., Williams, G.M., & Spray, D.C. (1993). Effect of tumor promoting stimuli on gap
junction permeability and connexin43 expression in ARL18 rat liver cell line. Arch. Toxicol.
67, 565-572.

Chang, C.C., Jone, C., Trosko, J.E., Peterson, A.R., & Sevanian. A. (1988a). Effect of cholesterol
epoxides on the inhibition of intercellular communication and on mutation induction in Chinese hamster V79 cells. Mutation Res. 206, 471-478.

Chang, C.C., Jone, C., Trosko, J.E., Peterson, A.R., & Sevanian, A. (1988b). Effect of cholesterol
epoxides on the inhibition of intercellular communication and on mutation induction in Chinese hamster V79 cells. Mutation Res. 206, 471-478.

Charles, A.C., Merrill, J.E., Dirksen, E.R., & Sanderson, M.J. (1991). Intercellular signaling in glial
cells: calcium waves and oscillations in response to mechanical stimulation and glutamate.
Neuron 6, 3-992.

Charles, A.C., Naus, C.C.G., Zhu, D., Kidder, G.M., Dirksen, E.R., & Sanderson, M.J. (1992). Intercellular calcium signaling via gap junctions in glioma cells. J. Cell Biol. 118, 195-201.

Clairmont, A., & Sies, H. (1997). Evidence for a posttranscriptional effect of retinoic acid on connexin43 gene expression via the 3'-untranslated region. FEBS Lett. 419, 268-70.

Cohen, S.M., & Ellwein, L.B. (1991). Genetic errors, cell proliferation, and carcinogenesis. Cancer Res. 51, 6493-6505.

Crow, D.S., Beyer, E.C., Paul, D.L., Kobe, S.S., & Lau, A.F. (1990). Phosphorylation of connexin43 gap junction protein in uninfected and Rous sarcoma virus-transformed mammalian fibroblasts. Mol. Cell. Biol. 10, 1754-1763.

Dahl, G., Levine, E., Rabadan, D.C., & Werner, R. (1991). Cell/cell channel formation involves disulfide exchange. Eur. J. Biochem. 197, 141-144.

De Feijter, A.W., Ray, J.S., Weghorst, C.M., Klaunig, J.E., Goodman, J.I., Chang, C.C., Ruch, R.J., & Trosko, J.E. (1990). Infection of rat liver epithelial cells with v-Ha-ras: correlation between oncogene expression, gap junctional communication, and tumorigenicity. Mol. Carcinog. 3, 54-67.

De Feijter, A.W., Trosko, J.E., Krizman, D.B., Lebovitz, R.M., & Lieberman, M.W. (1992). Correlation of increased levels of Ha-ras T24 protein with extent of loss of gap junction function in rat liver epithelial cells. Mol. Carcinog. 5, 205-212.

De Mello, W.C. (1984). Effect of intracellular injection of cAMP on the electrical coupling of mammalian cardiac cells. Biochem. Biophy. Res. Commun. 119, 1001-1007.

DeCaprio, J.A., Ludlow, J.W., Lynch, D., Furukawa, Y., Griffin, J., Piwnica-Worms, H., Huang, C.-M., & Livingston, D.M. (1989). The product of the retinoblastoma susceptibility gene has properties of a cell cycle regulatory element. Cell 58, 1085-1095.

Denning, C., & Pitts, J. D. (1997). Bystander effects of different enzyme-prodrug systems for cancer gene therapy depend on different pathways for intercellular transfer of toxic metabolites, a factor that will govern clinical choice of appropriate regimes [see comments]. Hum. Gene Ther. 8, 1825-1835.

Dermietzel, R., Traub, O., Hwang, T.K., Beyer, E., Bennett, M.V.L., Spray, D.C., & Willecke, K. (1989). Differential expression of three gap junction proteins in developing and mature brain tissues. Proc. Natl. Acad. Sci. U.S.A. 86, 10148-10152.

Dermietzel, R., Yancey, S.B., Traub, O., Willecke, K., & Revel, J.P. (1987). Major loss of the 28-kD protein of gap junction in proliferation hepatocytes. J. Cell Biol. 105, 1925-1934.

Dilber, M.S., Abedi, M.R., Christensson, B., Bjorkstrand, B., Kidder, G.M., Naus, C.C., Gahrton, G., & Smith, C.I. (1997). Gap junctions promote the bystander effect of herpes simplex virus thymidine kinase *in vivo*. Cancer Res. 57, 1523-8.

Dotto, G.P., El-Fouly, M.H., Nelson, C., & Trosko, J.E. (1989). Similar and synergistic inhibition of gap-junctional communication by ras transformation and tumor promoter treatment of mouse primary keratinocytes. Oncogene 4, 637-641.

Eghbali, B., Kessler, J.A., Reid, L.M., Roy, C., & Spray, D.C. (1991). Involvement of gap junctions in tumorigenesis: transfection of tumor cells with connexin 32 cDNA retards growth *in vivo*. Proc. Natl. Acad. Sci. U.S.A. 88, 10701-10705.

Eichner, R., Kahn, M., Capetola, R.J., Gendimenico, G.J., & Mezick, J.A. (1992). Effects of topical retinoids on cytoskeletal proteins: implications for retinoid effects on epidermal differentiation. J. Invest. Dermatol. 98, 154-161.

El-Sabban, M.E., & Pauli, B.U. (1991). Cytoplasmic dye transfer between metastatic tumor cells and vascular endothelium. J. Cell Biol. 115, 1375-1382.

Elias, P.M., Grayson, S., Caldwell, T.M., & McNutt, N.S. (1980). Gap junction proliferation in retinoic acid-treated human basal cell carcinoma. Lab. Invest. 42, 469-474.

Enomoto, T., Sasaki, Y., Shiba, Y. Kanno, Y., & Yamasaki, H. (1981). Tumor promoters cause a rapid and reversible inhibition of the formation and maintenance of electrical cell coupling in culture. Proc. Natl. Sci. U.S.A. 78, 5628-5632.

Enomoto, T., & Yamasaki, H. (1985). Rapid inhibition of intercellular communication between BALB/c 3T3 cells by diacylglycerol, a possible endogenous functional analogue of phorbol esters. Cancer Res. 45, 3706-3710.

Fairweather, N., Bell, C., Cochrane, S., Chelly, J., Wang, S., Mostacciuolo, M.L., Monaco, A.P., & Haites, N.E. (1994). Mutations in the connexin 32 gene in X-linked dominant Charcot-Marie-tooth disease. Hum. Mol. Genet. 3, 29-34.

Farber, E. (1984). Cellular biochemistry of the stepwise development of cancer with chemicals. Cancer Res. 44, 5463-5474.

Filson, A.J., Azarnia, R., Beyer, E.C., Loewenstein, W.R., & Brugge, J.S. (1990). Tyrosine phosphorylation of a gap junction protein correlates with inhibition of cell-to-cell communication. Cell Growth Differen. 1, 661-668.

Fitzgerald, D.J., Knowles, S.E., Ballard, F.J., & Murray, A.W. (1983). Rapid and reversible inhibition of junctional communication by tumor promoters in a mouse cell line. Cancer Res. 43, 3614-3618.

Fitzgerald, D.J., Mesnil, M., Oyamada, M., Tsuda, H., Ito, N., & Yamasaki, H. (1989). Changes in gap junction protein (connexin 32) gene expression during rat liver carcinogenesis. J. Cell Biochem. 41, 97-102.

Flaxman, B.A., & Cavoto, F.V. (1973). Low-resistance junctions in epithelial outgrowths from normal and cancerous epidermis in vitro. J. Cell Biol. 58, 219-223.

Fletcher, W.H., Byus, C.V., & Walsh, D.A. (1987a). Receptor-mediated action without receptor occupancy: a function for cell-cell communication in ovarian follicles. Adv. Exp. Med. Biol. 219, 299-323.

Fletcher, W.H., Shiu, W.W., Ishida, T.A., Haviland, D.L., & Ware, C.F. (1987b). Resistance to the cytolytic action of lymphotoxin and tumor necrosis factor coincides with the presence of gap junctions uniting target cells. J. Immunol.139, 956-962.

Flodstrom, S., Warngard, L., Ljungquist, S., & Ahlborg, U.G. (1988). Inhibition of metabolic cooperation in vitro and enhancement of enzyme altered foci incidence in rat liver by the pyrethroid insecticide fenvalerate. Arch. Toxicol. 61, 218-223.

Fraser, S.E., Green, C.R., Bode, H.R., & Gilula, N.B. (1987). Selective disruption of gap junctional communication interferes with a patterning process in hydra. Science 237, 49-55.

Friedman, E.A., & Steinberg, M. (1982). Disrupted communication between late-stage premalignant human colon epithelial cells by 12-O-tetradecanoylphorbol-13 acetate. Cancer Res. 42,5096-5105.

Fürstenberger, G., Rogers, M., Schnapke, R., Bauer, G., Höfler, P., & Marks. F. (1989). Stimulatory role of transforming growth factors in multistage skin carcinogenesis: possible explanation for the tumor-inducing effect of wounding in initiated NMRI mouse skin. Int. J. Cancer 43, 915-921.

Gibson, D.F.C., Hossain, M.Z., Goldberg, G.S., Acevedo, P., & Bertram, J.S. (1994). The mitogenic effects of transforming growth factor-B1 and 2 in 10T1/2 cells occur in the presence of enhanced gap junctional communication. Cell Growth Differen. 5,687-696.

Gimlich, R.L., Kumar, N.M., & Gilula, N.B. (1990). Differential regulation of the levels of three gap junction mRNAs in Xenopus embryos. J. Cell Biol. 110, 597-605.

Goldberg, G.,& Bertram, J.S. (1994). Correspondence re: Z. Ram et al., In situ retroviral-mediated gene transfer for the treatment of brain tumors in rats. Cancer Res. 53, 83-88, 1993; Cancer Res. 54, 3947-3948.

Gordon, R.E., Lane, B.P., & Marin, M. (1982). Regeneration of rat tracheal epithelium: changes in gap junctions during specific phases of the cell cycle. Exp. Cell Res. 3, 47-56.

Guo, H., Acevedo, P., Parsa, D.F., & Bertram, J.S. (1992). The gap-junctional protein connexin 43 is expressed in dermis and epidermis of human skin: differential modulation by retinoids. J. Invest. Dermatol. 99, 460-467.

Hamada, J., Takeichi, N., & Kobayashi, H. (1988). Metastatic capacity and intercellular communication between normal cells and metastatic cell clones derived from a rat mammary carcinoma. Cancer Res. 48, 5129-5132.

Hard, T., Kellenbach, E., Boelens, R., Maler, B.A., Dahlman, K., Freedman, L.P., Carlstedt-Duke, J., Yamamoto, K.R., Gustaffsson, J.-A., & Kaptein, R. (1990). Solution structure of the glucocorticoid receptor DNA-binding domain. Science 249, 157-160.

Hartman, T.G., & Rosen, J.D. (1986). The effect of anti-promoters and calcium antagonists on V-79 Chinese hamster lung fibroblasts exposed to phorbol myristate acetate. Carcinogenesis 7, 361-364.

Hayashi, T., Nomata, K., Chang, C.C., Ruch, R.J., &Hayashi, T., Nomata, K., Chang, C.C., Ruch, R.J., & Trosko, J.E. (1998). Cooperative effects of v-myc and c-Ha-ras oncogenes on gap junctional intercellular communication and tumorigenicity in rat liver epithelial cells. Cancer Lett. 128, 145-54.

Hendrix, E.M., Lomneth, C.S., Wilfinger, W.W., Hertzberg, E.L., Mao, S.J.T., Chen, L, & Larsen, W.J. (1992). Quantitative immunoassay of total cellular gap junction protein connexin32 during liver regeneration using antibodies specific to the COOH-terminus. Tissue Cell 24, 61-73.

Hennemann, H., Suchyna, T., Lichtenberg-Fraté, H., Jungbluth, S., Dahl, E., Schwarz, J., Nicholson, B.J., & Willecke, K. (1992). Molecular cloning and functional expression of mouse connexin40, a second gap junction gene preferentially expressed in lung. J. Cell Biol. 117, 1299-1310.

Herschman, H.R. & Brankow, D.W. (1986). Ultraviolet irradiation transforms C3H10T1/2 cells to a unique, suppressible phenotype. Science 234, 1385-1388.

Herschman, H.R., & Brankow, D.W. (1987). Colony size, cell density and nature of the tumor promoter are critical variables in expression of a transformed phenotype (focus formation) in co-cultures of UV-TDTx and C3H10T1/2 cells. Carcinogenesis 8, 993-998.

Hirschi, K.K., Xu, C.E., Tsukamoto, T., & Sager, R. (1996). Gap junction genes Cx26 and Cx43 individually suppress the cancer phenotype of human mammary carcinoma cells and restore differentiation potential. Cell Growth Differ. 7, 861-870.

Hixson, D.C., McEntire, K.D., & Obrink, B. (1985). Alterations in the expression of a hepatocyte cell adhesion molecule by transplantable rat hepatocellular carcinomas. Cancer Res. 45, 3742-3749.

Holden, P.R., McGuire, B., Stoler, A., Balmain, A., & Pitts, J.D. (1997). Changes in gap junctional intercellular communication in mouse skin carcinogenesis. Carcinogenesis 18, 15-21.

Hong, W.K., Lippman, S.M., Itri, L.M., Karp, D.D., Lee, J.S., Byers, R.M., Schantz, S.P., Kramer, A., Lotan, R., Peters, L.J., Dimery, I.W., Brown, B.W., & Goepfert, H. (1990). Prevention of second primary tumors with isotretinoin in squamous-cell carcinoma of the head and neck. N. Eng. J. Med. 323, 795-800.

Hooper, M.L. (1982). Metabolic co-operation between mammalian cells in culture. Biochim. Biophys. Acta 651, 85-103.

Hossain, M.Z., & Bertram, J.S. (1994). Retinoids suppress proliferation, induce cell spreading, and up-regulate connexin43 expression only in postconfluent 10T1/2 cells: implications for the role of gap junctional communication. Cell Growth Differen. 5, 1253-1261.

Hossain, M.Z., Wilkens, L.R., Mehta, P.P., Loewenstein, W., & Bertram, J.S. (1989). Enhancement of gap junctional communication by retinoids correlates with their ability to inhibit neoplastic transformation. Carcinogenesis 10, 1743-1748.

Jamakosmanovic, K.E., & Loewenstein, W.R. (1968). Intercellular communication and tissue growth. III. Thyroid cancer. J. Membr. Biol. 38, 556-561.

Jansen, L.A.M., Mesnil, M., & Jongen, W.M.F. (1996). Inhibition of gap junctional intercellular communication and delocalization of the cell adhesion molecule E-cadherin by tumor promoters. Carcinogenesis 17, 1527-1531.

Jongen, W.M., Fitzgerald, D.J., Asamoto, M., Piccoli, C., Slaga, T.J., Gros, D., Takeichi, M., & Yamasaki, H. (1991). Regulation of connexin 43-mediated gap junctional intercellular communication by Ca$^{2+}$ in mouse epidermal cells is controlled by E-cadherin. J. Cell Biol. 114, 545-555.

Jursnich, V.A., Fraser, S.E., Held,Jr., L.I., Ryerse, J., & Bryant, P.J. (1990). Defective gap-junctional communication associated with imaginal disc overgrowth and degeneration caused by mutations of the dco gene in Drosophila. Dev. Biol. 140, 413-429.

Jyonouchi, H., Hill, R.J., Tomita, Y., & Good , R.A. (1991). Studies of immunomodulating actions of carotenoids. I. Effects of beta carotene and astaxanthin on murine lymphocyte functions and cell surface marker expression in in vitro culture system. Nutr. Cancer 16, 93-105.

Kalimi, G.H., Hampton, L.L., Trosko, J.E., Thorgeirsson, S.S., & Huggett, A.C. (1992). Homologous and heterologous gap-junctional intercellular communication in v-raf-, v-myc-, and v-raf/v-myc-transduced rat liver epithelial cell lines. Mol. Carcinog. 5, 301-310.

Kalimi, G.H., & Sirsat, S.M. (1984). Phorbol ester tumor promoter affects the mouse epidermal gap junctions. Cancer Lett. 22, 343-350.

Kam, E., & Pitts, J.D. (1989). Tissue-specific regulation of junctional communication in the skin of mouse fetuses homozygous for the repeated epilation (Er) mutation. Development 107, 923-929.

Kam, E., Watt, F.M., & Pitts, J.D. (1987). Patterns of junctional communication in skin: studies on cultured keratinocytes. Exp. Cell Res. 173, 431-438.

Kamibayashi, Y., Oyamada, Y., Mori, M., & Oyamada, M. (1995). Aberrant expression of gap junction proteins (connexins) is associated with tumor progression during multistage mouse skin carcinogenesis in vivo. Carcinogenesis 16, 1287-1298.

Kaneko, S., Kagechika, H., Kawachi, E., Hashimoto, Y., & Shudo, K. (1991). Retinoid antagonists. Med Chem Res. 1, 220-225.

Kanno, Y., & Matsui, Y. (1968). Cellular uncoupling in cancerous stomach epithelium. Nature 218, 775-776.

Kelloff, G.J., Boone, C.W., Malone, W.F., & Steele, V.E. (1992). Chemoprevention clinical trials. Mutat. Res. Fundam. Mol. Mech. Mutagen. 267, 291-295.

Klann, R.C., Fitzgerald, D.J. Piccoli, C., Slaga, T.J., & Yamasaki, H. (1989). Gap-junctional intercellular communication in epidermal cell lines from selected stages of SENCAR mouse skin carcinogenesis. Cancer Res. 49, 699-705.

Klaunig, J.E., & Ruch, R.J. (1987a). Strain and species effects on the inhibition of hepatocyte intercellular communication by liver tumor promoters. Cancer Lett. 3, 161-168.

Klaunig, J.E., & Ruch. R.J. (1987b). Role of cyclic AMP in the inhibition of mouse hepatocyte intercellular communication by liver tumor promoters. Toxicol. Appl. Pharmacol. 91, 159-170.

Klaunig, J.E., & Ruch, R.J. (1990). Role of inhibition of intercellular communication in carcinogenesis. Lab. Invest. 62, 135-146.

Knudson, A.G., Jr. (1989). Hereditary cancers: clues to mechanisms of carcinogenesis. Br. J. Cancer 59, 661-666.

Kopan, R., Traska, G., & Fuchs, E. (1987). Retinoids as important regulators of terminal differentiation: examining keratin expression in individual epidermal cells at various stages of keratinization. J Cell Biol. 105, 427-440.

Kren, B.T., Kumar, N.M., Wang, S., Gilula, N.B., & Steer, C.J. (1993). Differential regulation of multiple gap junction transcripts and proteins during rat liver regeneration, J. Cell Biol. 123, 707-718.

Krutovskikh, V.A., Oyamada, M., & Yamasaki, H. (1991). Sequential changes of gap-junctional intercellular communications during multistage rat liver carcinogenesis: direct measurement of communication in vivo. Carcinogenesis 12, 1701-1706.

Kuerbitz, S.J., Plunkett, B.S., Walsh, W.V., & Kastan, M.B (1992). Wild-type p53 is a cell cycle checkpoint determinant following irradiation. Proc. Natl. Acad. Sci. U.S.A. 89, 7491-7495.

Laird, D.W., Yancey, S.B., Bugga, L., & Revel, J.-P. (1992). Connexin expression and gap junction communication compartments in the developing mouse limb. Devel. Dyn. 195, 153-161.

Larsen, W.J. (1977). Structural diversity of gap junctions. A review. Tissue Cell 9, 373-394.

Larsen, W.J. (1983). Biological implications of gap junction structure, distribution, and composition: a review. Tissue Cell 15, 645-671.

Larsen, W.J., Wert, S.E., & Brunner, G.D (1986). A dramatic loss of cumulus cell gap junctions is correlated with germinal vesicle breakdown in rat oocytes. Dev. Biol. 113, 517-521.

Larson, D.M., & Haudenschild, C.C. (1988). Junctional transfer in wounded cultures of bovine aortic endothelial cells. Lab. Invest. 59, 373-379.

Lau, A.F., Kanemitsu, M.Y., Kurata, W.E., Danesh, S., & Boynton, A.L. (1992). Epidermal growth factor disrupts gap-junctional communication and induces phosphorylation of connexin43 on serine. Mol. Biol. Cell 3, 865-874.

Lawrence, T.S., Rehemtulla, A., Ng, E.Y., Wilson, M., Trosko, J.E., & Stetson, P.L. (1998). Preferential cytotoxicity of cells transduced with cytosine deaminase compared to bystander cells after treatment with 5-flucytosine. Cancer Res. 58, 2588-93.

Lee, S., Gilula, N.B., & Warner, A.E. (1987). Gap junctional communication and compaction during preimplantation stages of mouse development. Cell 51, 851-860.

Lee, S.W., Tomasetto, C., Paul, D., Keyomarsi, K., & Sager, R. (1992). Transcriptional down-regulation of gap-junction proteins blocks junctional communication in human mammary tumor cell lines. J. Cell Biol. 118, 1213-1221.

Lee, S.W., Tomasetto, C., & Sager, R. (1991). Positive selection of candidate tumor-suppressor genes by subtractive hybridization. Proc. Natl. Acad. Sci. U.S.A. 88, 2825-2829.

Letourneau, R.J., Li, J.J., Rosen, S., & Villee, C.A (1975). Junctional specialization in estrogen-induced renal adenocarcinomas of the golden hamster. Cancer Res. 35, 6-10.

Levine, A.J., Momand, J., & Finlay, C.A. (1991). The p53 tumour suppressor gene. Nature 351, 453-456.

Ljungquist, S., & Warngard, L. (1990). Intercellular communication: A useful test for tumour promoters. Acta Physiol. 140, 41-45.

Lo, C.W. (1989). Communication compartments: a conserved role in pattern formation?. In: *Cell Interactions and Gap Junctions* (Sperelakis, N., & Cole, W.C., eds.), pp. 86-96. CRC Press, Boca Raton, FL.

Loch-Caruso, R., Caldwell, V., Cimini, M., & Juberg, D. (1990). Comparison of assays for gap junctional communication using human embryocarcinoma cells exposed to dieldrin. Fundam. Appl. Toxicol. 15, 63-74.

Loeb, L.A. (1989). Endogenous carcinogenesis: molecular oncology into the twenty-first century— presidential address. Cancer Res. 49, 5489-5496.

Loewenstein, W.R. (1979). Junctional intercellular communication and the control of growth. Biochim. Biophys. Acta 560, 1-65.

Loewenstein, W.R., & Kanno, Y. (1966). Intercellular communication and the control of tissue growth: lack of communication between cancer cells. Nature 209, 1248-1249.

Loewenstein, W.R., & Rose, B. (1992). The cell-cell channel in the control of growth. Sem. Cell Biol. 3, 59-79.

Loo, L.W.M., Berestecky, J.M., Kanemitsu, M.Y., & Lau, A.F. (1995). Pp60$^{src}$-mediated phosphorylation of connexin 43, a gap junction protein. J. Biol. Chem. 270, 12751-12761.

Madhukar, B.V., Oh, S.Y., Chang, C.C., Wade, M., & Trosko, J.E. (1989). Altered regulation of intercellular communication by epidermal growth factor, transforming growth factor-$\beta$ and peptide hormones in normal human keratinocytes. Carcinogenesis 10, 13-20.

Malcolm, A.R., & Mills, L.J. (1987). The potential role of bioactivation in tumor promotion: indirect evidence from effects of phenol, sodium cyclamate and their metabolites on metabolic cooperation *in vitro*. In: Biochemical Mechanisms and Regulation of Intercellular Communication

(Milman, H.A., & Elmore, E., eds.), pp. 237-23?. Princeton Scientific Publishing, Princeton, NJ.

Maldonado, P.E., Rose, B., & Loewenstein, W.R. (1988). Growth factors modulate junctional cell-to-cell communication. J. Membr. Biol. 106, 203-210.

Martin, W., Zempel, G., Hülser, D., & Willecke, K. (1991). Growth inhibition of oncogene-transformed rat fibroblasts by cocultured normal cells: relevance of metabolic cooperation mediated by gap junctions. Cancer Res. 51, 5348-5354.

McNutt, S.N. (1969). Carcinoma of the cervix: deficiency of nexus intercellular junctions. Science 165, 597-598.

McNutt, S.N., Hershberg, R.A., & Weinstein, R.S. (1971). Further observations on the occurrence of nexuses in benign and malignant human cervical epithelium. J. Cell Biol. 51, 805-825.

Mehta, P.P., Bertram, J.S., & Loewenstein, W.R. (1986). Growth inhibition of transformed cells correlates with their junctional communication with normal cells. Cell 44, 187-196.

Mehta, P.P., Bertram, J.S., & Loewenstein, W.R. (1989). The actions of retinoids on cellular growth correlate with their actions on gap junctional communication. J. Cell Biol. 108, 1053-1065.

Mehta, P.P., Hotz Wagenblatt, A., Rose, B., Shalloway, D., & Loewenstein, W.R. (1991). Incorporation of the gene for a cell-cell channel protein into transformed cells leads to normalization of growth. J. Membr. Biol. 124, 207-225.

Meiners, S., Xu, A., & Schindler, M. (1991). Gap junction protein homologue from Arabidopsis thaliana: evidence for connexins in plants. Proc. Natl. Acad. Sci. U.S.A. 88, 4119-4122.

Mesnil, M., Fitzgerald, D.J., & Yamasaki, H. (1988). Phenobarbital specifically reduces gap junction protein mRNA level in rat liver. Mol. Carcinog. 1, 79-81.

Mesnil, M., Krutovskikh, V., Piccoli, C., Elfgang, C., Traub, O., Willecke, K., & Yamasaki, H. (1995). Negative growth control of HeLa cells by connexin genes: connexin species specificity. Cancer Res. 55, 629-639.

Mesnil, M., Piccoli, C., Tiraby, G., Willecke, K., & Yamasaki, H. (1996). Bystander killing of cancer cells by herpes simplex virus thymidine kinase gene is mediated by connexins. Proc. Natl. Acad. Sci. U.S.A. 93, 1831-1835.

Mesnil, M., & Yamasaki, H. (1988). Selective gap-junctional communication capacity of transformed and non-transformed rat liver epithelial cell lines. Carcinogenesis 9, 1499-1502.

Michalopoulos, G.K. (1990). Liver regeneration: molecular mechanisms of growth control. FASEB J. 4, 176-187.

Mittnacht, S., & Weinberg, R.A. (1991). G1/S phosphorylation of the retinoblastoma protein is associated with an altered affinity for the nuclear compartment. Cell 65, 381-393.

Miyashita, T., Takeda, A., Iwai, M., & Shimazu, T. (1991). Single administration of hepatotoxic chemicals transiently decreases the gap-junction-protein levels of connexin 32 in rat liver. Eur. J. Biochem. 196, 37-42.

Mordan, L.J., Martner, J.E., & Bertram, J.S. (1983). Quantitative neoplastic transformation of C3H/10T½ fibroblasts: dependence upon the size of the initiated cell colony at confluence. Cancer Res. 43, 4062-4067.

Moreno, A.P., Fishman, G.I., & Spray, D.C. (1992). Phosphorylation shifts unitary conductance and modifies voltage dependent kinetics of human connexin43 gap junction channels. Biophys. J. 62, 51-53.

Muir, J.G., & Murray, A.W. (1986). Mimicry of phorbol ester responses by diacylglycerols. Differential effects on phosphatidylcholine biosynthesis, cell-cell communication and epidermal growth factor binding. Bioch. Biophys. Acta, 885, 176-184.

Murray, A.W., & Fitzgerald, D.J. (1979). Tumor promoters inhibit metabolic cooperation in cocultures of epidermal and 3T3 cells. Biochem. Biophys. Res. Comm. 91, 395-401.

Musil, L.S., Cunningham, B.A., Edelman, G.M., & Goodenough, D.A. (1990). Differential phosphorylation of the gap junction protein connexin43 in junctional communication-competent and -deficient cell lines. J. Cell Biol. 111, 2077-2088.

Musil, L.S., & Goodenough, D.A. (1993). Multisubunit assembly of an integral plasma membrane channel protein, gap junction connexin43, occurs after exit from the ER. Cell 74, 1065-1077.

Neveu, M.J., Babcock, K.L., Hertzberg, E.L., Paul, D.L., Nicholson, B.J., & Pitot, H.C. (1994a). Colocalized alterations in connexin32 and cytochrome P450IIB1/2 by phenobarbital and related liver tumor promoters. Cancer Res. 54, 12: 3145-3152.

Neveu, M.J., Hully, J.R., Babcock, K.L., Hertzberg, E.L., Nicholson, B.J., Paul, D.L., & Pitot, H.C. (1994b). Multiple mechanisms are responsible for altered expression of gap junction genes during oncogenesis in rat liver. J. Cell Sci. 107, 83-95.

Neveu, M.J., Hully, J.R., Babcock, K.L., Vaughan, J., Hertzberg, E.L. Nicholson, B.J., Paul, D.L., & Pitot, H.C. (1995). Proliferation-associated differences in the spacial and temporal expression of gap junction genes in rat liver. Hepatology 22(1), 202-212.

Neveu, M.J., Hully, J.R., Paul, D.L., & Pitot, H.C. (1990). Reversible alteration in the expression of the gap junctional protein connexin 32 during tumor promotion in rat liver and its role during cell proliferation. Cancer Commun. 2, 21-31.

Neveu, M.J., Sattler, C.A., Sattler, G.L., Hully, J.R., Hertzberg, E.L., Paul, D.L., Nicholson, B.J., & Pitot, H.C. (1994). Differences in the expression of connexin genes in rat hepatomas *in vivo* and *in vitro*. Mol. Carcinog. 11(3), 145-154.

Nicolson, G.L., Dulski, K.M., &Trosko, J.E. (1988). Loss of intercellular junctional communication correlates with metastatic potential in mammary adenocarcinoma cells. Proc. Natl. Acad. Sci. U.S.A. 85, 473-476.

Nicolson, G.L., Gallick, G.E., Dulski, K.M., Spohn, W.H., Lembo, T.M., & Tainsky, M.A. (1990). Lack of correlation between intercellular junctional communication, p21rasEJ expression, and spontaneous metastatic properties of rat mammary cells after transfection with c-H-rasEJ or neo genes. Oncogene 5, 747-753.

Nishino, H., Iwashima, A., Fujiki, H., & Sugimura, T. (1984). Inhibition by quercetin of the promoting effect of teleocidin on skin papilloma formation in mice initiated with 7,12-dimethylbenz(a)anthracene. Gann 75, 113-116.

Noda, K., Umeda, M., & Oso, T. (1981). Effects of various chemicals including bile acids and chemical carcinogens on the inhibition of metabolic cooperation. Gann 72, 772-776.

Oelze, I., Kartenbeck, J., Crusius, K., & Alonso, A. (1995). Human papillomavirus type 16 E5 protein affects cell-cell communication in an epithelial cell line. J. Virol. 69, 4489-4494.

Oyamada, M., Krutovskikh, V.A, Mesnil, M., Partensky, C., Berger, F., & Yamasaki, H. (1990). Aberrant expression of gap junction gene in primary human hepatocellular carcinomas: increased expression of cardiac-type gap junction gene connexin 43. Mol. Carcinog. 3, 273-278.

Park, J.Y., Elshami, A.A., Amin, K., Rizk, N., Kaiser, L.R., & Albelda, S.M. (1997). Retinoids augment the bystander effect *in vitro* and *in vivo* in herpes simplex virus thymidine kinase/ganciclovir-mediated gene therapy. Gene Ther. 4, 909-917.

Pauli, B.U., & Weinstein, R.S. (1981). Structure of gap junctions in cultures of normal and neoplastic bladder epithelial cells. Experientia 37, 1981.

Pelletier, D.B., & Boynton, A.L. (1994). Dissociation of PDGF receptor tyrosine kinase activity from PDGF-mediated inhibition of gap junctional communication. J. Cell. Physiol. 158, 427-434.

Pepper, M.S., Spray, D.C., Chanson, M., Montesano, R., Orci, L., & Meda, P. (1989). Junctional communication is induced in migrating capillary endothelial cells. J. Cell Biol. 109, 3027-3038.

Pitot, H.C., Dragan, Y.P., Neveu, M.J., Rizvi, T.A., Hully, J.R., & Campbell, H.A. (1991). Chemicals, cell proliferation, risk estimation, and multistage carcinogenesis. In: *Chemically Induced Cell Proliferation: Implications for Risk Assessment, Progress in Clinical and Biological Research,* pp. 517-532.

Pitts, J., Kam, E., Melville, L., & Watt, F.M. (1987). Patterns of junctional communication in animal tissues. Ciba Found. Symp.125, 140-153.

Powell, S.M., Zilz, N., Beazer-Barclay, Y., Bryan, T.M., Hamilton, S.R., Thibodeau, Vogelstein, B., & Kinzler, K.W. (1992). APC mutations occur early during colorectal tumorigenesis. Nature 359, 235-237.

Prutkin, L. (1975). Mucous metaplasia and gap junctions in the vitamin A acid-treated skin tumor, keratoacanthoma. Cancer Res. 35, 364-369.

Puranam, K.L., Laird, D.W., & Revel, J. (1993). Trapping an intermediate form of connexin43 in the golgi. Exp. Cell Res. 206, 85-92.

Radner, B., & Kennedy, A. (1984). Effects of agents known to antagonize the enhancement of *in vitro* transformation by 12-tetradecanoyl-phorbol-13-acetate (TPA) on the TPA suppression of metabolic cooperation. Cancer Lett. 25, 139-144.

Rae, R.S., Mehta, P.P., Chang, C.C., Trosko, J.E., & Ruch, R.J. (1998). Neoplastic phenotype of gap-junctional intercellular communication-deficient WB rat liver epithelial cells and its reversal by forced expression of connexin 32. Mol. Carcinog. 22, 120-7.

Revel, J.-P. (1988). The oldest multicellular animal and its junctions. In: *Gap Junctions* (Hertzberg, E., ed.), pp. 135-139. Alan R. Liss, New York.

Rheinwald, J.G., & Beckett, M.A. (1981). Tumorigenic keratinocyte lines requiring anchorage and fibroblast support cultured from human squamous cell carcinomas. Cancer Res. 41, 1657-1663.

Rivedal, E., Yamasaki, H., & Sanner, T. (1994). Inhibition of gap junctional intercellular communication in Syrian hamster embryo cells by TPA, retinoic acid and DDT. Carcinogenesis 5, 689-694.

Robenek, H., Rassat, J., & Themann, H. (1981). A quantitative freeze-fracture analysis of gap and tight junctions in the normal and cholestatic human liver. Vichows Arch 38, 39-56.

Rogers, M., Berestecky, J.M., Hossain, M.Z., Guo, H.M., Kadle, R., Nicholson, B.J., & Bertram, J.S. (1990). Retinoid-enhanced gap junctional communication is achieved by increased levels of connexin 43 mRNA and protein. Mol. Carcinog. 3, 335-343.

Rose, B., Mehta, P.P., & Loewenstein, W.R. (1993). Gap-junction protein gene suppresses tumorigenicity. Carcinogenesis 14, 1073-1075.

Ruch, R.J., Cesen-Cummings, K., & Malkinson, A.M.(1998). Role of gap junctions in lung neoplasia. Exp. Lung Res. 24, 523-539.

Ruch, R.J., & Klaunig, J.E. (1986). Effects of tumor promoters, genotoxic carcinogens and hepatocytotoxins on mouse hepatocyte intercellular communication. Cell Biol. 2, 469-483.

Ruch, R.J., & Klaunig, J.E. (1988a). Inhibition of mouse hepatocyte intercellular communication by paraquat-generated oxygen free radicals. Toxicol. Appl. Pharmacol. 94, 427-436.

Ruch, R.J., & Klaunig, J.E. (1988b). Kinetics of phenobarbital inhibition of intercellular communication in mouse hepatocytes. Cancer Res. 48, 2519-2523.

Ruch, R.J., Klaunig, J.E., Kerckaert, G.A., & LeBoeuf, R.A. (1990). Modification of gap junctional intercellular communication by changes in extracellular pH in Syrian hamster embryo cells. Carcinogenesis 11, 909-913.

Ruch, R.J., Madhukar, B.V., Trosko, J.E., & Klaunig, J.E. (1993). Reversal of ras-induced inhibition of gap-junctional intercellular communication, transformation, and tumorigenesis by lovastatin. Mol. Carcinog. 7, 50-59.

Rundhaug, J.E., Gimenez-Conti, I., Stern, M.C., Budunova, I.V., Kiguchi, K., Bol, D.K., Coghlan, L.G., Conti, C.J., DiGiovanni, J., Fischer, S.M., Winberg, L.D., & Slaga, T.J. (1997). Changes in protein expression during multistage mouse skin carcinogenesis. Mol. Carcinog. 20, 125-136.

Rutten, A.A., Jongen, W.M., de-Haan, L.H., Hendriksen, E.G., & Koeman, J.H. (1988). Effect of retinol and cigarette-smoke condensate on dye-coupled intercellular communication between hamster tracheal epithelial cells. Carcinogenesis 9, 315-320.

Sáez, J.C., Bennett, M.V.L., & Spray, D.C. (1987). Carbon tetrachloride hepatoxic levels blocks reversibly gap junctions between rat hepatocytes. Science 236, 967-969.

Sáez, J.C., Connor, J.A., Spray, D.C., & Bennett, M.V.L. (1989). Hepatocyte gap junctions are permeable to the second messenger, inositol 1,4,5-trisphosphate, and to calcium ions. Proc. Nat. Acad. Sci. U.S.A. 86, 2708-2712.

Sáez, J.C., Nairn, A.C., Czernik, A.J., Spray, D.C., Hertzberg, E.L., Greengard, P., & Bennett, M.V.L. (1990). Phosphorylation of connexin32, a hepatocyte gap-junctional protein, by cAMP-dependent protein kinase, protein kinase C and $Ca^{2+}$/calmodulin-dependent protein kinase II. Eur J. Cell Biol. 192, 263-273.

Sakamoto, H., Oyamada, M., Enomoto, K., & Mori, M. (1992). Differential changes in expression of gap junction proteins connexin 26 and 32 during hepatocarcinogenesis in rats. Jap. J. Cancer Res. 83, 1210-1215.

Salomon, D., Saurat, J.H., & Meda, P. (1988). Cell-to-cell communication within intact human skin. J. Clin. Invest. 82, 248-254.

Sandberg, K., Bor, M., Ji, H., Markwick, A., Millan, M.A., & Catt, K.J. (1990). Angiotensin II-induced calcium mobilization in oocytes by signal transfer through gap junctions. Science 249, 298-301.

Serras, F., & van-den, B.J. (1987). Is a mosaic embryo also a mosaic of communication compartments? Dev. Biol. 120, 132-138.

Shinoura, N., Chen, L., Wani, M.A., Kim, Y.G., Larson, J.J., Warnick, R.E., Simon, M., Menon, A.G., Bi, W.L., & Stambrook, P.J. (1996). Protein and messenger RNA expression of connexin43 in astrocytomas: Implications in brain tumor gene therapy. J. Neurosurg. 84, 839-845.

Simeone, A., Acampora, D., Arcioni, L., Andres, P.W., Boncinelli, E., & Mavilio, F. (1990). Sequential activation of HOX2 homeobox genes by retinoic acid in human embryonal carcinoma cells. Nature 763-766.

Stagg, R.B., & Fletcher, W.H. (1990). The hormone-induced regulation of contact-dependent cell-cell communication by phosphorylation. Endocr. Rev. 11, 302-325.

Stern, M.C., Gimenez-Conti, I.B., Budunova, I., Coghlan, L., Fischer, S.M., DiGiovanni, J., Slaga, T.J., & Conti, C.J. (1998). Analysis of two inbred strains of mice derived from the SENCAR stock with different susceptibility to skin tumor progression. Carcinogenesis 19, 125-132.

Stoker, M.G.P. (1967). Transfer of growth inhibition between normal and virus-transformed cells: autoradiographic studies using marked cells. J. Cell Sci. 2, 293-304.

Stutenkemper, R., Geisse, S., Schwarz, H.J., Look, J., Traub, O., Nicholson, B.J., & Willecke, K. (1992). The hepatocyte-specific phenotype of murine liver cells correlates with high expression of connexin32 and connexin26 but very low expression of connexin43. Exp. Cell Res. 201, 43-54.

Sugie, S., Mori, H., & Takahashi, M. (1987). Effect of *in vivo* exposure to the liver tumor promoters phenobarbital or DDT on the gap junctions of rat hepatocytes: a quantitative freeze-fracture analysis. Carcinogenesis 8, 45-51.

Swenson, K.I., Jordan, J.R., Beyer, E.C., & Paul, D.L. (1989). Formation of gap junctions by expression of connexins in Xenopus oocyte pairs. Cell 57, 145-155.

Swenson, K.I., Piwnica Worms, H., McNamee, H., & Paul, D.L. (1990). Tyrosine phosphorylation of the gap junction protein connexin 43 is required for the pp60$^{v-src}$-induced inhibition of communication. Cell Regulation 1, 989-1002.

Swierenga, S.H., Yamasaki, H., Piccoli, C., Robertson, L., Bourgon, L. Marceau, N., & Fitzgerald, D.J. (1990). Effects on intercellular communication in human keratinocytes and liver-derived cells of polychlorinated biphenyl congeners with differing *in vivo* promotion activities. Carcinogenesis 11, 921-926.

Tachikawa, T., Yamamura, T., & Yoshiki, S. (1984). Changes occurring in plasma membranes and intercellular junctions during the process of carcinogenesis and in squamous cell carcinoma. Virchows Arch 47, 1-15.

Takeda, A., Hashimoto, E., Yamamura, H., & Shimazu, T. (1987). Phosphorylation of liver gap junction protein by protein kinase C. FEBS Lett. 210, 169-172.

Takens-Kwak, B.R., & Jongsma, H.J. (1992). Cardiac gap junctions: three distinct single channel con-
    ductances and their modulation by phosphorylating treatments. Pflugers Archiv Eur. J. Phys-
    iol. 422, 198-200.
Tateno, C., Ito, S., Tanaka, M., Oyamada, M., & Yoshitake, A. (1994). Effect of DDT on hepatic gap
    junctional intercellular communication in rats. Carcinogenesis 15, 517-521.
Temme, A., Buchmann, A., Gabriel, H.D., Nelles, E., Schwarz, M., & Willecke, K. (1997). High inci-
    dence of spontaneous and chemically induced liver tumors in mice deficient for connexin32.
    Curr Biol 7, 713-716.
Tomasetto, C., Neveu, M.J., Daley, J., Horan, P.K., & Sager, R. (1993). Gap junctional intercellular
    communication among human mammary cells and connexin transfectants in culture. J. Cell
    Biol. 122, 157-167.
Tomomura, M., Kadomatsu, K., Matsubara, S., & Muramatsu, T. (1990). A retinoic acid-responsive
    gene, MK, found in the teratocarcinoma system. Heterogeneity of the transcript and the nature
    of the translation product. J. Biol. Chem. 265, 10765-10770.
Tong, C.C., & Williams, G.M. (1987). The effect of tumor promoters on cell-to-cell communication
    in liver epithelial cell systems. In: *Biochemical Mechanisms and Regulation of Intercellular
    Communication* (Milman, H.A., & Elmore, E., eds.), pp. 251-263. Princeton Scientific Pub-
    lishing, Princeton, NJ.
Trosko, J.E. (1998). Cooperative effects of v-myc and c-Ha-ras oncogenes on gap junctional intercel-
    lular communication and tumorigenicity in rat liver epithelial cells. Cancer Lett. 128, 145-154.
Trosko, J.E., Chang, C.C., & Madhukar, B.V. (1990a). Symposium: cell communication in normal
    and uncontrolled growth. Modulation of intercellular communication during radiation and
    chemical carcinogenesis. Radiat. Res. 123, 241-251.
Trosko, J.E., Chang, C.C., Madhukar, B.V., & Klaunig, J.E. (1990b). Chemical, oncogene and growth
    factor inhibition of gap junctional intercellular communication: an integrative hypothesis of
    carcinogenesis. Pathobiology 58, 265-278.
Trosko, J.E., Jone, C., & Chang, C.C. (1987). Inhibition of gap junctional-mediated intercellular com-
    munication *in vitro* by aldrin, dieldrin, and toxaphene: a possible cellular mechanism for their
    tumor-promoting and neurotoxic effects. Mol. Toxicol. 1, 83-93.
Trush, M.A., & Kensler, T.W. (1991). An overview of the relationship between oxidative stress and
    chemical carcinogenesis. Free Radic. Biol. Med. 10, 201-209.
Tsuda, H., Asamoto, M., Baba, H., Iwahori, Y., Matsumoto, K., Iwase, T., Nishida, Y., Nagao, S.,
    Hakoi, K., Yamaguchi, S., Ozaki, K., & Yamasaki, H. (1995). Cell proliferation and advance-
    ment of hepatocarcinogenesis in the rat are associated with a decrease in connexin 32 expres-
    sion. Carcinogenesis 16, 101-105.
Tsuda, H., & Okamoto, H. (1986). Elimination of metabolic cooperation by glycyrrhetinic acid, an
    anti-tumor promoter, in cultured Chinese hamster cells. Carcinogenesis 7, 1805-1807.
Upham, B.L., Kang, K.S., Cho, H.Y., & Trosko, J.E. (1997). Hydrogen peroxide inhibits gap junc-
    tional intercellular communication in glutathione sufficient but not glutathione deficient cells.
    Carcinogenesis 18, 37-42.
Van der Zandt, P.T.J., De Feijter, A.W., Homan, E.C., Spaaij, C., De Haan, L.H.J., Van Aelst, A.C.,
    & Jongen, W.M.F. (1990). Effects of cigarette smoke condensate and 12-O-tetradecanoylphor-
    bol-13-acetate on gap junction structure and function in cultured cells. Carcinogenesis 11,
    883-888.
Van Lieshout, E.M.M., Peters, W.H.M., & Jansen, J.B.M.J. (1996). Effect of oltipraz, α-tocopherol,
    β-carotene and phenethylisothiocyanate on rat oesophageal, gastric, colonic and hepatic glu-
    tathione, glutathione S-transferase and peroxidase. Carcinogenesis 17, 1439-1445.
Vanhamme, L., Rolin, S., & Szpirer, C. (1989). Inhibition of gap-junctional intercellular communica-
    tion between epithelial cells transformed by the activated H-ras-1 oncogene. Exp. Cell Res.
    180, 297-301.

Verma, A.K. (1987). Inhibition of both stage I and stage II mouse skin tumor promotion by retinoic acid and the dependence of inhibition of tumor promotion on the duration of retinoic acid treatment. Cancer Res. 47, 5097-5101.

Vrionis, F.D., Wu, J.K., Qi, P., Waltzman, M., Cherington, V., & Spray, D.C. (1997). The bystander effect exerted by tumor cells expressing the herpes simplex virus thymidine kinase (HSVtk) gene is dependent on connexin expression and cell communication via gap junctions. Gene Ther. 4, 577-85.

Wang, Y., & Mehta, P.P. (1995). Facilitation of gap-junctional communication and gap-junction formation in mammalian cells by inhibition of glycosylation. Eur. J. Cell Biol. 67, 285-296.

Wang, Y.J., & Rose, B. (1997). An inhibition of gap-junctional communication by cadherins. J. Cell Sci. 110, 301-309.

Warn-Cramer, B.J., Lampen, P.D., Kurata, W.E., Kanemitsu, M.Y., Loo, L.W.M., Eckhart, W., & Lau, A.F. (1996). Characterization of the mitogen-activated protein kinase phosphorylation sites on the connexin-43 gap junction protein. J. Biol. Chem. 271, 3779-3786.

Warner, A. (1992). Gap junctions in development - a perspective. Sem. Cell Biol. 3, 81-91.

Warner, A.E., Guthrie, S.C., & Gilula, N.B. (1984). Antibodies to gap-junctional protein selectively disrupt junctional communication in the early amphibian embryo. Nature 311, 127-131.

Weinberg, R.A. (1989). Oncogenes, antioncogenes, and the molecular bases of multistep carcinogenesis. Cancer Res. 49, 3713-3721.

Weinstein, R.S., Merk, F.B., & Alroy, J. (1976). The structure and function of intercellular junctions in cancer. Adv. Cancer Res. 23, 23-89.

Werner, R., Levine, E., Rabadan-Diehl, C., & Dahl, G. (1989). Formation of hybrid cell-cell channels. Proc. Natl. Acad. Sci. U.S.A. 86, 5380-5384.

Wert, S.E., & Larsen, W.J. (1990). Preendocytotic alterations in cumulus cell gap junctions precede meiotic resumption in the rat cumulus-oocyte complex. Tissue Cell 22, 827-851.

Wilgenbus, K.K., Kirkpatrick, C.J., Knuechel, R. Willecke, K., & Traub, O. (1992). Expression of Cx26, Cx32 and Cx43 gap junction proteins in normal and neoplastic human tissues. Int. J. Cancer 51, 522-529.

Willecke, K., Hennemann, H., Dahl, E., Jungbluth, S., & Heynkes, R. (1991). The diversity of connexin genes encoding gap junctional proteins. Eur. J. Cell Biol. 56, 1-7.

Winterhager, E., Grümmer, R., Jahn, E., Willecke, K., & Traub, O. (1993). Spatial and temporal expression of connexin26 and connexin43 in rat endometrium during trophoblast invasion. Dev. Biol. 157, 399-409.

Wolpert, L. (1978). Gap junctions: channels for communication in development. In: *Intercellular Junctions and Synapses* (Feldman, N.B.G., & Pitts, J.D., eds.), pp. 83-94. Chapman and Hall, London.

Wygoda, M.R., Wilson, M.R., Davis, M.A., Trosko, J.E., Rehemtulla, A., & Lawrence, T.S. (1997). Protection of herpes simplex virus thymidine kinase-transduced cells from ganciclovir-mediated cytotoxicity by bystander cells: the Good Samaritan effect. Cancer Res. 57, 1699-1703.

Yamasaki, H. (1990). Gap junctional intercellular communication and carcinogenesis. Carcinogenesis 11, 1051-1058.

Yamasaki, H., & Fitzgerald, F.J. (1988). The role of selective junctional communication in cell transformation. In: Tumor Promoters: *Biological Approaches for Mechanistic Studies and Assay Systems* (Langenbach, R., et al., eds.), pp. 131-147. Raven Press, New York.

Yamasaki, H., & Katoh, F. (1988). Further evidence for the involvement of gap-junctional intercellular communication in induction and maintenance of transformed foci in BALB/c 3T3 cells. Cancer Res. 48, 3490-3495.

Yancey, S.B., Easter, D., & Revel, J.-P. (1979). Cytological changes in gap junctions during liver regeneration. J. Ultrastruc. Res. 67, 229-242.

Yin, Y., Tainsky, M.A., Bischoff, F.Z., Strong, L.C., & Wahl, G.M. (1992). Wild-type p53 restores cell cycle control and inhibits gene amplification in cells with mutant p53 alleles. Cell 70, 937-948.

Yotti, L.P., Chang, C.C., & Trosko, J.E. (1979). Elimination of metabolic cooperation in Chinese hamster cells by a tumor promoter. Science 206, 1089-1091.

Yu, W., Dahl, G., & Werner, R. (1994). The connexin43 gene is responsive to estrogen. Proc. Royal Soc. (Lond.) B: Biological Sciences 255, 125-132.

Yuspa, S.H., & Poirier, M.C. (1988). Chemical carcinogenesis: from animal models to molecular models in one decade. Adv. Cancer Res. 50, 25-70.

Zeilmaker, M.J., & Yamasaki, H. (1986). Inhibition of junctional intercellular communication as a possible short-term test to detect tumor-promoting agents: results with none chemicals tested by dye transfer assay in Chinese hamster V79 cells. Cancer Res. 46, 6180-6186.

Zhang, J.-T., & Nicholson, B.J. (1989). Sequence and tissue distribution of a second protein of hepatic gap junctions, Cx26, as deduced from its cDNA. J. Cell Biol. 109, 3391-3401.

Zhang, L.-X., Cooney, R.V., & Bertram, J.S. (1991). Carotenoids enhance gap junctional communication and inhibit lipid peroxidation in C3H/10T1/2 cells: relationship to their cancer chemopreventive action. Carcinogenesis 12, 2109-2114.

Zhang, L., Cooney, R.V., & Bertram, J.S. (1992). Carotenoids up-regulate connexin43 gene expression independent of their pro-vitamin A or antioxidant properties. Cancer Res. 52, 5707-5712.

Zhu, D., Caveney, S., Kidder, G.M., & Naus , C.C.G. (1991). Transfection of C6 glioma cells with connexin 43 cDNA: analysis of expression, intercellular coupling, and cell proliferation. Proc. Natl. Acad. Sci. U.S.A. 88, 1883-1887.

Zhu, D., Kidder, G.M., Caveney, S., & Naus, C.C.G. (1992). Growth retardation in glioma cells co-cultured with cells over-expressing a gap junction protein. Proc. Natl. Acad. Sci. U.S.A. 89, 10218-10221.

# GAP JUNCTION FUNCTION

Paolo Meda and David C. Spray

|     |     |     |
|-----|-----|-----|
| I.   | Introduction | 264 |
| II.  | Investigating Physiological Roles for Gap Junctions | 265 |
|      | A. Electrical Measurements of Coupling Strength | 265 |
|      | B. Nonelectrical Methods for Determining Coupling Strength | 267 |
|      | C. Measurement of Gap Junction Coupling *In Vivo* | 269 |
| III. | What Gap Junctions Do | 270 |
| IV.  | Proposed Gap Junction Functions | 273 |
|      | A. Transmission of Hormonal Stimulation | 274 |
|      | B. Resistance to Viruses and Cytokines | 274 |
|      | C. Embryonic Development and Morphogenesis | 274 |
|      | D. Current Spread and Synchronization of Electrical Activity | 275 |
|      | E. Mechanical Synchronization | 277 |
|      | F. Compensation for Enzymatic Defects | 279 |
|      | G. The Bystander Effect | 279 |
|      | H. Nutrition | 280 |
|      | I. Dissipation of Extracellular and Intracellular Molecules and Ions | 281 |
|      | J. Differentiation | 281 |
|      | K. Cell Division | 284 |
|      | L. Migration | 285 |
|      | M. Secretion | 286 |
| V.   | Gap Junctional Communication in Somatic Disease States | 288 |
|      | A. Cardiovascular Diseases | 289 |

**Advances in Molecular and Cell Biology, Volume 30, pages 263-322.**
**Copyright © 2000 by JAI Press Inc.**
**All rights of reproduction in any form reserved.**
**ISBN: 0-7623-0599-1**

        B.    Tumorigenesis................................................291
        C.    Neurological Diseases.........................................292
        D.    Skin diseases................................................293
    VI.    Molecular Genetic Studies of Gap Junctions.........................294
    VII.   Gap Junction-Directed Therapeutics ................................297
    VIII.  Conclusions and Perspectives.......................................300

# I. INTRODUCTION

We will soon enter the fourth decade since the initial morphological identification of gap junctions (Robertson et al., 1963; Dewey and Barr, 1964; Barr et al., 1965) and the discovery that adjacent cells sharing these structures directly exchange current-carrying ions (Kuffler and Potter, 1964; Potter et al., 1966) and small cytoplasmic molecules (Loewenstein and Kanno, 1964; Furshpan and Potter, 1968; Crick, 1970; Gilula et al., 1972). In this period, much has been learned regarding the structure of gap junctions, their membrane topography, their tissue distribution, the characteristics of their constituent connexin proteins, and the molecular biology of the encoding genes (see the first three chapters in this volume). In parallel, our understanding of the biophysical characteristics of gap junction channels has also markedly progressed, and the functional regulation of the opening and closing of these channels has been investigated in a large number of systems and under variety of experimental conditions (see the chapters by Verselis and Veenstra, Lo and Gilula, Neveu and Bertram, aand Nagy and Dermietzel in this volume). Hence, it is now clear that gap junctions and the direct cell-to-cell communication which they mediate, are obligatory features of almost every cell type, throughout the phylogeny of multicellular animals. Furthermore, we have begun to appreciate the numerous variations of gap junction composition, structure, and regulation, which have developed with evolution, cell differentiation, and tissue morphogenesis. This unexpected diversity is now taken as a suggestion that gap junctions are likely to play multiple roles in different tissues (Bennett et al., 1991).

In contrast with this wealth of basic information, the specific physiological functions that gap junctions fulfill have been demonstrated in only a few selected cell types. Thus, it is now accepted that gap junctional communication plays a regulatory role in the electrical and mechanical coupling of different types of muscle cells (for reviews, see Spray and Burt, 1990; Garfield et al., 1992; Huizinga et al., 1992; Spray et al., 1994b), and increasing evidence also points to their involvement in the control of cell growth (Loewenstein and Rose, 1992) and differentiation in developing and adult systems (Allen et al, 1990; Warner, 1992), and in secretion of both endocrine and exocrine glands (Meda, 1984; Chanson and Meda, 1993). However, even in these cases, the proof that such roles are specifically and obligatorily fulfilled by gap junctions has not yet been provided, particularly under conditions of physiological relevance; that is, within intact tissues and organs *in vivo* (for exceptions, see Berga, 1984; Chanson et al., 1991; Segal and Bény, 1992; Little et al., 1995). Also, evidence is presently quite limited by which to understand why gap junctional communication is required for such functions. In this respect, the field is still shaped by ideas and hypotheses that were generated soon after the discovery of junctional coupling (Loewenstein, 1968a, 1968b; Subak-Sharpe et al., 1969; Sheridan, 1971). However, with the recent availability of new immunological and molecular biological tools with which to selectively perturb the formation and functioning of gap junction channels, including the availability of connexin knockout mice: (Reaume et al, 1995; Gong et al., 1997; Nicholson and Bruzzone, 1997; Simon et al., 1997; Gabriel et al., 1998; Willecke et al., 1998), we now appear to be on the threshold of a leap forward in understanding the role of gap junctions in physiological functions.

In this context, this chapter has two goals. The first is to consider the methodological requirements for investigating gap junction functions, the problems that complicate such studies, and the new tools and experimental strategies that may aid in this quest. Although voltage clamp and permeability measurements are considered in detail in the earlier chapter by Verselis and Veenstra in this volume, an overview is provided here to highlight the notion that what a gap junction does is to exchange ions and molecules between cells. The functions that these channels subserve in different tissues are determined by the diffusion of the most relevant molecules in those tissues, and methods of measurement may or may not assess these critical molecules. The second goal is to review some of the experimental evidence pointing to specific gap junction functions in a variety of normal and pathological organs and tissues. In many subheadings in this chapter, there is an overlap with other chapters in this volume. In those cases, we have attempted to paint the topic with broad strokes and refer the reader to the relevant chapters for detail.

## II. INVESTIGATING PHYSIOLOGICAL ROLES FOR GAP JUNCTIONS

The preparations chosen to study gap junction functions must necessarily balance the rigor of direct evaluation of junctional conductance and permeability against the desire for situations that most closely approximate those occurring *in vivo*. The paramount function of gap junction–mediated intercellular communication is to provide a pathway for the diffusional exchange of ions and small hydrophilic molecules from one cell to the next. The strength of coupling determines the efficiency of this exchange and any molecule to which junctional channels are permeant may be used to probe gap junction function.

### A. Electrical Measurements of Coupling Strength

Electrical measurements assess the resistance to passage of current-carrying ions. The earliest measurements of electrical coupling used microelectrodes inserted in two cells and passed current through one electrode while measuring voltage in both the injected and the recipient cell. During current pulses sufficiently long to completely charge membrane capacitance (so that voltages were recorded under steady state conditions), the ratio of voltages in the recipient and in the injected cell ($V_2/V_1$) was termed the coupling coefficient (k). In a two-cell system, alternating the application of current pulses to each cell ($I_1$, $I_2$) while measuring the resulting voltages allows for calculation of input and transfer resistances. From these values, conductances of the junctional ($g_j$) and nonjunctional membranes ($g_{nj}$) can be calculated using the equation $k = g_j/(g_j + g_{nj})$ by applying the pi-t transform to the equivalent circuit (Bennett, 1966). Because this measurement of coupling strength depends on both junctional and nonjunctional conductances, any treatment that changes cell input conductance will also cause an apparent change in cell coupling whether or not there is an actual effect on the junctional membrane itself. Thus, disruption of the cell membrane will uncouple cells, as may the activation of ligand- or voltage-activated channels in the nonjunctional membrane (see later discussion), regardless of whether junctional channels are affected. There are two additional limitations of the current-clamp method. Firstly, current pulses must be applied sequentially in order to obtain the four measurements from which the three equations of the pi-t transform can be solved. Thus, under conditions in which resistances may change rapidly, the measurements may not faithfully reflect these channels. The second limitation of current-clamp methods is the necessity of charging membrane capacitance before steady state voltages can be recorded. Finally, since different gap junction channels show distinct voltage- dependencies (Spray et al., 1991a; see also the earlier chapter by Verselis and Veenstra in this volume), the applied currents may induce the closure of some channels by eliciting voltage gradients between the cells.

A technique that avoids these problems is that of voltage clamping the nonjunctional membrane. Dual whole-cell voltage clamp was introduced in studies of large embryonic cells (Spray et al., 1979), where insertion of multiple electrodes presented no technical limitation. This technique remains the method of choice for evaluating junctional properties of connexins expressed in *Xenopus* oocytes (see the earlier chapter by Verselis and Veenstra in this volume), whereas for studies of mammalian cells, most workers now control voltage and measure currents using single patch-type electrodes on each cell, according to a modification of the originial technique (Neyton and Trautmann, 1985; White et al., 1985).

Direct measurements of junctional conductance cannot be performed in systems consisting of more than two cells. However, assumptions of symmetry can be used to model current spread in tissues of known geometry. For example, the application of cable analysis to strands of tissue containing gap junctions has a long history of separating internal longitudinal resistance, which is predominantly junctional in origin, from the resistance of the nonjunctional membrane (Weidmann, 1970). This technique is most commonly applied to cardiac tissue, and one use has been to separate pharmacological effects on conduction due to reduced membrane excitability from those due to reduced junctional conductance (Delmar et al., 1987; Allessie et al., 1989). In cell monolayers, current spread from a point source is described by a Bessel function when the space constant is much longer than cell diameter. In insect epidermis, for example, plotting the decay of the electrotonic current as a function of the logarithm of distance from source gives a slope that is directly proportional to junctional resistance (Caveney et al., 1986).

## B.  Nonelectrical Methods for Determining Coupling Strength

Various techniques have been devised to measure junctional permeance with a variety of tracer molecules. The earliest studies were performed using either radioactive precursors (Subak-Sharpe et al., 1966, 1969; Pitts and Sims, 1977) or fluorescent tracer molecules (Loewenstein 1968a, 1968b, 1979, 1981) and findings were reported as indicating "metabolic" or "dye" coupling. The strategy of metabolic coupling experiments is to incubate one population of cells in the presence of an excess of a radioactively labeled metabolite, to co-culture these cells in the presence of unlabeled cells, and then to determine the incidence of label transfer from loaded to unloaded cells as a function of time (Hooper and Subak-Sharpe, 1981). Variations on this theme include delivery of 2-deoxyglucose through laceration of a tissue sheet (Cole et al., 1985), "kiss of life" experiments, in which metabolite-rich cells rescue from cell death enzymatically deficient cells to which they are coupled, and "kiss of death" experiments in which recipient cells die after transfer of a noxious metabolite (Subak-Sharpe et al., 1966, 1969). The more recent version of the "kiss of death" experiment is the "bystander effect" (Bi et al., 1993), discussed later in this chapter. Also similar in philosophy are experiments in which cells deficient in $Na^+$-pump expression are rescued by cells in which pumps are well expressed (Ledbetter and Lubin, 1979). The functional role of gap junctions that is assayed with this method has also been described as "energy transfer" (Aslanidi et al., 1991).

The use of fluorescent tracers to map intercellular communication pathways was initially hampered by several inadequate properties of fluorescein derivatives. Fluorescein itself exhibits significant nonjunctional permeability, although negatively charged derivatives such as 6-carboxyfluorescein are more effectively retained by cells. Fluorescein esters, which are membrane permeant but are cleaved by cytoplasmic esterases to membrane impermeant, charged derivatives, have been used for studies in which one cell population is loaded, the loaded cells are paired with unlabeled ones, and subsequently dye transfer is measured by fluorescence microscopy of fluorescence activated cell sorting (El-Sabban and Pauli, 1991; Lee et al, 1992). A more serious limitation is that both fluorescein and derivatives exhibit significant photobleaching upon excitation, a property that has even been exploited in one type of assay, the fluorescence recovery after photobleaching (FRAP) technique, in which fluorescein-loaded cells in a monolayer are illuminated and recovery of their fluorescence after photobleaching is monitored as a function of time (Wade et al., 1986).

Several of these drawbacks were overcome by the synthesis of Lucifer yellow, which is roughly the same size as fluorescein but is essentially impermeant through the nonjunctional membrane. Lucifer yellow exhibits high quantum efficiency and is aldehyde fixable, thus allowing detailed observations after histological processing (Stewart, 1978). However, Lucifer yellow binds to nuclei and is rather poorly soluble in the presence of $K^+$ ions. In spite of these two problems, the use of this molecule has revolutionized the field, allowing unambiguous evaluation of dye transfer after either scrape-loading of cell monolayers (El Fouly et al., 1987) or intracellular injection through micropipettes (see later discussion). Among the earliest controversies that were resolved using this dye was whether junctional permeability of embryonic cells was lower than that of adults. Thus, it was shown that, in contrast with previous reports based on experiments with fluorescein derivatives (Bennett, 1973), Lucifer yellow diffusion occurs between coupled fish embryonic cells and that the rate of diffusion was not dramatically different from that observed in other cell types when the volume of the cell was taken into consideration (Sheridan, 1973; Bennett et al, 1978). Hence, geometrical issues such as cell volume can make results ambiguous when a threshold of detectability is used as the criterion of dye spread (Sheridan and Atkinson, 1985; Zimmerman and Rose, 1985). A drawback to the use of Lucifer yellow is that the size of this molecule is near the permeability limit of gap junction channels. In a plot compiling cardiac gap junction permeability to various molecules, the junctional permeability to Lucifer yellow was shown to be about four orders of magnitude less than that of $K^+$ (Imanaga, 1987). These data may be explicable on the basis of molecular size, as the Renkin equation describing permeability through a sieve predicts that permeability will decrease as the approximate third power of the diameter of the probe molecule (Spray, 1994b). Another drawback to the use of Lucifer yellow is its fixed negative charges, which initially were considered a desirable feature for aldehyde fixability and intracellular retention (Stewart, 1978). As discussed later (see also the earlier chapter by Veenstra and Verselis in this volume), many gap junction channels display selectivity for cations over anions. In this perspective, an advance in the detection of coupling is the use of biocytin and its derivative, Neurobiotin (Vaney, 1991). These molecules are significantly smaller than Lucifer yellow and are not negatively charged, so that their permeability across junctional channels tends to be higher than that of other fluorescent molecules currently used to determine the extent of coupling. These tracer molecules are thus enjoying popularity for studies of gap junction distribution in the central nervous system and the retina, where much more extensive domains of coupled cells are being revealed than were previously recognized (Peinado et al., 1993; Vaney, 1991).

Another category of studies quantitatively evaluated the ions permeating the junctional membrane (Verselis et al, 1986; Bodmer et al, 1988; see also the chapter by Verselis and Veenstra earlier in this volume) or determined relative permeabilities from single channel conductances or reversal potential measurements. Although not yet in widespread use, these experimental paradigms offer the possibility of comparing ionic selectivities among junctional channels formed by the different connexin isoforms. The direct measurement of tetra-alkylammonium permeability takes advantage of the pyramidal shape of molecules formed by symmetrical increase in chain length, to determine the limiting diameter of the junctional channel with more precision than has previously been achieved. In dye permeance studies, this determination has been complicated by the nonspherical nature of the tracer molecules. Channel selectivity can be calculated from unitary conductances measured under conditions in which the current carrying ions of both intracellular solutions are varied either symmetrically, allowing calculation of relative permeability, or asymmetrically, in which case reversal potentials allow calculation of absolute permeabilities. Substitution of Cl– with glutamate or gluconate was initially shown to reduce unitary conductances of Cx43 channels as expected from the larger size of the organic anion (e.g., Spray et al., 1991a), whereas similar substitutions for connexin37 (Cx37) and connexin40 (Cx40) channels exert less of an effect, indicating that these connexins are rather anion-permeant (Veenstra et al., 1994a, 1994b). Rigorous experiments utilizing a multiple of ionic substitutions over a range of concentrations (to control for possible channel saturation) are now providing more complete descriptions of ion permeability through junctional channels (Beblo and Veenstra, 1997; Wang and Veenstra, 1997).

## C.   Measurement of Gap Junctional Coupling *In Vivo*

Only a few studies thus far have attempted to assay coupling between cells *in vivo*. After the first report by Berga (1984), recent studies have evaluated whether the extent and modulation of coupling *in vivo* are consistent with tissue culture data and with measurements made in freshly explanted organs. For example, the vascular tissue in the hamster cheek pouch was used to evaluate electrical and dye coupling between endothelial and smooth muscle cells (Segal and Beny, 1992; Little et al., 1994, 1995), and rat pancreas has been studied to assess whether uncoupling correlated with changes in secretion *in vivo* (Chanson and Meda, 1993).

In addition to these studies on the whole animal, there are increasing attempts to approximate *in vivo* conditions using tissue slice techniques, which offer the advantages of maintaining cellular connectivity and extracellular matrix components. For example. one of the first studies using this approach was evaluated in dye injection and electronic coupling studies performed in liver lobules and showed that coupling was largely confined to single hepatic acini (Meyer et al., 1981), strengthening the notion of hepatic organization into coupled territories. Also, coupling of neurons within brain slices, as evidenced by Neurobiotin microinjection, is observed until the second postnatal week, and with the synchronous activation of neurons comprising cortical domains (Peinado et al., 1993), possibly under the control of $Ca^{2+}$ waves (Smith, 1992). Other studies have characterized junctional coupling in intact islets of Langerhans (Michaels and Sheridan, 1981; Meda et al., 1983), heart fragments (e.g., Allessie et al., 1989), and skin (Kam et al., 1986; Salomon et al., 1988). Together, these studies have shown that the basic *in vivo* characterization of coupling are not completely at odds with those derived from *in vitro* studies on simplified cell systems, with the noticeable exception of the compartmentation of the intercommunicating cells. Thus, whereas *in vitro*, most cell types appear able to form gap junctions with each other, *in vivo* the communicating network may be much more selective, and a number of instances in which two closely opposed cell types appear unable to establish functional gap junctions have been reported (Meda et al., 1983; Salomon et al., 1988; Segal and Beny, 1992; Beny, 1994). Also, whether the modulation of junctional channels is similar *in vitro* and *in vivo* is essentially unknown, as this question has been investigated only in one cell system involved in secretion (Chanson et al., 1991).

## III. WHAT GAP JUNCTIONS DO

The general purpose of gap junction channels is to permit the cell-to-cell diffusion of a variety of cytoplasmic components. Hence, besides the exogenous tracers mentioned earlier, a variety of endogenous ions and low molecular weight species have been shown to cross junctional channels. These include all current-carrying anions and cations, as well as glycolytic intermediates, vitamins, amino acids, and nucleotides (Pitts and Finbow, 1977). In view of the large size of the channel pore, it was surmised that morphogens could also pass connexin channels (Crick, 1970; Warner, 1992), although it is striking that morphogens having the requisite properties for molecules likely to permeate between cells have yet to be convincingly identified.

So far, most attention has been given to the exchange of second messengers, owing to the availability of molecules whose fluorescence is altered by interaction with specific ions. Although this interest has primarily been directed at the detection of $Ca^{2+}$, we anticipate that Cl–, $H^+$, and $Na^+$ indicators will also be more widely used to measure junctional permeability, as real time ratiometric methods become routinely applied to video images. With regard to $Ca^{2+}$ movement through gap junction channels, the first evidence came from studies on *Obelia,* where an endogenous $Ca^{2+}$-indicating molecule, which is expressed in a specific cell population, was shown to be excited after a spatially restricted illumination was applied to an adjacent photoreceptor (Dunlap et al., 1987; Brehm et al., 1989). In mammalian cells, $Ca^{2+}$ permeability of gap junctions was examined between rat hepatocytes after injection of Fura2 or incubation in the presence of its membrane permeant acetoxy ester (Sáez et al., 1989). These studies showed that Fura2, which has a size similar to that of Lucifer yellow, is gap junction permeant and that $Ca^{2+}$ injection into one cell leads to transfer of the cation into an adjacent cell, thus establishing a $Ca^{2+}$ gradient across the junctional membrane. The apparent intercellular diffusion of $Ca^{2+}$ was not affected by very low extracellular $Ca^{2+}$, indicating that the rise in the second cell was not due to activation of a $Ca^{2+}$ influx pathway and was completely and reversibly blocked by heptanol, a compound that inhibits current flow through gap junction channels (Sáez et al., 1989a). Moreover, intracellular injection of $IP_3$, which liberates $Ca^{2+}$ from intracellular stores, indicated that this second messenger or one of its metabolites, also diffused from one cell to another through gap junction channels (Sáez et al., 1989a). Observations of $Ca^{2+}$ waves flowing between coupled cells have more recently been extended to astroglia, tracheal epithelium, endothelial and smooth muscle cells, and many other cell types (reviewed in Sanderson et al., 1998; Steinberg et al., 1998) and even observed in intact liver, heart, and pancreatic islets (Valdeolmillos et al. 1993; Nathanson et al., 1995; Robb-Gaspers and Thomas, 1995). It is now also apparent that the gap junction–mediated transfer of $Ca^{2+}$ and $IP_3$ plays vital roles in the functions of many tissues. $Ca^{2+}$ waves may passively decay in concentration with distance from the source cell or, as appears to be more common, it may involve an active diffusion reaction scheme, in which $IP_3$-triggered $Ca^{2+}$ release may elicit regenerative, propagated $Ca^{2+}$ waves (Sneyd et al., 1994; Sanderson et al., 1998). The velocity of these propagated waves (20 to 50 μm/sec) is similar to that observed for the intracellular spread of $Ca^{2+}$- or $IP_3$-induced $Ca^{2+}$ waves, indicating that gap junctions are not a major permeability barrier for this process. One implication of these findings is that the repeatedly hypothesized $Ca^{2+}$-induced closure of gap junction channels (Loewenstein, 1979; Lazrak and Peracchia, 1993), requires $Ca^{2+}$ concentrations which should be significantly higher than those reached during the movement of the cation under most physiological conditions (Spray, 1994c; Spray and Bennett, 1985).

Although some studies, such as those in which glioma cells overexpressing connexin43 (Cx43) exhibited $Ca^{2+}$ wave spread whereas parentals did not have suggested that gap junctions play a critical role in spread of $Ca^{2+}$ waves between cells, other studies, such as those showing that $Ca^{2+}$ waves may propagate across areas devoid of astrocytes, indicate that extracellular signaling may also contribute substantially to the process (Enkvist and McCarthy, 1994; Hassinger et al., 1996). As originally shown in mast cells (Osipchuk and Cahallan, 1992), stimulation of one cell may cause adenosine triphosphate (ATP) release, which evokes a $Ca^{2+}$ entry sufficient to elicit a regenerative $Ca^{2+}$ release in nearby cells, over considerable distance. Studies comparing $Ca^{2+}$ wave propagation under conditions in which extracellular or intercellular communication pathways were selectively blocked revealed that both contribute in many cell types and that one or the other pathway may dominate in the absence of the other (Spray et al., 1998; Steinberg, et al., 1998).

One illustration of a case where direct transfer of second messenger molecules might be of importance is in the heterocellular communication between vascular endothelial and smooth muscle cells (Asada and Lee, 1992; Beny, 1994; Spray et al., 1994). The existence of such communication remains a controversial issue, due in part to the lack of adequate morphological evidence for myoendothelial gap junctions (Daniel et al., 1976; Davies, 1986; Segal and Bény, 1992). However, junctional communication has been demonstrated between endothelial and smooth muscle cells *in vitro*, as well as in small and large vessels *in vivo* (Sheridan and Larson 1982; Davies et al., 1985; Larson et al., 1987; Little et al., 1995), and intercellular coupling appears to allow for the spread of electrotonic potentials and vasomotor responses in at least some compartments of the vascular system (Segal and Duling, 1986, 1987, 1989; Beny, 1994). Such communication would seem to be an especially attractive mechanism for the transmission throughout muscular vessels of intracellular second messengers that may be generated by labile endothelial vasoactive substances such as nitric oxide (Christ et al., 1996).

Another system in which direct transfer of second messengers is believed to play a functional role is liver. Hepatocytes represent a heterologous population with phenotypic features that are graded from periportal to pericentral regions. Thus, hepatocytes with higher glucogenic activity are preferentially found close to the portal space (Gumuci and Miller, 1981; Jungerman and Katz, 1989), and expression of glucagon receptors is graded across the liver acinus, with the more active gluconeogenic cells having lower receptor numbers (Berthoud et al., 1992). In spite of this heterogeneity, hepatocytes are electrically coupled across relatively long distances (0.5 mm, or at least 20 cell diameters), which may account for the similarity in membrane potentials among periportal and pericentral hepatocytes, and for their synchronous glucagon-induced hyperpolarization throughout each acinus (Penn, 1966; Graf and Peterson, 1978). Moreover, glucagon-induced hyperpolarization is higher in periportal than in pericentral hepatocytes after treatment with the gap junction blocking agent octanol (Lee and Clemens, 1992) and heptanol-induced gap junction blockade has been shown to abolish metabolic and hemodynamic effects of nerve stimulation (Seseke et al., 1992). Therefore, it seems likely that gap junctions play a key role in transmitting the second messenger molecules that are generated by glucagon receptor occupancy, to distant and more glucogenic cells (Berthoud et al., 1992). This view is supported by the finding of reduced glucose mobilization in response to hepatic nerve stimulation in livers from mice lacking connexin32 (Cx32) expression (Willecke et al., 1996). The uptake, utilization and cell-to-cell exchange of glucose and some of its metabolites are also partly dependent on communication via connexin channels (Bosco et al., 1997; Giaume et al., 1997; Gabriel et al., 1998). In both astrocytes and pancreatic β-cells, some of these gap junction–permeant glucose derivatives act as second messengers by activating selective pathways in coupled cells (Bosco et al., 1997; Giaume et al., 1997).

Second messenger diffusion between astrocytes may also be functionally important. Recent studies have demonstrated that focal glutamate application, electrical stimulation or mechanical prodding leads to the propagation of $Ca^{2+}$ waves throughout glial cells (Charles et al., 1991; Enkvist and McCarthy, 1992; Hassiper et al., 1996; Veenstra et al., 1997; Scemes et al., 1998), and in many cases this propagation has been shown to be inhibited by agents that block gap junction channels. Thus, gap junction–mediated interactions between glial cells may actually play an active role in neuromodulation by propagating regenerative $Ca^{2+}$ signals, as a result of $Ca^{2+}$ or $IP_3$ induced $Ca^{2+}$ release (Usowicz et al., 1989; Smith, 1992). Such $Ca^{2+}$ signals might reflect moment-to-moment changes in glial $K^+$ buffering capacity or could even result in $K^+$ release from glial cells, thereby changing the level of neuronal excitability. Moreover, both intercellular and extracellular components of $Ca^{2+}$ wave propagation are facilitated under hyposmotic conditions, thereby potentially providing neuroprotective effects (Scemes and Spray, 1998).

Another second messenger molecule of enormous importance, namely cyclic adenosine monophosphate (cAMP), has been proposed to pass between cardiac myocytes based on inotropic and chronotropic responses of tissue to locally introduced cAMP (Gilula et al., 1972; Tsien and Weingart, 1976). Because cAMP is smaller than dye molecules, which are gap junction–permeant, diffusional exchange is predicted (Fletcher et al., 1987). Still missing, however, is the direct demonstration that should now be achievable using ratiometric imaging with cAMP-sensing molecules (Fletcher and Greenan, 1985). In cardiac myocytes, slow $Ca^{2+}$ waves occuring on a time scale 10,000 times less rapid than the electrical waves that mediate cardiac contraction, are also coordinated in part through diffusion gap junction–permeant second messengers. These waves may play a role in setting basal tone of normal heart and creating re-entrant circuits in ischemic myocardium (Spray et al., 1998).

## IV. PROPOSED GAP JUNCTION FUNCTIONS

A variety of specific functions have been proposed for the intercellular transfer of ions and molecules through gap junction channels. Several of these functions are still completely hypothetical and the few that have gained experimental support remain circumstantial, being defined under experimental conditions which are quite different from those we would like to ideally test, or in cell systems that do not faithfully represent the physiological conditions prevailing *in vivo*. In our view, these limitations do not undermine the interest of the studies cited here, which will certainly influence, if not inspire, the more direct investigations on gap junction functions that can now be envisaged *in situ*. Here, we summarize some of the gap junction functions that have recently received new experimental support, with a particular focus on those functions that have not been extensively discussed in previous reviews (Pitts, 1980; Hertzberg et al., 1981; Hooper and Subak-Sharpe, 1981; Sheridan and Atkinson, 1985; Dermietzel et al., 1990; Bennett et al., 1991).

## A. Transmission of Hormonal Stimulation

Cells devoid of appropriate receptors can respond to specific hormones, provided they contact cognate target cells in coculture (Lawrence et al., 1978; Murray and Fletcher, 1984; Hobbie et al., 1987). Most likely, this response is made possible by the transfer of second messengers that are generated within the receptor-bearing cells, into those cells which do not possess appropriate hormone receptors. Although there is little doubt that such a transfer depends on the establishment of gap junctions between the receptor bearing and nonbearing cells (Lawrence et al., 1978; Murray and Fletcher, 1984), it is still uncertain which second messengers are implicated. cAMP has received much attention (Lawrence et al., 1978; Murray and Fletcher, 1984; Spray and Burt, 1990; Stagg and Fletcher, 1990; De Mello, 1991), but the participation of other common mediators of hormonal action, such as $Ca^{2+}$ or $IP_3$ (Sáez et al., 1989a; Sanderson et al., 1990; Charles et al., 1992; Christ et al., 1992) cannot be excluded.

## B. Resistance to Viruses and Cytokines

A comparable mechanism may be implicated in the suggested role of gap junctions in the *in vitro* enhancement of cell resistance to both viral infections (Blalock and Stanton, 1980) and damage by cytokines (Fletcher et al., 1987b). In this case, gap junctions may increase the buffer capacity of the cell by effectively increasing the total cytoplasmic volume of the communication compartment hence diluting toxic agents.

## C. Embryonic Development and Morphogenesis

Several sets of data strongly support the involvement of gap junctional communication in the early development of embryos and in the subsequent morphogenesis of different organs. The first is the appearance of connexins at a critical stage of early embryonic development (Dulcibella et al., 1975; Lo and Gilula 1979; Nishi et al., 1991). The second set of data is provided by several studies at later stages of development, reporting that multiple connexins are differentially expressed in most ectodermal, mesodermal, and endodermal derivatives (Risek and Gilula, 1991; Risek et al., 1992; Yancey et al., 1992). A third set of evidence is provided by the early interruption of embryonic development in homozygous transgenic mice in which the gene coding for connexin26 (Cx26) has been knocked out (Gabriel et al., 1998) and by the clearcut and morphogenetic alterations that take place after knockout of Cx43 (Reaume et al., 1998). Interestingly, however, the knockout of other connexin genes has not led to the anticipated defects in embryological development in cell systems thought to express only the targeted connexin(s) (Charollais et al., 1998; Houghton et al., 1998; Simon et al., 1998). In at least one such case, however, these negative findings have led to the identification of other hitherto unsuspected connexins (Charollais et al., 1998). The reason why different tissues express distinct sets of connexins during morphogenesis and differentiation remains to be established. However, in some cases, the topographical distribution of gap junction proteins matches that of developmentally relevant territories (Kalimi and Lo, 1988; Lo, 1989; Bagnall et al., 1992; Ruangvoravat and Lo, 1992). Another set of data derives from experiments in *Xenopus* in which antibodies to connexin32 and antisense probes expected to hybridize to its cognate mRNA were used to interact with the functioning and formation of gap junctions, respectively (Warner et al., 1984; Warner, 1992). Under these conditions, newly hatched animals showed altered morphogenesis of anterior cephalic organs, possibly due to altered morphogen distribution. A comparable phenotype has been observed by expression in *Xenopus* of a hybrid connexin protein which interacts in a dominant negative fashion with multiple endogenous connexins. However, in this case, altered cell-to-cell adhesion has been implicated as the predominant cause of the developmental defects (Paul, 1995).

Further investigations of these findings are now conceivable using strategies such as reverse genetics, *in vivo* conditional connexin expression, and targeting to specific tissues of mutant or ectopic connexins. Such approaches may provide more direct evidence for or against the gap junction specificity of the provocative malformations that have been observed so far. Whatever the results of these future experiments, it remains to be shown whether a short-term, physiological modulation of junctional communication, which is likely to be much more subtle and fine-tuned than the chronic and massive over- and underexpression of connexins that can be achieved in the laboratory, is necessary for proper developmental control. In principle, gap junctional communication appears well suited to mediate the differential transmission of morphogens (Warner, 1992), most of which, however, still remain to be identified (Bryant and Gardiner, 1992; Jessell and Melton, 1992). These topics and data supporting them are considered more fully in the chapter by Lo and Gilula in this volume.

## D.  Current Spread and Synchronization of Electrical Activity

In excitable cells, such as neurons and cardiac myocytes, the primary role of gap junction channels is thought to allow for the passage of ions between nearby cells. This intercellular diffusion is driven by the electrochemical gradient and carries current flow from one cell to the next. Hence, the gap junction channel of excitable tissues functions primarily as an intercellular, and possibly also intracellular (Bergoffen et al., 1993) $K^+$ channel. Electrical measurements of this channel function are considered in Section IIA above.

One system in which coupling is probably of functional importance in this context is seen in the mollusc *Navanax*. In this voracious predator, a pool of coupled motoneurons innervates the radial pharyngeal musculature. Contact of the animal's lips with a prey excites the expansion motoneurons, causing the pharynx to inflate so that prey is ingested (Spira et al., 1980). Feedback from pharyngeal proprioceptors inhibits those expansion motoneurons which feature the largest motor fields and shunts currents flow between the cells (Spray et al., 1980). Expansion motoneurons with more restricted fields are thereby uncoupled, so that individual bands of pharyngeal muscle can expand. Together with the excitation of circumferential motoneurons, which have their own uncoupling circuitry (Spira et al., 1976), these contractions allow for the rhythmical peristaltic movement of the pharynx that forces the prey into the esophagus. In this behavior, uncoupling is entirely nonjunctional, being attributable to increased postsynaptic conductance of the motoneurons caused by the proprioceptive inputs. Current flow is effectively shunted so that cells can still be individually excited while being no longer coupled (Spira et al., 1980). The interplay between chemical and electrotonic synapses thus permits the same population of cells to exhibit synchronous and asynchronous activity. A homologous circuit exists in the inferior olive of the mammalian brain, suggesting that similar networks may be common throughout the central nervous system (Llinas et al., 1974).

Although gap junctions are difficult to locate between neurons of mammalian brain, the use of antibody and *in situ* hybridization techniques has revealed that connexin expression is much more widespread than previously appreciated (see the subsequent chapter by Nagy and Dermietzel in this volume; Micevych and Abelson, 1991; Dermietzel and Spray, 1993). Moreover, the use of novel lower molecular weight tracers has demonstrated significant functional coupling among neurons, which in mammalian cortex persists postnatally (Peinado et al., 1993). In neurons, the high input resistance of cells at rest allows for high coupling coefficients even when cells are connected by only a few gap junction channels (Dermietzel and Spray, 1993). The very rapid propagation through these "electrotonic synapses" is presumably the selective advantage favoring their occurrence in escape systems of both invertebrates (Furshpan and Potter, 1959) and vertebrates (Auerbach and Bennett, 1969; Korn and Faber 1979). Furthermore, even a very weak coupling may be sufficient to synchronize electrical activity among some homologous neurons, and to generate bursting behavior in neuronal clusters (Dermietzel and Spray, 1993).

Ionic exchange across cardiac tissue also allows for synchronous activation and isochronous wave front spread in working myocardium, as well as for propagation of the impulse through the conduction system. The heart can thus be viewed as an interconnected series of communication compartments (Spray et al., 1994c), each of which is specialized for pacemaking (sinoatrial and atrioventricular nodes), conduction (the Purkinje conduction system), or contraction (atrium, ventricles). Gap junctions and connexins vary qualitatively and quantitatively in these communication compartments. Junctional contacts are sparse in the sinoatrial and atrioventricular pacemaking regions, presumably so as not to weigh down the pacemaker currents. By contrast, gap junctions are both large and numerous in the conduction system and in the working myocardium. In these compartments, however, gap junctions differ in terms of connexin composition. Cx43 is the most abundant gap junction protein in ventricular myocardium of most mammals, whereas Cx40 may be more abundant in the conduction system or atrium, or both (Kanter et al., 1993a, 1993b), and connexin45 (Cx45) may be coexpressed in ventricle (Kanter et al., 1993b). Interestingly, the distribution of connexins is reversed in avian species. Thus, chick connexin42 (Cx42), which is the homolog of mammal Cx40, is the more abundant ventricular gap junction protein, whereas chick Cx43 predominates in the vasculature (Beyer et al., 1992; Minkoff et al., 1993). This differential distribution could have functional consequences if the conductile connexin had a larger unitary conductance (Kanter et al., 1993a, 1993b), or if it did not readily pair with the connexin of the ventricular mass. The larger unitary conductance reported for Cx40 as compared to Cx43 channels (see the earlier chapter by Veenstra and Verselis) might provide more efficient impulse propagation along the conduction system. However, one possible inconsistency for such a model is the finding that mammalian Cx40 and Cx43 do not form functional heterologous channels when expressed in either oocytes (Bruzzone et al., 1993) or communication-deficient mammalian cells (Willecke et al. 1994). If such absolute compartmentalization were the case in intact heart, the boundaries between the conduction and the contractile compartments could present paradoxical barriers to propagation. Another predicted interplay is that of heterologous hemichannels made of Cx45 and Cx43, which appear to pair avidly (Moreno et al., 1995). Because Cx45 is strongly voltage dependent, while Cx43 is not, a highly asymmetrical voltage dependence results from heterotypic pairing of cells that express these two proteins. Because of gating characteristics, expression of Cx45 in conductile tissue and of Cx43 in ventricle would favor propagation toward the ventricle and disfavor retrograde firing of the Purkinje system.

### E. Mechanical Synchronization

Regulation of contractile activity and its synchronization over long distances is a prerequisite for the proper functioning of smooth muscles throughout the digestive, respiratory, urinary, and vascular systems, and the genital tract. Although such control depends on the proper cell-to-cell propagation of action potentials, which helps to synchronize cytoplasmic $Ca^{2+}$ increases in multiple cells, the mechanism whereby such synchronization is achieved is still the subject of controversy (Daniel et al., 1976; Wray, 1993). Studies conducted on uterine and vascular smooth muscles have provided strong evidence that this synchronization involves gap junctional communication.

The uterus exhibits spontaneous contractile activity which, in both pregnant and nonpregnant females, is maintained relatively quiescent up to labor, at which time it markedly increases in strength, duration, and coordination to permit effective parturition (Garfield et al., 1980b; Wray, 1993). In several animal species, including humans, gap junctions and their constitutive connexins are sparsely represented among cells of prelabor myometrium (Garfield et al., 1980a, 1981b; Garfield and Hamayaski, 1991; Tabb et al., 1992), a tissue which, at this time, also shows very limited coupling, poor spatial propagation of action potentials, and asynchronous contractions (Miller et al., 1989). Gap junctions, electrical and metabolic coupling markedly increase, due to a differential modulation of different connexin isoforms between uterine muscle cells during labor and delivery (Garfield et al., 1977; Garfield and Hayashi, 1981; Blennerhassett and Garfield, 1991; Sakai et al., 1992; Tabb et al., 1992; Albrecht et al., 1996; Ou et al., 1997). At this time, action potentials and contractions spread across much larger distances, become synchronized and intensify (Wray, 1993). In vitro and *in vivo* studies have shown that the gap junction, coupling, and contraction changes are all controlled by similar modifications in steroid hormones, prostaglandins, and their receptors (Garfield et al., 1980a, 1987; MacKenzie and Garfield, 1986; Dookwah et al., 1992; Ou et al., 1997), a major indication for their likely endogenous regulation under physiologically relevant conditions. Moreover, transcription of Cx43 is upregulated by estrogen due to the presence of a specific responsive element on the promoter region of this gap junction protein (Yu et al., 1994). This impressive series of studies convincingly demonstrates that changes in the contractile ability of the myometrium correlate with changes in the gap junctional communication between its smooth muscle cells. What is not yet established is whether the mechanical changes are actually dependent on gap junctions and, if so, for what reason. It has been hypothesized that gap junctional coupling may indirectly control intracellular $Ca^{2+}$ levels, particularly during phasic contractions of myometrium (Garfield et al., 1992).

Functional gap junctions have also been demonstrated between smooth muscle cells of arterioles, in both resistance and conduit vasculature (Spray et al., 1994c), and numerous studies have documented the presence of Cx43 as well (e.g., Blennerhassett et al., 1987; Larson et al., 1990; Moore et al., 1991; Campos de Carvalho et al., 1993; Christ et al., 1993b; Little et al., 1995). Because vascular smooth muscle cells do not readily generate propagated action potentials (Mekata, 1981; Blennerhassett et al., 1987; Christ et al., 1993a), the coordination of contraction, relaxation, and other functions may be largely mediated by the intercellular exchange of second messengers, presumably through gap junctions (Christ et al., 1996). For example, smooth muscle cells of human corpus cavernosum, the relaxation of which leads to erection, are interconnected by gap junctions formed of Cx43 (Campos de Carvalho et al., 1993) and cultured corporal smooth muscle cells are well coupled electrically (Moreno et al., 1993). Moreover, the distribution of channel sizes and macroscopic junctional conductance are regulated by α-

and β-adrenergic receptor activation, as well as by physiologically relevant second messengers, such as cAMP- and cyclic guanosine monophosphate (cGMP)-dependent protein kinases as well as protein kinase C (Moreno et al., 1993). Pharmacological studies on isolated corporal tissue strips have indicated that gap junctions contribute significantly to pharmacomechanical coupling and syncytial tissue contraction during activation of α-adrenergic receptors (Christ et al., 1991). Because the innervation of corporal smooth muscle is sparse, and because $Ca^{2+}$ is freely diffusible between corporal smooth muscle cells in culture (Christ et al., 1992), a model has been proposed in which the diffusion of second messengers is hypothesized to underlie the spread of contraction and relaxation among cells (Christ et al., 1996).

## F. Compensation for Enzymatic Defects

Since the unraveling of a gap junction–mediated pathway by which cells equipped with hypoxanthine phosphoribosyltransferase permit the survival in cocultures of mutant cells deficient in this enzyme (Subak-Sharpe et al., 1966, 1969), the mechanism of metabolic coupling has been shown to have multiple implications. Thus, at least *in vitro*, several other enzymatic and cofactor defects have been shown to be compensated through the direct cell-to-cell exchange of appropriate metabolites (Hooper and Subak-Sharpe, 1981). However, to what extent such a compensation actually takes place *in vivo* remains to be demonstrated. This uncertainty is due in part to the difficulty of assessing metabolic coupling in intact tissues. This assessment has so far been possible in only a few, selected systems, such as the lens, where connections between epithelial and lens fiber cells appear responsible for metabolite supply throughout this avascular tissue (Goodenough, 1992).

## G. The Bystander Effect

Clinical trials against tumors have begun in which retroviral vectors expressing the herpes simplex virus thymidine kinase gene HSV tk have been introduced into tumor cells (Kolberg, 1994). Infected cells expressing HSV tk are then killed by exposure to ganciclovir, a guanosine analog, which is metabolized to a phosphorylated product by HSV but not mammalian thymidine kinase. This small (300-D) phosphorylated product is incorporated into nascent DNA molecules, stopping synthesis in proliferating cells. Remarkably, and a phenomenon which is fundamental to its therapeutic usefulness, as few as 70% of the tumor cells need to be infected in order to make the tumor sensitive to ganciclovir (Culver et al., 1992). This conferral of ganciclovir sensitivity to uninfected cells is termed the "bystander effect." Although it has been variably attributed to secretion of toxic factors, to disruption of blood supply to the tumor *in vivo*, to apoptosis followed by uptake of toxic metabolites, and to initiation of immune response in killed cells

(Kianmanesh et al., 1997), the bystander effect is now believed to also involve metabolic cooperativity (Bi et al., 1993; Kolberg, 1994). Indeed, recent evidence indicates that the 300-D cytoplasmic phosphorylated ganciclovir metabolite formed in HSV tk$^+$ cells can be transferred to adjacent, noninfected cells, thereby killing them as well (Bi et al., 1993; Denning and Pitts, 1997). Moreover, the bystander effect is much reduced in cells deficient in gap junction expression (Fick et al., 1995) and is conferred upon expression of Cx32 or Cx43 in communication- deficient cells (Elshami et al., 1996; Mesnil et al., 1996) and *in vivo* after Cx37 transfection (Vrionis et al., 1997). Nonetheless, in at least one model system the expression of gap junctions between infected and noninfected cells was found to protect cells from ganciclovir-induced death, a phenomenon termed the "Good Samaritan effect" (Wygoda et al., 1997). Also, whereas expression of gap junctions appears to mediate the bystander effect in most model systems analyzed using the ganciclovir/tk$^+$ system, other enzyme/drug systems do not appear to involve gap junctions (Lawrence et al., 1998).

A function of gap junctions that is analogous to the bystander effect, but involves ions rather than toxic molecules, is channel sharing, whereby ionic fluxes into only a fraction of the cell population can serve to modulate the functions of the entire population. Thus, with relevance to gene therapy, as few as 6% to 10% of cells lacking the cystic fibrosis conductance regulator protein (CFTR) need to be corrected in order to restore normal Cl- transport properties of epithelial sheets (Johnson et al., 1992). Moreover, the sychronized bursting of pancreatic islet cells in clusters, in contrast to the chaotic firing of isolated cells, has been suggested to result, in part, from the sharing of $Ca^{2+}$-sensitive $K^+$ channels which are open on average in less than one cell at a time (Atwater et al., 1983; Sherman et al., 1988).

## H.  Nutrition

Gap junctions are unusually abundant between the terminally differentiated fiber cells of the eye lens, which are organized in a compact avascular mass (Goodenough, 1992). The existence between such cells of extensive metabolic coupling, in spite of a remarkably low metabolism rate, has been taken as a suggestion that junctional coupling could provide a feeding pathway by which cells distant from blood supply, as are the fibers of the lens core, could receive the nutrients required for their rather long life time (Goodenough, 1992). Recent studies revealing cataractous changes in mice lacking the epithelial cell connexin (Cx43) or the connexins in fiber cells (connexin46 [Cx46] and connexin50 [Cx50]) are consistent with such a role (see later discussion).

An analogous feeding function of junctional communication could well be relevant to the functioning of other large avascular systems such as skin epidermis, where the uppermost layers of living keratinocytes are distant from the microvascular front. These metabolically active cells communicate with basal cells

(Salomon et al., 1988) via gap junctions which, in humans, are predominantly made up of Cx43 (Caputo and Pelluchetti, 1977; Guo et al., 1992; Wilgenbus et al., 1992; Salomon et al., 1994). Experimental and pathological conditions that lead to a marked thickening of epidermis, hence resulting in a further increase in the distance between blood supply and some keratinocytes, are associated with a marked increase in connexin expression, noticeably in that of the Cx26 isoform (Masgrau-Peya, et al., 1997; Labarthe, et al., 1998)

Junction-mediated transfer of nutrients has also been implicated in the ovarian follicle. In mammals, gap junctions form between the granulosa cells and the oocytes until the time of ovulation (Anderson and Albertini, 1976), and appear able to transfer to the female gamete metabolites that are necessary for its survival and growth (Gilula et al., 1978; Heller and Schultz, 1980; Brower and Schultz, 1982). Mice lacking Cx37, the connexin that fulfills this function in mammals, have been found to have reduced fertility (Simon et al., 1997).

## I.   Dissipation of Extracellular and Intracellular Molecules and Ions

It has long been proposed that gap junctions among glia provide a pathway for $K^+$ disposal into the perivascular compartment, thus reducing the concentration of this cation in the immediate environment of the neuron (Kuffler and Potter, 1964). In addition, coupling among astrocytes may directly promote $K^+$ buffering as a result of increasing the effective volume of the buffer sink. Thus, the function of glial cells in reducing $K^+$ activity in the vicinity of neurons is clearly enhanced by astroglial junctional interconnections (see the subsequent chapter by Nagy and Dermietzel in this volume). Within processes of a single cell, gap junctions may also function to dissipate intracellular gradients in ions or metabolites, as between cytoplasmic folds of Schwann cells at nodes of Ranvier (Bergoffen et al., 1993), where the presence of Cx32 is believed to provide a millionfold shorter diffusion distance from outer cytoplasm to the adaxonal portions of Schwann cells (Scherer, 1997). Also, gap junctions have been implicated in the dissipation of metabolites across coupled population of astrocytes (Giaume et al., 1997).

## J.   Differentiation

Several lines of work suggest a link between gap junctional coupling and cell differentiation. The first is that the 13 or more members of the connexin family are differentially distributed in distinct tissues and in different cell types of a same tissue (Bennett et al., 1991; Kumar and Gilula, 1996). Even though this complex pattern is not yet easily explainable, its consistency in different species does indicate that not all differentiated cells express the same set of connexins and, hence, do not intercommunicate via the same type of junctional channels. Moreover, this tissue-specific pattern of connexin distribution may be essential for establishment of proper communication compartments, leading to normal tissue differentiation.

A second line of evidence derives from screening connexin distribution in tissues where cells undergo programmed differentiation changes at precise times and locations. In human skin, for example, keratinocytes comprising the surface covering interfollicular epidermis express Cx43 (Guo et al., 1992; Wilgenbus et al., 1992; Salomon et al., 1994). However, this protein is barely detectable in the basal layer of the tissue where poorly differentiated keratinocytes form by mitotic division, sharply increases in the adjacent suprabasal layers where keratinocytes progressively acquire additional differentiation characteristics, decreases abruptly in the granular layer as keratinocytes progress toward full differentiation, and disappears in the uppermost layers of the epidermis that comprise dead cells (Guo et al., 1992; Wilgenbus et al., 1992; Salomon et al., 1994). In the upper layers, Cx26 appears to be highly expressed (Sawey et al., 1996) and several other connexins are also territorially distributed (Butterweck et al., 1994). Thus, in human skin, the heterogeneous distribution of Cx43 correlates with that of distinct differentiation compartments. Furthermore, keratinocytes forming glands and hair follicles undergo a quite distinct differentiation program from those comprising the interfollicular epidermis and, accordingly, also express a different connexin pattern (Salomon et al., 1994).

A third line of support is provided by observations that the pattern of connexins expressed by a given cell type may be qualitatively and quantitatively modified as a function of development and degree of (de)differentiation. In rodent brain, for example, some neuronal populations express Cx26 and Cx43 but no Cx32 during embryogenesis, whereas after birth they express Cx32, much lower levels of Cx43, and no Cx26 (Dermietzel et al., 1989). Interfollicular keratinocytes of human skin which, at variance with what is observed in rodents (Risek et al., 1992), do not normally feature Cx26 (Guo et al., 1992; Wilgenbus et al., 1992; Salomon et al., 1994), express appreciable levels of this protein following tumoral (Wilgenbus et al., 1992) or psoriatic transformation (Labarthe et al., 1998) or chronic retinoic acid treatment (Masgrau-Peya et al., 1997). Also, in a pancreatic acinar cell carcinoma line, exposure to dexamethasone modifies in parallel secretory differentiation and connexin pattern in such a way that the expression of amylase and Cx32 is increased whereas that of Cx26 is decreased. Eventually, several lines of insulin-producing cells which have lost one or more of the biochemical markers that characterize native pancreatic $\beta$-cells, do not retain Cx43 expression (Vozzi et al., 1995), the predominant connexin thus far identified in the islets of Langerhans (Meda et al., 1991). For at least one of these cell lines, a single passage *in vivo* rapidly restores the expression of both $\beta$-cell specific markers and Cx43 (Vozzi et al., 1997).

The fourth, more direct line of evidence has come from studies in which changes in the pattern of connexin types resulted in changes in cell-specific characteristics. Primary hepatocytes express Cx26 and Cx32 but not Cx43, both *in situ* and *in vitro* (Paul, 1986; Zhang and Nicholson, 1989), whereas a number of liver-derived cell lines and of immortalized embryonic hepatocytes express mostly, if not uniquely Cx43 (Stutenkemper et al., 1992; Neveu et al., 1994). Interestingly, the levels of this protein, relative to those of Cx26 and Cx32, is reversibly altered by manipulating the culture medium, and were found to be inversely related with the expression of hepatocyte-specific transcripts (Stutenkemper et al., 1992). Thus, at least in this system, a reversible change in connexin pattern is related to a parallel alteration in cell differentiation. However, it remains to be established whether the connexin modulation affects the hepatocyte-specific phenotype or vice versa, and what the molecular mechanism is linking these two events. Some clues to this question may be derived from recent studies on transgenic mice in which either one of the two genes that code for Cx46 and Cx50, the two connexin isoforms that form gap junctions between lens fibers, has been knocked out. In both cases, the functional loss of one gene resulted in an altered content of the fiber specific crystallins, even though the observed alterations were not the same in the two cases (Gong et al., 1997; White et al., 1998).

A fifth line of evidence is that cells cultured on tissue culture plastic in the presence of serum-supplemented medium, often lose their tissue-specific, differentiated phenotype over a rapid time course (Reid et al., 1988). This process can be slowed down or prevented by matching the culture conditions more closely to the *in vivo* environment; for example, by including hormones, growth factors, and matrix components. In at least a few systems, changes in gap junctions parallel these differentiation changes. For example, gap junctions increase markedly between rabbit corneal fibroblasts within 24 hours of plating on collagen (Nishida et al., 1988), as do rat and human granulosa cells at 24 and 72 hours after plating in a serum-containing or in a matrix-supplemented medium (Ben-Ze'ev and Amsterdam, 1986; Amsterdam et al., 1989). Furthermore, metabolic cooperation between Chinese hamster ovary cells is enhanced fourfold when cells are cultured on coverslips coated with matrix derived from embryonic fibroblasts (Willecke et al., 1983). Although the matrix molecules responsible for these effects were not identified, the studies indicate that either the structure of the matrix, or its binding of some regulatory factor can have marked effects on intercellular communication. The major components of the matrix, proteogylycans and glycosaminoglycans, stimulate dye and electrical coupling between hepatocytes (Spray et al., 1987). Heparin, a glycosaminoglycan, exerts transcriptional effects on Cx32 when added in the presence of glucagon (Rosenberg et al., 1993). In the absence of proteoglycans and glycosaminoglycans and in the presence of serum, gap junctions disappear within 12 hours between primary hepatocytes, leading to uncoupling (Spray et al., 1987). More recent experiments have extended these studies in rat hepatocytes in culture, demonstrating that Cx32 expression can be induced by treatment with epidermal growth factor, together with 2% dimethylsulfoxide (DMSO) and that Cx26 can be induced by DMSO plus glucagon (Kojima et al., 1995a, 1995b). Conversely, gap junction expression is prolonged about twofold by omitting serum and adding glucagon and linoleic acid, and coupling can be reinduced between cells that have lost it by the addition of heparin and dermatan sulfate. Coupling re-expression, however, is more dramatic when proteoglycans are added (Spray et al., 1987).

Signaling by matrix and soluble factors may stimulate coupling between heterologous cell types to a greater extent than between homologous cells. For example, coculture of hepatic endothelial cells with hepatocytes induces the production of fibronectin, type I collagen, and laminin, and selectively enhances gap junction expression between hepatocytes (Goulet et al., 1988).

## K.  Cell Division

At present, the possible involvement of gap junctional communication in controlling the growth of normal tissues has focused mostly on the mammalian oocyte. The meiotic division of this gamete starts during fetal life, is arrested at birth, and is normally completed only after puberty, at the time of ovulation (Lindner et al., 1974). The observations that the meiotic arrest of the oocyte depends on the maintenance of contacts with surrounding granulosa cells (Racowsky and Satterlie, 1985; Racowsky et al., 1989), and that completion of meiosis can be induced experimentally either by separating the oocytes and the granulosa cells (Edwards, 1965) or by pharmacologically blocking their gap junctions (Dekel and Piontkewitz, 1991), strongly suggest the gap junction–mediated transfer of some meiotic inhibitor, perhaps cAMP or $IP_3$ from granulosa cells into the oocyte (Sandberg et al., 1992). Interruption of this transfer would be expected to allow resumption of oocyte maturation. Certainly, a junction-mediated exchange of ions and molecules takes place between granulosa cells and the oocyte and can be modulated by hormones surging just prior to ovulation (Dekel and Beers, 1978; Gilula et al., 1978; Moor et al., 1980; Dekel et al., 1981; Larsen et al., 1981, 1986, 1987; Granot and Dekel, 1994). The critical question is to know whether this exchange also implicates the putative meiotic inhibitor(s), as the nature of these molecules is still elusive. Consistent with this possibility are the recent observations made in a transgenic mice line in which the gene coding for Cx37, a connexin that forms junctional channels between the oocyte and the granulosa cells of ovarian follicles, had been functionally suppressed. Homozygous knockout animals that do not express this connexin, no longer exhibit junctional coupling between the oocyte and the granulosa cells and are infertile, as a result of one or more alterations that may include lack of follicle maturation, premature luteal transformation or granulosa cells, or insufficient maturation of the meiotic competence of the oocyte, or a combination of all of these (Simon et al., 1997). Another knockout transgenic model suggests that loss of Cx50 affects the postnatal growth fibers in the mouse lens (White et al., 1998). However, it remains to be established why such a change was not observed when Cx46, the other lens fiber connexin, is no longer expressed (Gong et al., 1997). The development of double transgenic animals lacking both connexin isoforms may provide some insight into this question. Much more numerous have been the studies evaluating the role of gap junctional communication in the control of growth of tumoral cells. These studies are reviewed later in section V.B.

## L. Migration

Several important biological processes such as neurulation, morphogenesis of differentiated epithelia, and wound closure depend on the movement of cell sheets; that is, on the coordinated motion of confluent assemblies of tightly associated cells. A particularly important example is the coordinated movement of endothelial cell monolayers, which is essential to ensure a continuous, nonthrombogenic epithelial-like covering during both formation and renewal of blood vessels (Ross, 1993). Such movement can be reproduced in culture by mechanical wounding a confluent monolayer of endothelial cells (Sholley et al., 1977). In that case, the cells bordering the wound are induced to proliferate and to migrate into the denuded dish area. At least with cells derived from microvessels, this migration is associated with increased expression of Cx43 (Pepper et al., 1992) and a local enhancement of coupling, which is not observed between the quiescent cells that are distant from the wound edge (Pepper et al., 1989). A series of experiments indicates that conditions blocking migration, but not cell division, prevented the coupling increase and, conversely, that pharmacological blockers of gap junctions abolished the movement of endothelial monolayers, without blocking the migration of individual cells (Pepper et al., 1989). Thus, gap junctional communication may coordinate the movement of endothelial cells when their motion as a coherent sheet is required. Whether gap junctions actually play this role *in vivo* remains to be determined, particularly because endothelial cells then express a different connexin pattern. Indeed, whereas Cx43 (Larson et al., 1990; Pepper et al.,1992) and to a lesser degree Cx37 (Reed et al., 1993) appear to be the predominant endothelial connexins *in vitro*, *in vivo* these proteins may be downregulated relative to Cx40, in some but not all vascular compartments (Bruzzone et al., 1993). Nevertheless, the occurrence of extensive endothelial cell coupling *in situ* (Segal and Bény, 1992, Little et al., 1994) and the observation that bFGF, an angiogenic stimulus that is active under physiologically relevant conditions (Klagsbrun and D'Amore, 1991), colocalizes in part to Cx43 containing gap junctions (Kardami et al., 1991; Yamamoto et al., 1991) and increases coupling in monolayers of quiescent endothelial cells (Pepper and Meda, 1992), supports the possibility that gap junctional communication may coordinate the migration of endothelial cells. This coordination may also be relevant for the migration of other cell types. Thus, lineages of neural crest cells, which contribute to a wide range of embryonic tissues, also form functional gap junction channels made of Cx43, and an abnormally elevated expression of this connexin perturbs the development of tissues that form as a result of a coordinated, prenatal migration of neural crest subpopulations (Ewart et al., 1997; Lo et al., 1997).

## M.  Secretion

It is now well established that gap junctions and coupling are obligatory features of virtually every glandular epithelium, even after the morphogenetic and functional development of secretory cells is completed (Meda et al., 1984, 1995, 1996; Munari-Silem and Rousset, 1996). Often, gap junctions are unusually abundant in glandular tissues, as compared to nonsecretory epithelia (Friend and Gilula, 1972). This extensive development, together with the short half-life of connexins (Fallon and Goodenough, 1981; Yancey et al., 1981; Traub et al., 1989; Laird et al., 1991), suggests that the maintenance of gap junctions is important for proper functioning of adult glands. The following observations, made primarily in the pancreas, provide support for this idea.

Firstly, the insulin-producing β-cells, which represent the prominent cell type of the endocrine pancreas, secrete poorly following disruption of native cell-to-cell contacts. Restoration of such contacts promotes insulin secretion, protein biosynthesis, and the expression of the insulin gene (Salomon and Meda, 1986; Bosco and Meda, 1991; Philippe et al., 1992). Secondly, sustained stimulation of insulin secretion, *in vitro* and *in vivo*, increases gap junctions, expression of Cx43 and coupling of β-cells (Meda et al., 1979, 1980, 1983, 1990a; Kohen et al., 1983). Conversely, conditions inhibiting insulin secretion decrease or abolish β-to-β-cell coupling (Kohen et al., 1983). Thirdly, an acute pharmacological blockade of gap junctions markedly alters basal and stimulated insulin secretion of isolated islets of Langerhans and intact pancreas (Bruzzone and Meda, 1988; Meda et al., 1990b). These alterations appear to be specifically related to perturbation of coupling, as they could not be accounted for by detectable changes in the second messengers known to control insulin secretion, were rapidly and fully reversible after wash out of the uncoupling drug, and were not observed with single β-cells (Meda et al., 1990a). Fourthly, a number of tumoral and transformed insulin-producing cells that feature secretory defects do not express connexins, do not form gap junctions, and are virtually uncoupled *in vitro* (Vozzi et al., 1995). Induction of gap junction channels by transfection of Cx43 in these cell lines improved their limited secretion of insulin and the expression of the insulin gene *in vitro* (Vozzi et al., 1997). A similar promoting effect of coupling was observed *in vivo*. Thus, after Cx43 was expressed in tumoral insulin-producing cells to achieve a moderate increase in junctional coupling, insulin release was enhanced (Vozzi et al., 1997). In contrast, glucose-stimulated insulin secretion was found to be markedly reduced after the extent of β-cell coupling was strongly enhanced by overexpression of an ectopic connexin (White et al., 1996), suggesting the importance of critical levels of junctional communication in this system.

The participation of junctional coupling in secretion is also supported by experiments on the exocrine acinar cells of salivary glands (Sasaki et al., 1988; Kanno et al., 1991) and pancreas (Meda et al., 1988). The latter gland has been most studied and is now one of the nonelectrically excitable systems in which the function of gap junctional coupling is best characterized by several lines of experimental evidence. The first is that dispersed acinar cells secrete poorly under basal and stimulatory conditions and markedly improve their response to secretagogues following reestablishment of junctional contacts (Bosco et al., 1994). The second line of evidence is the finding that most physiological pancreatic secretagogues rapidly uncouple acinar cells in a fully reversible way, *in vitro* and *in vivo* (Iwatsuki and Petersen, 1978; Meda et al., 1986, 1987). At least for cholinergic neurotransmitters, which are the prominent physiological stimuli of the exocrine pancreas, it is now established that this effect is caused by a $Ca^{2+-}$, pH-, and PKC-independent gating of gap junction channels (Meda et al., 1990b). The third line of evidence is that several alkanols, which effectively uncouple acinar cells, reproduce the secretory stimulation induced by physiological secretagogues in various preparations of acinar cells, ranging from cell pairs to the intact gland (Meda et al., 1986, 1987; Bosco et al., 1988, 1994). The finding that such a stimulation is not observed when acinar cells are previously uncoupled by physical dispersion or cholinergic stimulation, is not associated with detectable alterations in nonjunctional conductances and in second messengers controlling acinar cell secretion, is prevented under conditions inhibiting exocytosis of pancreatic enzymes, and is fully and rapidly reversible following washout of the uncoupling drugs, indicate the specificity of the observed coupling-secretion relationship (Meda et al., 1986, 1987, 1990a; Bosco, et al., 1994). Further support for this specificity has come from the analysis of transgenic mice in which the gene coding for Cx32, the predominant connexin expression by acinar cells of mouse pancreas, had been knocked out by homologous recombination (Nelles et al., 1996). Knockout mice that no longer expressed this connexin and featured a much reduced acinar cell coupling, also showed an increase in the basal release of amylase, resulting in elevated circulating levels of the enzyme in spite of a normal responsiveness to cholinergic secretagogues (Chanson et al., 1998). The data extend to *in vivo* conditions, the observations previously made *in vitro*. Taken together, this set of observations also provides compelling evidence that coupling and secretion of pancreatic cells are causally linked. Clearly, however, the molecular mechanism underlying such a link remains to be unraveled, as is its functional significance. The identification of an intrinsic structural and functional heterogeneity of both β (Salomon and Meda, 1986, Bosco and Meda, 1991) and acinar cell populations (Bosco et al., 1994) certainly provides a first clue to conceptualization of the homeostatic manifestation of coupling likely to have important consequences in both the endocrine and exocrine secretion of the pancreas (Meda, 1995, 1996).

If the perspective that coupling provides a mechanism to compensate for cell heterogeneity is correct (see section VI), it is unclear why the β- and acinar cells of the pancreas express a completely different pattern of connexins, gap junctions, and coupling, and show an essentially opposite regulation of junctional communications during stimulation (Meda, 1995; Meda et al., 1990b). Screening of connexins in a variety of glands has revealed that endocrine cells always express Cx43 and usually not Cx32, as in insulin-producing β-cells, whereas exocrine cells express Cx32 and usually not Cx43, as in the pancreatic enzyme-producing acinar cells. This alternative connexin distribution was observed in glands derived from the three germ layers, irrespective of the spatial arrangement of secretory cells or whether these cells released peptides, glycoproteins, or lipids, suggesting that the expression of gap junction proteins may be differentially regulated depending on whether a secretory cell becomes part of an endocrine or of an exocrine gland (Meda et al., 1993). From this perspective, it is intriguing that the segregation of endocrine and exocrine cells into distinct glands occurred in the early vertebrates, at about the same time when a gene duplication permitted the divergence of Cx43, and other connexins of group II, from Cx26 and Cx32, and other connexins of group I (Bennett et al., 1991).

Recent studies further suggest that, in addition to participating in peptide secretion and gland morphogenesis, junctional coupling could also be involved in modulating the expression of gland-specific genes. Thus, in the AR4-2J cell line, derived from a pancreatic acinar cell carcinoma, dexamethasone increases amylase secretion in parallel with the expression of Cx32 and junctional coupling while decreasing that of Cx26. An analogous regulation may also apply to the insulin gene, whose glucocorticoid-mediated inhibition can be prevented by cAMP in coupled but not uncoupled insulin-producing cells (Philippe et al., 1992).

## V. GAP JUNCTIONAL COMMUNICATION IN SOMATIC DISEASE STATES

The occurrence of gap junctions in virtually every tissue, and the major, if still putative, influences the intercellular communication they mediate could have on multiple cell functions, raises the obvious possibility that alterations in connexins, gap junctions or coupling, or both, could participate in the pathogenesis of numerous disease states. Accordingly, such a participation has been hypothesized in a number of rather different clinical situations, including some resulting from alterations of myocardium (Saffitz et al., 1992; 1993; Campos de Carvalho et al., 1994), myometrium (Garfield et al., 1980b), and smooth muscle contraction (McGuire and Twietmeyer, 1985; Larson et al., 1990; Christ et al., 1993a, 1993b), as well as from abnormal vessel repair or function (Larson and Haudenschild, 1988; Pepper et al., 1989, 1992; Haefliger et al., 1997; 1999), cell differentiation and growth (Loewenstein, 1968a; 1968b; Loewenstein and Rose, 1992), and defective secretion (Meda, 1996). Only within the past few years has it been shown that diseases could be specifically linked to mutations of specific connexin isoforms (Bergoffen et al., 1993; Fairweather et al., 1994, Ionasescu et al., 1994;

Kelsell et al., 1997; Zelante et al., 1997; such hereditary diseases and phenotypic changes observed in connexin knockout mice are considered in the next section).

There are several reasons why the role of gap junctions in neural somatic diseases remains unclear. The functions, and thus the dysfunctions, of gap junctions are not easy to track, as these structures probably are responsible for pleiotropic roles, which may differ in different tissues. Furthermore, gap junctional coupling represents but one form of cell-to-cell communication, and it is most likely that vital systems have evolved mechanisms by which other forms of direct or indirect cell interactions may functionally compensate for a junctional defect. Nevertheless, it is important to point out that the most common illnesses of modern societies (i.e., cardiovascular diseases, cancer, and diabetes) are all diseases that are thought to represent disorders in communication between cells (Rasmussen, 1991). It is therefore essential to precisely define the initial derangements of intercellular signaling that begin the process of these chronic diseases. It would be rather surprising if, once the analysis of such derangements is completed using updated technological tools, no involvement of gap junctional communication were to be found.

## A. Cardiovascular Diseases

Mammalian survival on a minute-to-minute basis depends on delivery of blood, the force for which is provided by coordinated contractions of the heart. Because gap junctions provide the pathways of current flow between cells, clinically relevant cardiac dysfunction would be expected to result from conditions in which cardiac gap junction function, organization, or expression is altered. Conductance of gap junction channels is decreased by many of the factors that have been found to change in the failing or ischemic heart (Gettes and Cascio, 1992), where pH is decreased, $Ca^{2+}$ is elevated, fatty acids and free radicals are generated, ATP levels and protein kinase activities are reduced, and cells may be individually depolarized, hence generating transjunctional voltage gradients (Spray and Fishman, 1996). In addition to these acute changes, which may profoundly alter channel gating and hence affect both conduction velocity and pathway anisotropy, long-term changes in gap junction expression may also result in conduction changes (Spray et al., 1994). For example, resculpting of ventricular bundles due to collagen deposition splits the aging heart into a maze of meandering conduction pathways which hinders gap junction formation, hence reducing conductivity and synchronization (Spach et al., 1986). Similar resculpting and disorganization may occur at the infarct border zone immediately or chronically after the infarction incident, and be sustained as a chronic abnormality (Luke and Saffitz, 1991; Smith et al., 1991; Green and Severs, 1993; Spray et al., 1994c; Peters et al., 1997). As a consequence, conduction in the longitudinal direction may be favored over transverse conduction, or the total complement of junctional channels may be reduced (see Spach, 1997).

Chagas' disease, which is one of the prevalent causes of heart disease in South America, is caused by infection with *Trypanosoma cruzi*. In vitro infection of rat neonatal cardiac myocytes with this protozoan parasite is accompanied by marked disturbances in intercellular communication, which may underlie aberrant conduction patterns (Campos de Carvalho et al., 1992; 1994). Dye and electrical coupling are decreased between infected cells, and this uncoupling is consistent with a marked decrease of Cx43 in the appositional membrane regions of infected cells. These findings would predict that infection of the heart with this parasite should result in a slowing of cardiac conduction and in alterations of propagation patterns. Indeed, the synchrony and rhythmicity of contractions in infected myocyte cultures were found to be decreased compared with controls (Campos de Carvalho et al., 1994). Chronic Chagas' disease generally results in severe cardiomyopathy in which the integrity of both the myocardium and the conduction system is progressively compromised. Under these conditions, the spread of excitation through the heart is altered, leading to arrhythmias and other conduction disturbances. Likely, these disturbances result from decreased safety factor for intercellular communication, as focal tissue damage exaggerates the fragmentation of the conduction pathways that takes place within the aging heart (Spach et al., 1986; Spach, 1997).

In addition to pathological conduction and asynchrony in cardiac tissue, gap junction alterations are likely to be involved in dysfunctions of vasomotor tone (Segal and Duling, 1987; 1989; Christ et al., 1996). Thus, disturbed coordination and second messenger exchange among endothelial and smooth muscle cells could contribute to a number of vascular diseases such as coronary and cerebral vasospasm, Raynaud's syndrome, atherosclerosis (Blackburn et al., 1995), and cardiovascular complications of diabetes mellitus. As yet, however, most studies have focused on the possible participation of gap junctions in the pathogenesis and maintenance of the cardiovascular changes that take place in chronic hypertension. Thus, in different animal models of the disease that feature the pathological thickening of both endothelial and smooth muscle layers of elastic arteries, alterations in the appearance of gap junction plaques and in connexin expression have been reported (Huttner et al., 1982; Watts and Webb, 1996; Blackburn et al., 1997; Haefliger et al., 1997; 1999). A consistent finding has been the increased expression of Cx43 in the wall of arteries featuring altered isobaric distensibility (Watts and Webb, 1996; Blackburn et al., 1997; Haefliger et al., 1997; 1999), whereas either a decrease or no change in the levels of this connexin were observed when arteries retained a normal distensibility in spite of the elevation in blood pressure (Haefliger et al., 1999). Also, changes in Cx43 were not obligatorily observed in the hypertrophic myocardium of the same hypertensive animals that featured increased expression of this protein in arteries (Bastide et al., 1993; Sepp et al., 1996; Haefliger et al., 1997; 1999). These apparently contradictory findings have led to the suggestion that Cx43 may act as a mechanical sensor, transcriptionally regulatable by stretch of the vascular wall (Haefliger et al., 1997;

1999). Recent *in vitro* studies on the response of different vascular connexins to mechanical loads are consistent with this view (Cowan et al., 1998). The further observation that the gene coding for Cx43 cosegregates with body weight in the model of spontaneously hypertensive rats further suggests the noteworthy possibility that genetic alterations in the locus of this connexin could predispose to hypertension (Katsuya et al., 1995).

## B. Tumorigenesis

Soon after the identification of gap junctional coupling, it was hypothesized that this unique form of intercellular communication could participate in the control of cell growth (Furshpan and Potter, 1968; Loewenstein, 1968a, 1968b; Loewenstein and Rose, 1992). Since then, support for this idea has been examined mostly *in vitro*, using primary or permanent cell lines of normal, transformed, or transfected cells.

The simple hypothesis that tumor cells were devoid of intercellular channels (Loewenstein and Kanno, 1967) was soon shown not to be universally correct (Hulser, 1974; Sheridan and Johnson, 1975), but the possibility has remained that the strength of coupling or the abundance of connexins between cancer cells, or both, is reduced compared to that of the normal tissues from which they derive (Atkinson and Sheridan, 1985; Loewenstein and Rose, 1992). In hepatomas, for example, gap junction proteins, as shown by immunofluorescence, were found to be strikingly reduced compared to adjacent normal tissue (Yamasaki et al., 1987). Furthermore, it is possible that different connexins may offer properties that are more or less favorable for tumorigenesis. Thus, Cx32 expression appears reduced in rat liver carcinogenesis, whereas expression of Cx26 and Cx43 may be increased (Beer et al., 1988; Mesnil et al., 1988; Fitzgerald et al., 1989; Neveu et al., 1990). This different regulation could produce communication barriers between host and tumor cells, or could affect the permeability of the junctions between normal and tumor cells. However, it is still uncertain whether the effects on gap junctions are early epigenetic events leading to tumor formation. On the one hand, loss of Cx32 increases the incidence of chemically induced tumors, at least in liver (Temme et al., 1997). On the other hand, tumor promotion may be initiated by incorporation into the genome of the genetic information provided by an oncogenic virus or by the activation of either oncogenes or progenitor genes in the target cell. The role of reduced junctional communication in such tumor promotion has been hypothesized as permissive, so that initiated cells may continue to divide prior to the expression of additional genetic changes (Trosko and Chang, 1980).

One approach to evaluate the relationship between intercellular communication and cell growth has involved the introduction of Cx32 cDNA into tumorigenic cells in which communication is poor or absent. This condition is found, for example, in the highly metastatic human hepatoma line SKHep1. Studies comparing clones of SKHep1 cells stably transfected with either a neo vector alone (communication-deficient cells) or with a vector containing Cx32 cDNA driven by an SV40 promoter (coupled cells), revealed no difference in growth rate under tissue culture conditions (Eghbali et al., 1991). However, when these clones were injected into nude mice, the uncoupled cells (as well as a very poorly coupled clone of Cx32-transfectants) formed tumors that grew much more rapidly, in weight and volume, than those arising from well coupled clones (Eghbali et al., 1991). Together with other recent studies showing that some connexin-transfected clones of other cell types exhibit reduced growth rate (Mehta et al., 1991; Zhu et al., 1992; Naus et al., 1992; Bond et al., 1994; Vozzi et al., 1997), these data provide evidence that junctional communication may play a significant role in controlling the *in vivo* growth of tumoral cells. Other recent though less direct studies further suggest such a role. For example, glioma cells transfected with the oncogene neu (cerb-B2) have been shown to exhibit a major reduction in intercellular communication with no decrease in overall expression of Cx43 (Hofer et al., 1996). The change in the growth rate of these cells is interpreted as resulting from altered trafficking of the gap junction protein to the membrane. Moreover, one study has indicated that, compared to normal human breast cells, several breast tumor cell lines do not express Cx26 and do not show dye coupling, suggesting that, in this system, Cx26 may be a candidate tumor suppressor (Lee et al., 1992).

In summary, whereas growth rate may be influenced by changes in junctional communication in some cell types and under some conditions, it seems unlikely that gap junction deficiency per se is a necessary and sufficient condition for tumorigenesis (Omori et al., 1998). Obviously, this tentative conclusion does not dismiss the possible participation of gap junctions in cancer behavior.

## C. Neurological Diseases

Mutations in the Cx32 gene have now been clearly associated with the X-linked form of Charcot-Marie-Tooth syndrome (Bergoffen et al., 1993; Scherer, 1997), mutations of Cx26 have been associated with hereditary nonsyndromic sensorineural deafness (Kelsell et al., 1997; Zelante et al., 1997), and mutations of Cx31 have been associated with another autosomal dominant form of hearing disorder (Xia et al., 1998). In addition to these genetic disorders, discussed in the next section, there have been several studies showing gap junction changes in various nonhereditary brain pathologies (Spray and Dermietzel, 1995).

Several lesions of the central nervous system appear to result in alterations of gap junctional coupling. For example, lesion of the facial nerve results in a marked increase in coupling among astrocytes of its motor nucleus (Rohlmann et al., 1993), and an experimental wound of the visual cortex resulted in regional loss of astrocyte dye coupling (C. Muller, unpublished results). These alterations may be elicited by changes in extracellular matrix or in the transcriptional control of connexins, possibly influenced by the cell cycle (Dermietzel et al., 1987). Interestingly, a kainic acid injection into brain, which intially was expected to cause the loss of gap junctions in astrocytes (Vakelic et al., 1991), has now been shown to result in a masking of the connexin epitopes, without change in Cx43 abundance (Ochalski et al., 1995).

Parasitic infection also alters gap junction distribution in brain cells. *T. cruzi* and *Toxoplasma gondii*, which are associated with brain infection in immunosuppressed individuals, result in loss of coupling and reduced expression of Cx43-containing plaques in astrocytes (Campos de Carvalho et al., 1998). In leptomeningeal cells, both protozoans also reduce coupling and the expression of Cx26 and Cx43. Thus, the two parasites exert generalized junctional disruption, both with regard to type of host cell and of connexin. Because the changes in connexin distribution occur in parasitized cells without changes in the abundance of the proteins, a trafficking disorder appears responsible (Campos de Carvalho et al., 1994). Although not yet demonstrated, the consequence of parasite-induced loss of intercellular communication would be expected to disrupt both $K^+$ balance within glia and the pattern and occurrence of $Ca^{2+}$ waves passing between glial cells.

Increased neuronal excitability could arise from changes in extent of coupling between either neurons or astrocytes, as glial cells appear to mediate the efficient removal of extracellular $K^+$, which is required to prevent the constant depolarization of neurons. Previous evidence of increased coupling between hyperexcitable neurons and in immature cortex was interpreted as consistent with a participation of gap junctional communication in the lower seizure threshold of infants and adults (Peinado et al., 1993). Analysis of Cx43 mRNA levels in samples of cerebral cortex showed that this gap junction protein is markedly increased in epileptic patients (Naus et al., 1991; Elisevich et al., 1997). As most Cx43 found in brain is localized in astrocytes, an implication of this study is presumably that gap junction expression is altered in glial cells of epileptogenic tissue.

A curiosity that has puzzled neurobiologists is that strong mechanical or electrical stimuli delivered focally to brain or retina can results in a radially spreading wave of inhibition, which leaves the entire area refractory to further stimulation (Leao, 1944). The rate of propagation of this wave of depression, which can be observed visually in an eyecup preparation as a slowly moving change in opacity, is about 20 to 50 m/sec (Martins-Ferreira, 1994; Nedergaard, 1994). Because this velocity is similar to that measured for $Ca^{2+}$ waves among glial cells, in both culture and slice preparations, and because the phenomenon is inhibited by agents blocking gap junction channels (Martins-Ferreira and Riberao, 1995; Nedergaard et al., 1995), it is plausible that the spreading depression also results from a gap junction–mediated $Ca^{2+}$ exchange.

## D.  Skin Diseases

Mutations in Cx31 have been shown to be associated with ethrythrokeratodermia variabilis, an autosomal dominant genodermiatosus (Richard et al., 1998), and alterations in the levels of Cx26 have been reported in plaques of psoriatic patients (Labarthe et al., 1998).

## VI. MOLECULAR GENETIC STUDIES OF GAP JUNCTIONS

The application of molecular genetic strategies to the gap junction field is beginning to yield rich insights into the roles that individual gap junction proteins play in different tissues. Thus, an increasing number of studies have investigated the effects of eliminating connexin expression in mice by homologous recombination, whereas others have mapped human genetic diseases with pedigree analysis. One lesson that can be already be derived from these studies is that the degree to which gap junction distribution is similar or different in mouse and human can result in similarities or striking differences in the phenotype associated with the disturbed expression.

Mice lacking Cx26 die in utero, presumably as a result of failure of placental formation or function (Gabriel et al., 1998; Willecke et al., 1998), which appears to depend markedly on the expression of Cx26 in the trophoblast (Winterhager, et al., 1993). In several human kindreds with nonsyndromic hereditary deafness, Cx26 (which is expressed between auditory hair cells) appears to be the culprit gene (Kelsell et al., 1997; Zelante et al., 1997), and skin disease may also result from severe Cx26 coding region mutations (Richard et al., 1997).

Mice lacking Cx32 exhibit a reduced sympathetic-evoked mobilization of glucose from hepatocytes (Nelles et al., 1996), and an increased basal release of amylase from acivar cells (Chanson et al., 1998) presumably as a consequence of inadequate diffusion of second messenger throughout hepatic (see Sáez et al., 1993) and pancreatic acini, respectively. Cx32 knockout mice also display more frequent spontaneous hepatomas and are less resistant to hepatic but not to pancreatic chemical carcinogenesis (Temme et al., 1997). Finally, older Cx32 knockout mice exhibit slower peripheral nerve conduction velocities, Schwann cell onion bodies, and occasional abnormalities in gait (Anzini et al., 1997). These latter phenotypes are believed to be analogous to the abnormalities seen in patients with the X-linked form of Charcot-Marie-Tooth disease (CMTX), which has been associated with more than 70 distinct mutations in the coding region of the Cx32 gene in over 100 families (Bergoffen et al., 1993; Scherer, 1997). Cx32 localization in myelinating Schwann cells is believed to be between the cytoplasmic folds, which are restricted to paranodal regions, and at the Schmidt-Lantermann incisures. Why CMTX patients do not apparently display abnormal liver and pancreatic functions or a higher incidence of hepatomas is unexplained, as is the question of why the other connexins expressed in Schwann cells (Zhao et al., 1999) do not functionally compensate for the loss of Cx32. Although one explanation for the second issue could be found in the potentially negative-dominant action of Cx32 mutants (Omori et al., 1996), this would not explain the similarities found between Cx32 knockouts and CMTX patients.

Mice lacking Cx37 are less fertile than their littermates, apparently due to the loss of the major gap junction protein providing nourishment from granulosa cells to oocytes in the graafian follicle (Simon et al., 1997). Although polymorphisms in Cx37 DNA sequence are common (Krutovskikh et al., 1996), no disease state has yet been linked to Cx37 mutations.

In one recent report, mice lacking Cx40 were shown to display a slowed intraventricular conduction velocity (Willecke et al., 1998), which is consistent with the belief that this connexin preferentially contributes channels to the cardiac conduction system. However, such striking changes have not been detected in another study (Simon et al., 1998). Although linkage analysis initially suggested that Cx40 might be the culprit gene in a hereditary cardiomyopathy with conduction block (Kass et al., 1994), this view has now been reevaluated.

Mice lacking Cx43 die at birth, owing to a developmental defect that thus far appears to be limited to one region of the heart (Reaume et al., 1995; Lo et al., 1997). The right ventricular outflow tract is enlarged and the pulmonary artery is occluded, so that the animals are killed by closure of the cardiac shunts that at birth redirect circulation from placenta to lungs. Brains and most other organs of these animals are normal in appearance (Reaume et al., 1995; Charollini et al., 1998; Houghton et al., 1998), although an abnormality has been detected histologically in lens (Gao and Spray, 1998). In wild-type lens, epithelial cells are connected to one another by Cx43 gap junctions, whereas underlying lens fiber cells are connected by gap junctions formed of Cx46 and Cx50. In Cx43 knockout lenses, epithelial cells are more loosely apposed to one another and are largely separated from lens fiber cells. Within the lens, and especially in deep medullary areas, fiber cells are also separated by vacuolar spaces. These changes are similar to those encountered in early stages of osmotic cataract, reinforcing the hypothesis that Cx43 in epithelial cells is essential in providing nutrient and metabolite exchange from anterior chamber fluid to the inner, avascular lens.

Although the perinatal mortality of Cx43 knockout mice has limited the physiological studies of the animals, analysis of cells cultured from the heart, brain, and lens of these animals have revealed the presence of other connexins that may partially compensate for intercellular communication after loss of Cx43. Junctional conductance measured in pairs of Cx43 knockout cardiac myocytes is about 60% lower than in wild-type siblings, whereas that measured in astrocytes is 95% lower (Spray et al., 1998). Remarkably, however, dye coupling, as evaluated by the intercellular transfer of Lucifer yellow, is more common in these astrocytes than in cardiac myocytes, implying that the residual connexins are different. Single channel studies and molecular probes reveal that Cx40 is present in both cell types, whereas Cx46 is also expressed in astrocytes, and the difference in the anion permeability of the channels formed by these connexins may account for the difference. Strikingly, cardiac cells of Cx43 knockout cultures contract at a slower rate and in a less synchronized manner than wild-type controls, presumably because of decreased entrainment of cells by endogenous pacemakers. The propagation of slow $Ca^{2+}$ waves between cardiac myocytes and between astrocytes is also compromised in cells cultured from Cx43-deficient mice compared to wild-type siblings, although the degree of impairment is less than might be expected from the dramatic decrease in junctional conductance. These studies highlight the importance that extracellular mechanisms play in the propagation of $Ca^{2+}$ waves and emphasize that the safety factor for ensuring such intracellular waves is very high.

In culture, lens epithelial cells obtained from Cx43 knockout and wild-type siblings readily differentiate into lentoids, recapitulating the differentiation that occurs *in vivo* at the bow regions of the lens, and expressed Cx46, Cx50 and markers of numerous differentiation, such as MP26 and α-crystallin. These findings indicate that the presence of Cx43 is not obligatory for lens fiber differentiation, as is also the case in other cell systems *in vivo* (Charcollai et al., 1998, Hampton et al., 1998; Reaume et al., 1998).

Growth rates have been compared in astrocytes and fibroblast cell lines obtained from Cx43 knockout mice, with quite different results. Whereas lines of Cx43 knockout fibroblasts grow more rapidly than the wild-type lines (Martyn et al., 1997), Cx43 knockout astrocytes grow so slowly that care must be taken to study cultures at similar degrees of confluence (Naus et al., 1997; Spray et al., 1998). The extent to which this striking difference in the behavior of two cell types lacking Cx43 represents different requirements of intercellular signaling may provide insights about control of cell growth.

Heterozygous mice carrying only one allele of the mutated Cx43 null gene, feature a decreased Cx43 expression closely approximating the 50% reduction expected from gene dosage. Surprisingly, considering the high safety factor expected for ventricular action potential propagation, a reduced conduction velocity has been reported in these heterozygtes (Guerrero et al., 1997). Studies investigating whether the function of other organs (e.g., brain and pancreas) that appear to require Cx43 expression will also reveal functional abnormalities in the Cx43 heterozygotes are now very much needed.

Human genetic diseases involving Cx43 mutations might be expected to be common, given the widespread expression and functional importance of these channels in most tissues. Alternatively, the requirement for constitutive Cx43 expression might be such that any severe defect would be embryonically lethal. If this were the case, one might predict that subtle changes in Cx43 sequence would result in profound deficits. Consistent with this latter view, sequence polymorphism is much lower in Cx43 than in Cx37 (Sato et al., 1997). Point mutations of Cx43 in the consensus region of serine phosphorylation sites have been reported in a small subset of patients affected by visceroatrial heterotaxia (Britz-Cunningham et al., 1995), a clinically heterozygous syndrome characterized by laterality defects in major internal organs. Although extensive pedigree analysis by other investigators has failed to find additional examples of such changes (Penman-Splitt et al., 1997; Gebbia et al., 1998), it remains possible that the patients in which such mutations were reported represented a particular subset of the complex heterotoxic picture. Interestingly, laterality affects reminiscent of those observed in these patients are also seen in transgenic mice overexpressing Cx43 (Ewart et al., 1997).

Mice lacking Cx46, a major component of gap junctional communication between lens fiber cells, exhibit profound lens opacification that is similar to the nuclear cataracts seen in other experimental models and in human disease (Gong et al., 1997). This defect is associated with altered crystallin expression, which normally characterizes proper differentiation of fiber cells. Whether the remyelination that follows nerve crush injury, a process in which Cx46 expression between Schwann cells has been hypothesized to play a major role (Chandross et al., 1996), is altered in these animals awaits further experimentation. As yet, human genetic diseases involving Cx46 have not been reported.

Mice lacking Cx50 also exhibit nuclear cataracts and featured micro-opthalmia (White et al., 1998), two defects that may be predicted based on loss of a major pathway of intercellular communication between fiber cells. A recent report has suggested that a rare human nuclear cataractous condition referred to as zonular pulverulent cataract, may be linked to a point mutation in Cx50 (Steele et al., 1998). Whether such a mutation leads to alterations in the function or trafficking of connexin channels awaits the use of exogenous expression systems in which the mutated construct could be evaluated.

## VII.   GAP JUNCTION-DIRECTED THERAPEUTICS

The discussion in sections V and VI emphasizes that aberrations in gap junction expression or function may underlie certain pathological conditions, and it is thus of interest to consider the possibility of developing gap junction-targeted therapeutics, including genetic (Nabel et al., 1989; Wilson et al., 1989; Goodenough and Musil, 1993) and pharmacological approaches. In the latter perspective, it is worth recalling that a large number of drugs and of endogenous molecules (hormones, growth factors, morphogens, differentiation factors, etc.) have been shown to affect the expression of connexins, gap junctions, or coupling, or all of these (Spray and Bennett, 1985; Spray, 1994a). Nevertheless, most of these agents act with little tissue and channel specificity or have extremely variable effects on coupling depending on the experimental conditions. One of the most exciting possibilities that emerges from the extensive work devoted to the gating properties of gap junction channels is to identify agents that may act selectively on specific tissues and on selected types of connexins, and thus could be of therapeutic value.

It is generally believed that the tissue-specific distribution of the 15 connexins that have been cloned and sequenced reflects the specific properties of the channels these connexins form. For example, a highly voltage-dependent gap junction channel would have little functional utility in heart, where transjunctional voltage swings of 100 mV occur at least transiently. However, superimposed on these connexin-specific properties are regulatory mechanisms that, for the same connexin, will vary in different cell types. In this respect, the most illustrative examples relate to the phosphorylation-induced modulation of channel properties, where unitary conductance of Cx43 channels varies according to cell type (Spray, 1994a). Several other agents that also modulate the gating of gap junction channels have distinct specificity in different tissues.

The actions of certain agents on intercellular coupling can be distinguished on the basis of their time courses. In addition to the rapid gating mechanisms discussed earlier, the very rapid turnover of junctional proteins, which have a half-life of less than 3 to 5 hours, necessitates consideration that synthesis be another regulatory site of coupling strength. Although mRNA stability has been indicated to be profoundly affected by cAMP (Sáez, et al., 1989b; Kren et al., 1993), there is as yet no indication of stabilizing or degrading sequences in connexin mRNA. Most connexin genes with the exception of those coding for skate Cx35 (O'Brien et al., 1996) and Cx36 (Condorelli et al., 1997) have been shown to exhibit a common structure, with a single intron separating a noncoding upstream exon from a second downstream exon, which contains the entire coding sequence. For connexins 43, 40, 32, and 31, the basal promoter regions have been mapped upstream from the first exon (DeLeon et al., 1994; Bai et al., 1993; Yu, et al., 1994; Seul et al., 1997; Strobl et al., 1998). Canonical sites for action of gene regulatory elements have been identified (see earlier discussion and Miller et al., 1988), and studies are underway to resolve which of these factors are functional in transcriptional regulation. One set of exciting new data indicates the existence of a novel DNA-binding protein in cells expressing Cx32 (Bai et al., 1993). The turn-on and turn-off of these tissue specific regulatory proteins could thus allow for additional, and potentially highly selective modulation of coupling. Another new data set indicates that Cx32 has separate tissue-specific promoters with alternate splicing, thereby allowing differential regulation in different organs (Neuhaus et al., 1995; Sohl et al., 1996).

Targetted over- and underexpression of gap junction proteins to specific cell types has not yet been accomplished, although such studies will usher in a new age in the study of the physiological roles of gap junctions. Among the systems that are poorly coupled *in vivo* and in which overexpression might be insightful are the β-cells of the endocrine pancreas (which could be targeted in transgenic animals using an insulin promoter; White et al., 1996) and both oligodendrocytes and selected neuronal brain populations (for which selective promoters exist). Among the systems where insight would be gained from decreased coupling are hepatocytes, astrocytes, and cardiovascular cells. In addition, studies in which communication compartments may be disrupted (either by increasing expression at

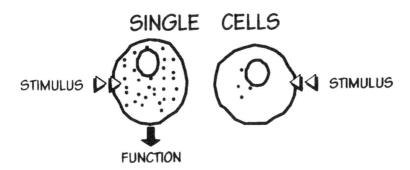

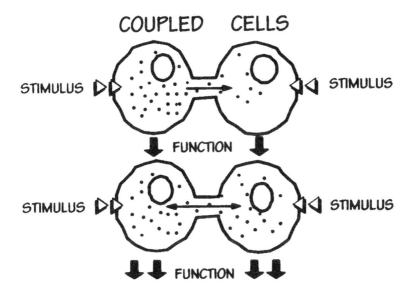

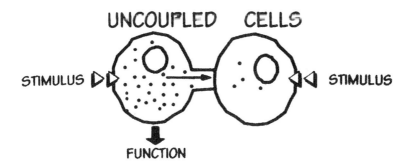

*Figure 1.*    Schematic view of a working hypothesis concerning the influence of gap junctional coupling on cell functioning. *Upper panel*: In functionally heterogeneous cell systems, stimuli result in a direct change of functioning of only those cells in which the cytoplasmic concentration of critical ion(s) or molecule(s) (represented as black dots), or both, reaches a threshold level. *Middle panel*: Establishment of gap junctional communications permits the diffusion-driven passage of the critical factor(s) from functioning into nonfunctioning cells. As a result, the latter cells become activated and synchronized with those already functioning. Under steady state conditions, the factors regulating cell functioning rapidly equilibrate across junctional channels, resulting in a change of their cytoplasmic concentrations within the coupled cells. This change in times alters the functioning of the cells equiped with gap junctions. *Lower panel*: Absence of gap junctions or closure of their channels abolishes the cell-to-cell exchange of cytoplasmic ions and molecules. As a result, uncoupled cells in contact function as if they were single.

---

margins or decreasing expression within the compartment) could also provide useful information.

## VIII.    CONCLUSIONS AND PERSPECTIVES

The experimental evidence summarized in the previous sections suggests that junctional coupling may participate in a variety of cell functions. However, how gap junctions ensure such diverse roles and what is (are) the advantage(s) of junctional communications for cell, tissue, and organ physiology are still open questions.

As compared to other forms of cell-to-cell communications, which involve the interaction of cells either with signal molecules (hormones, neurotransmitters, nitric oxide, etc.) diffusing in the intercellular spaces, or with adjacent cells (connected by cell adhesion molecules, via integrins, etc.), the intercellular communication mediated by gap junctions is unique in that it is driven by diffusion, and thus provides a direct mechanism to equilibrate ionic and molecular electrochemical gradients between coupled cells. In such a system, the increase of cytoplasmic ions or molecules less than about 900-D into one cell may be followed by its diffusion-driven passage into nearby cells connected by functional gap junction channels. At steady state, this passage will lead to the equilibration of electrochemical concentrations on the two sides of the channels (Figure 1). If the resulting concentration reaches a threshold level for activation or inhibition of an effector mechanism, functioning will be modified not only in the cell in which the ionic and molecular change first occurred, but also in all other cells coupled to it. In this case, junctional coupling would result in the functional recruitment of cells that could not be directly activated (or inhibited) in the absence of junctional channels or when such channels are temporarily closed. Experimental support for such a recruiting role of coupling has been obtained in different systems (Salomon and Meda, 1986; Mehta et al., 1991; Saffitz et al., 1992; Bosco et al., 1991). In some of these cases, it was further noticed that the activity of the coupled cells was actually improved as compared to that which the same cells show when they function as single, independent units (Salomon and Meda, 1986; Bosco et al., 1993). It is therefore likely that the equilibration induced by coupling also optimizes, at the level of a cell population, the threshold concentration of factors that are required for stimulation (or inhibition) of different effector mechanisms. Eventually, the coupling-induced equilibration of cytoplasmic constituents would be expected to synchronize those functions that are modulated by the factors exchanged through gap junctions. This expectation has also been clearly verified in some systems (Salomon and Meda, 1986; Mehta et al., 1991; Saffitz et al., 1992; Bosco et al., 1993).

These considerations imply that junctional coupling may be of value in tissues composed of cells with substantial structural, metabolic, and functional differences. Although we usually assume, at least implicitly, that cells of a given type in a specified tissue are all alike and function in the same way under a certain set of conditions, increasing evidence shows that this simplistic situation is unlikely to prevail in most tissues. On the contrary, significant structural and functional heterogeneity has been documented in all cell systems in which such diversity has been investigated (Salomon and Meda, 1986; Bosco and Meda, 1991; Bosco et al., 1994; Spray and Fishman, 1996; Spray et al., 1994c). Disparities in intrinsic properties will conceivably result in the asynchronous function of individual cells. By equilibrating ionic and molecular gradients between coupled cells, junctional coupling could correct these localized disbalances, thus permitting distinct cell subpopulations to function simultaneously or at the same rate. This synchronization, however, may not necessarily extend throughout an entire tissue. Actually, in several embryonic and adult systems, coupled cells appear grouped in distinct communication territories, which may or not be functionally linked to each other (Meda et al., 1982, 1983; Warner and Lawrence, 1982; Kalimi and Lo, 1988; Lo, 1989; Lo and Gilula, 1989; Bagnall et al., 1992; Ruanguoravat and Lo, 1992; Dermietzel and Spray, 1993; Spray et al., 1994b). At least in the case of cardiac muscle, this compartmentalization of intercommunicating cells is crucial for the proper functioning of the tissue (Spray et al., 1994c). Even though several aspects of this scheme are still rather hypothetical, the overall picture provides at least a useful conceptual framework to think of specific contributions of coupling and of its potential advantages over other forms of cell-to-cell communication.

This chapter documents that the physiological function(s) of gap junctions remain, despite increasing investigation over recent years, a field largely open for future research. It is indeed obvious that, in virtually each cell system we have considered here, more work is required to progress in our knowledge of this essential aspect of cell biology. The types of studies most urgently needed in order for the field to evolve from a descriptive to an analytical stage include finding answers to the following questions: (1) Why are there so many connexins (15 so far identified in rodents)? Does the answer lie in the specific biophysical properties of the channels, in their sensitivity to gating by specific stimuli, in their affinities to pair with one another, or in their differential control at the transcriptional or post-transcriptional levels? (2) Are there pharmacological agents that will selectively modify gap junction function without affecting other ion channels or other cell properties? And will it be possible to engineer differential selectivity so that channels of a single connexin type in a selected cell type will be affected? (3) In diseases in which gap junction function may be a primary cause (e.g., X-linked dominant Charcot-Marie-Tooth disease, atriovisceral heterotaxia, etc.) or a primary pathophysiological consequence (e.g., infarct, parasitic infection, etc.), what is the detailed nature of the disorder? For example, are multiple tissues involved in the genetic diseases and do the cardiac conduction disorders result from alterations in connexin synthesis, or from disrupted connexin trafficking? (4) For each cell type, what is the ionic or molecular message that most importantly crosses the junctional membrane? And for each tissue, are the compartmental boundaries restricting the flow of such molecules?

A concluding question comes inevitably to mind: Do we really know so little about the reasons why, fairly early in the evolution of multicellular organisms, gap junctions became almost obligatory constituents of a large variety of cells? Relative to all that could be learned, and that we are eager to know, we believe that the answer still must be affirmative. Nevertheless, our perspective is one of optimism. Never before have so many focused questions been formulatable with so many tools available for a concerted effort. It is therefore with great enthusiasm that we look forward to the next decade of junctional research and the insights that will emerge from these studies.

## ACKNOWLEDGMENTS

Work of PM is supported by grants from the Swiss National Science Foundation (53720.98), the Juvenile Diabetes Foundation International (197124), and the Commission of the European Union (QLRT-1999-00516). Work of DCS was supported by the Beatrice P. Parvin Grant in Aid Award from the New York Chapter of the American Heart Association, a grant from the Muscular Dystrophy Association, and National Institutes of Health grants HL 38449, NS-07512, DK-41918, EY-08969, and Ns-34931.

## REFERENCES

Albrecht, J.L., Atal, N.S., Tadros, P.N., Orsino, A., Lye, S.J., Sadovsky, Y., & Beyer, E.C. (1996). Rat uterine myometrium contains the gap junction protein connexin45, which has a differing temporal expression pattern from connexin43. Am.J. Obstet. Gynecol. 175 (4 Pt. 1), 853-858.

Allen, F., Tickle, C., & Warner, A. (1990). The role of gap junctions in patterning and chick limb bud. Development 108, 623-634.

Allessie, M.A., Schalij, M.J., Kirchlof, C.J., Boersma, L., Huybers, M., & Hollen, J. (1989). Experimental electrophysiology and arrhythmogenicity. Anisotropy and ventricular tachycardia. Eur. Heart. J. 10 (Suppl E.), 2-8.

Amsterdam, A., Rotmensch, S., Furman, A., Venter, E.A., & Vlodavsky, I. (1989). Synergistic effect of human chorionic gonadotropin and extracellular matrix on *in vitro* differentiation of human granulosa cells: progesterone production and gap junction formation. Endocrinology 124, 1956-1964.

Anderson, E., & Albertini, P.F. (1976). Gap junctions between the oocyte and the companion follicle cells in the mammalian ovary. J. Cell Biol. 71, 680-686.

Anzini, P., Neuberg, D.H., Schachner, M., Nelles, E., Willecke, K., Zielasek, J., Toyka, K.V., Suter, U., & Martini, R. (1997) . Structural abnormalities and deficient maintenance of peripheral nerve myelin in mice lacking the gap junction protein connexin32. J. Neurosci. 17, 4545-4551.

Asada, Y., & Lee, T. J-F. (1992). Role of myoendothelial junctions in endothelium dependent vasodilation. FASEB. J. Abstr. Part I, A974.

Aslanidi, K.B., Boitsaca, L.J., Chailakhyan, L.M., Kublik, L.N., Marachova, I.I., Potopova, T.V., & Vinogradova, T.A. (1991). Energetic cooperation via ion-permeable junctions in mixed animal cell cultures. FASEB 283, 295-297.

Atkinson, M.M., & Sheridan, J.D. (1985). Reduced junctional permeability in cells transformed by different viral oncogenes. In: *Gap Junctions* (Bennett, M.V.L., & Spray, D.C., eds.), pp. 205-213. Cold Spring Harbor Laboratory, Cold Spring Harbor, New York.

Atwater, I., Rosario, L., & Rojas, E. (1983). Properties of the Ca-activated $K^+$ channel in pancreatic beta-cells. Cell Calcium 4, 451-461.

Auerbach, A.A., & Bennett, M.V.L. (1969). A rectifying electrotonic synapse in the central nervous system of a vertebrate. J. Gen. Physiol. 53, 211-237.

Bagnall, K.M., Sanders, E.J., & Berdan, R.C. (1992). Communication compartments in the axial mesoderm of the chick embryo. Anat. Embryol. 186, 195-204.

Bai, S., Spray, D.C., & Burk, R. (1993). Characterization of rat connexin32 gene regulatory elements. In: *Gap Junctions: Progress in Cell Research* (Hall, J.E., Zampighi, G.A., & Davis, R.M., eds), vol. 3, pp. 291-297. Elsevier, Amsterdam.

Baimbridge, K.G., Peet, M.J., McLennan, H., & Church, J. (1991). Bursting response to current evoked depolarization in rat $CA_1$ pyramidal neurons is correlated with Lucifer yellow dye coupling but not with the presence of calbindin D28k. Synapse 7, 269-277.

Barr, L., Dewey, M.M., & Berger, W. (1965). Propagation of action potentials and the structure of the nexus in cardiac muscle. J. Gen. Physiol. 48, 797-823.

Bastide, B., Neyses, L., Ganten, D., Paul, M., Willecke, K., & Traub, O. (1993). Gap junction protein connexin40 is preferentially expressed in vascular endothelium and conductive bundles of rat myocardium and is increased under hypertensive conditions. Circ. Res. 73, 1138-1149.

Beblo, D.A., & Veenstra, R.D. (1997). Monovalent cation permeation through the connexin40 gap junction channel. Cs, Rb, K, Na, Li, TEA, TMA, TBA and effects of anions Br, Cl, F, acetate, aspartate, glutamate, and $NO_3$. J. Gen. Physiol. 109, 509-522.

Beer, D.G., Neveu, M.J., Paul, D.L., Rapp, U.R., & Pitot, H.C. (1988). Expression of the c-raf protooncogene, γ-glutamyltranspeptidase, and gap junction protein in rat liver neoplasms. Cancer Res. 48, 1610-1617.

Ben-Ze'ev, A., & Amsterdam, A. (1986). Regulation of cytoskeletal proteins involved in cell contact formation during differentiation of granulosa cells on extracellular matrix. Proc. Natl. Acad. Sci. U.S.A. 83, 2894-2898.

Bennett, M.V.L. (1966). Physiology of electrotonic junctions. Ann. N.Y. Acad. Sci. 37, 509-539.

Bennett, M.V.L. (1973). Function of electrotonic junctions in embryonic and adult tissues. Federation Proceedings 32, 65-75.

Bennett, M.V.L. (1994). Connexins in disease. Nature 368, 18-19.

Bennett, M.V.L., Barrio, L.C., Bargiello, T.A., Spray, D.C., Hertzberg, E.L., & Sáez, J.C. (1991). Gap junctions: new tools, new answers, new questions. Neuron 6, 305-320.

Bennett, M.V.L., Spira, M.E., & Spray, D.C. (1978). Permeability of gap junctions between embryonic cells of Fundulus; a re-evaluation. Dev. Biol. 65, 114-125.

Beny, J.L. (1994). Bidirectional electrical communication between smooth muscle and endothelial cells in the pig coronary artery. Am. J. Physiol. 266, H1465-H1472.

Berga, S.A. (1984). Electrical potentials and cell-to-cell dye movement in mammary gland during lactation. Am. J. Physiol. 247, C20-C25.

Bergoffen, J., Scherer, S.S., Wang, S., Oronzi-Scott, M. Bone, L. Paul. D.L., Chen, K., Leusch, M.W., Chance, P.F., & Fishbeck, K.H. (1993). Connexin mutations in X-linked Charcot-Marie-Tooth Disease. Science 262, 2039-2042.

Berthoud, V.M., Ivanij, V., Garcia, A.M., & Sáez, J.C. (1992). Connexins and glucagon receptors during development of the rat hepatic acinus. Am. J. Physiol. 263, G650-G658.

Beyer, E.C., Reed, K.E., Westphale, E.M., Kanter, H.L., & Larson, D.M. (1992). Molecular cloning and expression of rat connexin40, a gap junction protein expressed in vascular smooth muscle. J. Membr. Biol. 127, 69-76.

Bi, W.L., Parysek, L.M., Warnik, R., & Stambrook, P.J. (1993). In vitro evidence that metabolic cooperation is responsible for the bystander effect observed with HSV tk retriviral gene therapy. Human Gene Therapy 4, 725-731.

Blackburn, J.P., Connat, J.L., Severs, N.J., & Green, C.R. (1997). Connexin43 gap junction levels during development of the thoracic aorta are temporally correlated with elastic laminae deposition and increased blood pressure. Cell Biol. Int. 21(2), 87-97.

Blalock, J.E., & Stanton, J.D. (1980). Common pathways of interferon and hormonal action. Nature 283, 406-408.

Blennerhassett, M.G., & Garfield, R.E. (1991). Effect of gap junction number and permeability on intercellular coupling in rat myometrium. Am. J. Physiol. 261, C1001-C1009.

Blennerhassett. M.G., Kannan, M.S., & Garfield, R.E. (1987). Functional characterization of cell-to-cell coupling in cultured rat aortic smooth muscle. Cell Physiol. 21, C555-569.

Bodmer, R., Verselis, V., Levitan, I., & Spray, D.C. (1988). Electrotonic synapses in Aplysia neurons in situ and in culture: characteristics and permeability measurements. J. Neurosci. 8, 1656-1670.

Bond, S.L., Bechberger, J.F., Khoo, N.K., & Naus, C.C. (1994). Transfection of C6 glioma cells with connexin32: the effects of expression of a nonendogenous gap junction protein. Cell Growth Differ. 5, 179-186.

Bosco, D., Chanson, M., Bruzzone, R., & Meda, P. (1988). Visualization of amylase secretion from individual pancreatic acini. Am. J. Physiol. 254, G664-G670.

Bosco, D., & Meda, P. (1991). Actively synthetizing β-cells secrete preferentially during glucose stimulation. Endocrinology 129, 3157-3166.

Bosco, D. & Meda, P. (1997). Reconstructuring islet function in vitro Adv. Exp. Med. Biol. 426, 285-298.

Bosco, D., Soriano Mollá, J.V., Chanson, M., & Meda, P. (1994). Heterogeneity and contact-dependent regulation of amylase release by individual acinar cells. J. Cell. Physiol. 160, 378-388.

Brehm, P., Lechleiter, J., Smith, S., & Dunlap, K. (1989). Intercellular signaling as visualized by endogenous calcium-dependent bioluminescence. Neuron 3, 191-198.

Britz-Cunningham, S.H., Shah, M.M., Zuppan, C.W., & Fletcher, W.H. (1995). Mutations of the Connexin43 gap-junction gene in patients with heart malformations and defects of laterality. N. Engl. J. Med. 332, 1323-1329.

Brower, P.T., & Schultz, R.M. (1982). Intercellular communication between granulosa cells and mouse oocytes: existence and possible nutritional role during oocyte growth. Devel. Biol. 90, 144-153.

Bruzzone, R., & Meda, P. (1988). The gap junction: A channel for multiple functions? Eur. J. Clin. Invest. 18, 444-453.

Bruzzone R., White, T.W., Scherer, S.S., Fischbeck, K.H., & Paul, D.L. (1994). Null mutations of connexin32 in patients with X-linked Charcot-Marie-Tooth disease. Neuron 13, 1253-1260.

Bryant, S.V., & Gardiner, D.M. (1992). Retinoic acid, local cell-cell interactions, and pattern formation in vertebrate limbs. Dev. Biol. 152, 1-25.

Butterweck A., Elfang, C., Willecke, K., & Traub, O. (1994). Differential expression of the gap junction proteins connexin45, -43, -40, -31, and -26 in mouse skin. Eur. J. Cell Biol. 65, 152-163.

Campos de Carvalho, A.C., Masuda, M.O., Goldenberg, R.C.S., Tanowitz, H.B., & Spray, D.C. (1994). Conduction defects and arrhythmias in Chagas' disease: possible role of gap junctions and auto-immune mechanisms. J. Cardiovasc. Electrophys. 5, 686-698.

Campos de Carvalho, A.C., Roy, C., Christ, G.J., Moreno, A.P. Hertzberg, E.L., Melman, A., & Spray, D.C. (1993). Gap junctions formed of connexin43 are found between smooth muscle cells of human corpus cavernosum. J. Urol. 149, 1568-1575.

Campos de Carvalho, A.C., Roy, C., Hertzberg, E.L., Tanowitz, H.B., Kessler, J.A., Weiss, L.M., Wittner, M., Gao, Y., & Spray, D.C. (1998). Gap junction disappearance in astrocytes and leptomeningeal cells as a consequence of protozoan infection. Brain Res. 790, 304-314.

Campos de Carvalho, A.C., Tanowitz, H.B., Wittner, M., Dermietzel, R., Roy, C., Hertzberg, E.L., & Spray, D.C. (1992). Gap junction distribution is altered between cardiac myocytes infected with *Trypanosoma cruzi*. Circ. Res. 70, 733-742.

Caputo, R., & Peluchetti, D. (1977). The junctions of normal human epidermis. A freeze-fracture study. J. Ultrastruct. Res. 61, 44-61.

Caveney, S., Berdan, R.C., Blennerhassett, M.G., & Safranyos, R.G.A. (1986). Cell-to-cell coupling via membrane junctions, methods that show its regulation by a developmental hormone in an insect epidermis. In: Techniques in Cell Biology. (Kurstak, Ed.), vol. 2, pp.1-23. Elsevier, Amsterdam.

Chandross, K.J., Kessler, J.A., Cohen, R.I., Simburger, E., Spray, D.C., Bieri, P., & Dermietzel R. (1996). Altered connexin expression after peripheral nerve injury. Mol. Cell. Neurosci. 7, 501-518.

Chanson, M., & Meda, P. (1993). Rat pancreatic acinar cell coupling: comparison of extent and modulation *in vitro* and *in vivo*. Prog. Cell Res. 3, 199-205.

Chanson, M., Orci, L., & Meda, P. (1991). Extent and modulation of junctional communication between pancreatic acinar cells *in vivo*. Am. J. Physiol. 261, G28-G36.

Chanson, M., Fanjul, M., Bosco, D., Nelles, E., Suter, S., Willecke, K., & Meda, P. (1998). Enhanced secretion of amylase from exocrine pancreas of connexin32-deficient mice. J. Cell. Biol. 141, 1267-1275.

Charles, A.C., Merrill, J.E., Dirksen, E.R., & Sanderson, M.J. (1991). Intercellular signalling in glial cells: calcium waves and oscillations in response to mechanical stimulation and glutamate. Neuron 6, 983-992.

Charles, A.C., Naus, C.C.G., Zhu, D., Kidder, G.M., Dirksen, E.R., & Sanderson, M.J. (1992). Intercellular calcium signaling via gap junctions in glioma cells. J. Cell Biol. 118, 195-201.

Charollais, A., Serre, V., Mock, C., Cogne, F., Bosco, D., & Meda, P. (1999). Loss of alpha 1 connexin does not alter the prenatal differentiation of pancreatic beta cells and leads to the identification of another islet cell connexin. Dev. Genet. 24, 13-26.

Christ, G.J., Brink, P.R., Melman, A., & Spray, D.C. (1993a). The role of gap junctions and ion channels in the modulation of electrical and chemical signals in human corpus cavernosum smooth muscle. Int. J. Impot. Res. 5, 77-96.

Christ, G.J., Brink, P.R., Zhao, W., Moss, J., Gondre, C.M., Roy, C., & Spray, D.C. (1993b). Gap junctions modulate tissue contractility and alpha 1 adrenergic agonist efficacy in isolated rat aorta. J. Pharmacol. Exper. Ther. 266, 1054-1065.

Christ, G.J., Moreno, A.P., Melman, A., & Spray, D.C. (1992). Gap junction-mediated intercellular diffusion of $Ca^{2+}$ in cultured human corporal smooth muscle cells. Am. J. Physiol. 263, C373-C383.

Christ, G.J., Moreno, A.P., Parker, M.E., Gondre, C.M., Valcic, M., Melman A., & Spray, D.C. (1991). Intercellular communication through gap junctions: a potential role in pharmacomechanical coupling and syncytial tissue contraction in vascular smooth muscle isolated from the human corpus cavernosum. Life Sciences 49, PL195.

Christ, G.J., Spray, D.C., El-Sabban, M., Moore, L.K., & Brink, P.R. (1996). Gap junctions in vascular tissues: evaluating the role of intercellular communication in the modulation of vasomotor tone. Circ. Res. 79, 631-646.

Cole, W.C., Garfield, R.E., & Kirkaldy, J.S. (1985). Gap junctions and direct intercellular communication between rat uterine smooth muscle cells. Am. J. Physiol. 249, C20-C31.

Condorelli, D., Parent, R., Spinella, F., Salinaro, A.T., Belluardo, N., Cardile, V., & Cicirata, F. (1998). Cloning of a new gap junction gene (Cx36) highly expressed in mammalian brain neurons. Eur. J. Neurosci. 10, 1202-1208.

Cowan, D.B., Lye, S.J., & Langille, B.L. (1998). Regulation of vascular connexin43 gene expression by mechanical loads. Circ. Res. 82, 786-793.

Crick, F. (1970). Diffusion in embryogenesis. Nature 225, 420-422.

Culver, K.W., Ram, Z., Walbridge, S., Ishii, H., Oldfield, E.H., & Blaese, R.M. (1992). In vitro gene transfer with retroviral vector-producer cells for treatment of experimental brain tumors. Science 256, 1550-1152.

Daniel, E.E., Daniel, V.P., Duchon, G., Garfield, R.E., Nichols, M., Malhotra, S.K., & Oki, M. (1976). Is the nexus necessary for cell-to-cell coupling of smooth muscle? J. Membr. Biol. 28, 207-239.

Davies, P.F. (1986). Biology of disease. Vascular cell interaction with special reference to the pathogenesis of atherosclerosis. Labor. Invest. 55, 5-24.

Davies, P.F., Ganz , P., & Diehl, P.S. (1985). Reversible microcarrier-mediated junctional communication between endothelial and smooth muscle cell monolayers: an in vitro model of vascular cell interactions. J. Lab Invest. 53, 710-718.

Dekel, N., & Beers, W.H. (1978). Rat oocyte maturation in vitro: Relief of cyclic AMP inhibition by gonadotropins. Proc. Natl. Acad. Sci. U.S.A. 75, 4369-4373.

Dekel, N., Lawrence, T.S., Gilula, N.B., & Beers, W.H. (1981). Modulation of cell to cell communication in the cumulus-oocyte complex and the regulation of oocyte maturation by LH. Devel. Biol. 86, 356-362.

Dekel, N., & Piontkewitz, Y. (1991). Induction of maturation of rat oocytes by interruption of communication in the cumulus-oocyte complex. Bull. Ass. Anat. 75, 51-54.

De Leon, J.R., Buttrick, P.M., & Fishman, G.I. (1994). Functional analysis of the connexin43 gene promoter in vivo and in vitro. J. Mol. Cell. Cardiol. 26(3), 379-389.

Delmar, M., Michaels, D.C., Johnson, T., & Jalife, J. (1987). Effects of increasing intercellular esistance on transverse and longitudinal propagation in sheep epicardial muscle. Circ. Res. 60, 780-785.

De Mello, W.C. (1991). Cyclic AMP and junctional communication viewed through a multi-biophysical approach. In: Biophysics of Gap Junction Channels. (Peracchia, C., Ed.), pp. 229-239. CRC Press, Boca Raton, FL.

Denning, C. & Pitts, J.D. (1997). Bystander effects of different enzyme-prodrug systems for cancer gene therapy depend on different pathways for intercellular transfer of toxic metabolites, a factor that will govern clinical choice of appropriate regimes. Hum. Gene Ther. 8, 1825-1835.

Dermietzel, R., Hwang, T.K., & Spray, D.C. (1990). The gap junction family: structure, function and chemistry. Anat. Embryol. 182, 517-528.

Dermietzel, R., & Spray, D.C. (1993). Gap junctions in brain: how many, what type and why? Trends in Neurosicence 16, 186-192.

Dermietzel, R., Traub, O., Hwang, T.K., Beyer, E., Bennett, M.V.L., Spray, D.C., & Willecke, K. (1989). Differential expression of three gap junction proteins in developing and mature brain tissue. Proc. Nat. Acad. Sci. U.S.A. 88, 10148-10152.

Dermietzel, R., Yancey, S.B., Traub, O., Willecke, R., & Revel, J.P. (1987). Major loss of the 28-kD protein of gap junction in proliferating hepatocytes. J. Cell Biol. 105, 1925-1934.

Dewey, M.M., & Barr, L. (1964). A study of the structure and distribution of the nexus. J. Cell Biol. 23, 553-585.

Dookwah, H.D., Barhoumi, R., Narasimhan, T.R., Safe, S.H., & Burghardt, R.C. (1992). Gap junctions in myometrial cell cultures: evidence for modulation by cyclic adenosine 3':5'-monophosphate. Biol. Reprod. 47, 397-407.

Dulcibella, T., Albertini, D.F., Anderson, E., & Biggers, J.D. (1975). The preimplantation mammalian embryo: characterization of intercellular junctions and their appearance during development. Dev. Biol. 45, 231-250.

Dunlap, K., Takeda, K., & Brehm, P. (1987). Activation of a calcium-dependent photoprotein by chemical signaling through gap junctions. Nature 325, 60-62.

Edwards, R.G. (1965). Maturation *in vitro* of mouse, sheep, cow, pig, rhesus monkey and human ovarian oocytes. Nature 206, 349-351.

Eghbali, B., Kessler, J.A., Reid, L.M., Roy, C., & Spray, D.C. (1991). Involvement of gap junctions in tumorigenesis: transfection of hepatoma cells with connexin32 cDNA retards growth *in vivo*. Proc. Natl. Acad. Sci. U.S.A. 88, 10701-10705.

El-Fouly, M.H., Trosko, J.E., & Chang, C.C. (1987). Scrape-loading and dye transfer. A rapid and simple technique to study gap junctional intercellular communication. Exp. Cell. Res. 168, 422-431.

Elisevich, K., Rempel, S.A., Smith, B., & Allar, N. (1997). Connexin43 mRNA expression in two experimental models of epilepsy. Mol. Chem. Neuropathol. 32(1-3), 75-88.

El-Sabban, M.E., & Pauli, B.U. (1991). Cytoplasmic dye transfer between metastatic tumor cells and vascular endothelium. J. Cell. Biol. 115, 1375-1382.

Elshami, A.A., Saavedra, A., Zhang, H., Kucharczuk, J.C., Spray, D.C., Fishman, G.I., Kaiser, L.R., & Albelda, S.M. (1996). Gap junctions play a role in the "bystander effect" of the herpes simplex virus thymidine kinase/ganciclovir system *in vitro*. Gene Therapy 3, 85-92.

Enkvist, M.O.K., & McCarthy, D.D. (1992). Activation of protein kinase C blocks astroglial gap junction communication and inhibits the spread of calcium waves. J. Neurochem. 59, 519-526.

Enkvist, M.O., & McCarthy, D.D. (1994). Astroglial gap junction communication is increased by treatment with either glutamate or high K$^+$ concentration. J. Neurochem 62, 489-495.

Ewart, J.L., Cohen, M.F., Meyer, R.A., Huang, G.Y., Wessels, A., Gourdie, R.G., Chin, A.J., Park, S.M., Lazatin, B.O., Villabon, S., & Lo, C.W. (1997). Heart and neural tube defects in transgenic mice overexpressing the Cx43 gap junction gene. Development 124, 1281-1292.

Fairweather, N., Bell, C., Cochrane, S., Chelly, J., Wang, S., Mostacciuolo, M.L., Monaco, A.P., & Haites, N.E. (1994). Mutations in the connexin32 gene in x-linked Charcot-Marie-Tooth disease (CMTX1). Human Mol. Gen. 3, 29-34.

Fallon, R.F., & Goodenough, D.A. (1981). Five-hour half-life of mouse liver gap-junction protein. J. Cell Biol. 90, 521-526.

Fick, J., Barker II, F.G., Dazin, P., Westphale, E.M., Beyer, E.C., & Israel, M.A. (1995). The extent of heterocellular communication mediated by gap junctions is predictive of bystander tumor cytotoxicity *in vitro*. Proc. Natl. Acad. Sci. U.S.A. 92, 11071-11075.

Fitzgerald, D.J., Mesnil, M., Oyamada, M., Tsuda, H., Ito, N., & Yamasaki, H. (1989). Changes in gap junction protein (connexin 32) gene expression during rat liver carcinogenesis. J. Cell. Biochem. 41, 97-102.

Fletcher, W.H., Byus, C.V., & Walsh, D.A. (1987a). Receptor-mediated action without receptor occupancy: a function for cell-cell communication in ovarian follicles Adv. in Exp. Med. Biol. 219, 99-323.

Fletcher, W.H., & Greenan, J.R. (1985). Receptor mediated action without receptor occupancy. Endocrinol. 116, 1660-1662.

Fletcher, W.H., Shiu, W.W., Ishida, T.A., Haviland, D.L., & Ware, C.F. (1987b). Resistance to the cytolytic action of lymphotoxin and tumor necrosis factor coincides with the presence of gap junctions uniting target cells. J. Immunol. 139, 956-962.

Friend, D.S., & Gilula, N.B. (1972). Variations in tight and gap junctions in mammalian tissues. J. Cell Biol. 53, 758-776.

Furshpan, E.J., & Potter D.D. (1959). Transmission at the giant motor synapses of the crayfish. J. Physiol. 145, 289-325.

Furshpan, E.J., & Potter, D.D. (1968). Low-resistance junctions between cells in embryos and tissue culture. Curr. Top. Dev. Biol. 3, 95-127.

Gabriel, H.D., Jung, D., Butzler, C., Temme, A., Traub, O., Winterhager, E., & Willecke, K. (1998). Transplacental uptake of glucose is decreased in embryonic lethal connexin26-deficient mice. J. Cell Biol. 140, 1453-1461.

Gao, Y., & Spray, D.C. (1998). Gap junction expression in lens and cornea of wildtype (WT) and Cx43 deficient (KO) mice: cataractous changes in lens but no striking defect in cornea. In: Gap Junctions. (Werner, R., Ed.), pp. 314-318. IOS Press, Netherlands.

Garfield, R.E. & Hayashi, R.H. (1981). Appearance of gap junctions in the myometrium of women during labor. Am. J. Obstet. Gynecol. 140(3), 254-260.

Garfield, R.E., Hayashi, R.H., & Harper, M.J. (1987). In vitro studies on the control of human myometrial gap junctions. Int. J. Gyn. Obst. 25, 241-248.

Garfield, R.E., Kannan, M.S., & Daniel, E.E. (1980a). Gap junction formation in myometrium: control by estrogens, progesterone, and prostaglandins. Am. J. Physiol. 238, C81-C89.

Garfield, R.E., Merrett, D., & Grover, A.K. (1980b). Gap junction formation and regulation in myometrium. Am. J. Physiol. 239, C217-C228.

Garfield, R.E., Sims, S., & Daniel, E.E. (1977). Gap junctions: their presence and necessity in myometrium during parturition. Science 198, 958-960.

Garfield, R.E., Thilander, G., Blennerhassett, M.G., & Sakai, N. (1992). Are gap junctions necessary for cell-to-cell coupling of smooth muscle?: an update. Can. J. Physiol. Pharmacol. 70, 481-490.

Gebbia, M. Towbin, J.A., & Casey, B. (1996). Failure to detect connexin43 mutations in 38 cases of sporadic and familial heterotaxy. Circulation 94(8), 1909-1912.

Gettes, L.S., & Cascio, W.E. (1992). Effect of acute ischemia on cardiac electrophysiology. In: The Heart and Cardiovascular System, 2nd ed. (Fozzard, H. et al., Eds.), p. 2021. Raven Press, New York.

Giaume, C., Tabernero, A., & Medina, J.M. (1997). Metabolic trafficking through astrocytic gap junctions. Glia 21, 114-123.

Gilula, N.B., Epstein, M.C., & Beers, W.N. (1978). Cell-to-cell communication and ovulation. A study of the cumulus-oocyte complex. J. Cell Biol. 78, 58-75.

Gilula, N.B., Reeves, O.R., & Steinback, A. (1972). Metabolic coupling, ionic coupling and cell contacts. Nature 235, 262-265.

Gong, X., Li, E., Klier, G., Huang, Q., Wu, Ying, Lei, Hong, Kumar, N.M., Horwitz, H., & Gilula, N.B. (1997). Disruption of α3 connexin gene leads to proteolysis and cataractogenesis in mice. Cell 91, 833-843.

Goodenough, D.A. (1992). The crystalline lens. A system networked by gap junctional intercellular communication. Cell Biol. 3, 49-58.

Goodenough, D.A., & Musil, L.S. (1993). Gap junctions and tissue business: problems and strategies for developing specific functional reagents J. Cell Science 17, 133-138.

Goulet, F., Normand, C., & Morin, O. (1988). Cellular interactions promote tissue-specific function, biomatrix deposition and junctional communication of primary cultured hepatocytes. Hepatology 8, 1010-1018.

Graf, J., & Petersen, O.H. (1978). Cell membrane and resistance in liver. J. Physiol. (Lond.) 284, 105-126.

Granot, I., & Dekel, N. (1994). Phosphorylation and expression of connexin-43 ovarian gap junction protein are regulated by luteinizing hormone. J. Biol. Chem. 269, 30502-30509.

Green, C.R., & Severs, N.J. (1993). Distribution and role of gap junctions in normal myocardium and human ischaemic heart disease. Histochem. Cell Biol. 99, 105-120.

Guerrero, P.A., Schuessler, R.B., Davis, L.M., Beyer, E.C., Johnson, C.M., Yamada, K.A., & Saffitz, J.E. (1997). Slow ventricular conduction in mice heterozygous for a connexin43 null mutation. J. Clin. Invest. 15, 1991-1998.

Gumuci, J.J., & Miller, D.L. (1981). Functional implications of liver cell heterogeneity. Gastroenterology 80, 393-403.

Guo, H., Acevedo, P., Parsa, F.D., & Bertram, J.S. (1992). Gap-junctional protein connexin 43 is expressed in dermis and epidermis of human skin: Differential modulation by retinoids. J. Invest. Dermatol. 99, 460-467.

Haefliger, J.A., Castillo, E., Waeber, G., Bergonzelli, G.E., Aubert, J.F., Sutter, E., Nicod, P., Waeber, B., & Meda, P. (1997). Hypertension increases connexin43 in tissue-specific manner. Circulation 95, 1007-1014.

Haefliger, J.A., Meda, P., Formenton, A., Wiesel, P., Zanchi, A., Brunner, H.R., Nicod, P., & Hayoz, D. (1999). Aortic connexin43 is decreased during hypertension induced by inhibition of nitric oxide synthases. Arterisc. Thromb. Vasc. Biol. 19, 1615-1622.

Hassinger, T.D., Guthrie, P.B., Atkinson, P.B., Bennett, M.V.L., & Kater, S.B. (1996). An extracellular signaling component in propagation of astrocytic calcium waves. Proc. Natl. Acad. Sci. U.S.A. 93, 13268-13273.

Heller, D.T., & Schultz, R.M. (1980). Ribonucleoside metabolism in mouse oocyte: metabolic cooperativity between the fully-grown oocyte and cumulus cells. J. Exp. Zool. 214, 355-364.

Hertzberg, E.L., Lawrence, T.S., & Gilula, N.B. (1981). Gap junctional communication. Annu. Rev. Physiol. 43, 479-491.

Hobbie, L., Kingsley, D.M., Kozarsky, K.F., Jackman, R.W., & Krieger, M. (1987). Restoration of LDL receptor activity in mutant cells by intercellular junctional communication. Science 235, 69-73.

Hofer, A., Sáez, J.C., Chang, C.C., Trosko, J.E., Spray, D.C., & Dermietzel, R. (1996). C-erb2/neu transfection induces gap junctional communication-incompetence in glial cells. J. Neurosci. 16, 4311-4321.

Hooper, M.L., & Subak-Sharpe, J.H. (1981). Metabolic co-operation between cells. Int. Rev. Cytol. 69, 46-104.

Houghton, J.L., Davison, C.A., Kuhner, P.A., Torossov, M.T., Strogatz, D.S., & Carr, A.A. (1998). Heterogeneous vasomotor responses of coronary conduit and resistance vessels in hypertension. J. Am. Coll. Cardiol. 31(2), 374-382.

Huizinga, J.D., Liu, L.W.C., Blennerhassett, M.G., Thuneberg, L., & Molleman, A. (1992). Intercellular communication in smooth muscle. Experientia 48, 932-941.

Hulser, D.F. (1974). Ionic coupling between nonexcitable cells in culture. Methods Cell Biol. 8, 289-317.

Huttner, I., Costabella, P.M., De Chastonay, C., & Gabbiani, G. (1982). Volume, surface and junctions of rat aortic endothelium during experimental hypertension: a morphometric and freeze fracture study. Lab. Inves. 46, 489-504.

Imanaga, I. (1987). Cell-to-cell coupling studies by diffusional methods in myocardial cells. Experientia 43, 1080-1083.

Ionasescu, V., Searby, C., & Ionasescu, R. (1994). Point mutations of the connexin32 (GJB1) gene in X-linked dominant Charcot-Marie-Tooth neuropathy. Hum. Mol. Genet. 3, 355-358.

Iwatsuki, N., & Petersen, O.H. (1978). Electrical coupling and uncoupling of exocrine acinar cells. J. Cell Biol. 79, 533-545.

Jessell, T.M., & Melton, D.A. (1992). Diffusible factors in vertebrate embryonic induction [review]. Cell 68, 257-270.

Johnson, L.G., Olsen, J.C., Sarkadi, B., Moore, K.L., Swanstrom, R., & Boucher, R.C. (1992). Efficiency of gene transfer for restoration of normal airway epithelial function in cystic fibrosis. Nat. Genet. 2, 21-25.

Jungerman, K., & Katz, K. (1989). Functional specialization of different heypatocyte populations. Physiol. Rev. 69, 708-764.

Kalimi, G.H., & Lo, C.W. (1988). Communication compartments in the gastrulating mouse embryo. J. Cell Biol. 107, 241-255.

Kam, E., Melville, E.L., & Pitts, J.D. (1986). Patterns of junctional communication in skin. J. Invest. Dermatol. 87, 748-753.

Kanno, Y., Sasaki, Y., Hirono, C., & Shiba, Y. (1991). Membrane receptor and signal transduction on the regulation of gap junction permeability in rat submandibular gland. Biomed. Res. 12(suppl. 2), 93-94.

Kanter, H.L., Laing, J.G., Beau, S.L., Beyer, E.C., & Saffitz, J.E. (1993a). Distinct patterns of connexin expression in canine Purkinje fibers and ventricular muscle. Circ. Res. 72, 1124-1131.

Kanter, H.L., Laing, J.G., Beyer, E.C., Green, K.G., & Saffitz, J.E. (1993b). Multiple connexins colocalize in canine ventricular myocyte gap junctions. Circ. Res. 73, 344-350.

Kardami, E., Stoski, R.M., Doble, B.W., Yamamoto, T., Hertzberg, E.L., & Nagy, J.I. (1991). Biochemical and ultrastructural evidence for the association of basic fibroblast growth factor with cardiac gap junctions. J. Biol. Chem. 266, 19551-19557.

Kass, S., MacRae, C., Graber, H.L., et al. (1994). A gene defect that causes conduction system disease and dilated cardiomyopathy maps to chromosome 1p1-1q1, Nat. Genet. 7, 546-551.

Katsuya, T., Takami, S., Higaki, J., Serikawa, T., Mikami, H., Miki, T., & Oghihara, T. (1995). Gap junction protein locus on chromosome 18 cosegregates with body weight in the spontaneously hypertensive rat. Hypertens. Res. 18, 63-67.

Kelsell, D.P., Dunlop, J., Stevenes, H.P., Lench, N.J., Liang, J.N., Parry, G., Mueller, R.F., & Leigh, I.M. (1997). Connexin26 mutations in hereditary non-syndromic sensorineural deafness. Nature 387, 80-83.

Kianmanesh, A.R., Perrin, H., Panis, Y., Fabre, M., Nagy, H.J., Houssin, D., & Klatzmann, D. (1997). A "distant" bystander effect of suicide gene therapy: Regression of nontransduced tumors together with a distant transduced tumor. Human Gene Ther. 8, 1807-1814.

Klagsbrun, M., & D'Amore, P.A. (1991). Regulators of angiogenesis. Ann. Rev. Physiol. 53, 217-239.

Kohen, E., Koen, C., & Rabinovitch, A. (1983). Cell-to-cell communication in rat pancreatic islet monolayer cultures is modulated by agents affecting islet cell secretory activity. Diabetes 32, 95-98.

Kojima, T., Mitaka, T., Paul, D.C., Mori, M., & Mochizuki, Y. (1995a). Reappearance and long-term maintenance of connexin32 in proliferated adult rat hepatocytes: Use of serum-free L-15 medium supplemented with EGF and DMSO. J. Cell Sci. 108, 1347-1357.

Kojima, T., Mitake, T., Shibata, Y., & Mochizuki, Y. (1995b). Induction and regulation of connexin26 by glucagon in primary cultures of adult rat hepatocytes. J. Cell Sci. 108, 2771-2780.

Kolberg, R. (1994). The bystander effect in gene therapy: great, but *how* does it work? J. NIH Res. 6, 62-64.

Korn, H., & Faber, D.S. (1979). Electrical interactions between vertebrate neurons: field effects and electrotonic coupling. In: The Neurosciences: Fourth Study Program. (Schmitt, F.O., & Worden, F.G., Eds.), pp.333-358. MIT Press, Cambridge, MA.

Kren, B.T., Kumar, N.M., Wang, S.Q., Gilula, N.B., & Steer, C.J. (1993). Differential regulation of multiple gap junction transcripts and proteins during rat liver regeneration. J. Cell. Biol. 123, 707-718.

Krutovskikh, V., Mironov, N., & Yamasaki, H. (1996). Human connexin37 is polymorphic but not mutated in tumours. Carcinogenesis 17, 1761-1763.

Kuffler, S.W., & Potter, D.D. (1964). Glia in the leech central nervous system: physiological properties and neuron-glia relationships. J. Neurophysiol. 27, 290-320.

Kumar, N. M., Gilula, N. B. (1996). The gap junction channel. Cell 84, 381-388.

Labarthe, M-P, Bosco, D., Saurat, J-H, Meda, P., & Salomon, D. (1998). Upregulation of connexin26 between keratinocytes of psoriatic lesions. J. Invest. Dermat. 111, 72-76.

Laird, D.W., Puranam, K.L., & Revel, J.P. (1991). Turnover and phosphorylation dynamics of connexin43 gap junction protein in cultured cardiac myocytes. Biochem. J. 273, 67-72.

Larsen, W.J., Tung, H.N., & Polking, C. (1981). Response of granulosa cell gap junctions to human chorionic gonadotropin (hCG) at ovulation. Biol. Reprod. 25, 1119-1134.

Larsen, W.J., Wert, S.E., & Brunner, G.D. (1986). A dramatic loss of cumulus cell gap junctions is correlated with germinal vesicle breakdown in rat oocytes. Devel. Biol. 113, 517-521.

Larsen, W.J., Wert, S.E., & Brunner, G.D. (1987). Differential modulation of follicle cell gap junction populations at ovulation. Dev. Biol. 122, 61-71.

Larson, D.M., Carson, M.P., & Haudenschild, C.C. (1987). Junctional transfer of small molecules in cultured bovine brain microvascular endothelial cells and pericytes. Microv. Res. 34, 184-199.

Larson, D.M., & Haudenschild, C.C. (1988). Junctional transfer in wounded cultures of bovine aortic endothelial cells. Lab. Invest. 59, 373-379.

Larson, D.M., Haudenschild, C.C., & Beyer, E.C. (1990). Gap junction messenger RNA expression by vascular wall cells. Circ. Res. 66, 1074-1080.

Lawrence, T.S., Beers, W.H., & Gilula, N.B. (1978). Transmission of hormonal stimulation by cell-to-cell communication. Nature 272, 501-506.

Lawrence, T.S., Rehemtulla, A., Ng, E.Y., Wilson, M., Trosko, J.E., & Stetson, P.L. (1998). Preferential cytotoxicity of cells transduced with cytosine deaminase compared to bystander cells after treatment with 5-flucytosine. Cancer Res. 58, 2588-2593.

Lazrak, A., & Peracchia, C. (1993). Gap junction gating sensitivity to physiological internal calcium regardless of pH in Novikoff hepatoma cells. Biophys. J. 65, 2002-2012.

Ledbetter, ML. & Lubin M. (1979). Transfer of potassium. A new measure of cell-cell coupling. J. Cell Biol. 80, 150-165.

Leao, A.A.P. (1944). Spreading depression in the cerebral cortex. J. Neurophysiol. 7, 359-390.

Lee, S.-M., & Clemens, M.G. (1992). Subacinar distribution of hepatocyte membrane potential response to stimulation of glucogenesis. Am. J. Physiol. 263, G319-G326.

Lee, S.W., Tomasetto, C., Paul, D., Keyomarski, K., & Sager, R. (1992). Transcriptional downregulation of gap-junction proteins blocks junctional communication in human mammary tumor cell lines. J. Cell. Biol. 118, 1213-1221.

Lindner, H.R., Tsafriri, A., Lieberman, M.E., Zor, U., Koch, Y., Bauminger, S., & Barnea, A. (1974). Gonadotrophin action on cultured Graafian follicles: induction of maturation division of the mammalian oocyte and differentiation of the luteal cell. Rec. Prog. Horm. Res. 30, 79-138.

Little, T.L., Beyer, E.C., & Duling, B.R. (1995). Connexin43 and connexin40 gap junctional proteins are present in arteriolar smooth muscle and endothelium *in vivo*. Am. J. Physiol. 268, H729-H739.

Little, T.L., Xia, J., & Duling, B.R. (1994). Dye tracers define differential endothelial and smooth muscle coupling patterns within the arteriolar wall. Circ. Res. 76, 498-504.

Llinas, R., Baker, R., & Sotelo, C. (1974). Electrotonic coupling between neurons in the cat inferior olive. J. Neurophysiol. 37, 560-571.

Lo, C.W. (1989). Communication compartments: a conserved role in pattern formation? In: Cell Interactions and Gap Junctions. (Sperelakis, N., & Cole, W.C., Eds.), pp. 85-96. CRC Press, Boca Raton, FL.

Lo, C.W., Cohen, M.F., Huang, G-Y, Lazatin, B.O., Patel, N., Sullivan, R., Pauken, C., & Park, S.M.J. (1997). Cx43 gap junction gene expression and gap junctional communication in mouse neural crest cells. Dev. Genet. 20, 119-132.

Lo, C.W., & Gilula, N.B. (1979). Gap junctional communication in the post-implantation mouse embryo. Cell 18, 411-422.

Lo, C.W., & Gilula, N.B. (1989). Communication compartments: a conserved role in pattern formation? CRC Rev. Gap Junct. 1, 85-96.

Loewenstein, W.R. (1968a). Communication through cell junctions. Implications in growth control and differentiation. Dev. Biol. 190(suppl. 2), 151-183.

Loewenstein, W.R. (1968b). Some reflections on growth and differentiation. Perspect. Biol. Med. 11, 260-272.

Loewenstein, W.R. (1979). Junctional intercellular communication and the control of growth. Biochim. Biophys. Acta 560, 1-65.

Loewenstein, W.R. (1981). Junctional intercellular communication: the cell to cell membrane channel. Physiol. Rev. 61, 829-913.

Loewenstein, W.R., & Kanno, Y. (1964). Studies on an epithelial (gland) cell junction. I. Modification of surface membrane permeability. J. Cell Biol. 22, 565-586.

Loewenstein, W.R., & Kanno, Y. (1967). Intercellular communication and tissue growth. I: Cancerous growth. J. Cell. Biol. 33, 225-234.

Loewenstein, W.R., & Rose, B. (1992). The cell-cell channel in the control of growth. Sem. Cell Biol. 3, 59-79.

Luke, R.A., & Saffitz, J.E. (1991). Remodeling of ventricular conduction pathways in healed canine infarct border zones. J. Clin. Invest. 87, 1594-1602.

MacKenzie, L.W., & Garfield, R.E. (1986). Effects of tamoxifen citrate and cycloheximide on estradiol induction of rat myometrial gap junctions. Can. J. Physiol. Pharmacol. 64, 703-706.

Martins-Ferreira, H. (1994). Spreading depression: a neurohumoral reaction. Brazil J. Med. 27, 851-863.

Martins-Ferreira, H., & Ribeiro, L.J. (1995). Biphasic effects of gap junctional uncoupling agents on the propagation of retinal spreading depression. Braz. J. Med. Biol. Res. 28, 991-994.

Martyn, K.D., Kurata, W.E., Warn-Cramer, B.J., Burt, J.M., TenBroek, E., & Lau, A.F. (1997). Immortalized connexin43 knockout cell lines display a subset of biological properties associated with the transformed phenotype. Cell Growth Diff. 8, 1015-1027.

Masgrau-Peya, E., Salomon, D., Saurat, J-H, & Meda, P. (1997). In vivo modulation of connexins 43 and 26 of human epidermis by topical retinoic acid treatment. J. Histochem. Cytochem. 45, 1207-1215.

McGuire, P.G., & Twietmeyer, T.A. (1985). Aortic endothelial junctions in developing hypertension. Hypertension 7, 483-490.

Meda, P. (1995). Junctional coupling of pancreatic β-cells. In: Pacemaker Activity and Intercellular Communication. (Huizinga, J.D., Ed.), pp. 275-291. CRC Press, Boca Raton, FL.

Meda, P. (1996). The role of gap junction membrane channels in secretion and hormonal action. J. Bioenerg. Biomembr. 28, 369-377.

Meda, P., Bosco, D., Chanson, M., Giordano, E., Vallar, L., Wollheim, C., & Orci, L. (1990a). Rapid and reversible secretion changes during uncoupling of rat insulin-producing cells. J. Clin. Invest. 86, 759-768.

Meda, P., Bosco, D., Giordano, E., & Chanson, M. (1990b). Junctional coupling modulation by secretagogues in two-cell pancreatic systems. In: Biophysics of Gap Junction Channels. (Peracchia, C., Ed.), pp. 191-208. CRC Press, Boca Raton, FL

Meda, P., Bruzzone, R., Chanson, M., & Bosco, D. (1988). Junctional coupling and secretion of pancreatic acinar cells. In: Modern Cell Biology. (Hertzberg, E., & Johnson, R.G., Eds.), vol. 7, pp. 353-364. Alan Liss, New York.

Meda, P., Bruzzone, R., Chanson, M., Bosco, D., & Orci, L. (1987). Gap junctional coupling modulates secretion of exocrine pancreas. Proc. Natl. Acad. Sci. U.S.A. 84, 4901-4904.

Meda, P., Bruzzone, R., Knodel, S., & Orci, L. (1986). Blockage of cell-to-cell communication within pancreatic acini is associated with increased basal release of amylase. J. Cell Biol. 103, 475-483.

Meda, P., Chanson, M., Pepper, M., Giordano, E., Bosco, D., Traub, O., Willecke, K., El Aoumari, E., Gros, D., Beyer, E., Orci, L., & Spray, D.C. (1991). *In vivo* modulation of connexin 43 gene expression and junctional coupling of pancreatic β-cells. Exp. Cell Res. 192, 469-480.

Meda, P., Halban, P., Perrelet, A., Renold, A.E., & Orci, L. (1980). Gap junction development is correlated with insulin content in pancreatic β-cells. Science 208, 1026-1028.

Meda, P., Michaels, R.L., Halban, P.A., Orci, L., & Sheridan, J.D. (1983). *In vivo* modulation of gap junctions and dye coupling between β-cells of the intact pancreatic islet. Diabetes 32, 858-868.

Meda, P., Pepper, M.S., Traub, O., Willecke, K., Gros, D., Beyer, E., Nicholson, B., Paul, D., & Orci, L. (1993). Differential expression of gap junction connexins in endocrine and exocrine glands. Endocrinology 133, 2371-2378.

Meda, P., Perrelet, A., & Orci, L. (1979). Increase of gap junctions between pancreatic β-cells during stimulation of insulin secretion. J. Cell Biol. 82, 441-448.

Meda, P., Perrelet, A., & Orci, L. (1984). Gap junctions and cell-to-cell coupling in endocrine glands. In: Modern Cell Biology. (Satir, B.H., Ed.), vol. 3, pp. 131-196. Alan Liss, New York.

Mehta, P.P., Hotz-Wagenblatt, A., Rose, B., Shalloway, D., & Loewenstein, W.R. (1991). Incorporation of the gene for a cell-cell channel protein into transformed cells leads to normalization of growth. J. Membr. Biol. 124, 207-225.

Mekata, F. (1981). Electrical current induced contraction in the smooth muscle of the rabbit aorta. J. Physiol. 317, 149-161.

Mesnil, M., Fitzgerald, D.J., & Yamasaki, H. (1988). Phenobarbital specifically reduces gap junction protein mRNA level in rat liver. Mol Carcinogen 1, 79-81.

Mesnil, M., Piccoli, C., Tiraby, G., Willecke, K., & Yamasaki, H. (1996). Bystander killing of cancer cells by herpes simplex virus thymidine kinase gene is mediated by connexins. Proc. Natl. Acad. Sci. U.S.A. 93, 1831-1835.

Meyer, D.J., Yancey, S.B., & Revel, J.P. (1981). Intercellular communication in normal and regenerating rat liver: A quantitative analysis. J. Cell. Biol. 91(2 Pt. 1), 505-523.

Micevych, P.E., & Abelson, L. (1991). Distribution of mRNA coding for liver and heart gap-junction proteins in the rat central nervous system. J. Comp. Neurol. 305, 96-118.

Michaels, R.L., & Sheridan, J.D. (1981). Islets of Langerhans: dye coupling among immunocytochemically distinct cell types. Science 214, 801-803.

Miller, S.M., Garfield, R.E., & Daniel, E.E. (1989). Improved propagation in myometrium associated with gap junctions during parturition. Am. J. Physiol. 256, C130-C141.

Miller, T., Dahl, G., & Werner, R. (1988). Structure of a gap junction gene: rat connexin32. Biosci. Rep. 8, 455-464.

Minkoff, R., Rundus, V.R., Parker, S.B., Beyer, E.C., & Hertzberg, E.L. (1993). Connexin expression in the developing avian cardiovascular system. Circ. Res. 73, 71-78.

Moor, R.M., Smith, M.W., & Dawson, R.M.C. (1980). Measurement of intercellular coupling between oocytes and cumulus cells using intracellular markers. Exp. Cell Res. 126, 15-29.

Moore, L.K., Beyer, E.C., & Burt, J.M. (1991). Characterization of gap junction channels in A7r5 vascular smooth muscle cells. Am. J. Physiol. 260 (Cell Physiol. 29), C975-981.

Moreno, A.P., Campos de Carvalho, A.C., Christ, G., & Spray, D.C. (1993). Gap junctions between human corpus cavernosum smooth muscle cells: gating properties and unitary conductance. Amer. J. Physiol. 33, C80.

Moreno, A.P., Fishman, G.I., Beyer, E.C., & Spray, D.C. (1994). Voltage dependent gating and single channel analysis of heterotypic gap junction channels formed of Cx45 and Cx43. In: Gap Junctions: Progress in Cell Research. (Kanno, Y., Ed.), vol. 4, pp. 405-408. Elsevier, Amsterdam.

Moreno, A.P., Laing, J.G., Beyer, E.C., & Spray, D.C. (1995). Properties of gap junction channels formed of connexin45 endogenously expressed in human hepatoma (SKHep1) cells. Am. J. Physiol. 268(2 Pt. 1), C356-C365.

Munari-Silem, Y., & Rousset, B. (1996). Gap junction-mediated cell-to-cell communication in endocrine glands- molecular and functional aspects: a review. Eur. J. Endocrin. 135, 251-264.

Murray, S.A., & Fletcher, W.H. (1984). Hormone-induced intercellular signal transfer dissociates cyclic AMP-dependent protein kinase. J. Cell Biol. 98, 1710-1719.

Nabel, E.G., Plautz, G., Boyce, F.M., Stanley, J.C., & Nabel, G.J. (1989). Recombinant gene expression in vivo within endothelial cells of the arterial wall. Science 244, 1342-1344.

Nathanson, M.H., Burgstahler, A.D., Mennone, A., Fallon, M.B., Gonzalez, C.B., & Sáez, J.C. (1995). $Ca^{2+}$ waves are organized among hepatocytes in the intact organ. Am. J. Physiol. 269, G167-G171.

Naus, C.C.G., Bechberger, J.F., & Paul, D.L. (1991). Gap junction gene expression in human seizure disorder. Exp. Neurol. 111, 198-203.

Naus, C.C., Bechberger, J.F., Zhang, Y., Venance, L., Yamasaki, H., Juneja, S.C., Kidder, G.M., & Giaume, C. (1997). Altered gap junctional communication, intercellular signaling, and growth in cultured astrocytes deficient in connexin43. J. Neurosci. Res. 49, 528-540.

Naus, C.G., Elisevich, K., Zhu, D., Belliveau, D.J., & Del Maestro, R.F. (1992). In vivo growth of C6 glioma cells transfected with connexin43 cDNA. Cancer Res. 52, 4208-4213.

Nedergaard, M. (1994). Direct signalling from astrocytes to neurons in cultures of mammalian brain cells. Science 263, 1768-1771.

Nedergaard, M., Cooper, A.J., & Goldman, S.A. (1995). Gap junctions are required for the propagation of spreading depression. J. Neurobiol. 28, 433-444.

Nelles, E., Butzler, C., Jung, D., Temme, A., Gabriel, H.D., Dahl, U., Traub, O., Stumpel, F., Jungermann, K., Zielasek, J., Toyka, K.V., Dermietzel, R., & Willecke, K. (1996). Defective propagaqtion of signals generated by sympathetic nerve stimulation in the liver of connexin32-deficient mice. Proc. Natl. Acad. Sci. U.S.A. 93, 9565-9570.

Neuhaus, I.M., Dahl, G., & Werner, R. (1995). Use of alternate promoters for tissue-specific expression of the gene coding for connexin32. Gene 158, 257-262.

Neveu M.J., Hully J.R., Paul D.L., & Pitot H.C. (1990). Reversible alteration in the expression of the gap junctional protein connexin 32 during tumor promotion in rat liver and its role during cell proliferation. Cancer Comm. 2, 21-31.

Neveu, M.J., Sattler, C.A., Sattler, G.L., Hully, J.R., Hertzberg, E.L., Paul, D.L., Nicholson, B.J., & Pitot, H.C. (1994). Differences in the expression of connexin genes in rat hepatomas in vivo and in vitro. Mol. Carcinog. 11, 145-154.

Neyton, J., & Trautmann, A. (1985). Single-channel currents of an intercellular junction. Nature 317, 331-335.

Nicholson, S.M., & Bruzzone, R. (1997). Gap junctions: getting the message through. Current Biol. 7, R340-R344.

Nishi, M., Kumar, N.M., & Gilula, N.B. (1991). Developmental regulation of gap junction gene expression during mouse embryonic development. Dev. Biol. 146, 117-130.

Nishida, T., Ueda, A., Fukuda,M., Mishima, H., Yasumoto, K., & Otori, T. (1988). Interactions of extracellular collagen and corneal fibroblasts: morphological and biochemical changes of rabbit corneal cells cultured in a collagen matrix. In Vitro Cell Dev. Biol. 24, 1009-1014.

O'Brien, J., al-Ubaidi, M.R., & Ripps, H. (1996). Connexin35: a gap-junctional protein expressed preferentially in the skate retina. Mol. Biol. Cell. 7, 233-243.

Ochalski, P.A., Sawchik, M.A., Hertzberg, E.L., & Nagy, J.I. (1995). Astrocytic gap junction removal, connexin43 redistribution, and epitope masking at excitatory amino acid lesion sites in rat brain. GLIA 14, 279-294.

Omori, Y., Duflot-Dancer, A., Mesnil, M., Yamasaki, H. (1998). Dominant-negative inhibition of connexin-mediated tumor suppression by mutant connexin genes. In: *Gap Junctions* (Werner, R., ed.), pp. 377-381. IOS Press, Netherlands.

Omori, Y., Mesnil, M., & Yamasaki, H. (1996). Connexin32 mutations from X-linked Charcot-Marie-Tooth disease patients: functional defects and dominant negative effects. Mol. Biol. Cell. 7, 907-916.

Osipchuk, Y., & Cahalan, M. (1992). Cell-to-cell spread of calcium signals mediated by ATP receptors in mast cells. Nature 359, 241-244.

Ou, C.W., Orsino, A., & Lye, S.J. (1997). Expression of connexin43 and connexin26 in rat myometrium during pregnancy and labor is differentially regulated by mechanical and hormonal signals. Endocrinology 138, 5398-5407.

Paul, D.L. (1986). Molecular cloning of cDNA for rat liver gap junction protein. J. Cell Biol. 103, 123-134.

Paul, D.L. (1995). New functions for gap junctions. Curr. Opin. Cell Biol. 7(5), 665-672.

Peinado, A., Yuste, R., & Katz, L.C. (1993). Extensive dye coupling between rat neocortical neurons during their period of circuit formation. Neuron 10, 103-114.

Penman-Splitt, M., Tsai, M.Y., Burn, J., & Goodship, J.A. (1997). Absence of mutations in the regulatory domain of the gap junction protein connexin43 in patients with visceroatrial heterotaxy. Heart 77, 369-370

Penn, R.D. (1966). Ionic communication between liver cells. J. Cell Biol. 29, 171-174.

Pepper, M.S., & Meda, P. (1992). Basic fibroblast growth factor increases junctional communication and connexin 43 expression in microvascular endothelial cells. J. Cell. Physiol. 153, 196-205.

Pepper, M.S., Montesano, R., El Aoumari, A., Gros, D., Orci, L., & Meda, P. (1992). Coupling and connexin 43 expression in microvascular and large vessel endothelial cells. Am. J. Physiol. 262, C1246-C1257.

Pepper, M.S., Spray, D.C., Chanson, M., Montesano, R., Orci, L., & Meda, P. (1989). Junctional communication is induced in migrating capillary endothelial cells. J. Cell Biol. 109, 3027-3038.

Peters, N.S., Coromilas, J., Severs, N.J., & Wit, A.L. (1997). Disturbed connexin43 gap junction distribution correlates with the location of reentrant circuits in the epicardial border zone of healing canine infarcts that cause ventricular tachycardia. Circulation 18, 988-996.

Philippe, J., Giordano, E., Gjinovci, A., & Meda, P. (1992). Cyclic adenosine monophosphate prevents the glucocorticoid-mediated inhibition of insulin gene expression in rodent islet cells. J. Clin. Invest. 90, 2228-2233.

Pitts, J.D. (1980). The role of junctional communication in animal tissues. In Vitro 16, 1049-1056.

Pitts, J.D., & Finbow, M. (1977). Junctional permeability and its consequences. In: *Intercellular Communication* (DeMello, W.C., ed.), pp. 61-68. Plenum Press, New York.

Pitts, J.D., & Simms, J.W. (1977). Permeability of junctions between animal cells. Intercellular transfer of nucleotides but not of macromolecules. Exp. Cell. Res. 104, 153-163.

Potter, D.D., Furshpan, E.J., & Lennox, E.S. (1966). Connections between cells of the developing squid as revealed by electrophysiological methods. Proc. Natl. Acad. Sci. U.S.A. 55, 328-336.

Racowsky, C., Baldwin, K.V., Larabell, C.A., DeMarais, A.A., & Kazilek, C.J. (1989). Downregulation of membrana granulosa cell gap junctions is correlated with irreversible commitment to resume meiosis in golden Syrian hamster oocytes. Eur. J. Cell Biol. 49, 244-251.

Racowsky, C., & Satterlie, R.A. (1985). Metabolic, fluorescent dye and electrical coupling between hamster oocytes and cumulus cells during meiotic maturation *in vivo* and *in vitro*. Dev. Biol. 108, 191-202.

Rasmussen, H. (1991). Disordered cell communication as the basis of human disease: implications for 21st-century medicine. In: *Biology and Medicine Into the 21st Century* (Hardy, M.A., & Kinne, R.K.H., eds.), pp. 33-68. Karger, Basel, Switzerland.

Reaume, A.G., de Sousa, P.A., Kulkarni, S., Lowell Langile, B., Zhu, D., Davies, T.C., Juneja, S.C.,
    Kidder, G.M., & Rossant, J. (1995). Cardiac malformation in neonatal mice lacking
    connexin43. Science 267, 1831-1834.

Reed, K.E., Westphale, E.M., Larson, D.M., Wang, H.-Z. Veenstra, R.D., & Beyer, E.C. (1993).
    Molecular cloning and functional expression of human connexin37, an endothelial cell gap
    junction protein. J. Clin. Invest. 91, 997-1004.

Reid, L., Abreu, S.L., & Montgomery, K. (1988). Extracellular matrix and hormonal regulation of syn-
    thesis and abundance of messenger RNAs in cultured liver cells. In: The Liver: Biology and
    Pathobiology, 2nd Ed. (Arias, I.M., Jakoby, W.B., Popper, H., Schachter, D., & Shafritz, D.A.,
    eds.), pp. 717-737. Raven Press, New York.

Richard, G., Lin, J.P., Smith, L., Whyte, Y.M., Itin, P., Wollina, U., Epstein, E. Jr, Hohl, D., Giroux,
    J.M., Charnas, L., Bale, S.J., & DiGiovanna, J.J. (1997). Linkage studies in erythrokeratoder-
    mias: Fine mapping, genetic heterogeneity and analysis of candidate genes. J. Invest. Derma-
    tol. 109, 666-671

Richard, G., Smith, L.E., Bailey, R.A., Itin, P., Hohl, D., Epstein, E.H., Jr., DiGiovanna, J.J., Comp-
    ton, J.G., & Bale, S.J. (1998). Mutations in the human connexin gene GJB3 cause erythrokera-
    todermia variabilis. Nat. Genet. 20, 366-369.

Risek, B., & Gilula, N.B. (1991). Spatiotemporal expression of three gap junction gene products
    involved in fetomaternal communication during rat pregnancy. Development 113, 165-181.

Risek, B., Klier, F.G., & Gilula, N.B. (1992). Multiple gap junction genes are utilized during rat skin
    and hair development. Development 116, 639-651.

Robb-Gaspers, L.D., & Thomas, A.P. (1995). Coordination of $Ca^{2+}$ signaling by intercellular propa-
    gation of $Ca^{2+}$ waves in the intact liver. J. Biol. Chem. 270, 8102-8107.

Robertson, J.D., Bodenheimer, T.S., & Stage, D.E. (1963). The ultrastructure of Mauthner cell syn-
    apses and nodes in goldfish brains. J. Cell Biol. 19, 159-199.

Rohlmann, A., Laskawi, R., Hofer, A., Dobo, E., Dermietzel, R., & Wolff, J.R. (1993). Facial nerve
    lesions lead to increased immunostainng of the astrocytic gap junction protein (connexin43) in
    the corresponding facial nucleus of rats. Neurosc. Lett.154, 206-208.

Rosenberg, E., Spray, D.C., & Reid, L.M. (1993). Matrix regulation of gap junctions and of excitabil-
    ity in cells. In: Extracellular Matrix: Chemistry, Biology and Pathobiology with Emphasis on
    Liver. (Zern, M.A., & Reid, L.M., Eds.), pp. 449-462. Marcel Dekker, New York.

Ross, R. (1993). The pathogenesis of atherosclerosis: a perspective for the 1990s. Nature 362,
    801-809.

Ruangvoravat, C.P., & Lo, C.W. (1992). Connexin 43 expression in the mouse embryo: Localization
    of transcripts within developmentally significant domains. Devel. Dyn. 194, 261-281.

Sáez, J.C., Berthoud, V.M., Moreno, A.P., & Spray, D.C. (1993). Gap junctions: multiplicity of con-
    trols in differentiated and undifferentiated cells and possible functional implications. In:
    Advances in Second Messenger and Phosphoprotein Research. (Shenolikar, S., & Nairn, A.,
    Eds), vol. 27, pp. 163-198. Raven Press, New York.

Sáez, J.C., Connor, J.A., Spray, D.C., & Bennett, M.V.L. (1989a). Hepatocyte gap junctions are per-
    meable to the second messenger, inositol 1,4,5-triphosphate, and to calcium ions. Proc. Natl.
    Acad. Sci. U.S.A. 86, 2708-2712.

Sáez, J.C., Gregory, W.A., Dermietzel, R., Hertzberg, E.L., Watanabe, T., Reid, L.M., Bennett,
    M.V.L., & Spray, D.C. (1989b). cAMP delays disappearance of gap junctions between pairs of
    rat hepatocytes in primary culture. Amer. J. Physiol. 257, C1-C11.

Saffitz, J.E., Corr, P.B., & Sobel, B.E. (1993). Arrhythmogenesis and ventricular dysfunction after
    myocardial infarction. Is anomalous cellular coupling the elusive link? Circulation 87,
    1742-1745.

Saffitz, J.E., Hoyt, R.H., Luke, R.A., Kanter, H.L., & Beyer, E.C. (1992). Cardiac myocyte intercon-
    nections at gap junctions. Role in normal and abnormal electrical conduction. Trends Cardio-
    vasc. Med. 2, 56-60.

Sakai, N., Tabb, T., & Garfield, R.E. (1992). Modulation of cell-to-cell coupling between myometrial cells of the human uterus during pregnancy. Am. J. Obstet. Gynecol. 167, 472-480.

Salomon, D., Masgrau, E., Vischer, S., Ullrich, S., Dupont, E., Sappino, P., Saurat, J.-H., & Meda, P. (1994). Topography of mammalian connexins in human skin. J. Invest. Dermatol. 103, 240-247.

Salomon, D., & Meda, P. (1986). Heterogeneity and contact-dependent regulation of hormone secretion by individual β-cells. Exp. Cell Res. 162, 507-520.

Salomon, D., Saurat, J.H., & Meda, P. (1988). Cell-to-cell communication within intact human skin. J. Clin. Invest. 82, 248-254.

Sandberg, K., Iida, H.J., & Catt, K.J. (1992). Intercellular communication between follicular angiotensin receptors and *Xenopus laevis* oocytes—mediation by an inositol 1,4,5 triphosphate-dependent mechanism. J. Cell. Biol. 117, 157-167.

Sanderson, M.J., Charles, A.C., & Dirksen, E.R. (1990). Mechanical stimulation and intracellular communication increases intracellular $Ca^{2+}$ in epithelial cells. Cell Regulation 1, 585-596.

Sanderson, M.J., Paemeleire, Strahonja, A., & Leybaert, L. (1998). Intercellular $Ca^{2+}$ signaling between glial and endothelial cells. In: Gap Junctions. (Werner, R., Ed.), pp. 261-265. IOS Press, Netherlands.

Sasaki, Y., Shiba, Y., & Kanno, Y. (1988). Suppression of intercellular communication in acinar cells from rat submandibular gland by cholinergic and adrenergic agonists. Japan. J. Physiol. 38, 531-543.

Sato, K., Gratas, C., Lampe, J., Biernat, W., Kleihues, P., Yamasaki, H., & Ohgaki, H. (1997). Reduced expression of the P2 form of the gap junction protein connexin43 in malignant meningiomas. J. Neuropathol Exp Neurol. 56, 835-839.

Sawey, M.J., Goldschmidt, M.H., Risek, B., Gilula, N.B., & Lo, C.W. (1996). Perturbation in connexin 43 and connexin26 gap-junction expression in mouse skin hyperplasia and neoplasia. Mol. Carcinog. 17, 49-61.

Scemes, E., Dermietzel R., & Spray, D.C. (1998). Calcium waves between astrocytes from Cx43 knockout mice. Glia, 24, 65-73.

Scemes, E., & Spray, D. C. (1998) Increased intercellular communication in mouse astrocytes exposed to hyposmotic shocks. Glia, 24, 74-84.

Scherer, S.S. (1997). The biology and pathobiology of Schwann cells. Curr. Opin. Neurol. 10(5), 386-397.

Segal, S.S., & Bény, J.-L. (1992). Intracellular recording and dye transfer in arterioles during blood flow control. Am. J. Physiol. 263, H1-H7.

Segal, S.S., & Duling, B.R. (1986). Flow control among microvessels coordinated by intercellular conduction. Science 234, 868-870.

Segal, S.S., & Duling, B.R. (1987). Propagation of vasodilation in resistance vessels of the hamster: development and review of a working hypothesis. Circ. Res. 61(suppl. II), 20-25.

Segal, S.S., & Duling, B.R. (1989). Conduction of vasomotor responses in arterioles: a role for cell-to-cell coupling? Am. J. Physiol. 256, H838-H845.

Sepp, R., Severs, N.J., & Gourdie, R.G. (1996). Altered patterns of cardiac intercellular junction distribution in hypertrophic cardiomyopathy. Heart 76, 412-417.

Seseke, F.G., Gardeman, A., & Jungermann, K. (1992). Signal propagation via gap junctions, a key step in the regulation of liver metabolism by the sympathetic hepatic nerves. FEBS Lett. 301, 265-270.

Seul, K.H., Tadros, P.N., & Beyer, E.C. (1997). Mouse connexin40: gene structure and promoter analysis. Genomics 46, 120-126.

Sheridan, J.D. (1971). Dye movement and low resistance junctions between reaggregated embryonic cells. Dev. Biol. 26, 627-636.

Sheridan, J.D. (1973). Functional evaluation of low resistance junctions: influence of cell shape and size. Amer. Zool. 13, 1119-1129.

Sheridan, J.D., & Atkinson, M.M. (1985). Physiological roles of permeable junctions: some possibilities. Ann. Rev. Physiol. 47, 337-353.

Sheridan, J.D., & Johnson, R.G. (1975). Cell junctions and neoplasia. In: Molecular Pathology. (Good, R.A. et al., Eds.), pp.354-378. Charles C. Thomas, Springfield, IL.

Sheridan, J.D., & Larson, D.M. (1982). Junctional communication in the peripheral vasculature. In: The Functional Integration of Cells in Animal Tissues. (Pitts, J.D., & Finbow, M.E., Eds), pp. 263-283. Cambridge University Press, Cambridge.

Sherman, A., Rinzel, J., & Keizer, J. (1988). Emergence of organized bursting in clusters of pancreatic beta-cells by channel sharing. Biophys. J. 54, 411-425.

Sholley, M.M., Grimbrone, A.M., & Cotran, R.S. (1977). Cellular migration and replication in endothelial migration. A study using irradiated endothelial cultures. Lab. Invest. 36, 18-25.

Simon, A.M., Goodenough, D.A., Li, E., & Paul, D.L. (1997). Female infertility in mice lacking connexin 37. Nature 385, 525-529

Simon, A.M., Goodenough, D.A., & Paul, D.L. (1998). Role of Cx40 in cardiac conduction. In: *Gap Junctions* (Werner, R., ed.), pp. 299-303. IOS Press, Netherlands.

Smith, J.H., Green, C.R., Peters, N.S., Rothery, S., & Severs, N.J. (1991). Altered patterns of gap junction distribution in ischemic heart disease. An immunohistochemical study of human myocardium using laser scanning confocal microscopy. Am. J. Path. 139, 801.

Smith, S.J. (1992). Do astrocytes process neural information? Prog. Brain Res. 94, 119-136.

Sneyd, J., Charles, A.C., & Sanderson, M.J. (1994). A model for the propagation of intercellular calcium waves. Am. J. Physiol. 266(1 Pt. 1), C293-C302.

Sohl, G., Gillen, C., Bosse, F., Gleichmann, M., Muller, H. W., & Willecke, K. (1996). A second alternative transcript of the gap junction gene connexin32 is expressed in murine Schwann cells and modulated in injured sciatic nerve. Eur. J. Cell Biol. 69, 267-275.

Spach, M.S. (1997). Discontinuous cardiac conduction: its origin in cellular connectivity with long-term adaptive changes that cause arrythmias. In: Discontinuous Conduction in the Heart. (Spooner, P.M., Joyner, R.W., & Jalife, J., Eds.), pp. 5-51. Futura Publishing, Armonk, NY.

Spach, M.S., & Dolber, P.C. (1986). Relating extracellular potentials and their derivatives to anisotropic propagation at a microscopic level in human cardiac muscle. Evidence for electrical uncoupling of side-to-side fiber connections with increasing age. Circ. Res. 58, 356-371.

Spira, M.E., Spray, D.C., & Bennett, M.V.L. (1976). Electrotonic coupling: effective sign reversal by inhibitory neurons. Science 194, 1065-1067.

Spira, M.E., Spray, D.C., & Bennett, M.V.L. (1980). Synaptic organization of expansion motoneurons of Navanax inermis. Brain Res. 195, 241-269.

Spray, D.C. (1994a). CMTX1: a gap junction genetic disease. Lancet 343, 1111-1112.

Spray, D.C. (1994b). Physiological and pharmacological regulation of gap junction channels. In: Molecular Mechaisms of Epithelial Cell Junctions: From Development to Disease, pp. 195-215. Medical Intelligence Unit, RG Landes, Co, Biomedical Publishers, Austin, TX.

Spray, D.C., Bai, S., & Burk, R.D. (1994a). Regulation and function of liver gap junctions and their genes. Prog. Liver Dis. 12, 1-18.

Spray, D.C., & Bennett, M.V.L. (1985). Physiology and pharmacology of gap junctions. Ann. Rev. Physiol. 47, 281-303.

Spray, D.C., Bennett, M.V.L., Campos de Carvalho, A.C., Eghbali, B., Moreno, A., & Verselis, V. (1991a). Voltage dependence of junctional conductance. In: Biophysics of Gap Junction Channels. (Peracchia, C., Ed.), pp. 97-116. CRC Press, Boca Raton, FL.

Spray, D.C., & Burt, J.M. (1990). Structure activity relations of the cardiac gap junction channel. Am. J. Physiol. 258, C195-C205.

Spray, D.C., Chanson, M.A., Moreno, A.P., Dermietzel, R., & Meda. P. (1991b). Distinctive types of gap junction channels connect WB cells, a clonal cell line derived from rat liver. Am. J. Physiol. 260 (Cell Physiol. 29), C513-C527.

Spray, D.C., & Dermietzel, R. (1995). X-Linked Charcot-Marie Tooth disease and other potential gap-junction diseases of the nervous system. Trends Neurosci. 18, 256-262.

Spray, D.C., & Fishman, G.I. (1996). Physiological and molecular properties of cardiac gap junctions. In: Molecular Physiology of Cardiac Ion Channels and Transporters. (Morad, M., Ebashi, E., Trautwein, W., & Kurachi, Y., Eds.), pp. 209-221. Kluwer Academic Publishers, Dordrecht, Netherlands.

Spray, D.C., Fujita, M., Sáez, J.C., Choi, H., Watanabe, T., Hertzberg, E.L., Rosenberg, L.C., & Reid, L.M. (1987). Glycosaminoglycans and proteoglycans induce gap junction synthesis and function in primary liver cultures. J. Cell. Biol. 105, 541-551.

Spray, D.C., Harris, A.L., & Bennett, M.V.L. (1979). Voltage dependence of junctional conductance in early amphibian embryos. Science 204, 432-434.

Spray, D.C., Rook, M.B., Moreno, A.P., Sáez, J.C., Christ, G.J., Campos de Carvalho, A.C., & Fishman, G.I. (1994c). Cardiovascular gap junctions: gating properties, function and dysfunction. In: Ion Channels in the Cardiovascular System: Function and Dysfunction. (Spooner, P.M., Brown, A.M., Catterall, W.A., Kaczorowski, G.J., & Strauss, H.C., Eds.), pp. 185-217. Futura Publishing, Mt. Kisco, NY.

Spray, D.C., Sáez, J.C., Hertzberg, E.L., & Dermietzel, R. (1994c). Gap junctions in liver: Composition, function and regulation. In: The Liver: Physiology and Pathophysiology. (Arias, I. et al., eds.), pp. 951-967. Raven Press, New York.

Spray, D.C., Spira, M.E., & Bennett, M.V.L. (1980). Synaptic connections of buccal mechanosensory neurons in the opisthobranch mollusc, Navanax inermis. Brain Res. 182, 271-286.

Spray, D.C., & Vink, M.J. (1995). Cardiac gap junctions as $K^+$ (and $Ca^{2+}$) channels. In: Potassium Channels in Normal and Pathological Conditions. (Vereecke, J., Verdonck, F., & van Bogaert, P.-P., Eds.), pp. 424-427. Leuven University Press, Leuven, Belgium.

Spray, D.C., Vink, M.J., Scemes, E., Suadicani, S., Fishman, G.I., & Dermietzel, R. (1998). Characteristics of coupling in cardiac myocytes and CNS astrocytes cultured from wildtype and Cx43-null mice. In: Gap Junctions. (Werner, R., Ed.), pp. 281-285. IOS Press, Netherlands.

Stagg, R.B., & Fletcher, W.H. (1990). The hormone-induced regulation of contact-dependent cell-cell communication by phosphorylation. Endocrine Rev. 11, 302-325.

Steele, E.C., Jr., Lyon, M.F., Glenister, P.H., Buillot, P.V., & Church, R.L. (1998). Identification of a mutation in the connexin 50 (Cx50) gene of the NO2 cataractous mouse mutant. In: Gap Junctions. (Werner, R., Ed.), pp. 289-293. IOS Press, Netherlands.

Steinberg, T.H., Civitell, R., Beyer, E.C., Jorgensen, N.R., Cao, D., Geist, S.T., & Lin, G. (1998). Multiple mechanisms for intercellular calcium waves. In: Gap Junctions. (Werner, R., Ed.), pp. 271-275. IOS Press, Netherlands.

Stewart. W.W. (1978). Functional connections between cells as revealed by dye-coupling with a highly fluorescent naphthalimide tracer. Cell 14, 741-759.

Strobl, B., Hellman, P., Bittner, R., & Winterhager, E. (1998). Structural and functional analysis of the rat connexin31 promoter. In: Gap Junctions. (Werner, R., Ed.), pp. 335-339. IOS Press, Netherlands.

Stutenkemper, R., Geisse, S., Schwarz, H.J., Look, J., Traub, O., Nicholson, B.J., & Willecke, K. (1992). The hepatocyte-specific phenotype of murine liver cells correlates with high expression of connexin32 and connexin26 but very low expression of connexin43. Exp. Cell Res. 201, 43-54.

Subak-Sharpe, H., Bürk, R.R., & Pitts, J.D. (1966). Metabolic co-operation by cell to cell transfer between genetically different mammalian cells in tissue culture. Heredity 21, 342-343.

Subak-Sharpe, H., Bürk, R.R., & Pitts, J.D. (1969). Metabolic co-operation between biochemically marked mammalian cells in tissue culture. J. Cell Sci. 4, 353-367.

Tabb, T., Thilander, G., Grover, A., Hertzberg, E., & Garfield, R. (1992). An immunochemical and immunocytologic study of the increase in myometrial gap junctions (and connexin 43) in rats and humans during pregnancy. Am. J. Obstet. Gynecol. 167, 559-567.

Temme, A., Buchmann, A., Gabriel, H.D., Nelles, E., Schwarz, M., Willecke, K. (1997). High incidence of spontaneous and chemically induced liver tumors in mice deficient for connexin32. Curr. Biol. 7, 713-716.

Traub, O., Look, J. Dermietzel, R., Brümmer, F., Hülser, D., & Willecke, K. (1989). Comparative characterization of the 21-kD and 26-kD gap junction proteins in murine liver and cultured hepatocytes. J. Cell Biol. 108, 1039-1051.

Trosko, J.E., & Chang. C.C. (1980). An integrative hypothesis linking cancer, diabetes and atherosclerosis. The role of mutations and epigenetic changes. Med. Hypoth. 7, 455-468.

Tsien, R., & Weingart, R. (1976). Inotropic effect of cyclic AMP in calf ventricular muscle studied by a cut-end method. J. Physiol. 260, 117-141

Usowicz, M.M., Gallo, V., & Cull-Candy, S.G. (1989). Multiple conductance channels in type-2 cerebellar astrocytes activated by excitatory amino acids. Nature 339, 380-383.

Valdeolmillos, M., Nadal, A., Soria, B., & Garcia-Sancho, J. (1993). Fluorescence digital image analysis of glucose-induced $Ca^{2+}i$ oscillations in mouse pancreatic islets of langerhans. Diabetes 42, 1210-1214.

Vaney, D.I. (1991). Many diverse types of retinal neurons show tracer coupling when injected with biocytin or Neurobiotin. Neurosci. Lett. 125, 187-190.

Veenstra, R.D., Wang, H.Z., Beyer, E.C., & Brink, P.R. (1994a). Selective dye and ionic permeability of gap junction channels formed by connexin45. Circ. Res. 75, 483-490.

Veenstra, R.D., Wang, H.Z, Beyer, E.C., Ramanan, S.V., & Brink, P.R. (1994b). Connexin37 forms high conductance gap junction channels with subconductance state activity and selective dye and ionic permeabilities. Biophys. J. 66, 1915-1928.

Venance, L., Stella, N., Glowinski, J., & Giaume, C. (1997). Mechanism involved in initiation and propagation of receptor-induced intercellular calcium signaling in cultured rat astrocytes. J. Neurosci. 17, 1981-1992.

Verselis, V., White, R.L., Spray, D.C., & Bennett, M.V.L. (1986). Gap junctional conductance and permeability are linearly related. Science 234, 461-464.

Vozzi, C., Bosco, D., Dupont, E., Charollais, A., & Meda, P. (1997). Hyperinsulinemia-induced hypoglycemia is enhanced by overexpression of connexin43. Endocrinology 138, 2879-2885.

Vozzi, C., Ullrich, S., Charollais, A., Philippe, J., Orci, L., & Meda, P. (1995). Adequate connexin-mediated coupling is required for proper insulin production. J. Cell Biol. 131, 1561-1562.

Vrionis, F.D., Wu, J.K., Qi, P., Waltzman, M., Cherington, V., Spray, D.C. (1997). The bystander effect exerted by tumor cells expressing the herpes simplex virus thymidine kinase (HSVtk) gene is dependent on connexin expression and cell communication. Gene Ther. 4, 577-585.

Wade, M.H., Trosko, J.E., & Schindler, M. (1986). A fluorescence phtobleaching assay of gap junction-mediated communication between human cells. Science 232, 525-528.

Wang, H-Z., & Veenstra, R.D. (1997). Monovalent ion selectivity sequences of the rat connexin43 gap junction channel. J. Gen. Physiol. 109, 491-507.

Warner, A. (1992). Gap junctions in development - a perspective. Sem. Cell Biol. 3, 81-91.

Warner, A.E., Guthrie, S.C., & Gilula, N.B. (1984). Antibodies to gap-junctional protein selectivity disrupt junctional communication in the early amphibian embryo. Nature 311, 127-131.

Warner, A.E., & Lawrence, P.A. (1982). Permeability of gap junctions at the segmental border in insect epidermis. Cell 28, 243-259.

Watts, S.W., & Webb, R.C. (1996). Vascular gap junctional communication is increased in mineralocorticoid-salt hypertension. Hypertension 28, 888-893.

Weidmann S. (1970). Electrical constants of trabecular muscle from mammalian heart. J. Physiol. 264, 341-365.

White, T.W., Chanson, M., Huarte, J., & Meda, P. (1996). Generation of transgenic mice expressing connexin32 in pancreatic β cells. Mol. Biol. Cell 7, 461a.

White T.W., Goodenough, D.A., & Paul, D.L. (1998). Ocular abnormalities in connexin50 knockout mice. Europ. J. Cell. Biol. 48, 23.

White, R.L., Spray, D.C., Campos de Carvalho, A.C., Wittenberg, B.A., & Bennett, M.V.L. (1985). Some physiological and pharmacological properties of gap junctions between cardiac myocytes dissociated from adult rat. Amer. J. Physiol. 249 (Cell Physiol. 18), C447-C455

Wilgenbus, K.K., Kirkpatrick, C.J., Knuechel, R., Willecke and K., Traub, O. (1992). Expression of Cx26, Cx32 and Cx43 gap junction proteins in normal and neoplastic human tissues. Int. J. Cancer 51, 522-529.

Willecke, K., Buchman, A., Butzler, C., Heinz-Dieter, G., Hagendorff, A., Jung, D., Kirchloff, S., Kruger, O., Nelles, E., Schwarz, M., Temme, A., Traub, O., & Winterhager, E. (1998). Biological functions of gap junctions revealed by targeted inactivation of mouse connexin32, -26 and -40 genes. In: Gap Junctions. (Werner, R., Ed.), pp. 304-308. IOS Press, Netherlands.

Willecke, K., Elfgang, C., Lichtenberg-Frate, H., Butterweck, A., & Traub, O. (1994). Analysis of murine gap junction channels formed by compatible and incompatible connexins [abstract]. Workshop on Intercerllular Communication, Puschino, Russia.

Willecke, K., Muller, D., Druge, P.M., Frixen, U., Schafer, R., Dermietzel, R., & Hulser, D. (1983). Isolation and characterization of Chinese hamster cells defective in cell-cell coupling by gap junctions. Exp. Cell. Res. 144, 95-113.

Wilson, J.M., Birinyi, L.K., Salomon, R.N., Libby, P., Callow, A.D., & Mulligan, R.C. (1989). Implantation of vascular grafts lined with genetically modified endothelial cells. Science 244, 1344-1346.

Winterhager, E., Grummer, R., Jahn, E., Willecke, K., & Traub, O. (1993). Spatial and temporal expression of connexin26 and connexin43 in rat endometrium during trophoblast invasion. Dev. Biol. 157, 399-409.

Wray, S. (1993). Uterine contraction and physiological mechanisms of modulation. J. Physiol. 264, C1-C18.

Wygoda, M.R., Wilson, M.R., Davis, M.A., Trosko, J.E., Rehemtulla, A., & Lawrence, T.S. (1997). Protection of herpes simplex virus thymidine kinase-transduced cells from ganciclovir-mediated cytotoxicity by bystander cells: the Good Samaritan effect. Cancer Res. 57, 1699-703.

Xia, J.H., Liu, C.Y., Tang, B.S., Pan, Q., Huang, L., Dai, H.P., Zhang, B.R., Xie, W., Hu, D.X., Zheng, D., Shi, X.L., Wang, D.A., Xia, K., Yu, K.P., Liao, X.D., Feng, Y., Yang, Y.F., Xiao, J.Y., Xie, D.H., & Huang, J.Z. (1998). Mutations in the gene encoding gap junction protein beta-3 associated with autosomal dominant hearing impairment. Nat. Genet. 20, 370-373.

Yamamoto, T., Kardami, E., & Nagy, J.I. (1991). Basic fibroblast growth factor in rat brain: localization to glial gap junctions correlates with connexin43 distribution. Brain Res. 554, 336-343.

Yamasaki, H., Hollstein M., Mesnil M., Martel N., & Aguelon A.M. (1987). Selective lack of interecellular communication between transformed and nontransformed cells as a common property of chemical and oncogene transformation of BALB/c 3T3 cells. Cancer Res. 47, 5658-5664.

Yancey, S.B., Biswal, S., & Revel, J.-P. (1992). Spatial and temporal patterns of distribution of the gap junction protein connexin43 during mouse gastrulation and organogenesis. Development 114, 203-212.

Yancey, B.S., Nicholson, B.J., & Revel, J.-P. (1981). The dynamic state of liver gap junctions. J. Supramol. Struct. Cell Biochem. 16, 221-232.

Yu, W., Dahl, G., & Werner, R. (1994). The connexin43 gene is responsive to oestrogen. Proc. Royal Soc. Lond. Series B: Biol. Sci. 255, 125-132.

Zelante, L., Gasparini, P., Estivill, X., Melchionda, S., D'Agruma, L., Govea, N., Mila, M., Monica, M.D., Lutfi, J., Shohat, M., Mansfield, E., Delgrosso, K., Rappaport, E., Surrey, S., & Fortina, P. (1997). Connexin26 mutations associated with the most common form of non-syndromic neurosensory autosomal recessive deafness (DFNB1) in Mediterraneans. Human Mol. Genet. 6, 1605-1609.

Zhao, S., Fort, A., & Spray, D.C. (1999). Gap junctions in Schwann cells from wildtype and connexin-deficient mice. Ann. N.Y. Acad. Sci. (In press.)

Zhang, J.T., & Nicholson, B.J. (1989). Sequence and tissue distribution of a second protein of hepatic gap junctions, Cx26, as deduced from its cDNA. J. Cell Biol. 109, 3391-3401.

Zhu, D., Kidder, G.M., Caveney, S., & Naus, C.C. (1992). Growth retardation in glioma cells cocultured with cells overexpressing a gap junction protein. Proc. Nat. Acad. Sci. U.S.A. 89, 10218-10221.

Zimmerman, A.L., & Rose, B. (1985). Permeability properties of cell-to-cell channels: Kinetics of fluorescent tracer diffusion through a cell junction. J. Membr. Biol. 84, 269-283.

# GAP JUNCTIONS AND CONNEXINS IN THE MAMMALIAN CENTRAL NERVOUS SYSTEM

James I. Nagy and Rolf Dermietzel

I.  Introduction . . . . . . . . . . . . . . . . . . . . . . . . . . . . . . . . . . . . . . . . . . . . . . 324
II.  Intercellular Communication between Neurons . . . . . . . . . . . . . . . . . . . . . . 326
    A.  Occurrence of Neuronal Gap Junctions. . . . . . . . . . . . . . . . . . . . . . . . . 326
    B.  Connexin Expression in Mature Neurons . . . . . . . . . . . . . . . . . . . . . . . 330
    C.  Possible Connexin-Related Proteins in Neurons. . . . . . . . . . . . . . . . . . . 332
    D.  Regulation of Conductance State . . . . . . . . . . . . . . . . . . . . . . . . . . . . 333
    E.  Functional Considerations. . . . . . . . . . . . . . . . . . . . . . . . . . . . . . . . . . 334
III.  Junctional Communication between Astrocytes . . . . . . . . . . . . . . . . . . . . . . 337
    A.  Astrocytic Gap Junctions. . . . . . . . . . . . . . . . . . . . . . . . . . . . . . . . . . . 337
    B.  Connexin Expression by Astrocytes . . . . . . . . . . . . . . . . . . . . . . . . . . . 338
    C.  Connexin43 in Astrocytes *In Vivo* . . . . . . . . . . . . . . . . . . . . . . . . . . . . 338
    D.  Connexin30 in Astrocytes *In Vivo* . . . . . . . . . . . . . . . . . . . . . . . . . . . . 340
    E.  Astrocytic Syncytial Compartments . . . . . . . . . . . . . . . . . . . . . . . . . . . 342
    F.  Astrocytic Coupling and Connexin *In Vitro*. . . . . . . . . . . . . . . . . . . . . . 343
    G.  Interastrocytic Calcium Waves . . . . . . . . . . . . . . . . . . . . . . . . . . . . . . 344
    H.  Calcium Waves and Neuronal/Glial Interactions. . . . . . . . . . . . . . . . . . . 346
    I.  Regulation of Astrocytic Connexin Expression . . . . . . . . . . . . . . . . . . . . 347
    J.  Regulation of Astrocytic Coupling . . . . . . . . . . . . . . . . . . . . . . . . . . . . 347

**Advances in Molecular and Cell Biology, Volume 30, pages 323-396.**
**Copyright © 2000 by JAI Press Inc.**
**All rights of reproduction in any form reserved.**
**ISBN: 0-7623-0599-1**

|     | K.  | Regulation of Connexin43 by Phosphorylation | 350 |
|     | L.  | Astrocytic Gap Junctions in Response to Injury | 351 |
| IV. |     | Junctional Communication between Myelinating Cells | 354 |
|     | A.  | Oligodendrocyte Gap Junctions | 354 |
|     | B.  | Localization of Oligodendrocytic Connexins *In Vivo* | 355 |
|     | C.  | Oligodendrocytic Connexins and Functional Coupling | 356 |
|     | D.  | Oligodendrocytic Coupling *In Vitro* | 357 |
|     | E.  | Schwann Cells and the Charcot-Marie-Tooth Type X1 Syndrome | 359 |
| V.  |     | Junctions between Astrocytes and Oligodendrocytes | 360 |
|     | A.  | Heterologous Astro/Oligo Gap Junctions *In Vivo* | 360 |
|     | B.  | Connexins at Heterotypic Astro/Oligo Junctions | 361 |
| VI. |     | Retinal Gap Junctions and Connexins | 362 |
| VII.|     | Gap Junctions and Connexins during Central Nervous System Development | 363 |
|     | A.  | Neuronal Gap Junctions | 363 |
|     | B.  | Connexin Expression | 365 |
|     | C.  | Connexin43 Localization | 366 |
|     | D.  | Connexin30 Localization | 368 |
|     | E.  | Connexin26 and Connexin36 Localization | 369 |
| VIII.|    | Gap Junctions between Other Central Nervous System Cell Types | 369 |
|     | A.  | Leptomeningeal Cells | 370 |
|     | B.  | Ependymal Cells | 371 |
|     | C.  | Pericyte and Endothelial Cells | 371 |
|     | D.  | Pinealocytes | 372 |
| IX. |     | Conclusions and Prospects | 373 |

# I. INTRODUCTION

Gap junctions have been found in virtually all tissues except mature skeletal muscle. Consistent with their role of providing a pathway for intercellular communication in peripheral tissues, it is generally well accepted that gap junctions mediate intercellular electrotonic and metabolic coupling between neurons and between various other cell types in the central nervous system (CNS) (for reviews, see Bennett, 1974; Bennett and Goodenough, 1978; Sotelo and Korn, 1978; Peracchia, 1980; Hertzberg et al., 1981; Loewenstein, 1981; Sotelo and Triller, 1981; Dudek et al., 1983; Dermietzel et al., 1990; Bennett et al., 1991). Progress on the elucidation of the prevalence of neuronal gap junctions and the functional contributions of these as well as glial gap junctions to the operation of mammalian CNS has been slow because studies of gap junction function in such systems are technically difficult. In the strictest sense, they must involve direct measurements of electrical or metabolic coupling between cells. Indirect methods to assess function are also limited because, unlike chemical transmission via neurotransmitters, gap junctions have no substrates or enzymes that can be measured and lack receptor components as specific targets for ligand binding. There are currently no useful pharmacological tools that could be employed to interfere specifically with gap junctional communication *in vivo*. These limitations have been partly circumvented with approaches afforded by the identification of connexins which are, as described in other chapters of this volume, the structural proteins of gap junctions.

It has been known for some time that gap junction channels are composed of six subunits of proteins that form hemichannels or connexons contributed by each of the coupled cells. Because gap junctions are abundant in liver and heart, the search for the molecules forming these channels utilized primarily these tissues, leading ultimately to the elucidation of primary sequence and membrane topologies of three gap junction proteins, namely connexin32 (Cx32) and connexin26 (Cx26) from liver, and connexin43 (Cx43) from heart as well as several more connexins from other tissues (see other chapters of this volume). Most recently, low stringency hybridization and polymerase chain reaction (PCR) cloning from genomic libraries have increased the number of putative mammalian connexins to about 15. All family members possess similar gene structure (but see section II.B), exhibit about 50% sequence homology at the amino acid level, and show diverse patterns of tissue distribution (for reviews see Beyer et al., 1990; Dermietzel et al., 1990; Bennett et al., 1991, and references therein).

With the discovery of each new connexin, many reports have included mammalian brain in routine Western or Northern blot screening of tissues in which these are expressed. To date nine connexins have been detected in adult and developing CNS by blotting techniques. These include connexin45 (Cx45), Cx43, connexin40 (Cx40), connexin37 (Cx37), connexin36 (Cx36), Cx32, connexin31 (Cx31), connexin30 (Cx30), and Cx26 in rat brain (Dermietzel et al., 1989; Aoumari et al., 1990; Naus et al., 1990; Dupont et al., 1991; Dermietzel, 1996, Nagy et al., 1998). The mRNA of cloned Cx37 and Cx40 have also been detected in whole brain homogenates (Willecke et al., 1991; Hennemann et al., 1992). Although the present review is focused on mammalian species, it appears that molecular probes and antibodies developed for connexins expressed in mammalian tissues may also be useful for the analysis of connexins and gap junctions in lower vertebrates (Yamamoto et al., 1989a; Minkoff et al., 1991). It should be noted that failure to detect a particular connexin in CNS does not necessary mean its absence in this tissue nor does the detection of connexin mRNA by *in situ* hybridization histology (ISIHI) always imply assembly of the gene product into gap junctions. In analyses of whole brain, for example, expression of a connexin or its mRNA in a small population of cells in relatively small brain nuclei may go undetected. Moreover, such analyses do not provide information on the cell types that express connexins that are detected. Owing to the cellular diversity and morphological heterogeneity of CNS tissue, cellular localization requires detailed examination, ultimately by immunolabeling techniques at the electron microscope (EM) level, to confirm connexin localization at gap junctions. These points are further discussed below in relation to connexins found in particular cells of the CNS.

In this review, we attempt to integrate classical work on gap junctions in the mammalian CNS with more recent information derived from biochemical and anatomical studies of connexins in neural tissues, with a focus on gap junctions between neurons and between glial cells in adult and developing systems. Our aim is to provide a broad survey of literature that will serve as background for anticipated advances in understanding of CNS gap junctions. Numerous reviews on gap junctions and connexins have appeared in recent years (Bruzzone et al., 1996a, 1996b; Goodenough et al., 1996; Kumar and Gilula, 1996), including an excellent series in the *Journal of Bioenergetics and Biomembranes,* vol. 28, 1996, and a monograph co-edited by one of the authors of this chapter (Spray and Dermietzel, 1996). There are also several reviews that summarize topics relevent to this chapter (Dermietzel and Spray, 1993, Cook and Becker, 1995; Fulton, 1995; Giaume and Venance, 1995; Wolburg and Rohlmann, 1995; Vernadakis, 1996; Bruzzone and Ressot, 1997; Giaume et al., 1997).

## II.  INTERCELLULAR COMMUNICATION BETWEEN NEURONS

### A.  Occurrence of Neuronal Gap Junctions

In the CNS of lower vertebrates, the existence of electrical synapses formed by gap junctions is well documented (for review, see Bennett et al., 1991 and references therein) and there is a large body of evidence indicating that electrotonic transmission between particular assemblies of neurons constitutes an important mode of intercellular communication. In the sensory and motor components of the electrocommunication and electrolocation system in weakly electric fish, for example, it appears that the rapid transmission and synchronous neuronal activity afforded by gap junctional coupling may be indispensable for the particular types of integration and information transfer required at certain CNS levels (Carr and Maler, 1986; Yamamoto et al., 1989a). In contrast, although gap junctions between neurons were first demonstrated in the CNS of lower vertebrates several decades ago (Furshpan and Potter, 1959), the extent of their occurrence and the nature of their contribution to interneuronal communication in mammalian neural systems are still largely matters of speculation with many basic issues remaining to be resolved.

Evidence for the existence of gap junctions between mammalian CNS neurons has been obtained by several different strategies. More traditional studies have involved direct visualization of gap junctions between neurons by conventional or freeze fracture electron microscopy where junctions can be recognized as characteristic heptalaminar structures or paracrystalline arrays of particles and pits, respectively (Brightman and Reese, 1969; Dermietzel, 1974). These ultrastructural approaches allow definitive identification of neuronal gap junctions as well as the types of neurons contributing to junction formation in cases where the junctional elements such as dendrites can be identified as belonging to a particular class of neuron (for review, see Sotelo and Korn, 1978). Direct demonstrations of electrical coupling between neurons are provided by simultaneous intracellular recordings from pairs of coupled cells utilizing brain slice or tissue culture preparations (Bennett, 1977; Perez-Velasques, et al., 1994; Llinás and Yarom, 1981; MacVicar and Dudek, 1981). However, this approach is difficult to achieve in mammalian brain *in vivo*. Consequently, electrophysiological evidence for electrotonic transmission has been obtained by indirect tests involving the assumption that intracellularly recorded short latency depolarizations in antidromically activated neurons represent electrotonic coupling when the depolarizations are not blocked by a preceding orthodromic spike in the impaled neuron (Korn and Farber, 1979). Another approach, based on the intercellular exchange of small molecules up to 1000 D through gap junctions (Loewenstein, 1981) is termed dye coupling or tracer-transfer and involves injection of a fluorescent dye into junctionally coupled cells resulting in the cell-to-cell diffusion of dye or tracer, presumably via gap junctions (Dudek et al., 1983). Until recently the most common tracer for the determination of dye transfer has been Lucifer yellow (LY). It has been successfully exploited in slice preparations and tissue culture (Connors et al., 1983; Murphy et al., 1983). The abundance of coupled neurons estimated on the basis of intracellular LY injections (for critical discussion, see Peinado et al., 1993a, 1993b) has apparently been under estimated. Limitation of LY may be partly related to its size (Mr 430) and charge, but more so to its propensity to bind proteins (Stewart, 1978, 1981). Its mobility within the cytoplasm and permeability between cells are thus low. LY transfer is generally undetectable if junctional conductance is below 1 to 2 nS (D. Spray, personal communication). Although somewhat larger than LY and not widely tested as yet, the highly fluorescent cyanine dyes with extinction coefficients up to six times greater than LY may hold some promise for dye transfer studies (Hossain et al., 1995). The biotinylated tracers Biocytin (Horikawa and Armstrong, 1988) and Neurobiotin (Kita and Armstrong, 1991) revealed that interneuronal tracer-coupling, for instance in the retina (Hampson et al., 1992) and in postnatal neocortical pyramidal cells (Peinado et al., 1993a), is much more extensive than expected on the basis of LY injections (Vaney, 1991). The application of dye and tracer transfer has demonstrated transfer in about 20% of the solitary complex neurons in dorsal medullary slices (Dean et al., 1997; Huang et al., 1997) and in 30% of suprachiasmatic nucleus neurons (Jiang et al., 1997). We expect a similar outcome when these new probes become more widely used in slice preparations from other brain regions. The possibility of intercellular propagation of calcium waves among neurons (Charles et al., 1996) may also have potential utility as an assay of gap junction activity, depending on the extent to which this process involves gap junctions (see Hassinger et al., 1996).

**Table 1.** Mammalian Neural Structures in Which Evidence for Gap Junctions Between Neuronal Elements Has Been Reported in the References Listed

| CNS Structure | References |
|---|---|
| 1. Between the cell bodies of primary afferent neurons in the trigeminal mesencephalic nucleus | Baker and Llinás, 1971; Hinrichsen and Larramendi, 1968 |
| 2. Between the perikarya or dendrites of inhibitory interneurons in the molecular layer, or both, and between the axonal processes of basket cells in the Purkinje cell layer in the cerebellum | Sotelo and Llinás, 1972 |
| 3. Between axon terminals and giant cells of Deiters in the lateral vestibular nucleus of rat | Korn et al., 1973; Sotelo and Palay; 1970; Wylie, 1973 |
| 4. Between dendrites in glomeruli and between granule cells in olfactory bulb of rat, gerbil, and marmoset | Pinching and Powell, 1971; Landis et al., 1974; Reyher et al., 1991; Miragall et al., 1996 |
| 5. Between dendrites in glomeruli within the inferior olive of rat, guinea pigs, opossum, cat, and monkey | Llinás et al., 1974; Sotelo et al., 1974; King, 1976; Gwyn et al., 1977; Rutherford and Gwyn, 1977; Sotelo et al., 1986; Benardo and Foster, 1986; DeZeeuw et al., 1989 |
| 6. Between photoreceptors, between horizontal cells, and between amacrine cells as well as between other cell types in the retina of rabbit, cat, and monkey | Raviola and Gilula, 1973, 1975; Kolb, 1979; Nishimura and Rakic, 1985; Dacheux and Raviola, 1986; Strettoi et al., 1990; Vaney, 1991; Hampson et al., 1992 |
| 7. Between dendrites or dendrites and somata of neurons in the cerebral cortex of rat, guinea pig, and monkey | Smith and Moskowitz, 1979; Peters, 1980; Gutnick and Prince, 1981; Gutnick et al., 1985; Sloper, 1972, 1973; Sloper and Powell, 1978 |
| 8. Between dendrites and between perikarya and dendrites of neurons in the anterior ventral cochlear nucleus and between stellate neurons in the dorsal cochlear nucleus of rat | Sotelo et al., 1976; Wouterlood et al., 1984 |
| 9. Between dopaminergic neurons in the substantia nigra | Grace and Bunney, 1983 |
| 10. Between magnocellular neurons in supraoptic and paraventricular hypothalamic nuclei | Andrew et al., 1981; Belin and Moos, 1986; Cobbett et al., 1985, 1987; Dudek et al., 1982; Hatton; 1983; Hatton et al., 1987; Renaud, 1987; Theodosis et al., 1981; Theodosis and Poulain, 1984, 1987 |

11. Between dendrites and somata of both pyramidal cells and granule cells Andrew et al., 1982; Jefferys and Haas, 1982; Knowles et al., 1982; Kosaka, as well as between dendrites of nonpyramidal cells in the hippocampus 1983a, 1983b; Kosaka and Hama, 1985; MacVicar and Dudek, 1980, 1981, of rat and guinea pig 1982; MacVicar et al., 1982; Schmalbruch and Jahnsen, 1981; Taylor and Dudek, 1982; Rao et al., 1986; Barnes et al., 1987; O'Beirne et al., 1987; Katsumaru et al., 1988a,b; Nunez et al., 1990; Baimbridge et al., 1991; Church and Baimbridge, 1991; Perez-Velazquez et al., 1994

12. Between primary afferent fibers in peripheral nerves of rat and cat Matthews and Holland, 1975; Matthews, 1976; Brenan and Matthews, 1983; Meyer et al., 1985; Meyer and Campbell, 1987, 1988

13. Between motoneurons in spinal cord and in the abducens nucleus of rat Nelson, 1966; Werman and Carlen, 1976; Rall et al., 1967; Collins and Erich- and cat sen 1988; Gogan et al., 1974; 1977; Matsumoto et al., 1988, 1989, 1991a, 1992; Zieglgansberger and Reiter, 1974; van der Want et al., 1998

14. In mixed synapses of neurons in spinal cord Rash et al., 1996

14. Between neurons in the striatum and accumbens of rat Cepeda et al., 1989; O'Donnell and Grace, 1993

15. Between neurons in the hypothalamic arcuate nucleus of rat Perez et al., 1990

Assessment of the importance of electrotonic coupling in the CNS of higher vertebrates has been difficult as each of the preceding approaches to the analysis of neuronal gap junctions does not readily permit quantitative estimates of the occurrence of these in neural tissues. Their distribution either throughout the CNS or presence in high concentrations in particular brain regions is a reasonable prerequisite that could be taken as at least a first indication of their legitimate contribution to interneuronal communication. Nevertheless, in the past two decades, the number of mammalian CNS structures and types of neurons that have been found to harbor neuronal gap junctions has steadily increased (Sotelo and Korn, 1978; Sotelo and Triller, 1981). Evidence for gap junctions between neurons has been obtained in various areas of mammalian CNS listed in Table 1. A detailed discussion and critical evaluation of findings in each of the neural systems listed is beyond the scope of the present review. However, it should be noted that although many of the reports cited could be considered to have presented strong or definitive evidence for gap junction–mediated interneuronal communication, the results of others are only suggestive. Moreover, the literature on gap junctional interactions among neurons in some brain areas contains inconsistencies and, in some cases, remains controversial.

## B.  Connexin Expression in Mature Neurons

A variety of immunohistochemical and *in situ* hybridization studies reported in the past few years have attempted to determine connexin expression in neurons. In brain, Dermietzel and colleagues (1989) suggested that some neurons express Cx32 on the basis of their finding that 20% of the cells in basal ganglia, thalamus, and brain stem that were immunopositive for Cx32 were also positive for neuron specific enolase. These authors found neurons elsewhere to be devoid of Cx32. Others have reported ISHH detection of Cx32 and Cx43 mRNA in unspecified cells in rat vestibular nuclei (Wachym et al., 1991). In a report on connexin localization in the olfactory bulb (Reyher et al., 1991), antibodies to Cx32 and Cx43 were found by light microscopy (LM) to produce similar immunolabelling patterns. Both antibodies labeled neuronal as well as glial gap junctions as seen by EM. In rat brain, immunohistochemical analyses have been conducted with an antibody (designated A893) against Cx32 (Nagy et al., 1988; Shiosaka et al., 1989; Yamamoto et al., 1989b). Despite precautions taken including affinity purification, preimmune and absorption controls, and demonstrations of its recognition of a single protein on Western blots corresponding to the apparent molecular weight of Cx32, this antibody labeled what can be classified as three distinct structures; namely, neuronal gap junctions, glial gap junctions, and a class of varicose axons throughout the brain whose identity has not yet been determined. With respect to the former of these, the reported findings were consistent with the preceding descriptions of gap junction localization between neurons in rat brain. However, possible recognition of other known or as yet unidentified members of the connexin family by this antibody (discussed later in this chapter) reduced its usefulness for studies of neuronal connexin expression. Nevertheless, the results suggested that Cx32 or a closely related protein may be a component of neuronal gap junctions.

In separate LM and EM work on Cx43, it should be noted that Cx43-immunoreactive neuronal gap junctions were never observed (Yamamoto et al., 1990b, 1990c), although observations to the contrary have been made immunohistochemically and by ISHH localization of Cx43 mRNA in mature rat neurons (Nadarajah et al., 1996; Simburger et al., 1997). Likewise, ISHH for Cx43 indicated its expression in mitral and tufted cells of olfactory bulb (Miragall et al., 1996). In a thorough investigation involving ISHH for Cx32 and Cx43 mRNA, Micevych and Abelson (1991) reported a wide range in the density of autoradiographic labeling for Cx32 mRNA in what appeared to be neurons throughout the brain. The greatest densities were detected in such areas as the mid-layers of the cerebral cortex, islands of Calleja, pyramidal and granule layers of the hippocampus, and lateral habenula, among others. The most general or parsimonious conclusion is that some antibodies with greater specificity for Cx32 may provide for its localization, although antibodies to several epitopes should be used for such studies. Furthermore, the presence of other connexins in neurons, including some perhaps not yet identified, is suggested by data obtained with some antibodies and by ISHH. In any case, it remains to be determined whether the connexins detected in neurons so far are authentic components of neuronal gap junctions.

Only recently has the first neuronal connexin been cloned, which shows an enriched expression in diverse nuclei and subpopulations of cells known to be electrically coupled (Condorelli et al., 1998, Söhl et al., 1998). This connexin (Cx36) is the homolog of the skate connexin35 (Cx35), which has been cloned by O'Brian and colleagues (1996). Apparently, Cx36 is exclusively expressed in the brain. According to *in situ* hybridization and neurotoxic lesioning (Condorelli et al., 1998), it seems to be confined to the neuronal compartment. Interestingly, this connexin is different from all other known connexin genes. Whereas the structure of the "conventional" connexin genes (α and β groups) consists of two exons (Ex1 and Ex2) with a long spanning intron separating the two exons and the complete coding region localized on Ex2 (see other chapters of this book), the gene structure of Cx36 contains an intron that interrupts the coding region. According to this unique feature, Cx36 likely belongs to a new subgroup of connexins which may be designated as the γ group. Unpublished data obtained by O'Brian and associates (Bruzzone, personal communication) indicate that this group consists of at least another member also expressed in fish retina. Some features of the protein sequence of this group, which has been described exclusively in neurons thus far, seem of interest. Cx36 as well as the Cx35 homolog have a long putative cytoplasmic loop, connecting TM2 and TM3, which is unique among the connexin family except for Cx45. Only the putative carboxy-terminus of Cx26 is shorter than that of Cx36 (Cx35). Of considerable interest in this context is the ball-and-chain model that has been proposed for the gating mechanism of the gap junction channel (Liu et al., 1993). According to this model, the putative carboxy-terminus serves as the ball and chain and the cytoplasmic hinge side as the receptor. Thus, the peculiar protein structure of Cx36 (Cx35) might imply some special features in the gating mechanisms of this neuronal gap junction channel.

## C.  Possible Connexin-Related Proteins in Neurons

A potential complication is the possibility that connexins or connexin-related proteins in the CNS may have other roles than the formation of typical gap junctions. For example, there has been speculation that structures analogous to gap junctions may be involved in vesicle exocytosis. Also, studies of intracellular calcium mobilization have suggested the possibility of gap junctional communication between intracellular organelles. It was previously reported that a monoclonal anti-Cx32 antibody and polyclonal antibody A893 immunohistochemically label not only gap junctions in liver and brain, but also intracellular structures that in motoneurons of rat and cat were identified as membranous elements referred to as subsurface cisterns (SSC) (Nagy et al., 1988; Yamamoto et al., 1990a, 1991a). The ultrastructural appearance of SSCs may be pertinent to observations of low stringency ISHH detection of strong signals in motoneurons with Cx32 (Micevych and Abelson, 1991) and possibly to the identification of novel neuronal connexins.

Subsurface cisterns were first found in neurons of rat cerebral cortex and spinal cord (Rosenbluth, 1962a, 1962b). They were described as having a flattened, sheetlike configuration separated from the plasma membrane by 5 to 8 nm and in continuity with the deeper endoplasmic reticulum (ER) at their lateral edges. Some authors have described the inner and outer cisternal membranes as being separated by a gap of 2 to 5 nm (Henkart et al., 1976). In nigral neurons of rat and cat, Le Beux (1972) drew an analogy between SSCs and the 2-nm extracellular gap of gap junctions, noting that such a gap may also occur between the cisternal membranes. As anti-Cx32 antibodies label SSCs and given the structural similarities between SSCs and gap junctions, it is possible that the mRNA detected in motoneurons by low stringency ISHH with connexin probes corresponds to the translated products at SSCs. SSCs have been found in neurons in a number of other brain regions as well as in various peripheral cell types, but are absent in glial cells (Herndon, 1964; Bunge et al., 1965; Peters et al., 1968; Siegesmund, 1968; Weis, 1968; Adinolfi, 1969; Takahashi and Wood, 1970; Anzil et al., 1971; Gulley and Wood, 1971; Bodian, 1972; Le Beux, 1972; Pappas and Waxman, 1972; Watanabe and Burnstock, 1976; Spoerri and Gleas, 1977; Hervas and Lafarga, 1979). Thus, it may be prudent to proceed cautiously in the interpretation of immunolabeling or ISHH results, as has also been emphasized by others (Micevych and Abelson, 1991), especially if further work points to a relationship between SSCs and gap junctions. It might also be noted in this context that SSCs in brain are about the size of most neuronal gap junctions and that immunolabeling of both these structures by LM yields largely punctate staining localized around the periphery of neurons. Thus, a punctate morphology of labeling associated with neurons, despite fulfillment of appropriate antibody specificity controls, should not be considered a sufficient criterion for designation of these puncta as representing connexin localization at neuronal gap junctions.

Further reasons to keep in mind the possible existence of as yet unidentified connexins or connexin-related proteins stem from work in two separate areas. Firstly, in pharmacological studies of cellular calcium mobilization, structures analogous to gap junctions have been repeatedly postulated to form intracellular calcium release or communication channels (Irvine and Moor, 1987; Merritt and Rink, 1987; Berridge and Taylor, 1988; Irvine et al., 1988). Moreover, relevant to the above descriptions of SSCs are models of calcium movements, which suggest calcium filling of some intracellular compartment directly from an extracellular source (Putney, 1986; Mullaney et al., 1987; Nahorski, 1988; Irvine, 1989). Because SSCs appear to be structurally analogous to the sarcoplasmic reticulum (SR) and triads in muscle (Rosenbluth, 1962b) and given the known role of the SR terminal cistern T-tubule complex in the process of calcium mobilization, it has been speculated that SSCs in neurons may also serve a function in this process (Rosenbluth, 1962b; McBurney and Neering, 1987; Yamamoto et al., 1990a, 1991a). Secondly, based on studies of vesicular secretion, there has been considerable speculation and some evidence that exocytosis in certain cell types may occur via initial formation of a vesicle/plasma membrane fusion pore similar to gap junctional channels (Zimmerberg et al., 1985; Breckenridge and Almers, 1987a, 1987b; Nagy et al., 1988; Lemos et al., 1989, 1991; Shiosaka et al., 1989; Spruce et al., 1990; Yamamoto et al., 1990a, 1991a; Chow et al., 1992; Lee et al., 1992; Yamamoto et al., 1993). These studies further emphasize the need to establish ultrastructurally the cellular elements that are recognized by anticonnexin antibodies *in situ.*

## D.  Regulation of Conductance State

Little is known as yet about mechanisms that regulate neuronal gap junctional communication and connexin expression. Junctional communication in other cell types is regulated by a variety of factors including calcium, pH, phosphorylation of junctional proteins, transmembrane voltage, neurotransmitters, growth factors, nitric oxide, arachidonic acid, and steroids (Spray et al., 1992; Spray and Bennett, 1985, Sáez et al., 1986; Maldonado et al., 1988; Randriamampita et al., 1988; Miyachi et al., 1994; Chandross et al., 1996; Rorig et al., 1996; Rorig and Sutor, 1996a, 1996b; Reuss et al., 1998; O'Donnell and Grace, 1997). This regulation may be achieved either by altered connexin production leading to increased numbers or sizes of gap junctions or by more rapid changes in the conductance state of intercellular communicating channels. Relevant to the identification of connexin molecules expressed by various cell types is evidence suggesting that the conductance states of different connexins not only differ, but are likely to also be regulated by different mechanisms. As discussed later in this chapter, dopamine regulates gap junctional coupling between horizontal cells in the retina of lower vertebrates (Lasater and Dowling, 1985; Neyton et al., 1985; Rogawski, 1987), as well as between amacrine cells in mammalian retina (Hampson et al., 1992). These findings raise the question of whether dopamine modulates gap junction permeability between neurons in brain. Although this remains to be determined, it may be pertinent here that lesions of the nigrostriatal dopaminergic system were shown to increase the percentage of dye coupling between striatal neurons by an order of magnitude (Cepeda et al., 1989). Moreover, given that dopaminergic nigral neurons appear to be gap junctionally coupled (Grace and Bunney, 1983) and that dendritic release of dopamine from these neurons is well established, the possibility that dopamine regulates coupling between nigral cells may be considered (O'Donnell and Grace, 1993). Sáez and colleagues (1991) revealed that norepinephrine increases gap junctional couplin between pinealocytes (see later discussion) when kept in culture. The molecular mechanisms underlying this regulative capacities of hormones and neurotransmitters remain to be established. A direct effect on junctional conductance, however, seems unlikely. Rather, it is anticipated that regulation involves second messenger effects as has been convincingly demonstrated for cyclic adenosine monophosphate (cAMP)-dependent phosphorylation of hepatic Cx32 (Sáez et al., 1986; Traub et al., 1987). Junctional conductance in leptomeningeal cells is also increased by a factor of two within 5 minutes after addition of membrane permeant cAMP or forskolin, and is strongly inhibited by addition of protein kinase C activated by phorbol esters (Spray et al., 1991). Besides neurotransmitter effects on neuronal coupling, androgen dependency of gap junctions and connexin expression has been demonstrated in spinal motoneurons of rat (Matsumoto et al., 1988, 1989, 1991a, 1992). Other factors of physiological relevance, for example, pH shift (Baimbridge et al., 1991), have also been considered to be influential on neuronal coupling.

## E.   Functional Considerations

Many years after the discovery of electrical neurotransmission and its morphological correlate, electrical synapses were still considered a primitive form of neural communication and therefore unlikely to be important in mammals. However, support for the proposition that gap junctions make a significant contribution to interneuronal communication in higher vertebrates is provided by investigations of electrical coupling between neurons in the inferior olive, the magnocellular hypothalamic nuclei, and retina. In the olivo-cerebellar system, it is known that there is a close temporal relationship between the activity of neurons in the inferior olivary (IO) nucleus and cerebellar Purkinje cells in which climbing fiber projections from olivary neurons produce complex spikes. Moreover, it has been shown that complex spikes among particular sets of rostrocaudally oriented rows of Purkinje cells are triggered almost simultaneously. This was inferred to indicate that IO neurons giving rise to this near synchronous activation of Purkinje cells must themselves form sets in which members discharge action potentials to the cerebellum in near unison (Sasaki et al., 1989). Evidence has been presented that the unitary activity of clusters of IO neurons is determined by the dynamic state of electrical coupling between these cells (Llinás, 1985; Llinás and Sasaki, 1989). Moreover it was suggested that the organization of cells into synchronously active clusters may be achieved by a process involving selective uncoupling of sets of neurons which are ordinarily extensively and isotropically coupled to their neighbors (Llinás and Yarom, 1981). In view of these findings and given that the patterns of activity exhibited by collections of Purkinje cells form in part the basis for most theories of cerebellar function, it would appear that gap junctions between IO neurons are an essential feature of the function of the olivo-cerebellar system as a whole.

In the hypothalamo-neurohypophysial system, numerous reports have appeared demonstrating synchronous activity, dye coupling and electrical coupling between magnocellular neurons in the supraoptic (SON) and paraventricular nuclei (PVN) (Belin and Moos, 1986; Yang and Hatton, 1988; for reviews, see Hatton et al., 1988; Hatton, 1990). The incidence of coupling between these neurons was found to be higher in female compared with male rats. In female rats, the numbers of dye coupled neurons was reported to be dependent on the physiological state of the animal; coupling was higher in lactating or maternally behaving virgin rats than in virgin animals and in both sexes was found to be regulated by gonadal steroids (Cobbett et al., 1987; Hatton et al., 1987; Modney et al., 1990; Hatton et al., 1992a,b). Equally significant was the observation that, despite the intermingled distribution of oxytocin- and vasopressin-containing cell bodies and more so their dendrites in the PVN, dye coupling between these neurons was homotypic; that is, dye transfer occurred only between neurons containing oxytocin and between those containing vasopressin, but not between neurons containing different peptides. On the basis of these results, it has been suggested that in the case of oxytocin cells, electrotonic coupling may serve to coordinate their activation during the periodic high-frequency bursts of action potentials that they exhibit in response to suckling and thereby cause sufficient oxytocin release to evoke milk ejection (Hatton, 1990). Although gap junctions between magnocellular hypothalamic neurons have not as yet been found, their presence is suggested by dye transfer results and ISHH demonstrating an increase of Cx32 mRNA in these neurons during lactation (Micevych et al., 1996). The synchronizing properties these gap junctions would confer may underlie the coordinated activity displayed by these neurons.

The retina presents a further model in which to appreciate the roles that gap junctions may play in neuronal networks. Electrotonic coupling has been known for some time between rod photoreceptors, between amacrine cells, and between

horizontal cells. Junctional coupling has been well characterized between horizontal cells in lower vertebrates (see Dowling, 1991). Coupling among horizontal
cells is believed to enlarge the size of the receptive fields, effectively increasing
convergence of the photoreceptor input onto this cell layer. Dopamine uncouples
these cells, thereby reducing the size of the receptive fields, and the mechanism
apparently involves cAMP-dependent protein kinase (Lasater and Dowling,
1985). Amacrine cells of the AII type are also connected by gap junctions and
they receive synaptic input from the dopaminergic amacrine cells. Injection of the
small biotinylated tracer Neurobiotin (see earlier discussion) in rabbit retina
revealed extensive dye transfer between these cells which could be dramatically
reduced by exogenous applied dopamine (Hampson et al., 1992). The uncoupling
effect is apparently initiated by a D1-like dopamine receptor that stimulates
cAMP production. The mechanism underlying dopaminergic uncoupling of the
rabbit retina is suggested to be substantially similar to that established for horizontal cells in lower vertebrates (Lasater, 1987; De Vries and Schwartz, 1989).
Effects of dopamine are both acute and long term and may account for acuity in
dark adaptation when dopamine levels in the retina are highest.

These three examples, in addition to demonstrating functional contributions of
electrical transmission, provide the basis for conceptualization of the types of
neural integrative processes that in mammalian CNS may be best served by electrical synapses. Although it may be hazardous to generalize at this stage of knowledge, it would nevertheless seem inconceivable that such a fundamental role of
electrotonic transmission as suggested by investigations on the three previously
discussed systems would not be represented in other neural systems of higher vertebrates. In this context, it is worth noting that although gap junctional coupling
between neurons in the CNS often occurs at their dendrites, some mammalian
brain regions do contain relatively large, classical gap junctions at axodendritic
and axosomatic contacts, forming what are termed mixed synapses (Sotelo and
Palay, 1970; Sotelo and Llinás, 1972; Korn et al., 1973; Sotelo, 1975; Sotelo and
Korn, 1978; Korn and Farber, 1979; Sotelo and Triller, 1981). Functionally, these
gap junctions might modulate presynaptic events or subserve metabolic communication, or both (see Rash et al., 1996). The effects of these junctions could also
spread beyond the confines of the synapse itself. In the lateral vestibular nucleus,
for example, electrical coupling occurs between neuronal cell bodies that are not
themselves gap junctionally connected, but whose dendrites form gap junctions
with branches of the same axons. Thus, it was emphasized in some of the aforementioned studies that a single axon simultaneously forming gap junctions with
two dendrites can mediate electrical coupling between the dendrites. This
axo-dendritic coupling will also allow dye transfer. Quantitative high-resolution
scanning EM studies of spinal cord indicate that neurons are commonly contacted
by nerve terminals forming very small gap junctions with their postsynaptic elements (Rash et al., 1996). Indeed, nearly all spinal neurons are contacted by hundreds of mixed synapses. These small gap junctions would be difficult to detect as

a classical heptalaminar junctional structure by standard transmission EM. This is perhaps why they have remained elusive until now, but their prevalance again suggests a current underestimation of the importance of gap junctions in adult mammalian neurotransmission and the role of gap junctions in widespread communication in the CNS.

## III.  JUNCTIONAL COMMUNICATION BETWEEN ASTROCYTES

### A.  Astrocytic Gap Junctions

Among other functions (Hansson, 1988; Walz, 1989), astrocytes form structural barriers at vascular surfaces, ensheathe neuronal elements and separate regions of dissimilar or fluctuating ionic composition, thereby physically or metabolically compartmentalizing various neuronal components of the CNS (Chan-Palay and Palay, 1972; Palay and Chan-Palay, 1977; Szentagothai, 1970; Landis and Reese, 1982; Walz and Hertz, 1983; Waxman and Black, 1984). Metabolically, astrocytes contribute to extracellular $K^+$ homeostasis through, in part, a process termed spatial buffering (Orkand et al., 1966; Orkand, 1977; Walz and Hertz, 1983; Ransom and Carlini, 1986; Walz, 1989), for which there is now evidence in the retina (Karwoski et al., 1989; Skatchkov et al., 1995; Newman, 1996). This involves the cell-to-cell redistribution of excess extracellular $K^+$ through the cytoplasm of vast networks of contiguous astrocytes coupled by gap junctions (Futamachi and Pedley, 1976; Bennett and Goodenough, 1978; Massa and Mugnaini, 1982; Dudek et al., 1983; Gardner-Medwin, 1986; Mugnaini, 1986; Galambos, 1989; Walz, 1989). It has been assumed ever since astrocytes were assigned the property of buffering the $K^+$ activity surrounding active neurons that gap junctions in this system would most likely provide a direct pathway to the site of $K^+$ disposal into the perivascular compartments (Newman, 1986; Orkand et al., 1966). Equally plausible, however was the related notion that coupling would increase the volume of the buffer sink, so that if a neuron in proximity to a glial cell were particularly active, the astrocytic compartment would share the load (Kuffler and Nicholls, 1977).

Ultrastructural, dye transfer, and electrophysiological studies (Sipe and Moore, 1976; Schwartzkroin and Prince, 1979; Gutnick et al., 1981; Kettenmann et al., 1983; Connors et al., 1984; Waxman and Black, 1984; Fischer and Kettenmann, 1985; Kettenmann and Ransom, 1988) have provided strong support for the generally accepted view that astrocytes are extensively coupled to form what are defined as "functional syncytia" (Futamachi and Pedley, 1976; Bennett and Goodenough, 1978; Massa and Mugnaini, 1982; Dudek et al., 1983; Mugnaini, 1986; Ransom and Carlini, 1986). There are, in addition, reports indicating that gap junctions are particularly numerous between astrocytic processes that ensheathe chemical synapses, glomeruli, and nodes of Ranvier, as well as between those at brain and vascular surfaces (Williams, 1975; Sipe and Moore, 1976; Waxman and Black, 1984; Mugnaini, 1986). Although these reports suggest that astrocytic gap junctions are ubiquitous in the CNS, the functional significance of the observed local heterogeneities has remained largely unclear. The global organization of astrocytic gap junctions has been difficult to deduce using traditional methods of junction visualization by EM, which allow the examination of only minute samples. The alternative of dye transfer and electrophysiological methods, which have been used to demonstrate coupling between glial cells in culture and tissue slices (Schwartzkroin and Prince, 1979; Gutnick et al., 1981; Kettenmann et al., 1983; Connors et al., 1984; Fischer and Kettenmann, 1985; Dudek et al., 1988; Kettenmann and Ransom, 1988), are usually qualitative, do not provide accurate topology of junctions among coupled cells, and are difficult to apply *in vivo* (Mugnaini, 1986). These and other methods have, nevertheless, provided a wealth of information relevant to functional issues of glial coupling. This early work has now been considerably extended by the discovery of connexins and identification of the particular connexin proteins expressed by astrocytes and other support cells in the CNS.

## B.   Connexin Expression by Astrocytes

On the basis of immunohistochemical, Western blot, and Northern blot analyses of neuronal and glial cultures, it was reported by some that astrocytes express Cx43 protein and mRNA and contain no detectable levels of Cx32 mRNA (Dermietzel et al., 1991; Giaume et al., 1991a; Naus et al., 1991b; Welsh and Reppert, 1996). In brain, Dermietzel and colleagues (1989) reported that only astrocytes are immunopositive for Cx43, and Yamamoto and associates (1990b, 1990c) have shown that Cx43 is localized to the majority if not all astrocytic gap junctions in rat brain. With respect to other cell types, EM observations of hippocampus and olfactory bulb failed to reveal Cx43-labeled neuronal gap junctions (Yamanoto et al., 1990c) despite the presence of gap junctionally coupled neurons in these areas (Pinching and Powell, 1971; Kosaka, 1983a, 1983b; Kosaka and Hama, 1985; Katsumaru et al., 1988a, 1988b; Shiosaka et al., 1989). Whereas Cx43 immunopositive gap junctions between oligodendrocytes and astrocytes have been occasionally observed (Ochalski et al., 1997; Nagy et al., 1998), the cytoplasmic membrane of the oligodendrocyte was seldom labeled. These junctions were thus asymmetrically stained, indicating heterologous coupling between connexins.

## C.   Connexin43 in Astrocytes *In Vivo*

In LM immunolocalization studies of connexins in sections of CNS tissue, fluorescence or peroxidase immunoreactivity often consists of small spots or aggregations or puncta. It is hazardous to assign these tiny puncta to particular cells types in the absence of ultrastructural data. By EM, immunolabeled gap junctions can often be identified on the basis of standard structural features and are often formed between thin astrocytic processes and between dendrites in the case of neuronal junctions. Immunolabeling for connexins *in situ* is only sometimes detected in glial or neuronal somata. Thus, it may be difficult by LM to attribute the localization of puncta to processes of a particular cell type. This is especially true when one considers that astrocytic processes bearing gap junctions are frequently found within a few microns of neuronal elements.

It has been found that Cx43 is widely distributed in rat brain, which is not surprising given the apparently ubiquitous CNS distribution of astrocytic gap junctions. However, levels of Cx43 detected by Western blotting or immunohistochemistry (IHC) were highly uneven among brain regions (Yamamoto et al., 1990b, 1990c; Nagy et al., 1992), suggesting considerable heterogeneity of either astrocytic gap junctions density, Cx43 protein incorporation into these junctions, or both as Cx43-immunoreactive puncta was correlated with amounts of Cx43 expressed among brain regions (Nagy et al., 1992). Cx43 mRNA as assessed by Northern blots was reported to be somewhat more homogeneous (Naus et al., 1990); however, the brain regions taken for analysis in this report were relatively large. In studies by ISHH, Cx43 mRNA was found to be ubiquitously expressed in rat CNS in what were concluded to be astrocytes and the noted regional heterogeneity (Micevych and Abelson, 1991) corresponded with that of Cx43 protein (Yamamoto et al., 1990c; Nagy et al., 1992). However, striking differences in Cx43 protein levels such as consistently found between the striatum (low) and globus pallidus (high) were apparently not seen at the Cx43 messenger level. These results suggest that Cx43 gene transcription may be related to Cx43 translation in some brain regions, but not in others, and point to regulatory control at various cellular levels.

EM studies have shown that Cx43 is associated with what can be clearly identified morphologically as astrocytic gap junctions (Yamamoto et al., 1990b, 1990c), and this has been confirmed by immunogold localization methods (Dermietzel et al., 1989; Miragall et al., 1992; Nagy et al., 1998). Relatively greater numbers of labeled gap junctions were found around neuronal elements such as those between astrocytic processes surrounding dendrites and synaptic glomeruli. Within cells, Cx43 was detected only at or in the vicinity of astrocytic gap junctions and no such junctions were found to be unlabelled for Cx43. As previously noted, virtually all labeling was punctate. From correlative LM and EM, it appears that the vast majority of puncta seen by LM correspond to individual or clusters of labeled gap junctions identified by EM. In some brain regions such as the cerebellum, it was possible to correlate patterns of Cx43 immunostaining with known patterns of anatomical organization. For example, the cerebellar cortex contains linear arrays of puncta projecting from the Purkinje cell layer to the cortical surface. Since Bergmann glial cells extend their process in this fashion through the molecular layer, since Cx43 is localized at gap junctions between

Bergmann glia, and since each puncta appears to correspond to a gap junction (Yamamoto et al., 1990c), it is possible now to interpret the LM view of the global organization of gap junctions between the processes of these cells.

Studies of Cx43 in spinal cord have revealed some further interesting points (Ochalski et al., 1997). Many gap junctions were observed between layers of the glia limitans, suggesting extensive intercellular communication around the circumference of the spinal cord just as occurs at the brain surface (Yamamoto et al., 1990c). Cx43 was distributed in all laminae, but most highly concentrated in the substantia gelatinosa. In stark contrast to observations in brain, astrocyte processes in grey matter often exhibited Cx43 labeling not only of their gap junctions, but also their nonjunctional membranes, which frequently ensheathed synaptic glomeruli forming what are observed by LM as Cx43-positive annular profiles. These have been seen in brain and their significance has been previously discussed (Yamamoto et al., 199cb). In white matter, Cx43 was localized to gap junctionally coupled fibrous astrocytic processes running parallel to myelinated axons, suggesting the creation of an extensive rostrocaudally running gap junctionally coupled syncytium. Thick radially directed astrocytic processes extending to and away from the glia limitans also formed extensive gap junctions with each other. Some of these processes formed either symmetrically or asymmetrically labeled gap junctions (i.e., stained for Cx43 on only one side). The asymmetrical staining suggests that one or more astrocytic subtypes peculiar to spinal cord may express an additional connexin that is neither Cx43 nor Cx30 (see later discussion), but is able to form heterotypic gap junctions with Cx43. Some radial processes also extended branches that contacted and often formed gap junctions with oligodendrocytic processes at the outer surface of myelinated fibers. These junctions were labeled for Cx43 only on the astrocytic side and were thus heterotypic. Further elucidation of the gap junctional relationships between spinal cord astrocytes together with identification of other connexins they may express is required to establish the organization of glial gap junctional intercellular communication (GJIC) pathways that prvide routes for movement or disposal of ions or small metabolites within the cord.

## D.    Connexin30 in Astrocytes *In Vivo*

Cx30 is among the most recent members of the family of connexin proteins to be identified in mammalian species. It is closely related to Cx26 with which it shares 77% amino acid sequence identity (Dahl et al., 1996) and is expressed at high levels in brain. Given its abundance in brain and its sequence similarity with Cx26, it is perhaps not surprising that some well-characterized anti-Cx26 antibodies were found to cross-react with Cx30 (Nagy et al., 1997b, 1998). Based on immunolabeling of astrocytic gap junctions and the detection of a 30-kD protein in brain with these antibodies, it was suggested that astrocytes produce a second connexin which was unlikely Cx26, but almost certainly Cx30 (Nagy et al.,

1997b). This was subsequently confirmed with Cx30-specific antibody that did not cross-react with Cx26 or any other connexin known to be expressed in brain (Nagy et al., 1998; Kunzelmann et al., 1999). Immunolabeling for Cx30 in adult rat brain was qualitatively similar to that of Cx43, with which it was found to be co-localized. Cx30 was restricted to astrocytes and to gap junctions formed by these cells. Interastrocytic gap junctional membranes were symmetrically labeled, oligodendrocytic-astrocytic junctional membranes were asymmetrically labeled only on the astrocyte side, and oligo-oligodendrocyte junctions were unlabeled. Immunolabeling for Cx30, as with Cx43, was found largely at gap junctions and their vicinity, which may indicate small cytoplasmic pools of the protein, rapid incorporation within junctional plaques following synthesis, or simply lower accessibility of antigenic sites outside intact junctions.

In contrast to regional Cx43 expression, diencephalic and hindbrain gray matter areas exhibited greater Cx30 levels than forebrain areas. This differential expression may simply be related to the developmental origin of these areas with astrocytes in phylogenetically older parts of brain expressing higher levels of Cx30. Thus, Cx30 and Cx43 may fulfill similar functional roles with a greater contribution of one over the other in regions of common developmental origin. Alternatively, Cx30 and Cx43 expression appear to be differentially regulated in adult brain, suggesting specific cellular requirements or different contributions to interglial gap junctional communication. It was also evident that subcortical blood vessels were more richly invested with perivascular astrocytic endfeet immunopositive for Cx30 than Cx43. Targeting of Cx30 to astrocytic endfeet throughout the brain, even areas with otherwise low levels of Cx30, suggests its special contribution to gap junctional interendfoot interactions that mediate redistribution of ions or metabolites entering or exiting brain.

A major difference between the two astrocytic connexins was the absent of Cx30 in white matter tracts, which generally contain an abundance of Cx43. Thus, Cx30 is expressed by gray, but not white matter astrocytes, thereby distinguishing these cells according to connexin phenotype, which corresponds to subpopulations defined by morphology and location, namely fibrous white matter and protoplasmic gray matter astrocytes. This differential Cx30 expression in white and gray matter, together with demonstrations that oligodendrocytes also exhibit distinctive localization of Cx32 along particular subpopulations of myelinated fibers (Li et al., 1997) raises questions about the nature of glial gap junctions at specific cellular loci, which may have important implications for the formation of communication compartments within the glial syncytium. Thus, depending on selectivity of ion and metabolite passage through glial connexin channels, the capability of various glial connexins to form functional heterotypic channels, and heterogeneity of connexin expression even by a particular glial cell type, different operationally defined compartments may emerge in neural tissue as a whole or separately in white or gray matter.

## E.  Astrocytic Syncytial Compartments

The highly heterogeneous distribution of both Cx43 and Cx30 in brain suggests that requirements for intercellular metabolic cooperation via gap junctions between astrocytes may be markedly different among brain regions. Indeed, it has been speculated (Yamamoto et al., 1990c) that nuclear and subnuclear regions demarcated by sharp borders of differential Cx43 expression may be endowed with specific glial GJIC capacities. In terms of ion spatial buffering, this may lead to restriction of ion redistribution to areas of similar buffering capacities resulting in intracellular communication pathways that may be somewhat functionally compartmentalized. Although no evidence for such compartmentation was found following tracer injections into individual astrocytes in the hippocampus (Konietzko and Muller, 1994), this structure is far less heterogeneous in Cx43 and Cx30 levels than, for example, the striatum compared with the adjacent globus pallidus (Yamamoto et al., 1990c; Nagy et al., 1992, 1998), which could thus serve as more critical tests of differential functional coupling among astrocyte populations *in vivo*. Of course, such speculation must take into account the possibility that astrocytes may express yet additional connexins.

Markedly different regional connexin levels in brain also suggests that their production by astrocytes may be locally regulated. As local demands for $K^+$ spatial buffering and the contributions of astrocytic gap junctions to this process are likely determined by the extracellular ionic milieu generated by active neurons, it is reasonable to consider that regulation of Cx43 and Cx30 expression in astrocytes, and hence establishment of putative GJIC compartments, may be exerted in part by neurons in their vicinity. This possibility is supported by observations that gap junctional coupling efficiency of astrocytes is modulated by interactions with neurons and their constituents *in vitro* (Fischer and Kettenmann, 1985; Anders and Woolrey, 1992), although meningeal cells co-cultured with astrocytic have also been reported to augment astrocytic GJIC (Anders and Salopek, 1989), raising the possibility that other cell types including oligodendrocytes might also impart such regulation. Further, it is noteworthy that brain nuclei such as the globus pallidus in which neurons are known to be tonically active contain among the highest levels of Cx43 and Cx30 (Yamamoto et al., 1990c; Nagy et al., 1998).

On a finer scale, detailed quantitative studies of gap junction deployment between astrocytes indicate that vast numbers of junctions connect not only processes of different cells, but also those of the same cell (Wolburg and Rohlmann, 1995; Wolff et al., 1998). Based on this extensive autocellular coupling or autocoupling, as it was termed, it was suggested that GJIC subserves pathways between astrocytic subcellular compartments as much as it does between cells. Although the purpose of such autocoupling is at present unclear, numerous gap junctions exist between the layers of astrocytic lamellar processes that typically envelope synaptic glomeruli, and it has been emphasized that patterns of Cx43 labeling throughout the brain and spinal cord appear to reflect in part concentration of Cx43 in these lamellae (Yamamoto et al., 1990c; Ochalski et al., 1997). Such lamellar junctions must invariably result in autocoupling if the enveloping lamellae arise from one or just a few astrocytes. Thus, by linking subcellular compartments as proposed (Wolburg and Rohlmann, 1995), autocoupling at least between lamellar processes around glomeruli could serve as a radially directed gap junctional short circuit for movement of substances from the glomerular mileau into the lamellar cytoplasm, through their attendant junctions and into the pool of surrounding astrocytes (Rohlmann and Wolff, 1996).

## F. Astrocytic Coupling and Connexins *In Vitro*

Some of the issues raised by the preceding observations *in vivo* have been investigated by studies of astrocyte coupling *in vitro*. It is intriguing, for example, that astrocytes cultured from different brain regions of neonatal rat exhibit markedly different levels of Cx43 expression and dye coupling after they have been grown to confluence (Batter et al., 1992; Lee et al., 1994). Interpretation of these findings, however, must take into account the time required for development of coupling between glial cells in culture (Venance et al., 1995a), raising the possibility that heterogeneity in coupling seen at a single time after cell plating may reflect its development at different rates in cells from different brain areas. Nevertheless, it is of note that brain regions from which astrocytes displayed the highest and lowest levels of Cx43 and Cx30 expression and coupling *in vitro* correspond roughly to those displaying high or low levels, respectively, of Cx43 in adult rats (Yamamoto et al., 1990c; Nagy et al., 1992). In this context, the pattern of Cx30 expression in primary cultured astrocytes is of interest. Consistent with its late onset of expression in the postnatal brain is the observation that Cx30 is expressed around 5 weeks after plating of postnatal astrocytes, where cultivation needs several passages before Cx30 commences (Kunzelmann et al., 1999). Inasmuch as astrocytic Cx43 and Cx30 expression and dye coupling between astrocytes in newborn rats is low (Binmoller and Muller, 1992; Yamamoto et al., 1992), these findings taken together raise the possibility that the communication competence of astrocytes in different brain regions is either preprogrammed during their differentiation or imparted by local environmental factors during some stage of their development. As noted earlier, such factors may include those derived from neurons and other cell types.

Astrocytes *in vitro* are classified as either type 1 or type 2 based on morphological and immunological criteria. Interestingly, it has been reported that Cx43 and dye coupling are absent in cultured type 2 astrocytes (Sontheiner et al., 1990; Belliveau and Naus, 1994), which may correspond to an astrocytic subpopulation that exhibits other peculiarities in ion channel expression (Sontheimer et al., 1991). The absence of coupling between type 2 astrocytes casts doubt on their debated correspondence to white matter astrocytes *in vivo*, which are known to express Cx43 (Yamamoto et al., 1990b, 1990c), but it is curious that this absence of coupling parallels the lack of Cx30 expression in white matter astrocytes. In any case, it remains to be determined whether these cells correspond to a noncoupled glial cell type *in vivo* or arise in cultures lacking appropriate signals for connexin expression. In studies of astrocytes cultured from Cx43 knockout mice (Naus et al., 1997), it was reported that the cells lacked dye coupling and exhibited reduced stimulus-evoked intercellular calcium waves, but were found to undergo normal differentiation. They also displayed reduced rates of proliferation compared to wild-type astrocytes, suggesting a role of astrocytic GJIC in promoting growth of these cells. Reports of residual GJIC between astrocytes cultured from these mice may be explained by Cx30 expression in these cells, particularly as the conductance properties of this residual coupling were different from those of channels formed by Cx43 (Spray et al., 1995). Cx45 and Cx43 together with Cx30 may account for the residual coupling prevalent in the Cx43 (–/–) mice (Spray et al., 1998, Kunzelmann et al., 1999). In organotypic slice cultures of brain from these mice, astrocyte morphology and electrical properties appeared normal, but the density of astrocytes in central regions of cultures was increased by over 25-fold compared to control slices, which was taken to indicate a possible regulatory role of GJIC in astrocyte migration (Velazquez et al., 1996). Transfections of Cx43 cDNA into glioma cells has been described, and these cells have been useful in studies of GJIC (Zhu et al., 1991; Naus et al., 1992a, 1992b).

## G.   Interastrocytic Calcium Waves

As in other cell types, agonist application to, or mechanical stimulation of, individual cells in a gap junctionally coupled syncytium of cultured astrocytes results in transient elevations in intracellular $Ca^{2+}$ levels in adjacent cells. By fluorescence imaging, these elevations are seen to propagate within cells and from cell to cell in a wavelike fashion at rates ranging from 10 to 50 $\mu$/sec over long distances (Cornell-Bell and Finkbeiner, 1991; Finkbeiner, 1993; Verkhratsky and Kettenmann, 1996). This phenomenon has become the focal interest of several laboratories and began with findings that such $Ca^{2+}$ wave propagation in astrocytes could be evoked by glutamate in pure cultures of astroctyes as well as those in organotypic preparations (Cornell-Bell et al., 1990; Glaum et al., 1990; Dani et al., 1992; Finkbeiner, 1992; Charles, 1998; Giaume and Venance, 1998). In individual astrocytes, such waves appear to originate from specific wave initiation loci (Yagodin et al., 1994). Evidence that these waves are in some cases propagated from cell to cell by gap junctional channels is suggested by their abolition following treatments with various agents such as halothane, heptanol, octanol, glycyrrhetinic acid, and phorbol esters that are known to inhibit GJIC in various systems. It should, however, be kept in mind that these may have other actions such as the reported intracellular acidification of astrocytes caused by octanol (Pappas et al., 1996). Stronger evidence is provided by observations of intercellular $Ca^{2+}$ waves in C6 cells that express gap junctions following transfection with

Cx43, but not in normal C6 cells that lack gap junctions (Charles et al., 1992). Calcium wave transmission has also been demonstrated in acutely isolated, but otherwise intact retina from mature rats, although gap junction mediation was not tested (Newman and Zahs, 1997).

The mechanism of cell-to-cell wave propagation is not fully understood, but probably involves signals similar to those underlying the generation of intracellular $Ca^{2+}$ waves that, in the case of coupled cells, are simply transmitted intercellularly via junctional channels linking the cytoplasm of one cell to that of another. Thus, $Ca^{2+}$ at a wave front may itself pass through these channels to elicit $Ca^{2+}$-stimulated $Ca^{2+}$ release from intracellular stores in adjacent cells (Sanderson et al., 1990). Alternatively, propagated waves may involve regenerative release of IP3 (Charles et al., 1993; Venance et al., 1997), which can permeate gap junctions in some systems (Sáez et al., 1989). It has also been emphasized that wave propagation appears to depend on several other simultaneously occurring processes in astrocytes, including ion fluxes and perhaps as yet unidentified diffusible factors (Kim et al., 1994). In fact, gap junctional coupling or even cell contact may not be a prerequisite for $Ca^{2+}$ wave transmission under some conditions as was recently reported in the case of cultured astrocytes that transmitted typical $Ca^{2+}$ waves between groups of cells separated by substantial cell-free lanes, implying mediation of waves by substances released by astrocytes into the extracellular space (Hassinger et al., 1996). Because a fraction of $Ca^{2+}$ waves can be blocked by suramin, a potent antagonist of purinergic receptors, it seem most likely that an extracellular paracrine mechanism that includes adenosine triphosphate (ATP) secretion contributes to the waves between astrocytes (Hassinger et al., 1996; Charles 1998; Scemes et al., 1998). This is consistent with observations of only partial reductions in $Ca^{2+}$ waves in cultured astrocytes derived from Cx43 knockout mice, although gap junction formation by compensatory connexins (Cx45, Cx40) that may be produced in these cells could not be excluded (Naus et al., 1997; Suadicani et al., 1998).

There has been much speculation concerning possible functions of interastrocytic $Ca^{2+}$ waves. Numerous cell types, including astrocytes exhibit $Ca^{2+}$ oscillations, which presumably mediate intracellular $Ca^{2+}$ signaling events and appear to be generated by mechanisms different form those underlying wave production (Charles et al., 1991). Because the transit of intercellular $Ca^{2+}$ waves influences agonist-evoked patterns of these oscillations, albeit under some conditions and in some cell types more than others (Jensen and Chiu, 1990; Cornell-Bell and Finkbeiner, 1991; van den Pol et al., 1992), it is conceivable that the waves may coordinate internal $Ca^{2+}$ fluctuations and thereby the responsiveness of a locally interconnected group of astrocytes to external factors whose signaling pathways are linked to cellular $Ca^{2+}$ levels, as has been suggested in the case of hepatocytes (Tordjmann et al., 1997). Hepatocytes, ordinarily coupled by Cx32, show a 78% reduction in noradrenaline (NA)-induced glucose mobilization from glycogen in Cx32 knockout mice, suggesting that GJIC spreads the NA signal (Nelles et al.,

1996). GJIC increases sensitivity of pancreatic islet cells to secretagogues, apparently by shifting their responsiveness to the most sensitive of a group of coupled cells, and was found to be essential for proper release of secretory products (Meda, 1996). These waves could similarly coordinate various astrocytic metabolic processes. One of many possible examples of this includes the proposed gap junction–dependent normalization of heterogeneous $Na^+$ levels observed in uncoupled astrocytes (Rose and Ransom, 1997). This may be initiated by the spread of $Ca^{2+}$ waves followed by $Ca^{2+}$-mediated actions on processes that govern internal $Na^+$ levels, rather than by direct flow of $Na^+$ through gap junctions.

## H.  Calcium Waves and Neuronal/Glial Interactions

Demonstrations of both induced and spontaneous $Ca^{2+}$ waves spreading throughout the astrocytic network have also led to speculation that glia and their gap junction–mediated interactions may actually play an active role in neuromodulation. According to this view, there may be coupled glial circuits *in vivo* that propagate regenerative $Ca^{2+}$ waves that might regulate moment-to-moment changes in glial $K^+$ buffering capacity or even cause $K^+$ release from glial cells, thereby influencing neuronal excitability. Indeed, besides neuron-to-glial signaling (Kriegler and Chiu, 1993; Murphy et al., 1993; Verkhratsky and Kettenmann, 1996), $Ca^{2+}$ waves in glial cultures have been reported to evoke transient $Ca^{2+}$ elevations in overlying neurons, and this was suggested to be mediated by glial to neuronal signaling via gap junctions between these two cell types (Charles, 1994; Nedergaard, 1994; Verkhratsky and Kettenmann, 1996; Froes et al., 1998) or by glutamate receptors (Parpura et al., 1994; Hassinger et al., 1995). Whatever the mechanism, glial-to-neuronal signaling appears to occur, at least in culture, and adds a new dimension to thinking about neuronal-glial interactions. In this context, the suppression of astrocytic GJIC by, for example, general anesthetics could have far-reaching implications (Mantz et al., 1993). Also of potential importance is that $Ca^{2+}$ waves share several features in common with spreading depression (SD), the characteristics of which include slow wavelike propagation of transient neuronal inactivity and loss of neuronal ion gradients. Such SD waves evoked in chick retina were shown to be inhibited by gap junction blockers (Martins-Ferreira and Ribeiro, 1995; Nedergaard et al., 1995; Largo et al., 1997), suggesting that waves of SD may be propagated in part by gap junctions between either neurons or glia, or both, and that SD may be a physiological manifestation of $Ca^{2+}$ waves. However, SD was not inhibited by the glial metabolic poison fluoroacetate (Largo et al., 1997), but it is not yet known whether treatment with this poison causes glial uncoupling. If it does not under the conditions studied, then these results would have some bearing on the role of glial gap junctions in pathology. For example, SD waves in focal ischemia contribute to the expansion of an ischemic core into the penumbra and may be a significant factor in the pathophysiology of migraine (Walz, 1997), thus implicating GJIC in these two processes.

Further, the characteristics of astrocytic $Ca^{2+}$ waves are susceptible to the effect of exogenously applied $Ca^{2+}$ buffers; the endogenous protein counterparts of such buffers are upregulated in glial cells after injury, which led to the suggestion that these proteins could play a role in the regulation of signaling events mediated by $Ca^{2+}$ waves (Wang et al., 1997).

## I.  Regulation of Astrocytic Connexin Expression

The uneven distribution of both Cx30 and Cx43 in normal CNS, as well as alterations in Cx43 levels in the various experimental and disease conditions described later in this chapter, indicate that expression of these astrocytic connexins is subject to long-term regulation. This may occur at the transcriptional or translational level involving connexin production or at the post-translational level involving protein trafficking and gap junction assembly. Very little is known as yet about these processes in astrocytes, but there have been suggestions of a potential role of basic fibroblast growth factor (FGF-2). This growth factor has been localized specifically to gap junctions between astrocytes in brain and between myocytes in heart (Kardami et al., 1991; Yamamoto et al., 1991b). These observations clearly indicate, at least in these two tissues, a physical association between a particular form of FGF-2 and gap junctions composed in part of Cx43. Others (Grothe et al., 1991; Iwata et al., 1991; Tooyama et al., 1991) have reported the same punctate pattern of immunolabeling for FGFs in brain as that observed for Cx43 (see, however, Reuss and Unsicker, 1998). In MPTP-induced lesions of the nigrostriatal system in mice, a transient increase in Cx43 levels was further augmented after local application of FGF-2 into the striatum (Rufer et al., 1996). Interestingly, no dye coupling was found between these cells in either control or treated animals, although this may be a function of channel number or selectivity, or both. Nonetheless, FGF-2 has been shown to influence GJIC and Cx43 expression in microvascular endothelial cells (Pepper and Meda, 1992). At present, functional relationships between FGF-2 and gap junctions or connexins remain to be clarified. Investigations of this issue are complicated by the existence of a family of FGFs as well as their receptors, which have multiple cellular locations and perhaps multiple functions (Rifkin and Moscatelli, 1989; Baird and Bohlen, 1990)

## J.  Regulation of Astrocytic Coupling

It is likely that astrocytic gap junctions are also regulated at the channel level involving moment-to-moment modulation of conductance state. Indeed, astrocytic coupling must be under stringent regulation if enhanced GJIC during neuronal activity is required to promote flow of ions and substances away from locally elevated levels, and if reduced GJIC following neuronal activity occurs to prevent spatial dissipation of local glial metabolic activation. Thus, it may be speculated that neurons or glial cells themselves produce factors that modulate astrocytic GJIC via various signalling pathways *in situ*. In the absence of agents such as GJIC blockers that can confidently be said to have specific actions at gap junctions, elucidation of how GJIC between astrocytes is regulated by endogenous factors in the CNS could represent the next best key to understanding how the astrocytic syncytium operates to support the activity of neurons under normal conditions, how it may contribute to limit neuronal injury in disease, and whether misdirected regulation may inadvertently occur and exacerbate or even cause disease. Some progress on such regulatory factors has been made in the past few years (Giaume and McCarthy, 1996).

That short-term modulation occurs is indicated by observations in other tissues. For example, cyclic guanosine monophophate (cGMP) has been shown to reduce channel conductance at cardiac myocyte gap junctions composed of Cx43 (Burt and Spray, 1988). Moreover, activation of guanylate cyclase in retinal cells reduces gap junction conductivity between these cells (DeVries and Schwartz, 1989). Positive effects on junctional communication have been shown by the application of membrane permeable cAMP derivatives in cultured hepatocytes (Sáez et al., 1986, Traub et al., 1987), and antagonistic depression of GJIC can be induced by protein kinase C stimulation via phorbol esters (Spray et al., 1991). In neural tissue, coupling between astrocytes is increased following optic nerve stimulation (Marrero and Orkand, 1996) and can be altered by substances that in most cases have been shown to be released by neurons, supporting the idea that neuronally released substances may dynamically regulate the strength of connectivity of the glial syncytium (Finkbeiner, 1992). Thus, it has been shown in cultured astrocytes (Giaume et al., 1991b) that the stimulation of β-adrenergic receptors by isoprotenerol in the presence of a cAMP phosphodiesterase inhibitor elicited a large increase of cAMP accumulation accompanied by an augmentation of dye coupling. Application of glutamate or $K^+$ also increases GJIC of cultured astrocytes (Guiame et al., 1991b; Enkvist and McCarthy, 1994). In contrast, norepinephrine reduced dye coupling via activation of α-adrenergic receptors, and a close relationship was found between the uncoupling effect and the extent of phospholipase C activation. As well, a number of other regulatory agents and signaling mechanisms that are extant in astrocytes (Murphy and Pearce, 1987; Bevan, 1990) have been tested, including ATP, endothelin-1, oleic acid, 2-chloroadenosine in conjunction with noradrenaline, nitric oxide or its peroxynitrite products, AMPA receptor activation, protein kinase C (PKC) activation, and products of phospholipase C and phospholipase A2 activation such as arachidonic acid (AA) and endogenous cannabinoids termed fatty-acid amides (anandamide, oleamide) (Giaume et al., 1991b, 1992; Enkvist and McCarthy, 1992; Glowinski et al., 1994; Konietzko and Muller, 1994; Venance et al., 1995b; Bolanos and Medina, 1996; Cravatt et al., 1996; Giaume and McCarthy, 1996; Muller et al., 1996; Lavado et al., 1997). All of these evoked inhibition of GJIC between astrocytes *in vitro*.

The role of increased astrocytic coupling under a few conditions (e.g., stimulated optic nerve, $K^+$, glutamate) has been much discussed and seems consistent with, for example, spatial buffering concepts. Less consideration has been given, however, to the physiological consequences and relevance of the plethora of substances that cause reduction or closure of GJIC between astrocytes. Given the proposed role of junctional coupling in $K^+$ spatial buffering, what purpose would be served by the counter-intuitive process of reduced astrocytic GJIC evoked by the above agents if this were also shown to occur *in vivo*? One possibility is as follows. Assuming that glial gap junctions channels are normally in a fully open state during neuronal quiescence, then a sphere of local closure may be generated around active neuronal elements. This area of closure, involving perhaps subcellular glial compartments discussed earlier rather than individual astrocytes, would promote ion redistribution to regions of neuronal inactivity via junctions that remain open with contiguous astrocytes outside the area of closure. If theories concerning electrophysiological mechanisms governing $K^+$ spatial buffering are correct, then local closure of junctions may reduce dissipation of $K^{+-}$induced glial cell depolarization over vast areas of the astrocytic syncytium, thereby generating more effective electrical gradients that drive $K^+$ movements across gap junctions that remain open between astrocyte subcellular compartments located outside the sphere of reduced conductance. An alternative scenario based on maintained closure of astrocytic GJIC during neuronal quiescence, but also involving dynamic regulation of junctions, has been proposed (McKhann et al., 1997). It was suggested, however, that polarization-mediated changes in intracellular astrocytic pH, which have been shown to affect astrocytic GJIC (Connors et al., 1984; Anders, 1988; Pappas et al., 1996; see however Cotrina et al., 1998), may be among factors that influence junctional conductance.

The many factors that reduce astrocytic GJIC can also be considered from a standpoint of gap junction–mediated metabolite exchange. Astrocyte metabolism and metabolite flow via gap junctions between these cells may be especially important in view of evidence that neurons derive a significant proportion of their energy needs from lactate produced via nonoxidative glycolysis in astrocytes (Tsacopoulos and Magistretti, 1996). Thus, junctional distribution of, for example, glucose or lactate may be promoted by local neuronal needs. In this context, it has been reported that uncoupled astrocytes in culture exhibit different intracellular $Na^+$ levels, and it was suggested that GJIC may serve to maintain ion balance by equalizing internal $Na^+$, thereby normalizing functions mediated by this ion among coupled astrocytes (Rose and Ransom, 1997). This process, however, may be more complex. On the one hand, $Na^+$/glutamate co-transport into astrocytes stimulates glycolysis and lactate production via $Na^+$ activation of Na/K ATPase (Pellerin and Magistretti, 1994), which could be optimized by local equalization of internal $Na^+$ via junctional exchange of this ion. Yet, it was through inhibition of astrocytic GJIC that various agents were considered to stimulate glucose uptake (Lavado et al., 1997) in a manner linked to Na/K ATPase activity and hence to Na$^+$ concentration (Yu et al., 1993; Tabernero et al., 1996; Tsacopoulos and Magistretti, 1996). Thus, it is not clear whether neuronal needs following activation would be best served by locally increased astrocytic GJIC to facilitate K$^+$ buffering and to disperse Na$^+$ mediated metabolic events over a population of astrocytes as suggested (Rose and Ransom, 1997), or by decreased GJIC to restrict Na$^+$ movement thereby enhancing its local activation of glycolysis. Interestingly, both purposes could be served if conductance at homotypic channels (Cx30/Cx30 and Cx43/Cx43) formed by the two different connexins in astrocytes were differentially regulated in response to neuronal activation (one closed, the other open), and if these channels exhibited differential permeability to ions compared with metabolites. This would also be true if Cx43 forms functional heterotypic channels with Cx30, which could allow for unidirectional conductance. These possibilities are worth considering given evidence that electrical conductance, ionic selectivity, and dye permeability can vary by an order of magnitude among channels formed by members of the connexin family (Veenstra, 1996).

## K.  Regulation of Connexin43 by Phosphorylation

One approach to investigation of Cx43 regulation in the CNS involves examination of its molecularly modified forms. Similar to many other proteins, Cx43 is subject to phosphorylation. Some cells contain both nonphosphorylated and phosphorylated forms of Cx43, which are detected, respectively, as 41-kD and 43-kD bands on polyacrylamide gels, whereas other tissues such as heart contain almost exclusively the phosphorylated 43-kD form (see Hossain et al., 1994a). Both forms have been detected in brain, but their relative levels as reported by various authors were highly variable (Dupont et al., 1991; Kadle et al., 1991; Miragall et al., 1992; Nagy et al., 1992). It was found, however, that in rats killed by cranial high-energy microwave irradiation leading to rapid inactivation of brain metabolism, Cx43 in several brain regions was present almost exclusively as the 43-kD phosphorylated form (Hossain et al., 1994a). This indicates that the appearance of the dephosphorylated form was a postmortem artifact arising from the rapid action of a brain phosphatase that is able to act on Cx43 in tissue prepared by more standard methods. This finding was confirmed using a monoclonal anti-Cx43 antibody that exhibits selective recognition of the unphosphorylated form the protein in both fixed sections and by Western blotting of brain and a number of other tissues (Nagy et al., 1997a; Li et al., 1998).

As described later in this chapter, changes in Cx43 phosphorylation state also occur after various forms of neural injury. The significance of this dephosphorylated state in relation to the coupling status of astrocytes is not entirely clear as yet, although such Cx43 dephosphorylation at intact gap junctions has been observed in spinal cord after mild sciatic nerve stimulation, indicating that this event is not restricted to pathological conditions, but occurs normally in response to neural activation (Nagy et al., 1996b). Cx43 phosphorylation clearly affects junctional communication-competence (Bruzzone et al., 1996a, 1996b; Goode-nough et al., 1996; Lau et al., 1996), has been correlated with altered astrocytic GJIC (Hofer et al., 1996), and appears to contribute to modulation of junctional channel conductance state (Moreno et al., 1992; Takens-Kwak and Jongsma, 1992). However, whereas significant recent progress has been made in identifying residues of Cx43 phosphorylated in response to protein kinase C (Sáez et al., 1997) and mitogen-activated protein (MAP) kinase (Warn-Cramer et al., 1998), the functional consequences of altered phosphorylation at any particular serine residue in the multiply phosphorylated Cx43 molecule has not been established.

The observation that a monoclonal antibody developed to a peptide correspond-ing to amino acids 360 to 376 of the Cx43 sequence recognizes only unphospho-rylated Cx43 (Nagy et al., 1997a), a sequence including ser-368 and ser-372 known to be phosphorylated in response to activation of PKC (Sáez et al., 1997), does suggest initial phosphorylation of at least one of these sites and that it is maintained in the phosphorylated state in assembled gap junctions. That treat-ments activating PKC appear to inhibit GJIC in astrocytes (Konietzko and Muller, 1994) might be inconsistent with a role for PKC in this hypothesized early and stable phosphorylation of Cx43 (360-376). A clearer mechanism is emerging for MAP kinase inhibition of GJIC. MAP-kinase phosphorylation of ser-255, 279, and 282 correlates with an inhibition of GJIC, substitution of alanine for all of these serine residues, whereas abolishing the MAP-kinase sensitivity of GJIC does not inhibit the assembly or other known functions of Cx43-containing gap junctions (Warn-Cramer et al., 1998). Clearly, a more detailed knowledge of the role of phosphorylation of specific amino acids is critical to interpretation of GJIC status following various *in vivo* or *in vitro* manipulations.

The role of the phosphatase implicated in postmortem dephosphorylation of Cx43 and whether it normally participates in either Cx43 metabolism or regula-tion of gap junction communication is in need of further investigation. In addition, given possible relationships between Cx43 phosphorylation state and coupling, it may be relevant that sites on Cx43 serve as substrates for phosphorylation by MAP kinase (erk kinase) (Warn-Cramer et al., 1996), that MAP kinase signaling cascades have been identified in astrocytes (Tournier et al., 1994; Bhat et al., 1995), and that these cascades are activated by agents such as endothelin (Kasuya et al., 1994) that influence GJIC between these cells. Thus, it is likely that signal-ing mechanisms associated with MAP kinase play a significant role in the regula-tion of astrocytic GJIC.

## L. Astrocytic Gap Junctions in Response to Injury

Astrocytic Cx43 responses have been investigated in various experimental par-adigms with a panel of sequence-specific anti-Cx43 antibodies that exhibit differ-ential epitope recognition of Cx43 depending of its cellular location, phosphorylation state, and perhaps functional state. Anti-Cx43 antibody 13-8300 detects only unphosphorylated Cx43, as noted earlier; antibodies 16A and 71-0500 produce elevated labeling when Cx43 is internalized, at least in neural tissue; and antibody 18A undergoes epitope masking, the functional significance of which remains to be established. In studies of excitotoxic amino acid lesions, Cx43 levels are elevated in regions surrounding intracerebral kainic acid (KA) injection sites, which is consistent with reports that reactive astrocytes possess numerous gap junctions (Alonso and Privat, 1993; Lafarga et al., 1993). However, the coupling state of these cells at various stages of their progression from a normal to a reactive state is unclear and requires detailed examination. For example, treatment conditions that cause a reactive state *in vivo* have been correlated with uncoupling *in vitro* (Anders et al., 1990). At latter times, Cx43 at KA lesion sites undergoes a transition such that some epitopes are hidden (Vukelic et al., 1991), whereas others are exposed. These changes in Cx43 immunorecognition are unlikely to be a result of altered sensitivity of Cx43 epitopes to tissue fixatives as similar results were obtained by the methods of *in situ* transblotting of fresh unfixed tissue (Sawchuk et al., 1995). Cx43 epitope masking may thus be due either to tight intra- or intermolecular interactions or to covalent modifications at or near the masked epitope. Despite the presence of Cx43 at such sites, gap junctions are no longer present, but begin to reappear after several weeks survival (Hossain et al., 1994c), which may explain observations of dye-coupled astrocytes at this time in KA-lesioned hippocampus (Burnard et al., 1990). Ultrastructural studies of KA-lesioned areas revealed that removal of astrocytic gap junctions occurred coincident with neuron loss and that Cx43 was sequestered intracellularly by internalization of gap junctional membranes (Ochalski et al., 1995).

In studies of rats subjected to ischemic brain injury, Cx43 was increased in animals with mild to moderate striatal damage, but reduced in regions severely depleted of neurons (Hossain et al., 1994b). Similar results were obtained in a spinal cord trauma model (Theriault et al., 1997), where severely damaged or necrotic areas proximal to a compression site displayed intensified Cx43 staining with antibodies that best recognize internalized Cx43, suggesting removal of gap junctions in these areas. In contrast, Cx43 sequestration at slightly more distal regions exhibiting a milder degree of injury was accompanied by Cx43 epitope masking. In a rat cerebral focal ischemia model with various ischemia/reperfusion times (Li et al., 1998), Cx43 at intact gap junctions undergoes reversible dephosphorylation after brief ischemia, whereas astrocytic gap junction internalization accompanied by elevated labeling of internalized Cx43 occurred in the ischemic core after longer periods of ischemia. Dephosphorylated Cx43 persisted at intact gap junctions confined to a thin corridor at the ischemic penumbra. Both the brain ischemia and spinal cord trauma models suggest that different Cx43 regulatory processes may be operative in areas with different degrees of neural damage, which may be correlated with areas containing cells undergoing different modes of cell death (e.g., apoptotic in the ischemic periphery and necrotic in the ischemic core). Thus, it is clear that Cx43 is subject to rapid dephosphorylation, membrane dispersal, and internalization, which proceed as a temporally and spatially ordered sequence of events in response to neural injury and culminate in differential patterns of Cx43 modification and sequestration at the lesion center and periphery.

More generally, these findings indicate that the astrocytic syncytium in the vicinity of injury undergoes remodeling, perhaps to suit altered tissue homeostatic requirements. Thus, it has been suggested (Hossain et al., 1994b, 1994c; Li et al., 1998) that injury-induced elevations in Cx43 levels and astrocytic gap junctions around neuronal elements exhibiting limited damage may represent induction of an enhanced glial coupling state to support neuronal survival. This idea is consistent with recent demonstrations that astrocytic GJIC in mixed neuronal-astrocyte cultures protects neurons from exposure to oxidative insults (Blanc et al., 1998). Conversely, in regions of severe neuronal damage, removal of gap junctions must certainly reflect reduced junctional communication between astrocytes, in analogy with reduced GJIC that occurs following energy depletion of these cells in culture (Vera et al., 1996). This may be an attempt by these cells to sever communication with their neighbors, thereby restricting flow of undesirable metabolites that could potentially cause remote neuronal damage. The state of astrocytic coupling shortly after injury (Li et al., 1998) as well as within the ischemic penumbra is currently uncertain. However, molecular alteration of gap junctions at the penumbra (Li et al., 1998) is consistent with morphological irregularities and smaller size of junctions observed in this area (Cuevas et al., 1984a). Penumbral neuronal survival hangs in a balance and is contingent on many factors, but it is conceivable that survival is in part dependent on the unique features exhibited by astrocytes and their attendant gap junctions at such borders.

Some of these views are supported by studies in which brain slices from 13- to 16-day old rats as well as astrocytes maintained in culture for a few weeks were exposed to ischemic conditions and metabolic inhibitors, respectively (Cotrina et al., 1998). It was found that GJIC between astrocytes in slices was reduced by 65% after ischemic treatments and by 70% in cultures after inhibition of metabolism. A caveat in this study is that Cx30 expression is very low in rat brain at this age (Nagy et al., 1998) and almost absent in astrocyte cultures at only 2 to 3 weeks after plating (Kunzelmann et al., 1999; Nagy, unpublished observations). Thus, further studies will need to take into account the regulatory responses of junctional channels composed of Cx30, which may be different than those composed of Cx43. Moreover, substantial increases in rat brain Cx43 expression occur in the third and fourth postnatal weeks, and this may be accompanied by maturation of regulatory mechanisms that determine responses of astrocytic GJIC to injury in adult as distinct from those in neonatal brain.

Alterations in astrocytic gap junctions that have been seen in other animal models as well as in human postmortem tissue may also be relevant to various pathological conditions. Lesions of the rat facial nerve were shown to cause increases in Cx43 immunostaining in not only the facial nucleus but also the cerebral cortex after very short survival times of 1 to 3 hours (Rohlmann et al., 1993, 1994; Wolburg and Rohlmann, 1995; Laskawi et al., 1997). Although it is uncertain whether this reflects altered epitope recognition of Cx43, as previously discussed, or de novo synthesis, it is clear that rapid neural alterations known to occur at sites remote from facial nerve lesions are accompanied by equally rapid responses in astrocytes and their gap junctions (Laskawi et al., 1997). In studies of human tissues, Cx43 mRNA expression and astrocytic GJIC were reported to be greater in epileptic foci or hyperexcitable tissue than in surrounding normal neocortical tissues removed from patients undergoing surgical treatment for intractable seizure disorder (Naus et al., 1991b; Lee et al., 1995). In a similar study taking into consideration the heterogeneity of Cx43 mRNA and protein seen in normal brain, no upregulation of these molecular species was seen in response to development of epileptogenicity (Elisevich et al., 1997). Elevated levels of Cx43 were observed within βA4-positive amyloid plaques in brain tissue of patients with Alzheimer's disease (Nagy et al., 1996c). This may reflect the invasion of these plaques by processes of surrounding reactive astrocytes that may be more highly coupled. Alternatively, by analogy with the reported induction of Cx43 in PC12 cells over-expressing a carboxy-terminal fragment of amyloid precursor protein (APP) (Lynn et al., 1995; Nagy et al., 1996a), regulation of Cx43 might somehow be tied to APP function. Consequently, abnormal APP breakdown products within plaques may cause an aberrant upregulation of astrocytic Cx43, leading to an imbalance in glial GJIC, which could exacerbate disease progression.

## IV. JUNCTIONAL COMMUNICATION BETWEEN MYELINATING CELLS

### A. Oligodendrocytic Gap Junctions

Oligodendrocytes provide the cellular source for the formation as well as maintenance of the CNS myelin sheath. This sheath is formed by a complex spiral wrapping and subsequent compaction mechanism that requires considerable mechanical and molecular efforts of activated oligodendrocytes (for review, see Bunge, 1965). How these processes are topologically coordinated *in situ* is still a mystery. Gap junctions may well serve a supportive function in the coordination process, because cohorts of oligodendrocytes must act in concert to establish the compact myelin sheath of single as well as adjacent internodal axonal elements. In peripheral myelin, it has been shown that wallerian regeneration leads to an upregulation of the gap junction complement, indicating an increase of communicative capacities during the onset and continuation of myelination (Tetzlaff, 1982). In the CNS, the existence of gap junctions between resident oligodendrocytes have been documented *in vivo* (Sotelo and Angaut, 1973; Dermietzel, 1978; Sandri et al., 1982). Based on recent studies (Li et al., 1997; Rash et al., 1997), oligodendrocytic gap junctions appear to be more prevalent than suggested by previous ultrastructural observations (Mugnaini, 1986), particularly in subcortical areas, brain stem, and spinal cord. The prevalence of interoligodendrocytic gap junctions together with the common occurrence of gap junctions between the two macroglial cell types provides reason to consider oligodendrocytes as part of a composite glial syncytium.

## B. Localization of Oligodendrocytic Connexins *In Vivo*

The distribution of Cx32 in the mammalian CNS has been examined by several groups or researchers. Cx32 was localized to oligodendrocyte cell bodies and processes in spinal cord, where it was absent along myelin sheaths but was found biochemically to be enriched in myelin fractions of brain stem and spinal cord (Dermietzel et al., 1989; Scherer et al., 1995). Others have observed Cx32 labeling along internodal regions of central myelin (Spray and Dermietzel, 1995; Li et al., 1997; Pastor et al., 1998) and the distribution of Cx32-positive oligodendrocyte cell bodies in rat CNS was reported to be somewhat heterogeneous. Ultrastructurally, gap junctions exhibited symmetrical labeling of homologous gap junctions between oligodendrocytic elements and asymmetrical labeling on the oligodendrocyte side of heterologous gap junctions between oligodendrocytic and presumptive astrocytic elements (Li et al., 1997).

The presence of oligodendrocytic Cx32 along myelinated fibers is consistent with evidence for numerous gap junctions between astrocytic and oligodendrocytic processes surrounding the outer turn of myelin (Rash et al., 1997). However, Cx32 was found by Li and co-workers (1997) to be localized to only a subpopulation of myelinated fibers, such as in the cerebellum, where it appears to be associated with myelin sheaths only on axons arising from Purkinje cells. Further, Cx32 was sparse along myelin in some white matter regions, whereas it was concentrated along individual fibers or small bundles passing through gray matter, suggesting that glial coupling requirements may differ in white and grey matter. Oligodendrocytes appear to be heterogeneous, raising the possibility that different subtypes may myelinate different caliber axons. Thus, it was suggested (Li et al., 1997) that Cx32 along certain fibers may arise from either differential Cx32 expression or targeting to discrete cellular sites, which may represent divergent responses of a uniform oligodendrocyte population to different properties of the axons they myelinate. Alternatively, oligodendrocyte subtypes may be preprogrammed to express particular phenotypic characteristics, including connexin expression, and in this way are predetermined to myelinate particular classes of axons.

Coupling between oligodendrocytic processes as well as between these and astrocytes along internodal regions of myelin may be important for ion and metabolite homeostasis during axonal activity at nodes of Ranvier. For example, ions may accumulate in paranodal loops, flow through gap junctions demonstrated to be present between these structures, collect in oligodendrocytic cytoplasm at the outer surface of myelin near internodes, and then move into astrocytic or other oligodendrocytic processes via gap junctions. If Cx32 along myelinated fibers subserves such communication pathways, then this might explain its selective localization along particular fiber types having a potentially greater demand for ion redistribution. It should be noted, however, that while oligodendrocytic paranodal loops form gap junctions in the CNS (Sandri et al., 1982), the rarity with which Cx32 has been observed at paranodal loops (Spray and Dermietzel, 1995; Li et al., 1997) together with selectivity of Cx32 for subclasses of fibers suggests either that not all nodal loops possess gap junctions or, if they do, that such junctions are composed primarily of another connexin, perhaps Cx45.

## C.  Oligodendrocytic Connexins and Functional Coupling

In addition to different cellular distribution of Cx32 in white and gray matter, functional analysis of the coupling efficiency between oligodendrocytes showed significant differences. Dye coupling studies in oligodendrocytes of gray and white matter of the developing rat spinal cord revealed that 18% of gray matter oligodendrocytes are strongly coupled whereas no dye coupling was evident in white matter oligodendrocytes. Lack of coupling is not only restricted to LY, but includes Neurobiotin (Pastor et al., 1998), which is a cationic low molecular weight tracer. Consequently, both types of oligodendrocytes seem to be furnished with differential coupling modalities. Two alternative explanations may account for the lack of coupling in white matter oligodendrocytes of the spinal cord. It may be that connexin expression serves a function exclusively for reflexive coupling of oligodendrocytic processes facilitating transcellular transport or signaling. This implies that the main function of white matter Cx32 gap junctions is to mediate signal propagation within the complex territory of individual myelinating oligodendrocytes. In the context of the recently proposed hetero- and autocontrol territories comprised by astrocytic gap junctions (Rohlmann and Wolff, 1996), the Cx32 complement of white matter oligodendrocytes could provide an exclusive autocontrol territory. This idea is consistent with the suggested function of Cx32 coupling in Schwann cells, which is considered to serve as a radially oriented autocellular route through the myelin sheath. As white matter oligodendrocytes are primarily myelinating cells in contrast to perineuronal oligodendrocytes of gray matter, which are more likely to perform satellite cell functions, a functional analogy between the peripheral and central myelinating cells with respect to Cx32 expression seems reasonable.

Alternatively, in contrast to Schwann cells, oligodendrocytes are embedded in a complex cellular environment which may require heterologous cross talk between different types of cells. This suggests that white matter oligodendrocytes establish heterologous routes of coupling where both hemichannels contributed by the coupled cells may be furnished with different permselectivities which exclude dye transfer from the oligodendrocytic side to the partner cell. This interpretation leaves open the class of cell to which oligodendrocytes are coupled. In view of the recent freeze fracture data by Rash and colleagues (1997), oligodendrocytes are exclusively coupled to astrocytes allowing the creation of a "panglial" syncytium. If true, this panglial syncytium must be provided with a high degree of asymmetry, especially in white matter. In the light of recent data (Dermietzel et al., 1997, Kunzelmann et al., 1997) indicating that oligodendrocytes express a second type of gap junction protein and that this connexin (Cx45) has functional properties that allow rectification in a physiological range (Moreno et al., 1995), the presence of extended asymmetrical coupling is likely. Unidirectional dye transfer between oligodendrocytes and astrocytes has been shown by Robinson and associates (1993) between retinal glial cells. A chimeric setting of connexins could occur in white matter oligodendrocytes of the spinal cord. Heterotypic combinations of Cx45 and Cx43 may account for a functional asymmetry of coupling in the glial compartment.

Homotypic Cx45 channels between both classes of glial cells provide a likely alternative to the heterotypic setting since astrocytes have been detected to express Cx45 as well (see earlier discussion). In this case, the lack of dye transfer in white matter oligodendrocytes can be explained by the limited conductance of the homotypic Cx45/Cx45 channel to negatively charged dyes such as LY (Steinberg et al., 1994; Koval et al., 1995). In this context, the lack of Neurobiotin transfer is less plausible since transmission of this tracer has been documented in Cx45-transfected HeLa cells (Elfgang et al., 1995). However, in the latter case transfection may have led to overexpression of this channel which could be responsible for the limited transfer of this tracer in the transfected cells. Continuation of this work with purified cultured adult oligodendrocytes as well as by studies *in vivo* corroborated this initial finding (Scherer et al., 1995; Li et al., 1997) and have shown co-expression of Cx32 with Cx45 in these cells (Dermietzel et al., 1997; Kunzelmann et al., 1997).

## D.  Oligodendrocytic Coupling *In Vitro*

Early studies have indicated gap junctional coupling between cultured oligodendrocytes (Kettenmann et al., 1983; Massa et al., 1984; Kettenmann et al., 1990). However, the low incidence of interoligodendrocytic gap junctions *in vitro* (Massa and Mugnaini, 1985) has been interpreted as an indication of coupling deficiency of this glial cell class. According to the data described earlier, it seems more likely that the *in vitro* conditions lead to a downregulation of connexin expression in cultured oligodendrocytes in consequence of the disruption of the intricate topology of the oligodendrocytic network and its structural relation to the myelin sheaths. Dye injection studies in brain slices also yielded controversial results with regard to the existence of interoligodendrocytic coupling. Although Berger and associates (1991) where unable to demonstrate dye transfer between oligodendrocytes after LY injection into corpus callosum slices, Ransom (personal communication) found evidence of dye coupling in oligodendrocytes of the optic nerve. These discrepancies may point to methodological problems, especially when LY is used as a fluorescent tracer (see earlier discussion) in a tissue of low-degree coupling. Alternatively, they may indicate either heterogeneity of coupling efficiency in a way similar to that discussed for astrocytes or the differentiation state of oligodendrocytes in culture (Blankenfeld et al., 1993; Venance et al., 1995a). These observations are consistent with electrophysiological studies, which showed weak coupling between cultured oligodendrocytes (Kettenmann and Ransom, 1988; Ransom and Kettenmann, 1990) and with a low level of intercellular calcium wave propagation between these cells (Takeda et al., 1995). By measuring the unitary conductance, oligodendrocytes revealed an accumulation of single channel events in the range of 100 pS. This is in agreement with the expected size for Cx32 channels. A second event of unitary conductance was found to cluster around 30 to 40 pS.

By a combination of *in situ* and *in vitro* approaches, Cx45 was finally identified as an additional gap junction protein of oligodendrocytes. This evidence was based on the following observations: (1) *In situ* immunostaining using two different anti-Cx45 antibodies revealed an expression of Cx45 in cells staining positively for myelinating cell markers. (2) Although the level of Cx45 expression in cultured oligodendrocytes proved beyond the limit of detectability when common immunochemical or Northern blotting techniques were applied, the more sensitive RT-PCR successfully identified Cx45 *in vitro*. (3) According to data obtained from transfected cell lines expressing the chicken Cx45 (Veenstra et al., 1994) and from the SKHep cell lines, which express Cx45 as an endogenous connexin (Moreno et al., 1995), the small unitary conductance in oligodendrocytes corresponds with the size of single channel conductance of this connexin. Additionally, the high degree of voltage sensitivity found between some cell pairs supports the conclusion that Cx45 is the second oligodendrocytic gap junction protein.

The existence of a second connexin gene expressed in oligodendrocytes is also suggested by the Charcot-Marie-Tooth type X1 syndrome (see next subsection), which correlates with mutations in the Cx32 gene (Bergoffen et al., 1993), and by transgenic Cx32 (–/–) mouse (Nelles et al., 1996, Anzini et al., 1997). The lack of central symptoms in both cases renders a compensation by a second connexin most likely. Alternatively, a dominant negative effect of the Cx32 mutations could for some unknown reason preferably affect myelinating Schwann cells while being compensated in oligodendrocytes. mRNA of Cx32 and its translation product were clearly identifiable, although at a much lower level than that found for Cx43 expression in astrocytes. Additional connexins for which cDNA probes are available (Cx26, Cx37, Cx40, Cx43) were not detectable in highly enriched oligodendrocytic cultures. These findings together with reports showing developmental increases of Cx32 protein and mRNA levels during myelination may indicate a role of Cx32 in myelin formation and maintenance (Dermietzel et al., 1989; Belliveau et al., 1991; Giaume and Venance, 1995; Scherer et al., 1995).

From the accumulating data it seems likely that the capacity of interoligoden-drocytic coupling is less extensive *in vitro* and *in situ* as compared with astro-cytes. Similar to the striking heterogeneity in astrocytic coupling as evident by the differences between protoplasmic and fibrillar astrocytes (as previously dis-cussed), oligodendrocytes represent a population of divergent cells that seem to differ in their coupling efficiency. Whether oligodendrocytic coupling is also altered in response to local functional requirements (i.e., de- and remyelination) awaits future experiments.

## E. Schwann Cells and the Charcot-Marie-Tooth Type X1 Syndrome

Myelinating Schwann cells in peripheral nerve, which are the counterparts of oligodendrocytes in white matter, form reflexive gap junctions under unchal-lenged conditions and express Cx32 at nodes of Ranvier and at Schmidt-Lanter-man incisures (Bergoffen et al., 1993; Paul, 1995; Scherer et al., 1995; Spray and Dermietzel, 1995). As previously mentioned, reflexive gap junctions are formed between apposed processes of the same cell, allowing intracellular or autocellular communication. Such reflexive junctions may provide shortcuts between adjacent processes in an extended cytoplasmic network. Peripheral myelin is composed of two domains: compact myelin and noncompact myelin. The latter is found at the periodic cytoplasmic interruptions in the compact myelin sheath in the form of Schmidt-Lantermann incisures and at the paranodal processes. The distinct local-ization of Cx32 at these sites suggested a reflexive mode of communication in myelinating Schwann cells (Scherer et al., 1995; Spray and Dermietzel, 1995), which would allow functional coupling between the peripheral cytoplasmic enve-lope, including the soma and the nuclei of the Schwann cell, and its adaxonal cytoplasmic collar. Such a shortcut may play a role for trophic support to the most peripheral extensions of the myelin sheath or provide spatial buffering of extracel-lular potassium during action potential activity (Orkand et al., 1966).

The idea of having intracellular avenues for rapid trafficking is an intriguing concept and had not been supported by experimental evidence until recently (Balic-Gordon et al., 1998). By injecting high and low molecular tracer into the outer cytoplasmic collar, these groups convincingly demonstrated a rapid dye transfer of low molecular weight tracer (5,6 carboxyfluorescein) from the periph-eral portion of the myelin sheath into the inner adaxonal loop. The occurrence of "double train track pattern," which can be explained by a filling of the adaxonal cytoplasm from the outer injected side, suggested fast accessibility of the inner collar from the Schwann cell periphery. High molecular weight tracer such as rhodamine dextran (molecular weight 3000 D) failed to pass through the gap junc-tional channels, and consequently did not show the "double train tracks." Func-tional loss or negative dominant effects of mutated Cx32 gene could, therefore, result in disturbance of these pathway with the consequence of "malnutrition" of the myelin sheath or defects in spatial buffering. In this context, the mutations of the Cx32 gene, as have been shown in the X-linked Charcot-Marie-Tooth syn-drome, can be explained by functional disruption of the radial diffusion pathways, which ultimately results in degeneration of the peripheral myelin sheath. Interest-ingly homozygote Cx32 (–/–) mouse do not develop a clinical phenotype, although the myelin sheaths of older animals show signs of a peripheral neuropa-thy (Anzini et al., 1997, Scherer et al., 1998). A reasonable explanation for the lack of an external phenotype in Cx32 (–/–) mice is provided by the species-spe-cific anatomy of the motoneurons. Because long motoneurons are primarily afflicted by the Charcot-Marie-Tooth type X1 (CMTX1) syndrome in humans, it seems feasible that deficits in coupling manifest at those sites where long dis-tances are traversed by the axon and, therefore, "malnutrition" is more likely to occur. In mice, these effects seem less stringent because of shorter motoneurons, which might be less prone for coupling deficiencies. However, mice given xenografts of nerve taken from human with Cx32 mutation did exhibit axonal abnormalities within the transplanted segment (Sahenk and Chen, 1998). Thus, the lack of a clinical phenotype in Cx32 null mice may be due to partial compen-sation for the loss of the Cx32 gene by expression of an alternative connexin.

Shift in connexin expression of Schwann cells has been demonstrated after peripheral nerve injury. After nerve crush, Cx32 is downregulated in a sequel con-curring with the degeneration events. Concomitantly with the axonal degenera-tion, expression of Cx46, not detectable in resting Schwann cells, increases during the phase of degeneration and ceases after the regeneration process is finished (around day 20 after the nerve crush). This differential expression of connexins in the resting and challenged Schwann cells is a clear indication of functional speci-fication of the gap junction channels. Apparently, the functional properties of the channels that are required for the resting Schwann cell are essentially different from those of the de- and remyelinating cell. This raises the intriguing possibility that there are coordinated changes in Schwann cell proliferation and connexin expression. Moreover, intercellular junctional coupling among Schwann cells may be important during injury responses. Consistent with this idea is the obser-vation that the strength of junctional coupling among cultured Schwann cells is modulated by a number of cytokines to which Schwann cells are exposed *in vivo* after nerve injury (Chandross et al., 1996; Chandross, 1998).

## V.  JUNCTIONS BETWEEN ASTROCYTES AND OLIGODENDROCYTES

### A.  Heterologous Astro/Oligo Gap Junctions *In Vivo*

Astrocytic processes frequently make contact with myelinated fibers, particu-larly in spinal cord where these contacts arise from fine branches of radial astro-cytes (Ochalski et al., 1997). Heterologous astro-oligodendrocytic gap junctions have been reported at these appositions (Massa and Mugnaini, 1982; Waxman and Black, 1984; Mugnaini, 1986), suggesting gap junctions between astrocyte and oligodendrocyte processes at the outer surface of myelin results in the incorporation of the latter into the extensive gap junctionally coupled astrocytic syncytium. Asymmetrical gap junctional labeling for Cx43 has been described in cases where oligodendrocytic cell bodies or processes contribute the unstained side to junctional appositions with astrocytes (Nagy et al., 1997b; Ochalski et al., 1997; Nagy et al., 1998). This is consistent with reports that oligodendrocytes commonly form gap junctions with astrocytes and that the former express Cx32 and Cx45 and not Cx43, as discussed earlier. Thus, at the surface of myelin, Cx43-positive astrocytic processes contacted the outermost loop of myelin and formed asymmetrically labeled gap junctions (Ochalski et al., 1997). The possibility of heterologous coupling arising from an interaction of astrocytic Cx30 with oligodendrocytic Cx32 may be likely. These studies also demonstrated astrocyte processes forming junctions with an oligodendrocyte and simultaneously with another astrocyte. Although the significance of this observation is at present unclear, it may be physiologically relevant to the rate and direction of intercellular ion and metabolite flow at these homologous and heterologous pairs of junctions. A more comprehensive discussion of heterologous and heterotypic coupling between glial cells and between other cells type in neural tissue has recently been presented (Zahs, 1998).

## B.  Connexins at Heterotypic Astro/Oligo Junctions

The expression of Cx30 and Cx43 in astrocytes and Cx32 and Cx45 in oligodendrocytes increases the possible combinations of heterotypic connexin pairings at glial gap junctions. In gray matter, these include interastrocytic gap junctional channels composed of Cx30/Cx30, Cx43/Cx43, or Cx30/Cx43; interoligodendrocytic channels composed of Cx32/Cx32, Cx45/Cx45, or Cx32/Cx45; and astrocytic-oligodendrocytic channels consisting of four combinations with either Cx30 or Cx43 on the astrocytic side and Cx32 or Cx45 on the oligodendrocytic side. The combinations that occur depend on the degree to which different connexins are targeted to single junctional plaques in individual cells and on the specific pairs of connexins that are able to form functional channels. For example, it has been reported that Cx43/Cx32 pairings are nonpermissive for channel formation (Elfgang et al., 1995), and it appears that most if not all interoligodendrocytic and astro-oligodendrocytic gap junctions contain Cx32 on the oligodendrocytic side (Li et al., 1997). This suggests that oligodendrocytic Cx32 may instead form channels with astrocytic Cx30, which is permissive (Dahl et al., 1996; White and Bruzzone, 1996). Further, since Cx30 and Cx43 are incorporated to various degrees at virtually all astrocytic gap junctions in gray matter (Yamamoto et al., 1990b, 1990c; Ochalski et al., 1997; Nagy et al., 1998), it appears that both of these connexins contribute to the formation of interastrocytic or astro-oligodendrocytic junctions in either a homotypic or heterotypic manner. Whether any of these connexins form heteromeric junctions, where individual connexons (hemichannels) contain a mixture of connexins (Stauffer, 1995; Jiang and Goodenough, 1996), remains to be determined.

As astrocytes in white matter express Cx43 but not Cx30 (Nagy et al., 1998), it appears that Cx30 does not participate in the formation of such junctions between astrocytes and oligodendrcytes in white matter. Thus, consistent with reports of functionally permissive pairing between Cx43 and Cx45 (Elfgang et al., 1995), it is likely that glial junctions in white matter may arise by pairing astrocytic Cx43 with oligodendrocytic Cx45 (Dermietzel et al., 1997; Kunzelmann et al., 1997).

## VI. RETINAL GAP JUNCTIONS AND CONNEXINS

The first successful attempt to isolate a neuronal connexin clone has been made by O'Brian and associates (1996) using a skate retinal expression library. In contrast to brain tissue, where the complexity of the neural tissue makes identification of neuron specific connexins a hard task, the retina offers an outstanding model for studying neuronal gap junctions. The retina constitutes a highly ordered laminar structure which comprises three compact layers of neurons separated by two synaptic layers. Glial cells are confined to specific radially oriented cells (the so-called Müller cells), which show extensive coupling in some lower vertebrates (Ball and McReynolds, 1998), but insignificant coupling in mammals, although unidirectional tracer-transfer from astrocytes to Müller cells has been observed in rat and rabbit retina (Robinson et al., 1993; Zahs and Newman, 1997) and these cells do exhibit gap junctions at least *in vitro* (Wolburg et al., 1990). Electrical coupling is found in all cell types that form the neuronal network of the retina.

The utilization of the cationic tracer Neurobiotin in the retina has led to the discovery of an unexpected incidence of coupled cells and diversity of coupling pattern so far unmatched by other part of the brain. In particular, horizontal cells constitute a population of retinal neurons that are endowed with extensive gap junction properties (see Dowling, 1991). Dual-recording experiments and dye transfer studies in parallel with freeze fracture investigation made horizontal cells by far the best-studied class of coupled neurons in the central nervous system. As a result of an extensive coupling through gap junctions, horizontal cells exhibit a receptive field that is much larger than their individual anatomical size. Thus, horizontal cells offer the rare opportunity to correlate a functional parameter like the receptive field size with the degree of coupling. The concerted effort of horizontal cell studies allowed researchers for the first time to unravel a neuromodulatory control of gap junction conductance in the central nervous system (for review, see Vaney, 1996; Weiler, 1996). Dopamine was identified as decreasing the junctional conductance of gap junctions in horizontal cells of various species (Hedden and Dowling, 1978; Piccolino et al., 1984; Hampson et al., 1992). Further studies showed that this effect is mediated by cAMP through activating adenylate cyclase (Watling and Dowling, 1979). Arachidonic acid has also been shown to inhibit coupling between horizontal cells, perhaps by a pathway that activates guanylate cyclase (Miyachi et al., 1994). The modulation of junctional conductance by a potent neurotransmitter is an indication that both modes of neuronal synaptic transmission did not evolve in parallel forms, but rather synergistically as interrelated mechanisms of intercellular signaling. A selective nature of neuronal gap junction–mediated coupling and their differential modifiability by second messenger has been shown for the amacrine AII cells (Mills and Massey, 1995). This study strongly suggests co-occurrence of muliple types of connexins within the neuronal populations of the retina.

Although knowledge of the connexin types expressed in the retina is still limited, recent cloning endeavors have been especially successful. By means of an antibody directed to Cx32 that cross-reacted with a retinal connexin, O'Brian and colleagues (1996) were able to isolate a new connexin from an expression library of skate retina. Although the exact localization of the Cx35 was not evaluated in their original work, Condorelli and associates (1998), using degenerate primers, were able to clone the murine homolog of the fish Cx35, which showed a putative molecular mass of about 36 kD. By *in situ* hybridization, Cx36 was demonstrated to be enriched in the inner plexiform layer and to a minor degree in the ganglion cell layer of the retina. As well, it was found in some brain regions known to contain electrically coupled neurons. Parallel studies by Söhl and colleagues (1998) using a cDNA probe of the skate Cx35 also successfully isolated the murine homolog of Cx35 and confirmed the preferential localization of Cx36 in the retina. According to its gene structure, Cx36 (Cx35) and a further connexin cloned from perch retina (Cx34.7, Bruzzone, personal communication), belong to a new subclass of connexins. It remains to be clarified whether this subclass comprises a specific component of the neuronal connexin complement or, like the other subclasses, is more generally expressed.

## VII.  GAP JUNCTIONS AND CONNEXINS DURING CENTRAL NERVOUS SYSTEM DEVELOPMENT

### A.  Neuronal Gap Junctions

There are indications that gap junctional coupling between neurons may be quite extensive during neuronal differentiation and morphogenesis. Evidence for gap junction–mediated interneuronal communication at various stages of CNS development has been obtained in the areas listed in Table 2. One intriguing aspect of functional relevance regarding electrical coupling early in brain development has been addressed by Lo Turco and Kriegstein (1991). They reported on the occurrence of clusters of coupled neuroblasts within the ventricular zone as

**Table 2.** Mammalian Neural Structures in Which Evidence for the Presence of Gap Junctions between Neuronal Elements during Development Has Been Reported in the References Listed

| | Neural Structure | References |
|---|---|---|
| 1. | Between columnar domains of neurons in the cortex of rat in the first three postnatal weeks | Connors et al., 1983; Yuste et al., 1992; Peinado et al., 1993a, 1993b; Mienville et al., 1994; Kasper et al., 1994; Kandler and Katz, 1995; Rorig et al., 1995, 1996; Rorig and Sutor, 1996a, 1996b |
| 2. | Between neuroblasts in embryonic neocortex of rat and between neuroepithelial cells in the neural tube of rat | Bergmann and Surchev, 1989; Lo Turco and Kriegsten, 1991 |
| 3. | Between neurons of neonatal rat locus coeruleus maintained in vitro | Christie et al., 1989 |
| 4. | Between neurons in slice preparations of neonatal rat striatum | Walsh et al., 1989 |
| 5. | Between motoneurons in the spinal cord of neonatal rat | Fulton et al., 1980; Arasaki et al., 1984; Walton and Navarrete, 1991; van der Want et al., 1998 |
| 6. | Between primary afferent neurons in sensory ganglia in embryonic chick | Pannese et al., 1977 |
| 7. | Between dorsal root ganglia or parasympathetic neurons and myotubes or myoblasts in culture | Bonner, 1988; Fischbach, 1972; Ennes et al., 1997 |
| 8. | Between dendrites of neurons in rat inferior olivary complex where gap junctions are formed during the second postnatal week | Gotow and Sotelo, 1987 |
| 9. | Between various cells types in the developing monkey and human retina | Fisher and Linberg, 1975; Nishimura and Rakic, 1985 |
| 10. | Between clusters of neuroblasts within the ventricular zone as early as embryonic day 14 | Lo Turco and Kriegstein, 1991 |
| 11. | Between granule cells and between granule cells and axo-axonic (chandelier) cells in the dentate gyrus of rat hippocampus during the first three postnatal weeks | Haring et al., 1997 |
| 12. | Between hilar cells in early postnatal hippocampal development | Strata et al., 1997 |
| 13. | Between mitral cells in the developing rat olfactory bulb | Paternostro et al., 1995 |

early as embryonic day 14. The number of coupled cells within the clusters decreased as neuronal emigration from the ventricular zone progressed. They speculated that the clusters may comprise neuronal compartments that serve the basis of columnar organization in the mature brain. Yuste and colleagues (1992) extended these observations by demonstrating large numbers of coupled neurons organized within small and often columnar regions in postnatal cerebral cortex. These "neuronal domains" exhibited simultaneous increases in intracellular free calcium in the absence of sodium action potentials. Because chemical synapses are undifferentiated and regenerative electrical activity fatigues easily at that time, the presence of extensive coupling via gap junctions may provide a mechanism by which groups of neurons exchange developmentally relevant signals and thereby contribute to the differentiation of functionally mature circuits. A further argument for a developmental role of neuronal gap junction coupling is provided by observations (Yuste et al., 1992, Peinado et al., 1993a) that coupling disappears after the third postnatal week. However, the loss of tracer-coupling does not necessarily indicate the absence of electrical coupling. As pointed out by Dermietzel and Spray (1993), the threshold for detection of dye coupling (in this case LY) appears to requires 20 times as many gap junction channels as are necessary for transmission of electrical signals. Anion versus cation selectivity of specific connexins further complicates interpretation of coupling in the absence of intercellular transfer of specific dyes. The application of dual-cell recording technique to neuronal cell pairs in brain slices should provide better evaluation whether electrical coupling persists in cortical neurons even after the critical postnatal period of cortical columnar imprinting. Electrophysiological detection provides the only way to definitely prove whether cortical coupling via gap junctions is a consistent mechanism of neuronal communication even in adult brains. The aforementioned cross-reactivity of molecular connexin probes (i.e., antibodies and mRNAs), with intracellular proteins that may not directly contribute to electrical synapses renders a clarification by means of morphological techniques difficult or even unlikely. It should be noted, however, that the recent identification of Cx36 as a neuronal connexin and demonstration of its relatively high abundance in brain and retina (Condorelli et al., 1998, Söhl et al., 1998) may yet provide probes of greater specificity for neuronal gap junctions.

## B.  Connexin Expression

Although several connexins have been detected in embryonic (E) and early postnatal (P) CNS tissues, their cellular localization is uncertain for most except that of Cx43, which is found in developing astrocytes and is discussed later in the context of glial gap junctions. By Northern and Western blot analysis, Cx32 protein and mRNA appear during the second postnatal week in developing rat brain, whereas Cx43 is detectable at E18 and increases somewhat or remain at relatively constant levels during postnatal development (Belliveau et al., 1991). Similar temporal patterns were observed by ISHH, except that these mRNAs were evident at earlier ages and exhibited regional heterogeneity in onset of expression (Belliveau and Naus, 1995). In the striatum of rats, the levels of Cx26 was reported to be high at E18 after which it decreased dramatically by P6, whereas Cx32 gradually increased from P14 to adulthood; the cell types expressing these connexins at these developmental ages were not identified (Dermietzel et al., 1989). Similar postnatal patterns of expression were more recently reported by others, although expression of Cx26 and Cx43 was detected at embryonic ages as early as E12 (Nadarajah et al., 1997). By ISHH, Cx32 mRNA was detected in the somata of neurons, glial cells, and ependymal cells in various regions of 2-day-old rat brain (Matsumoto et al., 1991b). As well, Cx37 has been detected in brain during development and its expression found to be up to fivefold higher in embryonic than in adult brain (Willecke et al., 1991). A regional specific expression has been described for Cx31 in the developing hind brain of the mouse. By *in situ* ISHH, this connexin has been found to be expressed in a segmental pattern in the three and four rhombomers, which indicates a role in morphogenetic regulation of hindbrain development.

As in adult brain, it was suggested that message for Cx43 and Cx32 was localized to astrocytes and oligodendrocytes, respectively, which is consistent with immunohistochemical studies of these proteins in developing brain (Dermietzel et al., 1989; Yamamoto et al., 1992). *In vitro* data using immortalized mouse neuroblasts confirmed the early occurrence of Cx43 in undifferentiated neuronal precursor cells (Rozental et al., 1998). More interesting was the finding that with the progression of differentiation, a switch to Cx40 and Cx33 was detectable. Apparently, the pattern of connexin expression in differentiating neuroblasts is under the control of a morphogenetic program that regulates the expression of gap junction proteins in a temporal and spatial pattern.

## C.  Connexin43 Localization

In peripheral tissues, there is evidence that during development groups of cells are organized into what are called developmental fields or compartments containing cells coupled by gap junctions, while those at the compartment boundaries are not coupled with cells in adjacent compartments (Caveney, 1985). If such fields exist among astrocytes during development, it is then possible that the heterogeneities in Cx43 observed in adult rat brain may simply represent developmental patterns of glial cells that are maintained at the adult stage. This is unlikely at least as indicated by studies of Cx43 during development of rat brain. Biochemical results indicate that although both Cx43 protein and mRNA are detectable at late embryonic and early postnatal stages in rat brain (Belliveau et al., 1991), the levels of these increase markedly at 1 to 2 weeks after birth, which is a relatively late stage of development. Immunohistochemical analysis of Cx43 in the developing olfactory system of mouse also indicates the presence of Cx43, but at very low levels, as early as E9 with expression gradually increased to reach highest levels in postnatal and adult animals (Miragall et al., 1992). In a quantitative study of Cx43 in developing rat brain, it was reported that the number of Cx43-immunoreactive puncta in the striatum decrease somewhat 2 to 6 days after birth and thereafter remains at relatively constant levels up to adulthood (Dermietzel et al., 1989). In order to establish relationships between GJIC and cell proliferation in neural cells, Cx43 expression in conjunction with BrdU-labeling was examined in the rostral migratory stream of mouse telencelphalon and found to be inversely correlated (Miragall et al., 1997), which is consistent with similar findings in other tissues.

Coupling has also been examined in embryonic mouse ventricular zone where clusters of coupled cells consisted of radial glia and neural precursor cells, but not differentiating or migrating neurons (Bittman et al., 1997). Clusters of coupled cells were concentrated around individual radial glial cells, which is consistent with the observation that radial glia express Cx43 early during development (see later discussion, Yamamoto et al., 1992), and it was suggested that these cells may not only be an organizing center for cell migration, but also for propagation of signals via gap junctions formed among cells in the cluster. In contrast to the preceding results in the mouse telencephalon, analysis of relations between cell cycle and coupling indicated that proliferating cells were more likely to be junctionally coupled. Based on this finding together with observations of dynamic changes in coupling at different stages of the cell cycle, it was suggested that coupling in the ventricular zone early in development may promote cell cycle synchronization of adjacent cells within a cluster (Bittman et al., 1997).

Anatomical observations (Yamamoto et al., 1992) on postnatally developing rat brain indicate a clear developmental sequence in the emergence of immunolabeling for Cx43. In most brain areas, Cx43 first appeared as continuous labeling of radial glial cell processes, which was particularly striking in Bergmann glial cells. Immunoreactivity in radial glia disappeared with age and was replaced with immunolabeling of astrocytic processes as well as to a much lesser extent astrocytic cell bodies. As the numbers and immunolabeling intensity of such processes increased, the staining pattern transformed from stellate to exclusively punctate profiles, which is the pattern seen throughout the adult brain. This sequence occurred at different ages in different brain regions and was evident at earlier ages in structures that exhibited the greatest levels of Cx43 in adult brain. In other studies, it was reported that no dye coupling exits between radial glial cells in the neocortex of fetal rats (Lo Turco and Kriegstein, 1991) and that the number of gap junctions between processes forming the glial limitans does not reach adult values until several weeks after birth (Landis and Reese, 1981). Perhaps most revealing was an analysis of the appearance of dye coupling between astrocytes in the developing visual cortex, where it was found that coupling among these cells is first detected at P11 (Binmoller and Muller, 1992). Thus, it appears likely that Cx43 is produced in radial glia in preparation for the formation of gap junctions at their latter developmental stage as mature astrocytes. Moreover, these observations suggest that the regionally heterogeneous distribution of Cx43 observed in adult rat brain (Yamamoto et al., 1990b, 1990c), which begins to appear in most areas at about P10 to P15, does not reflect the earlier formation of gap junctionally coupled developmental fields between glial cells or their immediate precursors.

It appears rather that Cx43 production by astrocytes and the emergence of the characteristic distribution seen in adult brain may be geared to some other developmental event(s) than cell proliferation or early morphogenesis. Because the punctate staining seen in adult brain can be taken to represent the organization of Cx43 in fully functional astrocytic gap junctions for reasons elaborated elsewhere (Yamamoto et al., 1990c; Nagy et al., 1992), it was suggested that the developmental stage at which a preponderance of particularly punctate labeling first appears in a region reflects to some degree the functional maturity of that region. Based on the proposition (Yamamoto et al., 1990b; Nagy et al.,1992) that the expression of Cx43 by astrocytes in adult brain is dependent on neurons and regulated by them in accordance with local neuronal needs for astrocytic metabolic coupling and $K^+$ spatial buffering, it is reasonable that demands for glial GJIC would be greatest in areas with high levels of neuronal electrical activity. Thus, in general terms it was suggested (Yamamoto et al., 1992) that the time of onset of punctate Cx43 immunoreactivity in particular brain regions may reflect the maturation of local electrical activity, which includes not only the capacity of neurons to generate action potentials but also other transmembrane inhibitory or excitatory ionic currents. Based on this supposition, the early appearance of Cx43 puncta in, for example, the amygdaloid complex, septohypothalamic nucleus, preoptic hypothalamus, zona incerta, and subfornical organ suggests the early functional maturation of these areas.

## D.  Connexin30 Localization

The pattern of Cx30 development in brain parenchyma contrasts with the spatial and temporal characteristics of early Cx43 expression. Thus, Cx30 mRNA levels were reported to be very low for the first 2 postnatal weeks in developing mouse brain and did not reach adult levels until the end of the fourth postnatal week (Dahl et al., 1996). This is consistent with observations that Cx30 levels were still low at P15 in rat brain and increased by about 20-fold from this age to adulthood in thalamic and brain stem regions (Nagy et al., 1998). Further, unlike the diffuse labeling exhibited by Cx43 during the first postnatal week, Cx30 immunolabeling was already punctate at the time of its emergence. These differences provide reasons to consider that the Cx43 and Cx30 may have different functional roles at gap junctions between astrocytes in gray matter where Cx30 is expressed. As previously discussed (Nagy et al., 1998), such roles for Cx30 could be related to various developmental events that occur around the time that Cx30 appears, as follows:

1. Myelination, which is underway or progressed substantially by P15 and might require astrocytic Cx30 if astrocytic-oligodendrocytic gap junctions involving postmyelinating oligodendrocytes in gray matter contribute to the maturation of this process.

2. Formation of blood-brain barrier, which is complete at about P10 to P15. Although the barrier is not structurally composed of astrocytic endfeet, its establishment is coincident with the appearance of endfeet on blood vessels and roughly with the appearance of Cx30 at gap junctionally coupled astrocytic endfeet, where it may contribute to ion or metabolite dispersal.

3. Maturation of brain energy metabolism, which undergoes a transition from a reliance on ketone bodies in addition to glucose during the first 2 or 3 postnatal weeks to a primary reliance on aerobic glycolysis. These biochemical processes may require interglial exchange of particular metabolites, such as lactate, which may serve as major energy sources after this transition.

4. Compaction of brain extracellular space, which occurs maximally between P10 and P20 in rat (Lehmenkuhler et al., 1993) and places greater burdens on homeostatic processes mediated by astrocytic GJIC. It also coincides with development of astrocytic GJIC, as noted elsewhere (Muller et al., 1996) as well as with maximally increasing levels of Cx30.

### E.  Connexin26 and Connexin36 Localization

Expression of Cx26 has been described in the early embryonic brain around day E12 in the periventricular layer (Dermietzel et al., 1989). Postnatally, Cx26 decreases and remains confined to non-neuronal tissues (leptomeningeal cells, ependyma) and some specific neurosecretory areas (pinealocytes; see later discussion). In terms of temporal expression patterns of connexins in the developing brain, the peak of postnatal neuronal coupling (Peinado et al., 1993a, 1993b; Kandler and Katz, 1998) is of special interest. Because the strengthing of neuronal coupling in the poastnatal brain has been suggested to be involved in the modular organization of the neocortex, identification of the connexin(s) responsible for this coupling effect is of prime importance. From all connexins studied so far during this critical phase of neocorticogenesis, the recently cloned Cx36 seems to be the most likely candidate responsible for the columnar neuronal coupling as it shows a pattern of postnatal expression that matches perfectly with course of junctional coupling in the postnatal brain (Söhl et al., 1998).

## VIII. GAP JUNCTIONS BETWEEN OTHER CENTRAL NERVOUS SYSTEM CELL TYPES

Other cell types between which gap junctions have been described include tanycytes (Hatton and Ellisman, 1982), ependymal cells, leptomeningeal cells including pial as well as arachnoid cells (Dermietzel, 1975; Zenker et al., 1994; Haseg-awa et al., 1997; Nagy et al., 1997b), choroid epithelial cells (Dermietzel et al., 1977), and pinealocytes (Sáez et al., 1991). Cx43 appears in meninges and pitu-itary gland early in mouse development (Belliveau and Naus, 1995; Dahl et al., 1995). A thorough *in vitro* study using similar approaches as has been described here for astrocytes and oligodendrocytes was performed on leptomeningeal cells (Spray et al., 1991) and on pinealocytes (Sáez et al., 1991; Berthoud and Sáez, 1993). We will concentrate on these cellular constituents as further examples of well-characterized junctional coupling in brain tissues.

## A. Leptomeningeal Cells

Leptomeningeal cells in culture are even more strongly coupled than astrocytes. Spray and associates (1991) have suggested that these cells might provide a useful preparation in which to pursue the biochemical analysis of connexin biogenesis. Gap junctions between these cells do not show prominent voltage dependence, but display at least two types of junctional single channel events. This latter obser-vation is consistent with the cells' expression of two junctional proteins, Cx43 and Cx26. Various treatments with pharmacological agents and manipulation of the physiological conditions affecting junctional coupling have been performed on cultured leptomeningeal cells. Because these experiments are of general inter-est concerning metabolic regulation of junctional communication, we will describe them in more detail.

Conductance between pairs of leptomeningeal cells and dye coupling in conflu-ent cultures were decreased by exposure to $CO_2$-equilibrated saline, heptanol, or halothane. The mechanism of action of $CO_2$ is presumably through cytoplasmic acidification, which uncouples cells of most tissues. Although the pH sensitivity in this preparation was not quantified by measuring intracellular pH and gj simul-taneously, the variable effect seen in reversibility after total uncoupling is consis-tent with that expected for preparations expressing Cx43. Variable uncoupling in response to $CO_2$ exposure could also arise from highly efficient proton pumping by the cells; such pumping has been reported in glial cells and has been proposed as an explanation for lack of uncoupling in oligodendrocytes exposed to organic acids (Kettenmann et al., 1990).

Heptanol and halothane rapidly and reversibly uncoupled leptomeningeal cells, although the time course of the reversibility of heptanol action was longer than in some other tissues (e.g., hepatocytes, adult cardiac myocytes). The rapid and reversible uncoupling action of halothane in intact leptomeninges suggests that it may prove useful in determining possible roles of gap junctions in leptomeningeal cell function, assuming that its specificity for gap junction channels is as precise as in heart cells, where excitability is not grossly altered at concentrations that totally block impulse propagation. The potent uncoupling by halothane prompts the question of whether toxicity might be associated with gap junction channel closure during halothane anesthesia. Uncoupling might thus underlie the increased cerebrospinal pressure associated with use of halothane as an anesthetic, an effect that could arise from osmotic imbalance between uncoupled leptomeningeal cells. Junctional conductance between leptomeningeal cells is increased by a factor of two within 5 minutes after addition of membrane permeant cAMP or forskolin, and is strongly inhibited by addition of the PKC-activating phorbol esters. Junctional response to both sorts of stimuli are accompanied by rapid changes from flattened to rounder morphology, but whether this correlation is relevant to the junctional effects or represents an epiphenomenon is unknown.

## B. Ependymal Cells

Cx43 mRNA and protein has been localized to ependymal cells in brain and spinal cord (Dermietzel et al., 1989; Belliveau and Naus, 1995; Yamamoto et al., 1990b, 1990c; Ochalski et al., 1997). Most ependymal cells lining the spinal central canal displayed Cx43-positive gap junctions and apical cytoplasmic immunoreactivity. However, a small subset of these were devoid of Cx43 and formed asymmetrically labeled junctions with their immunostained neighbors. These unlabeled cells were presumed to be tanycytes based on the reported characteristics of this subpopulation of ependymal cells (Ochalski et al., 1997). The lack of labeling for Cx43 in endfeet abutting subependymal blood vessels is also consistent with the absence of this connexin in tanycytes. Although Cx43 has been observed in tanycytes of rat brain (Yamamoto et al., 1992), another connexin may be expressed in spinal cord tanycytes and contribute to their formation of gap junctions. The emerging perspective, supported as well by grid-mapped freeze fracture analysis, is one of a functional and perhaps regulated gap junction–supported syncytium from ependymal cells to pial glial limitans involving astrocytes and oligodendrocytes supporting neurons (Rash et al., 1997).

## C. Pericyte and Endothelial Cells

There is as yet little information on the nature and extent of GJIC between cells forming the brain vasculature. Although pericytes, the brain counterpart of peripheral vascular smooth muscle cells, and endothelial cells are largely separated by basal lamina in brain capillaries, these cells have been observed to make contact at some points in human brain and gap junctions have been found at these locations (Cuevas et al., 1984b). Gap junctions have also been observed between pericytes and endothelial cells in the basal forebrain of rat embryos where they have been suggested to play a role in the development of cerebral microcirculation (Fujimoto, 1995). Both Cx43 and Cx40 have been localized to rat brain arteriole endothelial and smooth muscle cells by immunohistochemistry (Little et al., 1995).

## D.   Pinealocytes

Pinealocytes are secretory active cells responsible for melatonin production and discharge of the pineal gland. Gap junctions have been morphologically identified between pinealocytes of various mammalian species (for review, see Vollrath, 1985). Because pinealocytes express the S-antigen (van Veen et al., 1986) and its expression has been first described in retinal photoreceptors, pinealocytes can be regarded as phylogenetic derivatives of photic receptors. Therefore, studies of cell-to-cell communication of this cell type give some clues, if not for neuronal coupling, for coupling between "nervelike" cells with neurosecretory activity. The most relevant observation of the study by Sáez and colleagues (1991) was that norepinephrine, the natural neuroeffector of the pineal gland, induces the expression of functional coupling between pinealocytes in culture. The pineal gland is an atypical neurosecretory system in which secretion appears to be controlled by the rate of synthesis of its hormone, melatonin. Gap junctions could allow pinealocytes to share metabolites and second messengers such as cAMP and $Ca^{2+}$, which are involved in regulating melatonin synthesis. Moreover, as not every pinealocyte receives independent innervation, intercellular communication through gap junctions among groups of pinealocytes could permit synchronization of secretory activity. The strength of nerve input, in this case the catecholaminergic tonus from the habenular nuclei and the basal forebrain, could thus coordinate the secretory efficacy of the gland. This phenomenon could also hold true for other neurosecretory nuclei in the brain. The actual outcome, however, of this neurotransmitter-related coupling effect may well depend on the type of connexins and neurotransmitters involved. As has been discussed for astrocytes (Giaume et al., 1991b) norepinephrine exerts an uncoupling effect in this cell type. Although the latter effect has been discussed to be linked to phospholipase C–dependent second messenger events and subsequent PKC activation or eicosanoid production, or both, this mechanism is unlikely to be the case for pinealocytes as Cx26 has no obvious phosphorylation sites (Zhang and Nicholson, 1989) and is not phosphorylated in hepatocytes (Traub et al., 1989) or in isolated liver junctions incubated with ATP and the catalytic subunit of cAMP-dependent protein kinase (Traub et al., 1989; Sáez et al., 1990), PKC, or $Ca^{2+}$/calmodulin-dependent protein kinase II. The upregulation of coupling between pinealocytes is more likely to rely on an increase of gap junction protein synthesis and not on gating processes, as blocking of the mRNA or protein synthesis prevents the positive effect of norepinephrine on the coupling strength. Although pineal astrocytes are abundant in rat by 7 days postpartum, only low levels of Cx43 are expressed at that time. That their levels achieve adult levels between 2 and 3 weeks after birth, which parallels innervation and functional maturation of the pineal gland, further suggests a role for these gap junctions in pineal gland activities (Berthoud and Sáez, 1993).

# IX.  CONCLUSIONS AND PROSPECTS

The conclusion that can be drawn from the preceding data and the data given in this chapter on the heterogeneity of connexin expression in brain tissues can be summarized as follows: (1) The different cell types of brain tissues exhibit a remarkable diversity in connexin expression. (2) Within a particular cell population, that is, astrocytes, neurons, and neuronal derivatives (i.e., pinealocytes), a high degree of quantitative as well as qualitative variability is prevalent which may correlate with the actual functional requirements of the particular cell types. (3) Connexin expression in neurons and glial cells is developmentally regulated. (4) The molecular diversity of connexin expression is suggested to reflect differences in metabolic and functional regulation of junctional conductivity, both at the level of biosynthesis and at the level of the channel sites. The cooperative utilization of various investigative methods, incorporating techniques from the areas of ultrastructural analysis as well as electrophysiology and molecular biology, provides promise in unveiling further details about gap junctional communication within brain tissues.

# REFERENCES

Adinolfi, A.M. (1969). The fine structure of neurons and synapses in the entopeduncular nucleus of the cat. J. Comp. Neurol. 135, 225-248.

Alonso, G., & Privat, A. (1993). Reactive astrocytes involved in the formation of lesional scars differ in the mediobasal hypothalamus and in other forebrain regions. J. Neurosci. Res. 34, 523-538.

Anders, J. (1988). Lactic acid inhibition of gap junctional intercellular communication in *in vitro* astrocytes as measured by fluorescence recovery after laser photobleaching. Glia 1, 371-379.

Anders, J., Niedermair, S., Ellis, E., & Salopek, M. (1990). Response of rat cerebral cortical astrocytes to freeze- or cobalt-induced injury: An immunocytochemical and Gap-FRAP study. Glia 3, 476-486.

Anders, J.J. & Salopek, M. (1989). Meningeal cells increase *in vitro* astrocytic gap junctional communication as measured by fluorescence recovery after laser photobleaching. J. Neurocytol. 18, 257-264.

Anders, J.J. & Woolrey, S. (1992). Microbeam laser-injected neurons increase *in vitro* astrocytic gap junctional communication as measured by fluorescence recovery after laser photobleaching. Lasers Surg. Med. 12, 51-62.

Andrew, R.D., MacVicar, B.A., Dudek, F.E., & Hatton, G.I. (1981). Dye transfer through gap junctions between neuroendocrine cells of rat hypothalamus. Science 211, 1187- 1189.

Andrew, R.D., Taylor, C.P., Snow, R.W., & Dudek, F.E. (1982). Coupling in rat hippocampal slices: dye transfer between CA1 pyramidal cells. Brain Res. Bull. 8, 211- 222.

Anzil, A.P., Blinzinger, K., & Matsushima, A. (1971). Dark cisternal fields: specialized formations of the endoplasmic reticulum in striatal neurons of a rat. Z. Zellforsch. 113, 553-557.

Anzini, P., Neuberg, D.H.-H., Schachner, M., Nelles, E., Willecke, K., Zielasek, J., Toyka, K.V., Suter, U., & Martini, R. (1997). Structural abnormalities and deficient maintenance of peripheral nerve myelin in mice lacking the gap junction protein connexin32. J. Neurosci. 17, 4545-4551.

Aoumari, A.E., Fromaget, C., Dupont, E., Reggio, H., Durbec, P., Briand, J.-P., Boller, K., Kreitman, B., & Gros, D. (1990). Conservation of a cytoplasmic carboxy-terminus domain of connexin43, a gap junctional protein, in mammal heart and brain. J. Membr. Biol. 115, 229-240.

Arasaki, K., Kudo, N., & Nakanishi, T. (1984). Firing of spinal motoneurones due to electrical interaction in the rat: An *in vitro* study. Exp. Brain Res. 54, 437-445.

Baimbridge, K.G., Peet, M.J., Mclennan, H., & Church, J. (1991). Bursting response to current-evoked depolariazation in rat CA1 pyramidal neurons is correlated with Lucifer yellow dye coupling but not with the presence of calbindin-D28K. Synapse 7, 269-277.

Baird, A., & Bohlen, B. (1990). Fibroblast growth factors. Handbook Exp. Pharm. 95/I, 369-418.

Baker, R., & Llinás, R. (1971). Electrotonic coupling between neurons in the rat mesencephalic nucleus. J. Physiol. (London) 212, 45-63.

Ball, A.K., & McReynolds, J.S. (1998). Localization of gap junctions and tracer coupling in retinal Muller cells. J. Comp. Neurol. 393, 48-57.

Balic-Gordon, R.J., Bone, L.J., Scherer, S.S. Functional gap junctions in Schwann cell myelin sheath. J. Cell Biol.142, 1095-1104.

Barnes, C.A., Rao, G., & McNaughton, B.L. (1987). Increased electrotonic coupling in aged rat hippocampus: a possible mechanism for cellular excitability changes. J. Comp. Neurol. 259, 549-558.

Batter, D.K., Corpina, R., Roy, C., Spray, D.C., Hertzberg, E.L., & Kessler, J.A. (1992). Heterogeneity in gap junction expression in astrocytes cultured from different brain regions. Glia 6, 213-221.

Belin, V., & Moos, F. (1986). Paired recordings from supraoptic and paraventricular oxytocin cells in suckled rats: recruitment and synchronization. J. Physiol. 377, 369-390.

Belliveau, D.J., Kidder, G.M., & Naus, C.C.G. (1991). Expression of gap junction genes during postnatal neural development. Dev. Gen. 12, 308-317.

Belliveau, D.J., & Naus, C.C.G. (1994). Cortical type 2 astrocytes are not dye coupled nor do they express the major gap junction genes found in the central nervous system. Glia, 12, 24-34.

Belliveau, D.J., & Naus, C.C.G. (1995). Cellular localization of gap junction mRNAs in developing rat brain. Dev. Neurosci. 17, 81-96.

Benardo, L.S., & Foster, R.E. (1986). Oscillatory behavior in inferior olive neurons: mechanism, modulation, cell aggregates. Brain Res. Bull. 17, 773-784.

Bennett, M.V.L. (1974). Flexibility and rigidity in electrotonically coupled systems. In: Synaptic Transmission and Neuronal Interaction. (Bennett, M.V.L., Ed.), pp. 153-178. Raven Press, New York.

Bennett, M.V.L. (1977). Electrical transmission: a functional analysis and comparison with chemical transmission. In: Handbook of Physiology, Section I: The Nervous System, Cellular Biology of Neurons (Kandel, E.R., ed.) vol. 1, pt. 1, pp. 357-416. American Physiological Society, Bethesda, MD.

Bennett, M.V.L., Barrio, T.A., Bargiello, T.A., Spray, D.C., Hertzberg, E., & Sáez, J.C. (1991). Gap junctions: new tools, new answers, new questions. Neuron 6, 305-320.

Bennett, M.V.L., & Goodenough, D.A. (1978). Gap junctions, electrotonic coupling and intercellular communication. Neurosci. Res. Prog. Bull. 16, 373-486.

Berger, T., Schnitzer, J., & Kettenmann, H. (1991). Developmental changes in the membrane current pattern, $K^+$-buffer capacity, and morphology of glial cells in the corpus callosum slice. J. Neurosci. 11, 3008-3024.

Bergmann, M., & Surchev, L. (1989). Freeze-Etching study of intercellular junctions in the rat developing neural tube. Acta Anat. 136, 12-15.

Bergoffen, J., Scherer, S.S., Wang, S., Scott, M.O., Bone, L.J., Paul, D.L., Chen, K., Lensch, M.W., Chance, P.F., & Fischbeck, K.H. (1993). Connexin mutations in X-linked Charcot-Marie-Tooth Disease. Science 262, 2039-2042.

Berridge, M.J., & Taylor, C.W. (1988). Inositol trisphosphate and calcium signaling. Cold Spring Harb. Symp. Quant. Biol. 53, 927-933.

Berthoud, V.M., & Sáez, J.C. (1993). Changes in connexin43, the gap junction protein of astrocytes, during development of the rat pineal gland. J. Pineal Res. 14, 67-72.

Bevan, S. (1990). Ion channels and neurotransmitter receptors in glia. Semin. Neurosci. 2, 467-481.

Beyer, E.C., Paul, D.L., & Goodenough, D.A. (1990). Connexin family of gap junction proteins. J. Membr. Biol. 116, 187-194.

Bhat, N.R., Zhang, P., & Hogan, E.L. (1995). Thrombin activates mitogen-activated protein kinase in primary astrocyte cultures. J. Cell. Physiol. 165, 417-424.

Binmoller, F.-J., & Muller, C.M. (1992). Postnatal development of dye-coupling among astrocytes in rat visual cortex. Glia 6, 127-137.

Bittman, K., Owens, D.F., Kriegstein, A.R., & LoTurco, J. (1997). Cell coupling and uncoupling in the ventricular zone of developing neocortex. J. Neurosci. 15, 7037-7044.

Blanc, E.M., Bruce-Keller, A.J., & Mattson, M.P. (1998). Astrocyte gap junctional communication decreases neuronal vulnerability to oxidative stress-induced disruption of $Ca^{2+}$ homeostasis and cell death. J. Neurochem. 70, 958-970.

Blankenfeld, G. von., Ransom, B.R., & Kettenmann, H. (1993). Development of cell-cell coupling among cells of the oligodendrocyte lineage. Glia 7, 322-328.

Bodian, D. (1972). Synaptic diversity and characterization by electron microscopy. In: Structure and Function of Synapses. (Pappas, G.D., & Purpura, D.P., Eds.), pp. 45-65. Raven Press, New York.

Bolanos, J.P., & Medina, J.M. (1996). Induction of nitric oxide synthase inhibits gap junction permeability in cultured rat astrocytes. J. Neurochem. 66, 2091-2099.

Bonner, P.H. (1988). Gap junctions form in culture between chick embryo neurons and skeletal muscle myoblasts. Dev. Brain Res. 38, 233-244.

Breckenridge, L.J., & Almers, W. (1987a). Currents through the fusion pore that forms during exocytosis of a secretory vesicle. Nature 329, 814-817.

Breckenridge, L.J., & Almers, W. (1987b). Final steps in exocytosis observed in a cell with giant secretory granules. Proc. Natl. Acad. Sci. U.S.A. 84, 1945-1949.

Brenan, A., & Matthews, B. (1983). Coupling between nerve fibres supplying normal and injured skin in the cat. J. Physiol. 334, 70-71P.

Brightman, M.W., & Reese, T.S. (1969). Junctions between intimately apposed cell membranes in the vertebrate brain. J. Cell Biol. 40, 648-677.

Bruzzone, B., & Ressot, C. (1997). Connexins, gap junctions and cell-cell signalling in the nervous system. Eur. J. Neurosci. 9, 1-6.

Bruzzone, B., White, T.W., & Paul, D.L. (1996a). Connections with connexins: the molecular basis of direct intercellular signalling. Eur. J. Biochem. 238, 1-27.

Bruzzone, B., White, T.W., & Goodenough, D.A. (1996b). The cellular internet: on-line with connexins. BioEssays 18, 709-718.

Bunge, R.P., Bunge, M.B., & Peterson, E.R. (1965). An electron microscope study of cultured rat spinal cord. J. Cell Biol. 24, 163-191.

Burnard, D.M., Crichton, S.A., & MacVicar, B.A. (1990). Electrophysiological properties of reactive glial cells in the kainate-lesioned hippocampal slice. Brain Res. 510, 43-52.

Burt, J.M., & Spray, D.C. (1988). Inotropic agents modulate gap junctional conductance between cardiac myocytes. Am. J. Physiol. 254, H1206-H1210.

Carr, C.E., & Maler, L. (1986). Electroreception in gymnotiform fish: central anatomy and physiology. In: Electroreception. (Bullock, T.H. & Heiligenberg, W., Eds.), pp. 319-373. John Wiley and Sons, New York.

Caveney, S. (1985). The role of gap junctions in development. Annu. Rev. Physiol. 47, 319-335.

Cepeda, C., Walsh, J.P., Hull, C.D., Howard, S.G., Buchwald, N.A., & Levine, M.S. (1989). Dye-coupling in the neostriatum of the rat: I. Modulation by dopamine-depleting lesions. Synapse 4, 229-237.

Chandross, K.J. (1998). Nerve injury and inflammatory cytokines modulate gap junctions in the peripheral nervous system. Glia 24, 21-31.

Chandross, K.J., Spray, D.C., Cohen, R.I., Kumar, N.M., Kremer, M., Dermietzel, R., & Kessler, J.A. (1996). TNF alpha inhibits Schwann cell proliferation, connexin46 expression, and gap junctional communication. Mol. Cell. Neurosci. 7, 479-500.195-201.

Chan-Palay, V., & Palay, S.L. (1972). The form of velate astrocytes in the cerebellar cortex of monkey and rat: high voltage electron microscopy of rapid Golgi preparations. Z. Anat. Entwickl.-Gesch. 138, 1-19.

Charles, A.C. (1994). Glia-neuron intercellular calcium signaling. Dev. Neurobiol. 16, 196-206.

Charles, A.C. (1998). Intercellular calcium waves in glia. Glia 24, 39-49.

Charles, A.C., Dirksen, E.R., Merrill, J.E., & Sanderson, M.J. (1993). Mechanisms of intercellular calcium signaling in glial cells studies with dantroline and thapsigargin. Glia 7, 134-145.

Charles, A.C., Kodali, S.K., & Tyndale, R.F. (1996). Intercellular calcium waves in neurons. Mol. Cell Neurosci. 7, 337-53.

Charles, A.C., Merrill, J.E., Dirksen, E.R., & Sanderson, M.J. (1991). Intercellular signaling in glial cells: calcium waves and oscillations in response to mechanical stimulation and glutamate. Neuron 6, 983-992.

Charles, A.C., Naus, C.C.G., Zhu, D., Kidder, G.M., Dirksen, E.R., & Sanderson, M.J. (1992). Intercellular calcium signaling via gap junctions in glioma cells. J. Cell Biol. 118,

Chow R.H., von Ruden, L., & Neher, E. (1992). Delay in vesicle fusion revealed by electrochemical monitoring of single secretory events in adrenal chromaffin cells. Nature 356, 60-63.

Christie, M.J., Williams, J.T., & North, R.A. (1989). Electrical coupling synchronizes subthreshold activity in locus coeruleus neurons *in vitro* from neonatal rats. J. Neurosci. 9, 3584-3589.

Church, J. & Baimbridge, K.G. (1991). Exposure to high-pH medium increases the incidence and extent of dye coupling between rat hippocampal CA1 pyramidal neurons *in vitro*. J. Neurosci. 11, 3289-3295.

Cobbett, P., Smithson, K.G., & Hatton, G.I. (1985). Dye-coupled magnocellular peptidergic neurons of the rat paraventricular nucleus show homotypic immunoreactivity. Neuroscience 16, 885-895.

Cobbett, P., Yang, Q.Z., & Hatton, G.I. (1987). Incidence of dye coupling among magnocellular paraventricular nucleus neurons in male rats is testosterone dependent. Brain Res. Bull. 18, 365-370.

Collins, W.F., & Erichsen, J.T. (1988). Direct excitatory interactions between rat penile motoneurons. Soc. Neurosci. Abst. 14, Part 1, p.181.

Condorelli, D.F., Parenti, R., Spinella, F., Salinaro, T., Belluardo, N., Cardile, V., & Cicirata. F. (1998). Cloning of a new gap junction gene (Cx36) highly expressed in mammalian neurons. Eur. J. Neurosci. 10, 1202-1208.

Connors, B.W., Benardo, L.S., & Prince, D.A. (1983). Coupling between neurons of the developing rat neocortex. J. Neurosci. 3, 773-782.

Connors, B.W., Benardo, L.S., & Prince, D.A. (1984). Carbon dioxide sensitivity of dye coupling among glia and neurons of the neocortex. J. Neurosci. 4, 1324-1330.

Cook, J.E., & Becker, D.L. (1995). Gap junctions in the vertebrate retina. Microsc. Res. Tech. 31, 408-419.

Cornell-Bell, A.H., & Finkbeiner, S.M. (1991). $Ca^{2+}$ waves in astrocytes. Cell Calcium 12, 185-204.

Cornell-Bell, A.H., Finkbeiner, S.M., Cooper, M.S., & Smith, S.J. (1990). Glutamate induces calcium waves in cultured astrocytes: long-range glial signaling. Science 247, 470-473.

Cotrina, M.L., Kang, J., Lin, J.H.-C., Bueno, E., Hansen, T.W., He, L., Liu, Y., & Nedergaard, M. (1998). Astrocytic gap junctions remain open during ischemic conditions. J. Neurosci. 18, 2520-2537.

Cravatt, B.F., Giang, D.K., Mayfield, S.P., Goger, D.L., Lerner, R.A., & Gilula, N.B. (1996). Molecular characterization of an enzyme that degrades neuromodulatory fatty-acid amides. Nature 384, 83-87.

Cuevas, P., Diaz, J.A., Reimers, D., Dujovny, M., Diaz, F.D., & Ausman, J.I. (1984a). Aspects of interastrocytic gap junctions in blood-brain barrier in the experimental penumbra area, revealed by transmission electron microscopy and freeze-fracture. Experientia 40, 5-7.

Cuevas, P., Gutierrez-Diaz, J.A., Reimers, D., Dujovny, M., Diaz, F.G., & Ausman, J.I. (1984b). Pericyte endothelial gap junctions in human cerebral capillaries. Anat. Embryol. 170, 155-159.

Dacheux, R.F., & Raviola, E. (1986). The rod pathway in the rabbit retina: a depolarizing bipolar and amacrine cell. J. Neurosci. 6, 331-345.

Dahl, E., Manthey, D., Chen, Y., Schwarz, H.-J., Chang, S., Lalley, P.A., Nicholson, B.J., & Willecke, K. (1996). Molecular cloning and functional expression of mouse connexin30, a gap junction gene highly expressed in adult brain and skin. J. Biol. Chem. 271, 17903-17910.

Dahl, E., Winterhager, E., Traub, O., & Willecke, K. (1995). Expression of gap junction genes, connexin40 and connexin43, during fetal mouse development. Anat. Embryol. (Berlin) 191, 267-278.

Dani, J.W., Chernjavsky, A., & Smith, S.J. (1992). Neuronal activity triggers calcium waves in hippocampal astrocyte networks. Neuron 8, 429-440.

Dean, J.B., Huang, R.Q., Erlichman, J.S., Southard, T.L., & Hellard, D.T. (1997). Cell-cell coupling occurs in dorsal medullary neurons after minimizing anatomical-coupling artifacts. Neuroscience 80, 21-40.

Dermietzel, R. (1974). Junctions in the central nervous system of the cat: Gap junctions and membrane-associated orthogonal particle complexes (MOPC) in astrocytic membranes. Cell Tissue Res. 149, 121-135.

Dermietzel, R. (1975). Junctions in the central nervous system of the cat. V. The junctional complex of the pia-arachnoid membrane. Cell Tissue. Res. 164, 309-329.

Dermietzel, R. (1978). The oligodendrocytic junctional complex. Cell Tissue Res. 193, 61-72.

Dermietzel, R. (1996). Molecular diversity and plasticity of gap junctions in the nervous system. In: Gap Junctions in the Nervous System. (Spray, D.C., & Dermietzel, R., Eds), pp.13-38. Landes, Austin.

Dermietzel, R., Farooq, M., Kessler, J.A., Althaus, H., Hertzberg, E.L., & Spray, D.C. (1997). Oligodendroytes express gap junction proteins connexin32 and connexin45. Glia 20, 101-114.

Dermietzel, R., Hertzberg, E.L., Kessler, J.A., & Spray, D.C (1991). Gap junctions between cultured astrocytes: immunocytochemical, molecular, and electrophysiological analysis. J. Neurosci. 11, 1421-1432.

Dermietzel, R., Hwang, T.K., & Spray, D.S. (1990). The gap junction family: structure, function and chemistry. Anat. Embryo. 182, 517-528.

Dermietzel, R., Meller, K., Tetzlaff, W., & Waelsch, M. (1977). *In vivo* and *in vitro* formation of the junctional complex in choroid epithelium. Cell Tiss. Res. 181, 427-441.

Dermietzel, R., & Spray, D.C. (1993). Gap junctions in the brain. Where, what type, how many, and why? TINS 16, 186-192.

Dermietzel, R., Traub, O., Hwang, T.K., Beyer, E., Bennett, M.V.L., Spray, D.C., & Willecke, K. (1989). Differential expression of three gap junction proteins in developing and mature brain tissues. Proc. Natl. Acad. Sci. U.S.A. 86, 10148-10152.

DeVries, S.H., & Schwartz, E.A. (1989). Modulation of an electrical synapse between solitary pairs of catfish horizontal cells by dopamine and second messengers. J. Physiol. 414, 351-375.

De Zeeuw, C.I., Holstege, J.C., Ruigrok, T.J.H., & Voogd, J. (1989). Ultrastructural study of the GABAergic, cerebellar, and mesodiencephalic innervation of the cat medial accessory olive: anterograde tracing combined with immunocytochemistry. J. Comp. Neurol. 284, 12-85.

Dowling, J.E. (1991). Retinal neuromodulation: the role of dopamine. Vis. Neurosci. 7, 87-97.

Dudek, F.E., Andrew, R.D. MacVicar, B.A., & Hatton, G.I. (1982). Intracellular electrophysiology of mammalian peptidergic neurons in rat hypothalamic slices. Fed. Proc. 41, 2953-2958.

Dudek, F.E., Andrew, R.D., MacVicar, B.A., Snow, R.W., & Taylor, C.P. (1983). Recent evidence for and possible significance of gap junctions and electrotonic synapses in the mammalian brain. In: Basic Mechanisms of Neuronal Hyperexcitability. (Jasper, H.H., & van Gelder, N.M., Eds.), pp. 31-73. Alan R. Liss, New York.

Dudek, F.E., Gribkoff, V.K., Olson, J.E., & Hertzberg, E.L. (1988). Reduction of dye coupling in glial cultures by microinjection of antibodies against the liver gap junction polypeptide. Brain Res. 439, 275-280.

Dupont, E., Aoumari, A.E., Fromaget, C., Briand, J.-P., & Gros, D. (1991). Affinity purification of a rat brain junctional protein, connexin43. Eur. J. Biochem. 200, 263-270.

Elfgang, C., Eckert, R., Lichtenberg-Frate, H., Butterweck, A., Traub, O., Klein, R.A., Hulser, D.F., & Willecke, K. (1995). Specific permeability and selective formation of gap junction channels in connexin-transfected Hela cells. J. Cell Biol. 129, 805-817.

Elisevich, K., Rempel, S.A., Smith, B.J., & Edvardsen, K. (1997). Hippocampal connexin43 expression in human complex partial seizure disorder. Exptl. Neurol. 145, 154-164.

Enkvist, M.O.K., & McCarthy, K.D. (1992). Activation of protein kinase C blocks astroglial gap junction communication and inhibits the spread of calcium waves. J. Neurochem. 59, 519-526.

Enkvist, M.O.K., & McCarthy, K.D. (1994). Astroglial gap junction communication is increased by treatment with either glutamate or high $K^+$ concentration. J. Nuerochem. 62, 489-495.

Ennes, H.S., Young, S.H., Raybould, H.E., & Meyer, E.A. (1997). Intercellular communication between dorsal root ganglion cells and colonic smooth muscle cells in vitro. Neuroreport 8, 733-737.

Finkbeiner, S.M. (1992). Calcium waves in astrocytes - filling in the gaps. Neuron 8, 1101-1108.

Finkbeiner, S.M. (1993). Glial calcium. Glia 9, 83-104.

Fischbach, G.D. (1972). Synapse formation between dissociated nerve and muscle cells in low density cell cultures. Dev. Biol. 38, 407-429.

Fischer, G., & Kettenmann, H. (1985). Cultured astrocytes form a syncytium after maturation. Exp. Cell Res. 159, 273-279.

Fisher, S.K., & Linberg, K.A. (1975). Intercellular junctions in the human embryonic retina. J. Ultrastruc. Res. 51, 69-78.

Froes, M.M. & Campos de Carvalho, A.C. (1998). Gap junction-mediated loops of neuronal-glial interactions. Glia 24, 97-107.

Fujimoto, K. (1995). Pericyte-endothelial gap junctions in developing rat cerebral capillaries: a fine structural study. Anat. Rec. 242, 562-565.

Fulton, B.P. (1995). Gap junctions in the developing nervous system [review]. Perspect. Dev. Neurobiol. 2, 327-334.

Fulton, B.P., Miledi, R., & Takahashi, T. (1980). Electrical synapses between motoneurons in the spinal cord of the newborn rat. Proc. R. Soc. Lond. B. 208, 115- 120.

Furshpan, E.J., & Potter, D.D. (1959). Transmission at the giant motor synapses of the crayfish. J. Physiol. 145, 289-325.

Futamachi. K.J., & Pedley, T.A. (1976). Glial cells and extracellular potassium: their relationship in mammalian cortex. Brain Res. 109, 311-322.

Galambos. R. (1989). Electrogenesis of evoked potentials. In: Springer Series in Brain Dynamics 2. (Basar, E., & Bullock, T.H., Eds.), pp. 13-25. Springer-Verlag, Berlin Heidelberg.

Gardner-Medwin, A.R. (1986). A new framework for assessment of potassium-buffering mechanisms. Ann. N.Y. Acad. Sci. 481, 287-301.

Giaume, C., Cordier, J., & Glowinski, J. (1992). Endothelins inhibit junctional permeability in cultured mouse astrocytes. Eur. J. Neurosci. 4, 877-881.

Giaume, C., Fromaget, C., Aoumari, A.E., Cordier, J., Glowinski, J., & Gros, D. (1991a). Gap junctions in cultured astrocytes: single-channel currents and characterization of channel-forming protein. Neuron 6, 133-143.

Giaume, C., Marin, P., Cordier, J., Glowinski, J., & Premont, J. (1991b). Adrenergic regulation of intercellular communications between cultured striatal astrocytes from the mouse. Proc. Natl. Acad. Sci. U.S.A. 88, 5577-5581.

Giaume, C., & McCarthy, K.D. (1996). Control of gap junctional communication in astrocytic networks. Trends Neurosci. 8, 319-325.

Giaume, C., Tabernero, A., & Medina, J.M. (1997). Metabolic trafficking through astrocytic gap junctions. Glia 21, 114-123.

Giaume, C., & Venance, L. (1995). Gap junctions in brain glial cells and development. Perspect. Dev. Neurobiol. 2, 335-345.

Giaume, C. & Venance, L. (1998). Intercellular calcium signaling and gap junctional communication in astrocytes. Glia 24, 50-64.

Glaum, S.R., Holzwarth, J.A., & Miller, R.J. (1990). Glutamate receptors activate $Ca^{2+}$ mobilization and $Ca^{2+}$ influx into astrocytes. Proc. Natl. Acad. Sci. U.S.A. 87, 3454-3458.

Glowinski, J., Marin, P., Tence, M., Stella, N., Giaume, C., & Premont, J. (1994). Glial receptors and their intervention in astrocyto-astrocytic and astrocyto-neuronal interactions. Glia 11, 201-208.

Gogan, P., Gueritaud, J.P., Horcholle-Bossavit, G., & Tyc-Dumont, S. (1974). Electrotonic coupling between motoneurones in the abducens nucleus of the cat. Exp. Brain Res. 21, 139-154.

Gogan, P., Gueritaud, J.P., Horcholle-Bossavit, G., & Tyc-Dumont, S. (1977). Direct excitatory interactions between spinal motoneurons of the cat. J. Physiol. (Lond.) 272, 755-767.

Goodenough, D.A., Goliger, J.A., & Paul, D.L. (1996). Connexins, connexons and intercellular communication. Ann. Rev. Biochem. 65, 475-502.

Gotow, T., & Sotelo, C. (1987). Postnatal development of the inferior olivary complex in the rat: IV. Synaptogenesis of GABAergic afferents, analyzed by glutamic acid decarboxylase immunocytochemistry. J. Comp. Neurol. 263, 526-552.

Grace, A.A., & Bunney, B.S. (1983). Intracellular and extracellular electrophysiology of nigral dopaminergic neurons—3. Evidence for electrotnic coupling. Neuroscience 10, 333-348.

Grothe, C., Zachmann, K., & Unsicker, K. (1991). Basic FGF-like immunoreactivity in the developing and adult rat brainstem. J.Comp.Neurol. 305, 328-336.

Gulley, R.L., & Wood, R.L. (1971). The fine structure of the neurons in the rat substantia nigra. Tissue and Cell 3, 675-690.

Gutnick, M.J., Connors, B.W., & Ransom, B.R. (1981). Dye-coupling between glial cells in the guinea pig neocortical slice. Brain Res. 213, 486-492.

Gutnick, M.J., Lobel-Yaakov, R., & Rimon, G. (1985). Incidence of neuronal dye-coupling in neocortical slices depends on the plane of section. Neuroscience 15, 659-666.

Gutnick, M.J., & Prince, D.A. (1981). Dye coupling and possible electrotonic coupling in the guinea pig neocortical slice. Science 211, 67-70.

Gwyn, D.G., Nicholson, G.P., & Flumerfelt, B.A. (1977). The inferior olivary nucleus of the rat: A light and electron microscopic study. J. Comp. Neurol. 174, 489-520.

Hampson, E.C.G.M., Vaney, D.I., & Weiler, R. (1992). Dopaminergic modulation of gap junction permeability between amacrine cells in mammalian retina. J. Neurosci. 12, 4911-4922.

Hansson, E. (1988). Astroglia from defined brain regions as studied with primary cultures. Prog. Neurobiol. 30, 369-397.

Haring, J.H., Yan, W., & Faber, K.M. (1997). Neuronal dye coupling in the developing rat fascia dentata. Devel. Brain Res. 103, 205-208.

Hasegawa, M., Yamashima, T., Kida, S., & Yamashita, J. (1997). Membranous ultrastructure of human arachnoid cells. J. Neuropathol. Exp. Neurol. 56, 1217-1227.

Hassinger, T.D., Atkinson, P.B., Strecker, G.J., Whalen, L.R., Dudek, R.E., Kossel, A.H., & Kater, S.B. (1995). Evidence for glutamate-mediated activation of hippocampal neurons by glial calcium waves. J. Neurobiol. 28, 159-170.

Hassinger, T.D., Guthrie, P.B., Atkinson, P.B., Bennett, M.V.L., & Kater, S.B. (1996). An extracellular signaling component in propagation of astrocytic calcium waves. Proc. Natl. Acad. Sci. U.S.A. 93, 13268-13273.

Hatton, G.I. (1983). The hypothalamic slice approach to neuroendocrinology. Quart. J. Exp. Physiol. 68, 483-489.

Hatton, G.I. (1990). Emerging concepts of structure-function dynamics in adult brain: the hypothalamo-neurohypophysial system. Prog. Neurobiol. 34, 437-504.

Hatton, J.D., & Ellisman, M.H. (1982). The distribution of orthogonal arrays in the freeze-fractured rat median eminence. J. Neurocytol. 11, 335-349.

Hatton, G.I., Modney, B.K., & Salm, A.K. (1992a). Increases in dendritic bundling and dye coupling of supraoptic neurons after the induction of maternal behavior. Ann. N.Y. Acad. Sci. 652, 142-155.

Hatton, G.I., Yang, Q.Z., & Cobbett, P. (1987). Dye coupling among immunocytochemically identified neurons in the supraoptic nucleus: increased incidence in lactating rats. Neuroscience 21, 923-930.

Hatton, G.I., Yang, Q.Z., & Koran, L.E. (1992b). Effects of ovariectomy and estrogen replacement on dye coupling among rat supraoptic nucleus neurons. Brain Res. 572, 291-295.

Hatton, G.I., Yang, Q.Z., & Smithson, K.G. (1988). Synaptic inputs and electrical coupling among magnocellular neuroendocrine cells. Brain Res. Bull. 20, 751-755.

Hedden, W.L., & Dowling, J.E. (1978). The interplexiform cell system. II. Effects of dopamine on goldfish retinal neurones. Proc. R. Soc. Lond. 201, 27-55.

Henkart, M., Landis, D.M.D., & Reese, T.S. (1976). Similarity of junctions between plasma membranes and endoplasmic reticulum in muscle and neurons. J. Cell Biol. 70, 338-347.

Hennemann, H., Suchyna, T., Lechtenberg-Frate, H., Jungbluth, S., Dahl, E., Schwarz, J., Nicholson, B.J., & Willecke, K. (1992). Molecular cloning and functional expression of mouse connexin40, a second gap junction gene preferentially expressed in lung. J. Cell Biol. 117, 1299-1310.

Herndon, R.M. (1964). Lamellar bodies, an unusual arrangement of the granular endoplasmic reticulum. J. Cell Biol. 20, 328-342.

Hertzberg, E.L., Lawrence, T.S., & Gilula, N.B. (1981). Gap junctional communication. Ann. Rev. Physiol. 43, 479-491.

Hervas, J.-P., & Lafarga, M. (1979). Subsurface cisterns in paraventricular nuclei of the hypothalamus of the rat. Cell Tiss. Res. 199, 271-279.

Hinrichsen, C.F.L., & Larramendi, L.M.H. (1968). Synapses and cluster formation of the mouse mesencephalic fifth nucleus. Brain Res. 7, 296-299.

Hofer, A., Sáez, J.C., Chang, C.C., Trosko, J.E., Spray, D.C., & Dermietzel, R. (1996). C-erbB2/neu transfection induces gap junctional communication incompetence in glial cells. J. Neurosci. 16, 4311-4321.

Horikawa, K., & Armstrong, W.E. (1988). A versatile means of intracellular labeling: injection of biocytin and its detection with avidin conjugates. J. Neurosci. Methods 25, 1-11.

Hossain, M.Z., Ernst, L.A., & Nagy, J.I. (1995). Utility of intensely fluorescent cyanine dyes (CY3) for assays of gap junctional communication by dye-transfer. Neurosci. Lett. 184, 71-74.

Hossain, M.Z., Murphy, L.J., Hertzberg, E.L., & Nagy, J.I. (1994a). Phosphorylated forms of connexin43 predominate in rat brain: Demonstration by rapid inactivation of brain metabolism. J. Neurochem. 62, 2394-2403.

Hossain, M.Z., Peeling, J., Sutherland, G.R., Hertzberg, E.L., & Nagy, J.I. (1994c). Ischemia-induced cellular redistribution of the astrocytic gap junctional protein connexin43 in rat brain, Brain Res. 652 311-322.

Hossain, M.Z., Sawchuk, M.A., Murphy, L.J., Hertzberg, E.L., & Nagy, J.I. (1994b). Kainic acid induced alterations in antibody recognition of connexin43 and loss of astrocytic gap junctions in rat brain, Glia 10, 250-265.

Huang, R.Q., Erlichman, J.S., & Dean, J.B. (1997). Cell-cell coupling between CO2-excited neurons in the dorsal medulla oblongata. Neuroscience 80, 41-57.

Irvine, R.F. (1989). How do inositol 1,4,5-trisphosphate and inositol 1,3,4,5-tetrakisphosphate regulate intracellular $Ca^{2+}$? Biochem. Soc. Trans. 17, 6-8.

Irvine, R.F., & Moor, R.M. (1987). Inositol (1,3,4,5) tetrakisphosphate-induced activation of sea urchin eggs requires the presence of inositol trisphosphate. Biochem. Biophys. Res. Commun. 146, 284-290.

Irvine, R.F., Moor, R.M., Pollock, W.K., Smith, P.M., & Wreggett, K.A. (1988). Inositol phosphates: proliferation, metabolism and function. Phil. Trans. R. Soc. Lond. B320, 181-198.

Iwata, H., Matsuyama, A., Okumura, N., Yoshida, S., Lee, Y., Imaizumi, K., & Shiosaka, S. (1991). Localization of basic FGF-like immunoreactivity in the hypothalamo-hypophyseal neuroendocrine axis. Brain Res. 550, 329-332.

Jefferys, J.G.R., & Haas, H.L. (1982). Synchronized bursting of CA1 hippocampal pyramidal cells in the absence of synaptic transmission. Science 300, 448-450.

Jensen, A.M., & Chiu, S.Y. (1990). Fluorescence measurements of changes in intracellular calcium induced by excitatory amino acids in cultured cortical astrocytes. J. Neurosci. 10, 1165-1175.

Jiang, J.X., & Goodenough, D.A. (1996). Heteromeric connexons in lens gap junction channels. Proc. Natl. Acad. Sci. U.S.A. 93, 1287-1291.

Jiang, Z.-G., Yang, Y.-Q., & Allen, C.N. (1997). Tracer and electrical coupling of rat suprachiasmatic nucleus neurons. Neuroscience 77, 1059-1066.

Kadle, R., Zhang, J.T., & Nicholson, B.J. (1991). Tissue-specific distribution of differentially phosphorylated forms of Cx43. Mol. Cell. Biol. 11, 363-369.

Kandler, K., & Katz, L.C. (1995). Neuronal coupling and uncoupling in the developing nervous system. Curr. Opin. Neurobiol. 5, 98-105.

Kandler, K., & Katz, L.C. (1998). Coordination of neuronal activity in developing visual cortex by gap junction-mediated biochemical communication. J. Neurosci. 18, 1419-1427.

Kardami, E., Stoski, R.M., Doble, B.W., Yamamoto, T., Hertzberg, E.L., & Nagy, J.I. (1991). Biochemical and ultrastructural evidence for the association of basic fibroblast growth factor with cardiac gap junctions. J. Biol. Chem. 226, 19551-19557.

Karwoski, C.J., Lu, H.-K., & Newman, E.A. (1989). Spatial buffering of light-evoked potassium increases by retinal Muller (glial) cells. Science 244, 578-580.

Kaspar, E.M., Larkman, A.U., Lubke, J., & Blakemore, C. (1994). Pyramidal neurons in layer 5 of the rat visual cortex. II. Development of electrophysiological properties. J. Comp. Neurol. 339, 475-494

Kasuya, Y., Abe, Y., Hama, H., Sakurai, T., Asada, S., Masaki, T., & Goto, K. (1994). Endothelin-1 activates mitogen-activated protein kinases through two independent signalling pathways in rat astrocytes. Biophys. Biochem. Res. Commun. 204, 1325-1333.

Katsumaru, H., Kosaka, T., Heizmann, C.W., & Hama, K. (1988a). Immunocytochemical study of GABAergic neurons containing the calcium-binding protein parvalbumin in the rat hippocampus. Exp. Brain Res. 72, 347-362.

Katsumaru, H., Kosaka, T., Heizmann, C.W., & Hama, K. (1988b). Gap junctions on GABAergic neurons containing the calcium-binding protein parvalbumin in the rat hippocampus (CA1 region). Exp. Brain Res. 72, 363-370.

Kettenmann, H., Orkand, R.K., & Schachner, M. (1983). Coupling among identified cells in mammalian nervous system cultures. J. Neurosci. 3, 506-516.

Kettenmann, H., & Ransom, B.R. (1988). Electrical coupling between astrocytes and between oligo-dendrocytes studied in mammalian cell cultures. Glia 1, 64-73.

Kettenmann, H., Ransom, B.R., & Schlue, W.R. (1990). Intracellular pH shifts capable of uncoupling cultured oligodendrocytes are seen in low HCO3- solution. Glia 3, 110-117.

Kim, W.T., Rioult, M.G., & Cornell-Bell, A.H. (1994). Glutamate-induced calcium signaling in astro-cytes. Glia 11, 173-184.

King, J.S. (1976). The synaptic cluster (glomerulus) in the inferior olivary nucleus. J. Comp. Neurol. 165, 387-400.

Kita, H., & Armstrong, W.E. (1991). A biotin-containing compound N-(2-aminoethyl) biotinamide for intracellular labeling and neuronal tracer studies: comparison with biocytin. J. Neurosci. Methods 37, 141-150.

Knowles, W.D., Funch, P.G., & Schwartzkroin, P.A. (1982). Electrotonic and dye coupling in hippoc-ampal CA1 pyramidal cells *in vitro*. Neuroscience 7, 1713-1722.

Kolb, H. (1979). The inner plexiform layer in the retina of the cat: electron microscopic observations. J. Neurocytol. 8, 295-329.

Konietzko, U., & Muller, C.M. (1994). Astrocytic dye coupling in rat hippocampus: topography, developmental onset, and modulation by protein kinase C. Hippocampus 4, 297-306.

Korn, H., & Farber, D.S. (1979). Electrical interactions between vertebrate neurons: field effects and electrotonic coupling. In: The Neurosciences: Fourth Study Program. (Schmitt, F.O., & Worden, F.G., Eds.), pp. 333-358. MIT Press, Cambridge.

Korn, H., Sotelo, C., & Crepel, F. (1973). Electrotonic coupling between neurons in the rat lateral ves-tibular nucleus. Exp. Brain Res. 16, 255-275.

Kosaka, T. (1983a). Gap junctions between non-pyramidal cell dendrites in the rat hippocampus (CA1 and CA3 regions) Brain Res. 271, 157-161.

Kosaka, T. (1983b). Neuronal gap junctions in the polymorph layer of the rat dentate gyrus. Brain Res. 277, 347-351.

Kosaka, T., & Hama, K. (1985). Gap junctions between non-pyramidal cell dendrites in the rat hippoc-ampus (CA1 and CA3 regions): a combined golgi-electron microscopy study. J. Comp. Neu-rol. 231, 150-161.

Koval, M., Geist, S.T., Westphale, E.M., Kemendy, A.E., Civitelli, R., Beyer, E.C., & Steinberg, T.H. (1995). Transfected connexin45 alters gap junction permeability in cells expressing endoge-nous connexin43. J. Cell Biol. 130, 987-995.

Kriegler, S., & Chiu, S.Y. (1993). Calcium signaling of glial cells along mammalian axons. J. Neuro-sci. 13, 4229-4245.

Kuffler, S.W., & Nicholls, J.G. (1977). *From Neuron to Brain*. Sinauer Associates, Sunderland, MA.

Kumar, N.M., & Gilula, N. (1996). The gap junction communication channel. Cell 84, 381-388.

Kunzelmann, P., Blümcke, I., Traub O., Dermietzel, R., & Willecke, K. (1997). Coexpression of connexin45 and -32 in oligodendrocytes of rat brain. J. Neurocytol. 26, 17-22.

Kunzelmann, P., Schröder, W., Traub, O., Steinhäuser, C., Dermietzel, R., Willecke, K. (1999). Late onset and increasing expression of the gap junction protein connexin30 in adult murine brain and long-term cultured astrocytes. Glia 25, 111-119.

Lafarga, M., Berciano, M.T., Saurez, I., Andres, M.A., & Berciano, J. (1993). Reactive astroglia-neu-ron relationships in the human cerebellar cortex: aquantitative, morphological and immunocy-tochemical study in Creutzfelt-Jakob disease. Int. J. Dev. Neurosci. 11, 199-213.

Landis, D.M.D., & Reese, T.S. (1981). Membrane structure in mammalian astrocytes: a review of freeze-fracture studies on adult, developing, reactive and cultured astrocytes. J. Exp. Biol. 95, 35-48.

Landis, D.M.D., & Reese, T.S. (1982). Regional organization of astrocytic membranes in cerebellar cortex. Neuroscience 7, 937-950.

Landis, D.M.D., Reese, T.S., & Raviola, E. (1974). Differences in membrane structure between exci-tatory and inhibitory components of the reciprocal synapse in the olfactory bulb. J. Comp. Neurol. 155, 67-92.

Largo, C., Tombaugh, G.C., Aitken, P.G., Herreras, O., & Somjen, G.G. (1997). Heptanol but not flu-oroacetate prevents the propagation of spreading depression in rat hippocampal slices. J. Neu-rophysiol. 77, 9-16.

Lasater, E.M. (1987). Retinal horizontal cell gap junctional conductance is modulated by dopamine through a cyclic AMP-dependent protein kinase. Proc. Natl. Acad. Sci. U.S.A. 84, 7319-7323.

Lasater, E.M., & Dowling, J.E. (1985). Electrical coupling between pairs of isolated fish horizontal cells is modulated by dopamine and cAMP. In: *Gap Junctions* (Bennett, M.V.L., & Spray, D.C., eds.), pp. 393-404. Cold Spring Harbor Laboratory, Cold Spring Harbor, NY.

Laskawi, R., Rohlmann, A., Landgrebe, M., & Wolff, J.R. (1997). Rapid astroglial reactions in the motor cortex of adult rats following peripheral facial nerve lesions. Eur. Arch. Oto. Rhino. Laryngol. 254, 81-85.

Lau, A.F., Kurata, W.E., Kanemitsu, M.Y., Loo, L.W.M., Warn-Cramer, B.J., Eckhart, W., & Lampe, P.D. (1996). Regulation of connexin43 function by activated tyrosine protein kinases. J. Bioenerg. Biomembr. 28, 359-368.

Lavado, E., Sanchez-Abarca, L.I., Tabernero, A., Bolanos, J.P., & Medina, J.M. (1997). Oleic acid inhibits gap junction permeability and increases glucose uptake in cultured rat astrocytes. J. Neurochem. 69, 721-728.

Le Beux, Y.J. (1972). Subsurface cisterns and lamellar bodies: particular forms of the endoplasmic reticulum in the neurons. Z. Zellforsch. 1333, 327-352.

Lee, C.J., Dayanithi, G., Nordmann, J.J., & Lemos, J.R. (1992). Possible role during exocytosis of a $Ca^{2+}$-activated channel in neurohypophysial granules. Neuron 8, 335-342.

Lee, S.H., Kim, W.T., Cornell-Bell, A.H., & Sontheimer, H. (1994). Astrocytes exhibit regional spec-ificity in gap junction coupling. Glia 11, 315-325.

Lee, S.H., Magge, S., Spencer D.D., Sontheimer, H., & Cornell-Bell, A.H. (1995). Human Epileptic astrocytes exhibit increased gap junction coupling. Glia 15, 195-202.

Lehmenkuhler, A., Sykova, E., Svoboda, J., Zilles, K., & Nicholson, C. (1993). Extracellular space parameters in the rat neocortex and subcortical white matter during postnatal development determined by diffusion analysis. Neuroscience 55, 339-351.

Lemos, J.R., Lee, O.C., Ocorr, K.A., Dayanithi, G., & Nordmann, J.J. (1991). Possible role for neuro-secretory granule channel that resembles gap junctions. Ann. N.Y. Acad. Sci. 635, 480-482.

Lemos, J.R., Ocorr, K.A., & Nordmann, J.J. (1989). Possible role for ionic channels in neurosecretory granules of the rat neurohypophysis. In: Secretion and Its Control. (Oxford, G., & Armstrong, C., Eds.), pp. 333-347. Rockefeller University Press, New York.

Li, J., Hertzberg, E.L., & Nagy, J.I. (1997). Connexin32 in oligodendrocytes and association with myelinated fibers in mouse and rat brain. J. Comp. Neurol. 379, 571-591.

Li, W.E.I., Ochalski, P.A.Y., Hertzberg, E.L., & Nagy, J.I. (1998). Immunorecognition, ultrastructure and phosphorylation of astrocytic gap junctions and connexin43 in rat brain after cerebral focal ischemia. Eur. J. Neurosci. 10, 2444-2463.

Little, T.L., Beyer, E.C., & Duling, B.R. (1995). Connexin 43 and connexin 40 gap junctional proteins are present in arteriolar smooth muscle and endothelium *in vivo*. Am. J. Physiol. 268, H729-H739.

Liu, S., Taffer, S., & Stoner, L. (1993). A structural basis for the unequal sensitivity of the major car-diac and liver gap junctions to intracellular acidification: the carboxy tail length. Biophys. J. 64, 1422-1433.

Llinás, R.R. (1985). Electronic transmission in the mammalian central nervous system. In: Gap Junc-tions. (Bennett, M.V.L., & Spray, D.C., Eds.), pp. 337-353. Cold Spring Harbor Laboratory, Cold Spring Harbor, NY.

Llinás, R., Baker, R., & Sotelo, C. (1974). Electrotonic coupling between neurons in cat inferior olive. J. Neurophysiol. 37, 560-571.

Llinás, R., & Sasaki, K. (1989). The functional organization of the olivo-cerebellar system as examined by multiple Purkinje cell recordings. Eur. J. Neurosci. 1, 587-602.

Llinás, R., & Yarom, Y. (1981). Electrophysiology of mammalian inferior olivary neurones *in vitro*. Different types of voltage-dependent ionic conductances. J. Physiol. (Lond.) 315, 549-567.

Loewenstein, W.R. (1981). Junctional intercellular communication: the cell-to-cell membrane channel. Physiol. Rev. 61, 829-913.

Lo Turco, J.J., & Kriegstein, A.R. (1991). Clusters of coupled neuroblasts in embryonic neocortex. Science 252, 563-566.

Lynn, B.D., Marotta, C.A., & Nagy, J.I. (1995). Propagation of intercellular calcium waves in PC12 cells overexpressing a carboxy-terminal fragment of amyloid precursor protein. Neurosci. Lett. 199, 21-24.

MacVicar, B.A., & Dudek, F.E. (1980). Dye-coupling between CA3 pyramidal cells in slices of rat hippocampus. Brain Res. 196, 494-497.

MacVicar, B.A., & Dudek, F.E. (1981). Electrotonic coupling between pyramidal cells: a direct demonstration in rat hippocampal slices. Science 213, 782-785.

MacVicar, B.A., & Dudek, F.E. (1982). Electrotonic coupling between granule cells of rat dentate gyrus: physiological and anatomical evidence. J. Neurophysiol. 47, 579-592.

MacVicar, B.A., Ropert, N., & Krnjevic, K. (1982). Dye-coupling between pyramidal cells of rat hippocampus *in vivo*. Brain Res. 238, 239-244.

Maldonado, P.E., Rose, B., & Loewenstein, W.R. (1988). Growth factors modulate junctional cell-to-cell communication. J. Membrane Biol. 106, 203-210.

Mantz, J., Cordier, J., & Giaume, C. (1993). Effects of general aesthetics on intercellular communication mediated by gap junctions between astrocytes in primary culture. Anesthesiology 78, 892-901.

Marrero, H., & Orkand, R.K. (1996). Nerve impulses increase glial intercellular permeability. Glia 16, 285-289.

Martins-Ferreira, H., & Ribeiro, L.J. (1995). Biphasic effects of gap junctional uncoupling agents on the propagation of retinal spreading depression. Braz. J. Med. Biol. Res. 28, 991-994.

Massa, P.T., & Mugnaini, E. (1982). Cell junctions and intramembrane particles of astrocytes and oligodendrocytes: a freeze-fracture study. Neuroscience 7, 523-538.

Massa, P.T., & Mugnaini, E. (1985). Cell-cell interactions and characteristic plasma membrane features of cultured rat glial cells. Neuroscience 14, 2, 695-709.

Massa, P.T., Szuchet, S., & Mugnaini, E. (1984). Cell-cell interactions of isolated and cultured oligodendrocytes: formation of linear occluding junctions and expression of peculiar intramembrane particles. J. Neurosci. 4, 3128-3139.

Matsumoto, A., Arnold, A.P., & Micevych, P.E. (1989). Gap junctions between lateral spinal motoneurons in the rat. Brain Res. 495, 362-366.

Matsumoto, A., Arnold, A.P., Zampighi, G.A., & Micevych, P.E. (1988). Androgenic regulation of gap junctions between motoneurons in the rat spinal cord. J.Neurosci. 8, 4177-4183.

Matsumoto, A., Arai, Y., Urano, A., & Hyodo, S. (1991a). Androgen regulates gap junction mRNA expression in androgen-sensitive motoneurons in the rat spinal cord. Neurosci. Lett. 131, 159-162.

Matsumoto, A., Arai, Y., Urano, A., & Hyodo, S. (1991b). Cellular localization of gap junction mRNA in the neonatal rat brain. Neurosci. Lett. 124, 225-228.

Matsumoto, A., Arai, Y., Urano, A., & Hyodo, S. (1992). Effect of androgen on the expression of gap junction and β-actin mRNA in adult rat motoneurons. Neurosci. Res. 14, 133-144.

Matthews, B. (1976). Coupling between cutaneous nerves. J. Physiol. 254, 37-38P.

Matthews, B., & Holland, G.R. (1975). Coupling between nerves in teeth. Brain Res. 98, 354-358.

McBurney, R.M., & Neering, I.R. (1987). Neuronal calcium homeostasis. Trends Neurosci. 10, 164-169.

McKhann, G.M., D'Ambrosio, R., & Janigro, D. (1997). Heterogeneity of astrocyte resting membrane potentials and intercellular coupling revealed by whole-cell and gramicidin-perforated patches from cultured neocortical and hippocampal slice astrocytes. J. Neurosci. 17, 6850-6863.

Meda, P. (1996). The role of gap junction membrane channels in secretion and hormonal action. J. Bioenerg. Biomemb. 28, 369-377.

Merritt, J.E., & Rink, T.J. (1987). Regulation of cytosolic free calcium in Fura-2-loaded rat parotid acinar cells. J. Biol. Chem. 262, 17362-17369.

Meyer, R.A., & Campbell, J.N. (1987). Coupling between unmyelinated peripheral nerve fibers does not involve sympathetic efferent fibers. Brain Res. 437, 181-182.

Meyer, R.A., & Campbell, J.N. (1988). A novel electrophysiological technique for locating cutaneous nociceptive and chemospecific receptors. Brain Res. 441, 81-86.

Meyer, R.A., Raja, S.N., & Campbell, J.N. (1985). Coupling of action potential activity between unmyelinated fibers in the peripheral nerve of monkey. Science 227, 184-187.

Micevych, P.E., & Abelson, L. (1991). Distribution of mRNA coding for liver and heart gap junction proteins in the rat central nervous system. J. Comp. Neurol. 305, 96-118.

Micevych, P.E., Popper, P., & Hatton, G.I. (1996). Connexin 32 mRNA levels in the rat supraoptic nucleus: up-regulation prior to parturition and during lactation. Neuroendocrinology 63, 39-45.

Mienville, J.-M., Lange, G.D., & Barker, J.L. (1994). Reciprocal expression of cell-cell coupling and voltage-dependent Na current during embryogenesis of rat telencephalon. Dev. Brain Res. 77, 89-95.

Mills, S.L., & Massey, S.C. (1995). Differential properties of two gap junctional pathways made by AII amacrine cells. Nature 377,734-737.

Minkoff, R., Parker, S.B., & Hertzberg, E.L. (1991). Analysis of distribution patterns of gap junctions during development of embryonic chick primordia and brain. Development 111, 509-522.

Miragall, F., Albiez, P., Bartels, H., de Vries, U., & Dermietzel, R. (1997). Expression of the gap junction protein connexin43 in the subependymal layer and the rostral migratory stream of the mouse: evidence for an inverse correlation between intensity of connexin43 expression and cell proliferation activity. Cell Tissue Res. 287, 243-253.

Miragall, F., Hwang, T.-K., Traub, O., Hertzberg, E.L., & Dermietzel, R. (1992). Expression of connexins in the developing olfactory system of the mouse. J. Comp. Neurol. 325, 359-378.

Miragall, F., Simburger, E., & Dermietzel, R. (1996). Mitral and tufted cells of the mouse olfactory bulb possess gap junctions and express connexin43 mRNA. Neurosci. Lett. 216, 199-202.

Miyachi, E., Kato, C., & Nakaki, T. (1994). Arachidonic acid blocks gap junctions between retinal horizontal cells. Neuroreport 5, 485-488.

Modney, B.K., Yang, Q.Z., & Hatton, G.I. (1990). Activation of excitatory amino acid inputs to supraoptic neurons. II. Increased dye-coupling in maternally behaving virgin rats. Brian Res. 513, 270-273.

Moreno, A.P., Fishman, G.I., & Spray, D.C. (1992). Phosphorylation shifts unitary conductance and modifies voltage dependent kinetics of human connexin43 gap junction channels. Biophys. J. 62, 51-53.

Moreno, A.P., Laing, J.G., Beyer, E.C., & Spray, D.C. (1995). Properties of gap junction channels formed of connexin45 endogenously expressed in human hepatoma (SDHep1) cells. Am. J. Physiol. 268, C356-365.

Mugnaini, E. (1986). Cell junctions of astrocytes, ependyma, and related cells in the mammalian central nervous system, with emphasis on the hypothesis of a generalized functional syncytium of supporting cells. In: Astrocytes. (Fedoroff, S. & Verndakis, A., Eds.), vol. 1, pp. 329-371. Academic Press, New York.

Mullaney, J.M., Chueh, S.-H., Ghosh, T.K., & Gill, D.L. (1987). Intracellular calcium uptake acti-
    vated by GTP: evidence for a possible guanine nucleotide-induced transmembrane conveyance
    of intracellular calcium. J. Biol. Chem. 262, 13865-13872.
Muller, T., Moller, T., Neuhaus, J., & Kettenmann, H. (1996). Electrical coupling among Bergmann
    glial cells and it modulation by glutamate receptor activation. Glia 17, 274-284.
Murphy, A.D., Hadley, R.D., & Kater, S.B. (1983). Axotomy-induced parallel increases in electrical
    and dye-coupling between identified neurons of Helisoma. J. Neurosci. 3, 1422-1429.
Murphy, S., & Pearce, B. (1987). Functional receptors for neurotransmitter on astroglial cells. Neuro-
    science 22, 381-394.
Murphy, T.H., Blatter, L.A., Wier, W.G., & Baraban, J.M. (1993). Rapid communication between
    neurons and astrocytes in primary cortical cultures. J. Neurosci. 13, 2672-2679.
Nadarajah, B., Jones, A.M., Evans, W.H., & Parnavelas, J.G. (1997). Differential expression of con-
    nexins during neocortical development and neuronal circuit formation. J. Neurosci. 17,
    3096-3111.
Nadarajah, B., Thomaidou, D., Evans, W.H., & Parnavelas, J.G. (1996). Gap junctions in the adult
    cerebral cortex: regional differences in their distribution and cellular expression of connexins.
    J. Comp. Neurol. 376, 326-342.
Nagy, J.I., Hossain, M.Z., Hertzberg, E.L., & Marotta, C.A. (1996a). Induction of connexin43 and gap
    junctional communication in PC12 cells overexpressing the carboxy terminal region of amy-
    loid precursor protein. J. Neurosci. Res. 44, 124-132.
Nagy, J.I., Li, W.E.I., Doble, B.W., Hochman, S., Hertzberg, E.L., & Kardami, E. (1996b). Detection
    of dephosphorylated Cx43 in brain, heart and in spinal cord after nerve stimulation. Soc. Neu-
    rosci. Abstr. 22, 1023.
Nagy, J.I., Li, W., Hertzberg, E.L., & Marotta, C.A. (1996c). Elevated connexin43 immunoreactivity
    at sites of amyloid plaques in Alzheimer's disease. Brain Res. 717, 173-178.
Nagy, J.I., Li, W.E.I., Roy, C., Doble, B.W., Gilchrist, J.S., Kardami, E., & Hertzberg, E.L. (1997a).
    Selective monoclonal antibody recognition and cellular localization of an unphosphorylated
    form of connexin43. Exptl. Cell Res. 236, 127-136.
Nagy, J.I., Ochalski, P.A.Y., Li, J., & Hertzberg, E.L. (1997b). Evidence for the co-localization of
    another connexin with connexin43 at astrocytic gap junctions in rat brain. Neuroscience 78,
    533-548.
Nagy, J.I., Patel, D., Ochalski, P.A.Y., & Stelmack, G.L. (1999). Connexin30 in rodent, cat and human
    brain: selective expression in gray matter astrocytes, co-localization with connexin30 at gap
    junctions and late developmental appearance. Neuroscience 88, 447-468.
Nagy, J.I., Yamamoto, T., Sawchuk, M.A., Nance, D.M., & Hertzberg, E.L. (1992). Quantitative
    immunohistochemical and biochemical correlates of connexin43 localization in rat brain. Glia
    5, 1-9.
Nagy, J.I., Yamamoto, T., Shiosaka, S., Dewar, K.M., Whittaker, M.E., & Hertzberg, E.L. (1988)
    Immunohistochemical localization of gap junction protein in rat CNS: a preliminary account.
    In: Modern Cell Biology. (Hertzberg, E.L., & Johnson, R.G., Eds.), vol. 7, pp. 375-389. Alan
    R. Liss, New York.
Nahorski, S.R. (1988). Inositol polyphosphates and neuronal calcium homeostasis. Trends Neurosci.
    11, 444-448.
Naus, C.C.G., Bechberger, J.F., Caveney, S., & Wilson, J.X. (1991a). Expression of gap junction
    genes in astrocytes and C6 glioma cells. Neurosci. Lett. 126, 33-36.
Naus, C.C.G., Bechberger, J.F., & Paul, D.L. (1991b). Gap junction gene expression in human seizure
    disorder. Exptl. Neurol. 111, 198-203.
Naus, C.C.G., Bechberger, J.F., Zhang, Y., Venance, L., Yamasaki, H., Juneja, S.C., Kidder, G.M., &
    Giaume, C. (1997). Altered gap junctional communication, intercellular signalling, and growth
    in cultured astrocytes deficient in connexin43. J. Neurosci. Res. 49, 528-540.

Naus, C.C.G., Belliveau, D.J., & Bechberger, J.F. (1990). Regional differences in connexin32 and connexin43 messenger RNAs in rat brain. Neurosci. Lett. 111, 297-302.

Naus, C.C.G., Elisevich, K., Zhu, D., Belliveau, D.J., & Del Maestro, R.F. (1992a). *In vivo* growth of C6 glioma cells transfected with connexin43 cDNA. Cancer Res. 52, 4208-4213.

Naus, C.C.G., Zhu, D., Todd, S.D.L., & Kidder, G.M. (1992b). Characteristics of C6 glioma cells overexpressing a gap junction protein. Cell. Mol. Neurobiol. 12, 163-175.

Nedergaard, M. (1994). Direct signaling from astrocytes to neurons in cultures of mammalian brain cells. Science 263, 1768-1771.

Nedergaard, M., Cooper, A.J.L., & Goldman, S.A. (1995). Gap junctions are required for the propagation of spreading depression. J. Neurobiol. 28, 433-444.

Nelles, E., Butzler., C., Jung, D., Temme, A., Gabreil, H.-D., Dahl, U., Traub, O., Stumpel, F., Jungermann, K., Zielasek, J., Toyka, K., Dermietzel, R., & Willecke, K. (1996). Defective propagation of signals generated by sympathetic nerve stimulation in the liver of connexin32-deficient mice. Proc. Natl. Acad. Sci. U.S.A. 93, 9565-9570.

Nelson, P.G. (1966). Interaction between spinal motoneurons of the cat. J. Neurophysiol. 29, 275-294.

Newman, E.A. (1986). High potassium conductance in astrocyte endfeet. Nature 233, 453-454.

Newman, E.A. (1996). Regulation of extracellular $K^+$ and pH by polarized ion fluxes in glial cells: the retinal Muller cell. Neuroscientist 2, 109-117.

Newman, E.A., & Zahs, K.R. (1997). Calcium waves in retinal glial cells. Science 275, 844-847.

Neyton, J., Piccolino, M., & Gershenfeld, H.M. (1985). Neurotransmitter-induced modulation of gap junction permeability in retinal horizontal cells. In: Gap Junctions. (Bennett, M.V.L. & Spray, D.C., Eds.), pp. 381-391. Cold Spring Harbor Laboratory, Cold Spring Harbor, NY.

Nishimura, Y., & Rakic, P. (1985). Development of the Rhesus monkey retina. I. Emergence of the inner plexiform layer and its synapses. J. Comp. Neurol. 241, 420- 434.

Nunez, A., Garcia-Austt, E., & Buno, W. (1990). *In vivo* electrophysiological analysis of lucifer yellow-coupled hippocampal pyramids. Exptl. Neurol. 108, 76-82.

O'Beirne, M., Bulloch, A.G.M., & MacVicar, B.A. (1987). Dye and electrotonic coupling between cultured hippocampal neurons. Neurosci. Lett. 78, 265-270.

O'Brian, J., Al-Ubaidi, M.R., & Ripps, H. (1996). Connexin35: a gap junctional protein expressed preferentially in the skate retina. Mol. Biol. Cell 7, 233-243.

Ochalski, P.A.Y., Frankenstein, U.N., Hertzberg, E.L., & Nagy, J.I. (1997). Connexin43 in rat spinal cord: localization in astrocytes and identification of heterotypic astro-oligodendrocytic gap junctions. Neuroscience 76, 931-945.

Ochalski, P.A.Y., Hossain, M.Z., Sawchuk, M.A., Hertzberg, E.L., & Nagy, J.I. (1995). Astrocytic gap junction removal, connexin43 redistribution and epitope masking at at excitatory amino acid lesion sites in rat brain. Glia 14, 279-294.

O'Donnell, P., & Grace, A.A. (1993). Dopaminergic modulation of dye coupling between neurons in the core and shell regions of the nucleus accumbens. J. Neurosci. 13, 3456-3471.

O'Donnell, P., & Grace, A.A. (1997). Cortical afferents modulate striatal gap junction permeability via nitric oxide. Neuroscience 76, 1-5.

Orkand, R.K. (1977). Glial cells. In: Handbook of Physiology. (Kandel, E.R., Ed.), section 1, vol. 1, pt. 2, pp. 855-875. Waverly Press, Baltimore.

Orkand, R.K., Nicholls, J.G., & Kuffler, S.W. (1966). Effect of nerve impulses on the membrane potential of glial cells in the central nervous system of amphibia. J. Neurophysiol. 29, 788-806.

Palay, S.L., & Chan-Palay, V. (1977). General morphology of neurons and neuroglia. In: Handbook of Physiology, Cellular Biology of Neurons. (Kandel, E.R., Ed.), vol. 1, pt. 1, pp. 5-37. American Physiological Society, Bethesda, MD.

Pannese, E., Luciano, L., Iurato, S., & Reale, E. (1977). Intercellular junctions and other membrane specializations in developing spinal ganglia: A freeze-fracture study. J. Ultrastruc. Res. 60, 169-180.

Pappas, C.A., Rioult, M.G., & Ransom, B.R. (1996). Octanol, a gap junction uncoupling agent, changes intracellular [H⁺] in rat astrocytes. Glia 16, 7-15.

Pappas, G.D., & Waxman, S.G. (1972). Synaptic fine structure-morphological correlates of chemical and electrotonic transmission. In: Structure and Function of Synapses. (Pappas, G.D., & Purpura, D.P., Eds.), pp. 1-43. Raven Press, New York.

Parpura, V., Basarsky, T.A., Liu, F., Jeftinija, K., Jeftinija, S., & Haydon, P.G. (1994). Glutamate-mediated astrocyte-neuron signaling. Nature 369, 744-747.

Pastor, A., Kremer, M., Möller, T., Kettenmann, H., & Dermietzel, R. (1998). Dye-coupling between spinal cord oligodendrocytes is restricted to gray matter: indication for functional autocellular territories in white matter oligodendrocytes. Glia 24, 108-120.

Paternostro, M.A., Reyher, C.K., & Brunjes, P.C. (1995). Intracellular injections of Lucifer yellow into lightly fixed mitral cells reveal neuronal dye-coupling in the developing rat olfactory bulb. Dev. Brain Res. 84, 1-10.

Paul, D.L. (1995). New functions for gap junctions. Curr. Opinion Cell Biol. 7, 665-672.

Peinado, A., Yuste, R., & Katz, L.C. (1993a). Extensive dye coupling between rat neocortical neurons during the period of circuit formation. Neuron 10, 103-114.

Peinado, A., Yuste, R., & Katz, L.C. (1993b). Gap junctional communication and the development of local circuits in neocortex. Cerebral Cortex 3, 488-498.

Pellerin, L., & Magistretti, P.J. (1994). Glutamate uptake into astrocytes stimulates aerobic glycolysis: a mechanism coupling neuronal activity to glucose utilization. Proc. Natl. Acad. Sci. U.S.A. 91, 10625-10629.

Pepper, M.S., & Meda, P. (1992). Basic fibroblast growth factor increases junctional communication and connexin43 expression in microvascular endothelial cells. J. Cell Physiol. 153, 196-205.

Peracchia, C. (1980). Structural correlates of gap junction permeation. Internatl. Rev. Cytol. 66, 81-146.

Perez, J., Tranque, P.A., Naftolin, F., & Garcia-Segura, L.M. (1990). Gap junctions in the hypothalamic arcuate neurons of ovariectomized and estradiol-treated rats. Neurosci. Lett. 108, 17-21.

Perez-Velazquez, J.L., Valiante, T.A., & Carlen, P.L. (1994). Modulation of gap junctional mechanisms during calcium-free induced field burst activity: a possible role for electrotonic coupling in epileptogenesis. J. Neurosci. 14, 4308-4317.

Peters, A. (1980). Morphological correlates of epilepsy: Cells in the cerebral cortex. In: Antiepileptic Drugs: Mechanisms of Action. (Glaser, G.H., Penry, J.K., & Woodbury, D.M., Eds.), pp. 21-48. Raven Press, New York.

Peters, A., Proskauer, C.C., & Kaiserman-Abramof, I.F. (1968). The small pyramidal neuron of the rat cerebral cortex. The axon hillock and initial segment. J. Cell Biol. 39, 604-619.

Piccolino, M., Neyton, J., & Gerschenfeld, H.M. (1984). Decrease of gap junction permeability induced by dopamine and cyclic adenosine3'5'-monophosphate in horizontal cells of turtle retina. J. Neurosci. 4, 2447-2488.

Pinching, A.J., & Powell, T.P.S. (1971). The neuropil of the glomeruli of the olfactory bulb. J. Cell Sci. 9, 347-377.

Putney, J.W., Jr. (1986). A model for receptor-regulated calcium entry. Cell Calcium 7, 1-12.

Rall, W., Burke, R.E., Smith, T.G., Nelson, P.G., & Frank, K. (1967). Dendritic location of synapses and possible mechanisms for the monosynaptic EPSP in motneurons. J. Neurophysiol. 30, 1169-1193.

Randriamampita, C., Giaume, C., Neyton, J., & Trautmann, A. (1988). Acetylcholine-induced closure of gap junction channels in rat lacrimal glands is probably mediated by protein kinase C. Pfugers Arch. 412, 462-468.

Ransom, B.R., & Carlini, W.G. (1986). Electrophysiological properties of astrocytes. In: Astrocytes. (Fedoroff, S., & Vernadakis, A., Eds.), vol. 2, pp. 1-49. Academic Press, New York.

Ransom, B.R., & Kettenmann, H. (1990). Electrical coupling, without dye coupling, between mammalian astrocytes and oligodendrocytes in cell culture. Glia 3, 258-266.

Rao, G., Barnes, C.A., & McNaughton, B.L. (1986). Intracellular fluorescent staining with carboxy-fluorescein: a rapid and reliable method for quantifying dye-coupling in mammalian central nervous system. J. Neurosci. Meth. 16, 251-263.

Rash, J.E., Dillman, R.K., Bilhartz, B.L., Duffy, H.S., Whalen, L.R., & Yasumura, T. (1996). Mixed synapses discovered and mapped throughout mammalian spinal cord. Proc. Natl. Acad. Sci. U.S.A. 93, 4235-4239.

Rash, J.E., Duffy, H.S., Dudek, F.E., Bilhartz, B.L., Whalen, L.R., & Yasumura, T. (1997). Grid-mapped freeze-fracture analysis of gap junctions in gray and white matter of adult rat central nervous system, with evidence for a "panglial syncytium" that is not coupled to neurons. J. Comp. Neurol. 388, 265-292.

Raviola, E., & Gilula, N.B. (1973). Gap junctions between photoreceptor cells in the vertebrate retina. Proc. Nat. Acad. Sci. U.S.A. 70, 1677-1681.

Raviola, E., & Gilula, N.B. (1975). Intramembrane organization of specialized contacts in the outer plexiform layer of the retina: a freeze-fracture study in monkeys and rabbits. J. Cell. Biol. 65, 192-222.

Renaud, L.P. (1987). Magnocellular neuroendocrine neurons: update on intrinsic properties, synaptic inputs and neuropharmacology. Trends Neurosci. 10, 498-502.

Reuss, B., Dermietzel, R., & Unsicker, K. (1998). Fibroblast growth factor2 (FGF-2) differentially regulates connexin (cx) 43 expression and function in astroglial cells from distinct brain regions. Glia 22, 19-30.

Reuss, B. & Unsicker, K. (1998). Regulation of gap junction communication by growth factors from non-neural cells to astroglia: brief review. Glia 24, 32-38.

Reyher, C.K.H., Lubke, J., Larsen, W.J., Hendrix, G.M., Shipley, M.T., & Baumgarten, H.G. (1991). Olfactory bulb granule cell aggregates: morphological evidence for interperikaryal electrotonic coupling via gap junctions. J. Neurosci. 11, 1485-1495.

Rifkin, D.B., & Moscatelli, D. (1989). Recent developments in the cell biology of basic fibroblast growth factor. J.Cell Biol. 109, 1-6.

Robinson, R.R., Hampson, E.C.G., Munro, M.N., & Vaney, D.I. (1993). Unidirectional coupling of gap junctions between neuroglia. Science 262, 1072-1074.

Rogawski, M.A. (1987). New directions in neurotransmitter action: dopamine provides some important clues. Trends Neurosci. 10, 200-205.

Rohlmann, A., Laskawi, R., Hofer, A., Dermietzel, R., & Wolff, J.R. (1994). Astrocytes as rapid sensors of peripheral axotomy in the facial nucleus of rats. Neuroport 36, 409-412.

Rohlmann, A., Laskawi, R., Hofer, A., Dobo, E., Dermietzel, R., & Wolff, J.R. (1993). Facial nerve lesions lead to increased immunostaining of the astrocytic gap junction protein (connexin43) in the corresponding facial nucleus of rats. Neurosci. Lett. 154, 206-208.

Rohlmann, A. & Wolff, J.R. (1996). Subcellular topography and plasticity of gap junction distribution on astrocytes. In: Gap Junctions in the Nervous System. (Spray, D.C., & Dermietzel, R., Eds.), pp 175-192. Landes, Austin.

Rorig, B., Klausa, G., & Sutor, B. (1995). Dye coupling between pyramidal neurons in developing rat prefrontal and frontal cortex is reduced by protein kinase A activation and dopamine. J. Neurosci. 15, 7386-7400.

Rorig, B., Klausa, G., & Sutor, B. (1996). Intracellular acidification reduced gap junction coupling between immature rat neocortical pyramidal neurons. J. Physiol. 490, 31-49.

Rorig, B., & Sutor, B. (1996a). Regulation of gap junction coupling in the developing neocortex. Mol Neurobiol. 12, 225-249.

Rorig, B., & Sutor, B. (1996b). Serotonin regulates gap junction coupling in the developing rat somatosensory cortex. Eur. J. Neurosci. 8, 1685-1695.

Rose, C.R., & Ransom, B.R. (1997). Gap junctions equalize intracellular $Na^+$ concentration in astrocytes. Glia 20, 299-307.

Rosenbluth, J. (1962a). The fine structure of acoustic ganglia in the rat. J. Cell Biol. 12, 329-359.

Rosenbluth, J. (1962b). Subsurface cisterns and their relationship to the neuronal plasma membrane. J. Cell Biol. 13, 405-421.

Rozental, R., Morales, M., Mehler, M.F., Urban, M., Kremer, M., Dermietzel, R., Kessler, J.A., & Spray, D.C. (1998). Changes in the properties of gap junctions during neuronal differentiation of hippocampal progenitor cells. J. Neurosci. 18, 1753-1762.

Rufer, M., Wirth, S.B., Hofer, A., Dermietzel, R., Pastor, A., Kettenmann, H., & Unsicker, K. (1996). Regulation of connexin43, GFAP, and FGF-2 in not accompanied by changes in astroglial coupling in MPTP-lesioned, FGF-2-treated parkinsonian mice. J. Neurosci. Res. 46, 606-617.

Rutherford, J.G., & Gwyn, D.G. (1977). Gap junctions in the inferior olivary nucleus of the squirrel monkey, *Saimiri sciureus*. Brain Res. 128, 374-378.

Sáez, J.C., Berthoud, V.M., Kadle, R., Traub, O., Nicholson, B.J., Bennett, M.V.L., & Dermietzel, R. (1991). Pinealocytes in rats: connexin identification and increase in coupling caused by norepinephrine. Brain Res. 568, 265-275.

Sáez, J.C., Connor, J.A., Spray, D.C., & Bennett, M.V.L. (1989). Hepatocyte gap junctions are permeable to the second messenger, inositol 1, 4, 5-triphosphate, and to calcium ions. Proc. Natl. Acad. Sci. U.S.A. 86, 2708-2712.

Sáez, J.C., Nairn, A.C., Czernik, A.J., Fishman, G.I., Spray, D.C., & Hertzberg, E.L. (1997). Phosphorylation of connexin43 and the regulation of neonatal rat cardiac myocyte gap junctions. J. Mol. Cell. Cardiol. 29, 2131-45.

Sáez, J.C., Spray, D.C., & Hertzberg, E.L. (1990). Gap junctions: biochemical properties and functional regulation under physiological and toxicological conditions. *In Vitro* Toxicol. 3, 69-86.

Sáez, J.C., Spray, D.C., Nairn, A.C., Hertzberg, E.L., Greengard, P., & Bennett, M.V.L. (1986). cAMP increases junctional conductance and stimulates phosphorylation of the 27-kDa principal gap junction polypeptide. Proc. Natl. Acad. Sci. U.S.A. 83, 2473-2477.

Sahenk, Z., & Chen, L. (1998). Abnormalities in the axonal cytoskeleton induced by a connexin32 mutation in nerve xenografts. J. Neurosci. Res. 51, 174-184.

Sanderson, M.J., Charles, A.C., & Dirksen, E.R. (1990). Mechanical stimulation and intercellular communication in creases intracellular $Ca^{2+}$ in epithelial cells. Cell Regul. 1, 585-596.

Sandri, C., Van Buren, J.M., & Akert, K. (1982). Membrane morphology of the vertebrate nervous system. Prog. Brain Res. 46, 201-265.

Sasaki, K., Bower, J.M., & Llinás, R. (1989). Multiple Purkinje cell recording in rodent cerebellar cortex. Eur. J. Neurosci. 1, 572-586.

Sawchuk, M.A., Hossain, M.Z., Hertzberg, E.L., & Nagy, J.I. (1995). *In situ* transblot and immunocytochemical comparisons of astrocytic connexin43 responses to NMDA and kainic acid in rat brain. Brain Res. 683, 153-157.

Scemes, E., Dermietzel, R., & Spray, D.C. (1998). Calcium waves between astrocytes from Cx43 knockout mice. Glia 24, 65-73.

Scherer, S. S., Deschênes, S.M., Xu, Y.-T., Grinspan, J.B., Fischbeck, K.H., & Paul, D.L. (1995). Connexin32 is a myelin-related protein in the PNS and CNS. J. Neurosci. 15, 8281-8294.

Scherer, S.S., Xu, Y.T., Nelles, E., Fischbeck, K., Willecke, K., & Bone, L.J. (1998). Connexin32-null mice develop demyelinating peripheral neuropathy. Glia 24, 8-20.

Schmalbruch, H., & Jahnsen, H. (1981). Gap junctions on CA3 pyramidal cells of guinea pig hippocampus shown by freeze-fracture. Brain Res. 217, 175-178.

Schwartzkroin, P.A., & Prince, D.A. (1979). Recordings from presumed glial cells in the hippocampal slice. Brain Res. 161, 533-538.

Shiosaka, S., Yamamoto, T., Hertzberg, E.L., & Nagy, J.I. (1989). Gap junction protein in rat hippocampus: correlative light and electron microscope immunohistochemical localization. J. Comp. Neurol. 281, 282-297.

Siegesmund, K.A. (1968). The fine structure of subsurface cisterns. Anat. Rec. 162, 187-196.

Simburger, E., Stang, A., Kremer, M., & Dermietzel, R. (1997). Expression of connexin43 in adult rodent brain. Histochem. Cell Biol. 107, 127-137.

Sipe, J.C., & Moore, R.Y. (1976). Astrocytic gap junctions in the rat lateral hypothalamic area. Anat. Rec. 185, 247-252.

Skatchkov, S., Vyklicky, L., & Orkland, R. (1995). Potassium currents in endfeet of isolated Muller cells from the frog retina. Glia 15, 54-64.

Sloper, J.J. (1972). Gap junctions between dendrites in the primate neocortex. Brain Res. 44, 641-646.

Sloper, J.J. (1973). An electron microscopic study of the neurons of the primate motor and somatic sensory cortices. J. Neurocytol. 2, 351-359.

Sloper, J.J., & Powell, T.P.S. (1978). Gap junctions between dendrites and somata of neurons in the primate sensori-motor cortex. Proc. R. Soc. Lond. B. 203, 39-47.

Smith, D.E., & Moskowitz, N. (1979). Ultrastructure of layer IV of the primary auditory cortex of the squirrel monkey. Neuroscience 4, 349-359.

Söhl, G., Degen, J., Teubner, B., & Willecke, K. (1998). The murine gap junction gene Connexin36 is highly expressed in mouse retina and regulated during development. FEBS Lett. 428, 27-31.

Sontheiner, H., Minturn, J.E., Black, J.A., Waxman, S.G., & Ransom, B.R. (1990). Specificity of cell-cell coupling in rat optic nerve astrocytes *in vitro*. Proc. Natl. Acad. Sci. U.S.A. 87, 9833-9837.

Sontheimer, H., Waxman, S.G., & Ransom, B.R. (1991). Relationship between $Na^+$ current expression and cell-cell coupling in astrocytes cultured from rat hippocampus. J. Neurophysiol. 65, 989-1001.

Sotelo, C. (1975). Morphological correlates of electrotonic coupling between neurons in mammalian nervous system. In: Golgi Centenial Symposium Proceedings. (Santini, M., Ed.), pp. 355-365. Raven Press, New York.

Sotelo, C., & Angaut, P. (1973). The fine structure of the cerebellar central nuclei in the cat. I. Neurons and neuroglial cells. Exp. Brain Res. 16, 410-430.

Sotelo, C., Gentschev, T., & Zamora, A.J. (1976). Gap junctions in ventral cochlear nucleus of the rat. A possible new example of electrotonic junctions in the mammalian CNS. Neuroscience 1, 5-7.

Sotelo, C., Gotow, T., & Wassef, M. (1986). Localization of glutamic-acid decarboxylase immunoreactive axon terminals in the inferior olive of the rat, with emphasis on anatomical relations between GABAergic synapses and dendrodendritic gap junctions. J. Comp. Neurol. 252, 32-50.

Sotelo, C., & Korn, H. (1978). Morphological correlates of electrical and other interactions through low-resistance pathways between neurons of the vertebrate central nervous system. Int. Rev. Cytol. 5, 67-107.

Sotelo, C., & Llinás, R. (1972). Specialized membrane junctions between neurons in the vertebrate cerebellar cortex. J. Cell Biol. 53, 271-289.

Sotelo, C., Llinás, R., & Baker, R. (1974). Structural study of inferior olivary nucleus of the cat: morphological correlates of electrotonic coupling. J. Neurophysiol. 37, 541-559.

Sotelo, C., & Palay, S.L. (1970). The fine structure of the lateral vestibular nucleus in the rat. II synaptic organization. Brain Res. 18, 93-115.

Sotelo, C., & Triller, A. (1981). Morphological correlates of electrical, chemical and dual modes of transmission. In: Chemical Neurotransmission, 75 Years. (Stjarne, L., Lagercrantz, H., Hedqvist, P., & Wennmalm, A., Eds.), pp. 13-28. Academic Press, New York.

Spoerri, P.E., & Gleas, P. (1977). Subsurface cisterns in the Cynomolgus retina. Cell Tissue Res. 182, 33-38.

Spray, D.C., & Bennett, M.V.L. (1985). Physiology and pharmacology of gap junctions. Ann. Rev. Physiol. 47, 2811-303.

Spray, D.C., & Dermietzel, R. (1995). X-linked dominant Charcot-Marie-Tooth disease and other potential gap junction diseases of the nervous system. Trends Neurosci. 18, 256-262.

Spray, D.C., & Dermietzel, R. (Eds.) (1996). Gap Junctions in the Nervous System. Landes, Austin.

Spray, D.C., Moreno, A.P., Eghbali, B., Chanson, M., & Fishman, G.I. (1992). Gating of gap junctions channels as revealed in cells stably transfected with wild type mutant connexin cDNAs. Biophys. J. 62, 48-50.

Spray, D.C., Moreno, A.P., Kessler, J.A., & Dermietzel, R. (1991). Characterization of gap junctions between cultured leptomeningeal cells. Brain Res. 568, 1-14.

Spray, D.C., Viera, D., El-sabban, M.E., Gao, Y., & Bennett, M.V.L. (1995). Gap junction properties in astrocytes from connexin43 (Cx43) knockout mice. Soc. Neurosci. Abstr. 21, 563.

Spray, D.C., Vink, C., Scemes, E., Suadicani, S.O., Fishman, G.I., & Dermietzel, R. (1998). Characteristics of coupling in cardiac myocytes and CNS astrocytes cultured from wildtype and Cx43-null mice. In: Gap Junctions. (Werner, R., ed.), pp. 281-285. IOS Press, Amsterdam.

Spruce, A.E., Breckenbridge, L.J., Lee, A.K., & Almers, W. (1990). Properties of the fusion pore that forms during exocytosis of a mast cell secretory vesicle. Neuron 4, 643-654.

Stauffer, K.A. (1995). The gap junction protein β1-connexin (connexin-32) and β2-connexin (connexin26) can form heteromeric hemichannels. J. Biol. Chem. 270, 6768-6772.

Steinberg, T.H., Civitelli, R., Geist, S.T., Robertson, A.J., Hick, E., Veenstra, R.D., Wang, H.Z., Warlow, P.M., Laing, J.G., & Beyer, E. (1994). Connexin43 and connexin45 form gap junctions with different molecular permeabilities in osteoblastic cells. EMBO J. 13, 744-750.

Stewart, W.W. (1978). Functional connections between cells as revealed by dye-coupling with a highly fluorescent naphthalimide tracer. Cell 14, 741-759.

Stewart, W.W. (1981). Lucifer dyes—highly fluorescent dyes for biological tracing. Nature 292, 17-21.

Strata, F., Atzori, M., Molnar, M., Ugolini, G., Tempia, F., & Cherubini, E. (1997). A pacemaker current in dye-coupled hilar interneurons contributes to the generation of giant GABAergic potentials in developing hippocampus. J. Neurosci. 17, 1435-46.

Strettoi, E., Dacheux, R.F., & Raviola, E. (1990). Synaptic connections of rod bipolar cells in the inner plexiform layer of the rabbit retina. J. Comp. Neurol. 295, 449-466.

Suadicani, S.O., Scemes, E., & Spray, D.C. (1998). Slow intercellular $Ca^{2+}$ waves in cardiac myocytes and CNS astrocytes: comparisons in cultures from wildtype (WT) and Cx43 null (Cx43 –/–) mice. Biophys. J. 74, A377.

Szentagothai, J. (1970). Glomerular synapses, complex synaptic arrangements, and their operational significance. In: The Neurosciences, A Second Study Program. (Quarton, G.C., Melnechuck, T., & Schmitt, F.O., Eds.), pp. 427-443. Rockefeller University Press, New York.

Tabernero, A., Giaume, C., & Medina, J.M. (1996). Endothelin-1 regulates glucose utilization in cultured astrocytes by controlling intercellular communication through gap junctions. Glia 16, 1187-1195.

Takahashi, K., & Wood, R.L. (1970). Subsurface cisterns in the Purkinje cells of cerebellum of syrian hamster. Z. Zellforsch. 110, 311-320.

Takeda, M., Nelson, D.J., & Soliven, B. (1995). Calcium signaling in cultured rat oligodendrocytes. Glia 14, 225-236.

Takens-Kwak, B.R., & Jongsma, H.J. (1992). Cardiac gap junctions: three distinct single channel conductances and their modulation by phosphorylating treatments. Pflugers Arch. 422, 198-200.

Taylor, C.P., & Dudek, F.E. (1982). A physiological test for electrotonic coupling between CA1 pyramidal cells in rat hippocampal slices. Brain Res. 235, 351-357.

Tetzlaff, W. (1982). Tight junction events and temporary gap junctions in the sciatic nerve fibers of the chicken, during Wallerian degeneration and subsequent regeneration. J. Neurocytol. 11, 839-858.

Theodosis, D.T., Poulain, D.A., & Vincent, J.-D. (1981). Possible morphological bases for synchronization of neuronal firing in the rat supraoptic nucleus during lactation. Neuroscience 6, 919-929.

Theodosis, D.T., & Poulain, D.A. (1987). Oxytocin-secreting neurones: a physiological model for structural plasticity in the adult mammalian brain. Trends Neurosci. 10, 426-430.

Theodosis, D.T., & Poulain, D.A. (1984). Evidence for structural plasticity in the supraoptic nucleus of the rat hypothalamus in relation to gestation and lactation. Neuroscience 11, 183-193.

Theriault, E., Frankenstein, U.N., Hertzberg, E.L., & Nagy, J.I. (1997). Connexin43 and astrocytic gap junctions in the rat spinal cord after acute compression injury. J. Comp. Neurol. 382, 199-214.

Tooyama, I., Hara, Y., Yasuhara, O., Oomura, Y., Sasaki, K., Muto, T., Suzuki, K., Hanai, K., & Kimura H. (1991). Production of antisera to acidic fibroblast growth factor and their application to immunohistochemical study in rat brain. Neuroscience 40, 769-779.

Tordjmann, T., Berthon, B., Claret, M., & Combettes, L. (1997). Coordinated intercellular calcium waves induced by noradrenaline in rat hepatocytes: dual control by gap junction permeability and agonist. EMBO J. 16, 5398-5407.

Tournier, C., Pomerance, M., Gavaret, J.M., & Pierre, M. (1994). MAP kinase cascade in astrocytes. Glia 10, 81-88.

Traub, O., Look, J., Dermietzel, R., Brummer, F., Hulser, D., & Willecke, K. (1989). Comparative characterization of the 21-kD and 26-kD gap junction proteins in murine liver and cultured hepatocytes. J. Cell Biol. 108, 1039-1051.

Traub, O., Look, J., Paul, D., & Willecke, K. (1987). Cyclic adenosine monophosphate stimulates biosynthesis and phosphorylation of the 26 kDa gap junction protein in cultured mouse hepatocytes. Europ. J. Cell Biol. 43, 48-54.

Tsacopoulos, M., & Magistretti, P.J. (1996). Metabolic coupling between glia and neurons. J. Neurosci. 16, 877-885.

Van den Pol, A.N., Finkbeiner, S.M., & Cornell-Bell, A.H. (1992). Calcium excitability and oscillations in suprachiasmatic nucleus neurons and glia *in vitro*. J. Neurosci. 12, 2648-2664.

van der Want, J.J.L., Gramsbergen, A., Ijkema-Paassen, J., de Weerd, H., & Liem, R.S.B. (1998). Dendro-dendritic connections between motoneurons in the rat spinal cord: an electron microscopic investigation. Brain Res. 779, 342-345.

van Veen, T., Elofsson, R., Hartwig, H.-G., Gery, I., Mochizuki, M., Cena, V., & Klein, D.C. (1986). Retinal S-antigen: immunocytochemical and immunochemical studies on distribution in animal photoreceptors and pineal organs. Exp. Biol. 45, 15-25.

Vaney, D.I. (1991). Many diverse types of retinal neurons show tracer coupling when injected with biocytin or Neurobiotin. Neurosci. Lett. 125, 187-190.

Vaney, D.I. (1996). Cell coupling in the retina. In: Gap Junctions in the Nervous System. (Spray, D.C., & Dermietzel, R., Eds.), pp 79-102. Landes, Austin.

Veenstra, R.D. (1996). Size and selectivity of gap junction channels formed from different connexins. J. Bioenerg. Biomembr. 28, 327-337.

Veenstra, R., Wang, H.Z., Beyer, E.C., & Brink, P.R. (1994). Selective dye and ionic permeability of gap junction channels formed by connexin45. Circ. Res. 75, 483-490.

Velazquez, J.L., Frantseva, M., Naus, C.C.G., Bechberger, J.F., Junja, S.C., Velumian, A., Carlen, P.L., Kidder, G.M., & Mills, L.R. (1996). Development of astrocytes and neurons in cultured brain slices from mice lacking connexin43. Dev. Brain Res. 97, 293-296.

Venance, L., Cordier, J., Monge, M., Zalc, B., Glowinski, J., & Giaume, C. (1995a). Homotypic and heterotypic coupling mediated by gap junctions during glial cell differentiation *in vitro*. Eur. J. Neurosci. 7, 451-461.

Venance, L., Piomelli, D., Glowinski, J., & Giaume, C. (1995b). Inhibition by anandamide of gap junctions and intercellular calcium signalling in striatal astrocytes. Nature 376, 590-594.

Venance, L., Stella, N., Glowinski, J., & Giaume, C. (1997). Mechanism involved in initiation and propagation of receptor-induced intercellular calcium signaling in cultured rat astrocytes. J. Neurosci. 15, 1981-1992.

Vera, B., Sanchez-Abarca, L.I., Bolanos, J.P., & Medina, J.M. (1996). Inhibition of astrocyte gap junction communication by ATP depletion is reversed by calcium sequestration. FEBS Lett. 392, 225-228.

Verkhratsky, A., & Kettenmann, H. (1996). Calcium signalling in glial cells. Trend Neurosci. 19, 346-352.

Vernadakis, A. (1996). Glia-neuron intercommunication and synaptic plasticity. Prog. Neurobiol. 49, 185-214.

Vollrath, L. (1985). Mammalian pinealocytes: ultrastructural aspects and innervation. In: Photoperiodism, Melatonin and the Pineal. (Evered, D., & Clark, S., Eds.), vol. 177, pp. 9-22. Ciba Foundation Symposium, Pitman, London.

Vukelic, J.I., Yamamoto, T., Hertzberg, E.L., & Nagy, J I. (1991). Depletion of connexin43-immunoreactivity in astrocytes after kainic acid-induced lesions in rat brain. Neurosci. Lett. 130, 120-124.

Wachym, P.A., Popper, P., Abelson, L.A., Ward, P.H., & Micevych, P.E. (1991). Molecular biology of the vestibular system. Acta Otolaryngol. 481, 141-149.

Walsh, J.P., Cepeda, C., Hull, C.D., Fisher, R.S., Levine, M.S., & Buchwald, N.A. (1989). Dye-coupling in the neostriatum of the rat: II. Decreased coupling between neurons during development. Synapse 4, 238-247.

Walton, K.D., & Navarrete, R. (1991). Postnatal changes in motoneurone electrotonic coupling studied in the *in vitro* rat lumbar spinal cord. J. Physiol. 433, 283-305.

Walz, W. (1989). Role of glial cells in the regulation of the brain ion microenvironment. Prog. Neurobiol. 33, 309-333.

Walz, W. (1997). Role of astrocytes in the spreading depression signal between ischemic core and penumbra. Neurosci. Biobehav. Rev. 21, 135-142.

Walz, W., & Hertz, L. (1983). Functional interactions between neurons and astrocytes. II. Potassium homeostasis at the cellular level. Prog. Neurobiol. 20, 133-183.

Wang, Z., Tymianski, M., Jones, D.T., & Nedergaard, M. (1997). Impact of cytoplasmic calcium buffering on the spatial and temporal characteristics of intercellular calcium signals in astrocytes. J. Neurosci. 17, 7359-7371.

Warn-Cramer, B.J., Cottrell, G.T., Burt, J.M., & Lau, A.F. (1998). Regulation of connexin-43 gap junctional intercellular communication by mitogen-activated protein kinase. J. Biol. Chem. 273, 9188-9196.

Warn-Cramer, B.J., Lampe, P.D., Kurata, W.E., Kanemitsu, M.Y., Loo, L.W.M., Eckhart, W., & Lau, A.F. (1996). Characterization of the mitogen-activated protein kinase phosphorylation sites on the connexin43 gap junction protein. J. Biol. Chem. 271, 3779-3786.

Watanabe, H., & Burnstock, G. (1976). Junctional subsurface organs in frog sympathetic ganglion cells. J. Neurocytol. 5, 125-136.

Watling, K.J., & Dowling, J.E. (1979). Dopamine receptors in the retina may all be linked to adenylate cyclase. Nature 281, 578-580.

Waxman, S.G., & Black, J.A. (1984). Freeze-fracture ultrastructure of the perinodal astrocyte and associated glial junctions. Brain Res. 308, 77-87.

Weiler, R. (1996). The modulation of gap junction permeability in the retina. In: Gap Junctions in the Nervous System. (Spray, D.C., & Dermietzel, R.,Eds.), pp. 104-121. Landes, Austin.

Weis, P. (1968). Confronting subsurface cisternae in chick embryo spinal ganglia. J. Cell Biol. 39, 485-488.

Welsh, D.K., & Reppert, S.M. (1996). Gap junctions couple astrocytes but not neurons in dissociated cultures of rat suprachiasmatic nucleus. Brain Res. 706, 30-36.

Werman, R., & Carlen, P.L. (1976). Unusual behavior of the Ia EPSP in cat spinal motoneurons. Brain Res. 112, 395-401.

White, T.W., & Bruzzone, R. (1996). Multiple connexin proteins in single intercellular channels: connexin compatibility and functional consequences. J. Bioenerg. Biomembr. 28, 339-350.

Willecke, K., Heynkes, R., Dahl, E., Stutenkemper, R., Hennemann, H., Jungbluth, S., Suchyna, T., & Nicholson, B.J. (1991). Mouse connexin37: Cloning and functional expression of a gap junction gene highly expressed in lung. J. Cell Biol. 114, 1049-1057.

Williams, V. (1975). Intercellular relationships in the external glial limiting membrane of the neocortex of the cat and rat. Am. J. Anat. 144, 421-432.

Wolburg, H., Reichelt, W., Stolzenburg, J.-U., Richter, W. & Reichenbach, A. (1990). Rabbit retinal Muller cells in cell culture show gap and tight junctions which they do not express *in situ*. Neurosci. Lett. 111, 58-63.

Wolburg, H., & Rohlmann, A. (1995). Structure-function relationships in gap junctions. Int. Rev. Cytol. 157, 315-373.

Wolff, J.R., Stuke, K., Missler, M., Tytko, H., Schwarz, P., Rohlmann, A. & Chao, T.I. (1998). Autocellular coupling by gap junctions in cultured astrocytes: a new view on cellular autoregulation during process formation. Glia 24, 121-140.

Wouterlood, F.G., Mugnaini, E., Osen, K.K., & Dahl, A.-L. (1984). Stellate neurons in rat dorsal cochlear nucleus studied with combined Golgi impregnation and electron microscopy: synaptic connections and mutual coupling by gap junctions. J. Neurocytol. 13, 639-664.

Wylie, R.M. (1973). Evidence of electrotonic transmission in the vestibular nuclei of the rat. Brain Res. 50, 179-183.

Yagodin, S.V., Holtzclaw, L., Sheppard, C.A., & Russell, J.T. (1994). Nonlinear propagation of agonist-induced cytoplasmic calcium waves in single astrocytes. J. Neurobiol. 25, 265-280.

Yamamoto, T., Hertzberg, E.L., & Nagy, J.I. (1990a). Epitopes of gap junctional proteins localized to neuronal subsurface cisterns. Brain Res. 527, 135-139.

Yamamoto, T., Hertzberg, E.L., & Nagy, J.I. (1991a). Subsurface cisterns in a-motoneurons of the rat and cat: Immunohistochemical detection with antibodies against connexin32. Synapse 8, 119-136.

Yamamoto, T., Hossain, M.Z., Hertzberg, E.L., Uemura, H., Murphy, L.J., & Nagy, J.I. (1993). Connexin43 in rat pituitary: localization at pituicyte and stellate cell gap junctions and within gonadotrophs. Histochemistry, 100, 53-64.

Yamamoto, T., Kardami, E., & Nagy, J.I. (1991b). Basic fibroblast growth factor in rat brain: localization to glial gap junctions correlates with connexin 43 distribution. Brain Res. 554, 336-343.

Yamamoto, T., Maler, L., Hertzberg, E.L., & Nagy, J.I. (1989a). Gap junction protein in weakly electric fish (Gymnotide): immunohistochemical localization with emphasis on structures of the electrosensory system. J. Comp. Neurol. 289, 509-536.

Yamamoto, T., Ochalski, A., Hertzberg, E.L., & Nagy, J.I. (1990b). LM and EM immunolocalization of the gap junction protein connexin43 in rat brain. Brain Res. 508, 313-319.

Yamamoto, T., Ochalski, A., Hertzberg, E.L., & Nagy, J.I. (1990c). On the organization of astrocytic gap junctions in rat brain as suggested by LM and EM immunohistochemistry of connexin43 expression. J. Comp. Neurol. 302, 853-883.

Yamamoto, T., Shiosaka, S., Whittaker, M.E., Hertzberg, E.L., & Nagy, J.I. (1989b). Gap junction protein in rat hippocampus: Light microscope immunohistochemical localization. J. Comp. Neurol. 281, 262-281.

Yamamoto, T., Vukelic, J., Hertzberg, E.L., & Nagy, J.I. (1992). Differential anatomical and cellular patterns of connexin43 expression during postnatal development of rat brain. Dev. Brain Res. 66, 165-180.

Yang, Q.Z., & Hatton, G.I. (1988). Direct evidence for electrical coupling among rat supraoptic nucleus neurons. Brain Res. 463, 47-56.

Yu, N., Martin, J.-L., Stella, N., & Magistretti, P.J. (1993). Arachidonic acid stimulates glucose uptake in cerebral cortical astrocytes. Proc. Natl. Acad. Sci. U.S.A. 90, 4042-4046.

Yuste, R., Peinado, A., & Katz, L.C. (1992). Neuronal domains in developing neocortex. Science 257, 665-669.

Zahs, K.R. (1998). Heterotypic coupling between glial cells of the mammalian central nervous system. Glia 24, 85-96.

Zahs, K.R. & Newman, E.A. (1997). Asymmetric gap junctional coupling between glial cells in the rat retina. Glia 20, 10-22.

Zenker, W., Bankoul, S., & Braun, J.S. (1994). Morphological indications for considerable diffuse reabsorption of cerebrospinal fluid in spinal meninges particularly in the areas of meningeal funnels. An electronmicroscopical study including tracing experiments in rats. Anat. Embryol. 189, 243-258.

Zhang, J.-T., & Nicholson, B.J. (1989). Sequence and distribution of a second protein of hepatic gap junctions, Cx26, as deduced from its cDNA. J. Cell Biol. 109, 3391-3401.

Zhu, D., Caveney, S., Kidder, G.M., & Naus, C.C.G. (1991). Transfection of C6 glioma cells with connexin43 cDNA: Analysis of expression, intercellular coupling, and cell proliferation. Proc. Natl. Acad. Sci. U.S.A. 88, 1883-1887.

Zieglgansberger, W., & Reiter, C. (1974). Interneuronal movement of procion yellow in cat spinal neurones. Exp. Brain Res. 20, 527-530.

Zimmerberg, J., Curran, M., Cohen, F.S., & Brodwick, M. (1985). Simultaneous electrical and optical measurements show that membrane fusion precedes secretory granule swelling during exocytosis of beige mouse mast cells. Proc. Natl. Acad. Sci. U.S.A. 84, 1585-1589.

# INDEX

A-type junctions, 149
acylation, post-translational, 117-118
adenosine triphosphate (ATP), 35, 271
AHF, selective communication of,
    229-230
alkali treatments, 37
annular gap junctions, 108
antibodies
    anti-connexin, 333
    anti-gap junction antibodies, 2
    gap junction, 200
antibody reagents, inhibiting gap junc-
        tional communication,
        208-209
assays of gap junction communication,
        235-237
astrocytes, 337-354
    astrocytic syncytial compartments,
        342-343
    calcium waves, 344-347
    connexins *in vitro*, 343-344
    coupling, 343-344, 347-350
    injury and astrocytic gap junctions,
        351-354
    junctions between oligodendrocytes
        and, 360-362
    pineal, 372

B-type junctions, 149
binding, 12, 35

blastomeres
    connexin sorting, 131
    voltage dependence, 162
Boltzmann equations, open/closed
        equilibria, 163
boxlike compartments in mouse
        embryos, 201-203
brain tissues (*See* central nervous sys-
        tem)
Bystander Effect, 246-248, 267,
        270-280

cadherins, 107, 238, 244
calcium waves, 344-347
cAMP (*See* cyclic adenosine mono-
        phosphate (cAMP))
cancers
    carcinogenic process, 222-225,
        229-234, 238, 239
    carotenoids and suppression,
        242-243
    cell differentiation and gap junc-
        tional communication
        225-227
    prevention, 241-243
    primary *vs.* secondary, 241
    retinoids and suppression, 241-242
carcinogenesis, 229-232
    carcinogen-induced transformants,
        233-234

human papilloma virus (HPV), 232
oxidative damage, 238-239
tobacco, carcinogens, 241-242
(*see also* oncogenesis: tumors)
cardiovascular cells, connexin expression, 17
cardiovascular diseases
cardiovascular anomalies, 80
Chagas' disease, 290
congenital heart diseases, 117
congenital heart malformations, 209-211
gap junctional communication, 289-294
carotenoids, cancer inhibition, 242-243
CD spectroscopy, 65
cell adhesion molecules, role of, 106-107
cell death
"kiss of life"/"kiss of death" experiments, 267
permeability, 155-156
cell differentiation
cancer, 225-227
evidence of gap junction functions, 281-284
gap junctional communication, 213
role of gap junctions, 194-200
cell division, gap junctional communication, 284-285
cell-to-cell communication (*See* intercellular communication)
cell-to-cell tracer diffusion, 137-145
central nervous system, 323-373
connexin expression. 365-366
development, 363-369
neurons, 330-332
subsurface cisterns, 332-333
Chagas' disease, 290
channel pores, 181-182
size, 142-145
voltage sensitivity of, 160
channels, 47-54
conductivity measurement, 131-137

gap junction proteins, formation and identification, 43-45
pores, 142-145, 147, 160, 181-182
selective permeability of, 205
size of, 141-145
unitary conductance of connexin specific, 176-181
(*see also* hemichannels)
chaperones, molecular, 119
Charcot-Marie-Tooth (CMTX) disease, 50, 72, 80, 175-176, 226-227, 292-295
inhibition of gap junctions, 212
myelinating Schwann cells, 10-11, 359-360
chemopreventive agents, intermediate markers of response, 245-246
chromosomal mapping of connexin genes, 8-11
cloning
and biochemical characterization of gap junction, 45-47
connexin genes, 2-7, 18-19
identifying first connexins, 40
neuronal connexin clones, 362
of neuronal connexins, 331-332
communication, gap junctional
assays of, 235-237
astrocytes, 337-341
carcinogen induced transformants, 234-235
cardiovascular diseases, 289-294
cell-cell signaling in development, 213
cell differentiation and, 213
cell division, 284-285
compounds that increase intercellular communication, 241-243
coupling blocked with antibody reagents, 208-209
in development and adult tissues, 225-227
in embryogenesis and development, 194-208

gene knockouts, 209-212

human papilloma virus (HPV),
    effects on, 232

inhibition and developmental disrup-
    tions, 208-212

mediates "Good Samaritan effect,"
    248

metastatic tumors, 239

molluscan embryos, 203-204

nematode embryos, 204-205

neurological diseases, 292-293

oncogenesis, 228-241

restoration and inhibition of onco-
    genesis, 243-244

restrictions, 205-208

skin diseases, 293-294

transgenic mouse models, 209-210

tumor promoters and modification,
    234-239

tumorigenesis, 291-292

communication compartments, 226

    placentation, 201

    synchronization of heart, 276-277

    therapeutics, 298-300

conductance

    junctional conductance ($g_j$), mea-
        surement of, 131-136

    leptomeningeal cells, 370

    measuring coupling strength,
        265-267

    multiple channel conductances,
        177-181

    permeability and, 137-158

    regulation of conductance states,
        333-334

    septates, 131

    single channel, 135-136

    unitary conductances, 143-145,
        150-152, 155-158, 176-181

connexin expression

    astrocytes, 338-341, 343-344, 347

    cardiovascular cells, 17

    central nervous system development,
        365-366

embryonic, 15

increased, 241-242

ISHH studies and, 330-332

keratinocytes, 17

multiple expression and unitary con-
    ductance, 152-154

nervous system development,
    363-369

neurons, mature, 330-332

nontransformed cells, 227-228,
    330-332

pregnancy and, 15

proliferation-associated changes,
    227-228

Schwann cells and shift in, 360

(*see also* connexin genes)

connexin genes, 2-12

    $\alpha$ and $\beta$ groups, 46

    amino acid sequences, 8

    cell type–specific expression, 16-18

    chromosomal locations, 8-11

    cloning, 2-7, 18-19

    embryonic expression, 15-16

    gene control regions, 12-14, 20

    heterotypic junctions of, 165-169

    inherited disorders, 10-11

    knockouts, 209-212, 226, 275,
        294-297

    mutations, 10-11

    nomenclature, 3-4

    polypeptide sequences, 2-6

    pregnancy and, 15

    primary structure, 6, 8

    recombinant, 67

    tissue-specific expression patterns,
        14-15

    topology, 7-8

    transcriptional and post-transcrip-
        tional control of expression,
        19-20

connexin sorting

    blastomeres, voltage gating, 131

    granulosa cells, 105

connexins, 33

cell adhesion molecules, role of, 106-107
compatibility, 163-164, 182-183
defined, 130
fertility, role in, 281, 295
gap junction assembly, 102-107
Golgi apparatus, 102-103
half life of, 119
heterotypic and heteromeric interactions, 47-54
heterotypic astro/oligo gap junctions, 361-362
heterotypic connexon channel formation, 226
identifying first, 40-41
localization of, 366-369
mammalian central nervous system, 323-373
multiple, evidence for, 40-41
nomenclature, 46
nonconnexin proteins, 41-43
oligodendrocytic, 355-359
oligomerizations of, 103-105
phosphorylation, 110-118
pIs and post-translational modifications, 118
plant origins, 47
plaque formation, 105-106
sorting, 105, 131
therapeutic applications, 297-300
trafficking, 102-107
transcripts, regulation of steady-state levels, 100-102
transfected, channel conductances of, 177
transfected cells, 18-19
*in vitro* translation of, 102
(*see also* connexin expression; connexin genes)
connexons, 33, 48
defined, 130
docking, models for, 73-75
favored by assembly pathways, 51
intercellular pairing, 105-106

oligomerization of connexins, 103-105
unitary conductance, 155-158
(*see also* hemichannels)
coupling, gap junctional
influence on cell function, 297-300
measurement *in vivo*, 269-270
cyclic adenosine monophosphate (cAMP), 240-245, 273, 279
connexin assembly, 106
cystic fibrosis, 280
cytokine resistance, 274
cytoskeletons, gap junctions and, 57

deafness, 80, 212, 292
degradation of gap junctions, 109-110
detergents, nonionic, 35-37, 38, 41
development, gap junctions and, 193-213, 225-227, 363-369
developmental compartments, 197-200
dissipation of molecules and ions, 281, 337
disulfide bridge formation, post-translational, 117-118
dominant negative inhibition of gap junctions, 212
dopamine, 157-158, 333-334
dual voltage clamp technique, 131-135, 266-267
ductin, 42-43

electrical coupling
in CNS of higher vertebrates, 330
early research, 32-35
measuring coupling strength, 265-267
electrical current spread and synchronization, 275-277
electron cryo-crystallography of gap junctions, 66-72
electron microscopy
atomic force microscopy, 60
3D, low resolution, 61-63
limitations, 60

projection density maps, low resolution, 59-63
scanning transmission electron microscopy (STEM), 60-61
variation in projection density maps, 61
electron paramagnetic resonance spectroscopy (EPR), 65-66
embryogenesis, gap junctional communication, 194-208
embryonic development, gap junction functions, 194-208, 274-275
endoplasmic reticulum, oligomerization in, 104
endothelial cells, 371
enzymes
  enzymatic binding, 35
  enzymatic defects, compensation for, 279
ependymal cells, 371
EPR (electron paramagnetic resonance spectroscopy), 65-66
expression patterns of connexin genes, 14-15
extracellular loops
  connexin compatibility, 182-183
  docking, 74-75

fast voltage dependence, 169-171
fatty acids, 55-56
fertility, role of connexins, 281, 295
Fourier transform infrared spectroscopy (FTIR), 65
free radicals, 238-239
freeze-fracture techniques, 32-33
friction, in pore models, 145, 146-149
FTIR (Fourier transform infrared spectroscopy), 65
functions of gap junctions
  Bystander Effect, 279-280
  cell differentiation, 281-284
  cell division, 284-285
  dissipation of molecules and ions, 281

  electrical coupling, 265-267
  electrical current spread and synchronization, 275-277
  embryonic development, 194-208, 274-275
  enzymatic defects, compensation for, 279
  hormonal stimulation transfer, 274
  mechanical synchronization, 277-279
  molecular genetic studies, 294-297
  morphogenesis, 274-275
  nutrition, 280-281
  physiological roles, 265-270
  secretion, 286-288
  skin diseases, 293-294
  somatic diseases, 288-294
  viral and cytokine resistance, 274

gap junction channels (*See* channels)
gap junctions
  annular gap junctions, 108
  biochemical characterization, 35-57
  cancers and, 222-225, 232
  central nervous system development, 363-369
  channel size, 142-145
  coupling, 297-300
  cytoskeletons, 57
  defined, 130
  development, 193-213
  growth, 105
  internalization and degradation, 107-110
  ion selectivity, 154-155
  isolation protocols, 2, 39
  lipid components, 55-56
  model described, 33
  morphological features, 32-35
  post-transcriptional events, 99-119
  post-translational modification, 54-55
  retinal, 362-363

(*see also* communication, gap junctional; functions of gap junctions; permeability; proteins; structure of gap junctions)
gating
models of gap junctions, 75-76
pore location and, 181-182
regulation, 76-78, 77, 78
voltage-dependent gating, 129-131, 158-173
gating regulation
functional sites, 76-78
pH-induced, 77
voltage, 78
gene knockouts, 226
gap junctional communication, 209-212
molecular genetic studies of gap junctions, 294-297
unsuspected connexin genes, 275
genes
gene knockouts, 209-212, 226, 275, 294-297
tumor suppressor and oncogenes, 20, 223-224
*wnt* genes, 207-208
(*see also* connexin genes)
glia, neuromodulation, 346-347
glycosylating enzymes, 103
Golgi apparatus
connexins in, 102-103
oligomerization of connexins, 103
trafficking, 118
Good Samaritan Effect, 248, 279-180
granulosa cells
connexin sorting, 105
internalization of gap junctions, 108-109
growth control and heterologous communication, 244
growth factors, modulating cell-cell communication, 113-114
growth regulatory signal hypothesis, 244-245

heart
communication compartments and synchronization, 276-277
connexin expression in cardiovascular cells, 17
multiple channel conductances, 177-180
unitary conductance in, 150-152
(*see also* cardiovascular diseases)
hemichannels, 38-39
cell surface, 105-106
heterologous hemichannels, 277
oligomerization of connexins, 103-105
oocyte systems, 38
protocols for isolating, 50-54
unitary conductance in native tissues, 150-158
voltage dependence, 166-169
(*see also* connexons)
heterocellular communication, second messengers, 272
heterologous astro/oligo gap junctions, 360-361
heterologous communication, oncogenesis and, 244
heterologous hemichannels, 277
heteromeric channels, 47-54
heterotypic channels, 47-54
heterotypic gap junctions
astro/oligo junctions, 361-362
coupling properties, 205-207
nonfunctional, 182
slow voltage dependence, 165-169
homotypic channels, 47-54
homotypic junctions, slow voltage dependence, 160-165
hormonal stimulation transfer, 274
hormones
connexin transcription, 278
oxcytocin, 335
human papilloma virus (HPV), 232
hypothalamo-neurohypophysial system, 335

injury responses, 351-354, 359-360
innexins, 42-43, 46
insects, 42-43, 46
    channel pore size, 143-145
    metamorphoses and gap junction
      degradation, 110
    pattern formation studies, 197-200
intercellular communication
    compounds that increase, 241-243
    detected in *Xenopus* embryo, 200
    gap junction functions, 264-267
    growth factors, modulating, 113-114
    between neurons, 326-337
    oncogenesis and, 233-239, 244
    pinealocytes, 372
    second messengers, 272
    signaling in development, 213
internalization of gap junctions,
    107-109
ion selectivity, 154-155
ionic coupling, 194-195
ISHH (*in situ* hybridization studies),
    connexin expression, 330-332

junctional conductance ($g_j$), measure-
    ment of, 131-136

keratinocytes, 17, 228
kinases, connexin phosphorylation,
    111-115
"kiss of life"/"kiss of death," 248, 267
knockouts (*See* gene knockouts)

lens gap proteins, 11-12
leptomeningeal cells, 14, 18, 370-371
lipids, gap junction components, 55-56
liver function, second messengers and,
    272
*Loligo,* size of channels, 141-142
loops, extracellular
    connexin compatibility, 182-183
    docking, 74-75

mammalian central nervous system,
    323-373
    evidence of gap junctions, 326-330
mammalian development, gap junction
    communication, 200-203
    boxlike compartments, 201-203
    placentation, 201
mechanical synchronization, 277-279
membrane topology of gap junctions,
    63-64
membranes, junctional, 137-140
metabolic coupling, 35
migration, gap junction functions, 285
MIP26 (main intrinsic protein 26),
    41-42
molecular genetic studies, gap junction
    functions, 294-297
molecular weight shifts, phosphoryla-
    tion of connexins, 110-111
molluscan embryos, gap junctional
    communication, 203-204
morphogenesis, 225-227
    gap junction functions, 194-200,
    274-275
mosacism and patterning, 203-204-
multiple channel conductances,
    177-180
    modulation of, 180-181
multiple connexin expression, 152-154
myelinating cells, junctional communi-
    cation, 354-362

native tissue studies, unitary conduc-
    tance, 150-152
nematode embryos, gap junctional
    communication, 204-205
neoplastic transformation, 221-248
neurological diseases
    demyelinating neuropathy, 10-11
    gap junctional communication,
    292-293
    parasitic infections, 293
neuronal gap junctions, 326-330
    development, 363-366

NMR (nuclear magnetic resonance) spectroscopy, 65-66
nonconnexin proteins, 41-43
nutrition, gap junctions as feeding pathway, 280-281

oligemeric channels, 33
oligodendrocytes
  coupling, 356-359
  junctions between astrocytes and, 360-362
  localization of oligodendrocytic connexins, 355-356
  oligodendrocytic gap junctions, 354-355
oligomerization, 103-105, 118
olivo-cerebellar system, 334-335
oncogenes
  gap junctional communication modifications, 234
  inhibition of gap junctional communication, 240-241
oncogenesis
  altered gap junctional communication, 228-233
  gap junctional intercellular communication, 233-239
  growth regulatory signal hypothesis, 244-245
  inhibited by restoration of gap junctional communication, 243-244
  mechanisms for inhibition of gap junctional communication, 240-241
  (see also carcinogenesis; tumors)
oncogenic transformation, 221-248
oxytocin, 335

P-N theory, 169
P87 proline, $V_j$ dependence, 176

parasitic infection, 293
pericytes, 371
peripheral neuropathies (See Charcot-Marie-Tooth (CMTX) disease)
permeability
  cell death, 155-156
  cell-to-cell tracers for measurement, 137-145
  conductance and, 137-158
  conductance and prediction of, 154-155
  friction effects, 145, 146-149
  measuring, 131, 136-137
  measuring gap junction coupling strength, 267-269
  $P_j/g_j$ ratio, 140-142
  selectivity of gap junction channels, 205
  solute charges, effect of, 146
  total permeability ($P_j$), 140-142
permeance (See permeability)
pharmocological agents, possible applications, 297-300, 302
phosphatases, role in regulation, 115
phosphorylation, 54-55, 227
  of connexins, 110-119
  functional role, post-translational, 115-117
  gating regulation sites, 76-77
  Golgi apparatus connexins, 103
  kinases involved in, 111-115
  regulation of Cx43, 350-351
  sites of, 111-115
pinealocytes, 372
pIs
  post-translational modifications of connexins, 118
  shift, 55
$P_j/g_j$ ratio, 140-142, 148-149
placentation, 207, 294

communication compartments and, 201

plants, connexin related proteins in, 11

plaques, junctional
  formation, 34, 105-106
  series access resistance, 150

polyoma transformed cells, 222

polypeptides sequences, of connexins, 5

pores
  channel pores and voltage sensitivity, 160
  location and voltage gating, 181-182
  models to predict size, 145-150
  size of channel pore, 130, 142-150
  structures, 72-73

post-transcriptional events, 99-119
  regulation of steady state levels, 100-102

post-translational modifications, 54-55
  (*see also* phosphorylation)

post-translational regulation, 110-118
  phosphatases, role of, 115

pregnancy
  expression of connexin genes, 15
  uterine muscle synchronization, 277-278
  projection density maps of gap junctions, 59-63
  variation in, 61

promoter/enhancer regions of connexin genes, 12-14

proteases, 37

proteins
  folding models, 52-53, 64
  gap junction, 2, 40-43, 56-57
  identifying criteria, functional assay, 43-45
  innexins, 42-43, 46
  lens gap, 11-12
  molecular cloning, 2-7
  in neurons, 332-333
  nomenclature, 225
  nonconnexin proteins, 41-43

post-translational modification, 54-55

α-, β-protein groups, 225

renewal, 107-108

recombinant genes
  connexins, 67
  dominant negative inhibition of gap junctions, 212

retina cells, 335-336

retinal gap junctions, 362-363

retinoids, 245
  cancer inhibitors, 231, 241-242
  regulating connexion gene expression, 20

Schwann cells, 10-11, 294-295, 359-360

second messengers, gap junction functions, 270-273

secondary structure models of gap junctions, 64-66

secretion
  gap junction functions, 286-288
  pinealocytes, 372

selectivity, of channel pores, 145-150

septates
  conductivity, 131
  septate junctions, 35

series assess resistance, effects, 149-150

signal transduction, 248

sites of phosphorylation, 111-115, 116

skin diseases, gap junction functions, 293-294

slow voltage dependence
  heterotypic junctions, 165-169
  homotypic junctions, 160-165

smooth muscle synchronization, 277-279

solute charge, effects on permeability, 146

somatic diseases, gap junctional communication, 288-294

spreading depression (SD) waves,
        346-347
structure analysis of gap junctions
    connexon docking, models for,
        73-75
    electron cryo-crystallography, high
        resolution, 66-72
    electron microscopy, 59-63
    gating models, 75-76
    membrane topology, 63-64
    pore structure, 72-73
    projection density maps, low resolu-
        tion, 59-63
    secondary structure models, 64-66
    x-ray electron density profiling,
        58-59
structure of gap junctions
    defining features, 32
    function and, 173-183
    structure analysis, 57-76
    voltage sensitivity, evidence of,
        158-160
subsurface cisterns, 332-333

TATA boxes, 12, 14
therapeutics, gap junction-directed,
        297-300
tissue morphogenesis, role of gap junc-
        tions, 194-200
tobacco, carcinogens, 241-242
total permeability ($P_j$), 140-142
toxicology as predictive assay, 246
TPA (C-kinase activator), 230-231,
        237-239, 240-243
transfected cells, connexin expression,
        18-19
transjunctional flux, 136-137
tumor promoters, 224-225, 234-239
    toxicology as predictive assay, 246
    TPA (C-kinase activator), 230-231,
        237-239, 240-243
tumors, 103
    Bystander Effect, 279-280

gap junctional communication,
        229-232, 291-292
growth regulatory signal hypothesis,
        244-245
liver, 229-230
metastatic, 239
mRNA levels, 101
primary vs. secondary, 241
retinoids and suppression, 241
retroviral treatments, 246-248
skin, 230-231
tumor promoters, 224-225, 229-231,
        234-239, 240-243
tumor suppressor genes, 20, 223-224
(see also cancers; carcinogenesis;
        oncogenesis)

unitary conductance, 21, 143-145
    of connexin specific channels,
        176-181
    native tissue studies, 150-152
unitary currents, channel pores,
        145-150
uterine muscle, 277-278

vascular diseases, 290-291
viral resistance, 274
visceroatrial heterotaxia (VAH), 210,
        212
vitamin A (See retinoids)
$V_j$ sensors, 173-176
$V_m$ or $V_{i-o}$ dependence, 171-173
voltage dependence
    fast vs. slow, 160
    slow, 160-169
    transduction, functional properties
        of P87, 176
voltage-dependent gating, 129-131,
        158-173
    definitions and descriptions,
        158-160
    $V_m$ or $V_{i-o}$ dependence, 171-173
voltage gating

blastomeres and connexin sorting,
 131
of gap junction channels, 78
structure and function studies,
 173-183
voltage transduction, 176

whole-cell patch clamp technique, 132,
 135-136
*wnt* genes, 207-208

x-ray electron density profiling of gap
 junctions, 58-59